k, k_s	spring constant, or elasticity constant
k	wave number, $2\pi/\lambda$, where λ is the wavelength
k_C	Coulomb constant
K	degrees Kelvin, or absolute
K	coefficient of thermal conductivity
l	mean free path
l	liter
l	orbital angular momentum quantum number. $L = \sqrt{l(l+1)}\hbar$
l_z	quantum number for z-component of orbital angular momentum. $L_z = l_z\hbar$
L	orbital angular momentum
L_z	z-component of orbital angular momentum
L	latent heat
m	mass
M	magnetization, or magnetic moment per unit volume. $M = \mu/V$
n	number of moles
n	integer
n	degree of degeneracy
n	chemical concentration in moles/liter
n	number of particles in a quantum state
$\bar{n}$	occupation number of a quantum state
N_A	Avogadro's number
N	number of particles in system
N_a, N_e, N_p	number of atoms, electrons, phonons
N_g, N_x	number in ground state, number in excited states
$\mathfrak{N}$	number of degrees of freedom in system
p	momentum. $p = mv$
p	probability that the criterion is satisfied
p	pressure
p	probability
q	probability that the criterion is *not* satisfied
q	an arbitrary coordinate, position, or momentum
q	heat energy per mole
Q	heat energy
Q_x	heat flux in x-direction = rate of heat flow per unit area
r	distance from origin, radius
R	radius of a sphere
R	gas constant
s	label for quantum state (e.g., state s)
s	entropy per mole

DATE DUE

SEP 21 '90			
SEP 21 90			
DEC 1 4 1990			
MAY 1 8 1994			
JAN 3 1 1995			
FEB 2 5 1995			
MAR 2 2 1995			
MAY 1 8 1995			
MAY 0 5 1998			
GAYLORD			PRINTED IN U.S.A.

INTRODUCTION TO
STATISTICAL MECHANICS AND
THERMODYNAMICS

INTRODUCTION TO STATISTICAL MECHANICS AND THERMODYNAMICS

KEITH STOWE

California Polytechnic State University

JOHN WILEY & SONS

New York • Chichester • Brisbane • Toronto • Singapore

Library of Congress Cataloging in Publication Data:

Stowe, Keith S., 1943–
Introduction to statistical mechanics and thermodynamics.

Includes indexes.
1. Thermodynamics.   2. Statistical mechanics.
I. Title.
QC311.S84   1984   536'.7   83-3617
ISBN 0-471-87058-7

Printed in the United States of America

10 9 8 7 6 5 4 3 2 1

PREFACE

The subject of thermodynamics developed on a postulatory basis long before we understood the nature or behavior of the elementary constituents of matter, such as electrons, atoms, or molecules. In fact, it was well into the twentieth century before we were familiar with these elementary constituents, and even then we were slow to place our trust in the "new" field of quantum mechanics, which was telling us that their behaviors could correctly and accurately be described using probabilities and statistics.

The influence of this historical sequence of events has lingered in our "traditional" thermodynamics curriculum. Until quite recently, we still tended to teach the first introductory course in thermodynamics using a more formal and abstract postulatory approach. Now, however, there is a growing feeling among instructors that the most effective way of introducing the subject is to first demonstrate the firm physical and statistical basis of the subject by showing how the properties of macroscopic systems are simple and direct consequences of the behaviors of their elementary constituents. An added advantage of this statistical approach is that it is easily extended to include the fundamentals of classical and quantum statistics in the first introductory course. It gives the student a broader spectrum of skills as well as a better understanding of their physical bases.

The need for an introductory text using the statistical approach is the reason I wrote this book. I have tried to introduce the subject as simply and succinctly as possible, with enough applications to indicate the relevance of the results, but not so many as might risk losing the student in details. The book introduces the subject to students who have never been exposed to thermodynamics before. There are many more advanced books of high quality on the market, which can help the interested and more knowledgeable student probe more deeply into the subject and its more specialized applications.

I have tried to structure the book in such a way as to maximize the freedom of the instructor to tailor the material to fit his/her particular interests and course outline. The fundamental concepts are developed slowly and thoroughly in the 10 chapters that are identified by asterisks (*).

Chapters 5–11: The basic ideas of conventional thermodynamics, including internal energy, entropy, and interactions between systems.
Chapter 12: Models and equations of state for various kinds of systems.
Chapter 16: Introduction to classical statistics.
Chapter 24: Introduction to quantum statistics.

These 10 chapters provide the basic skeleton of the course. Each one should be mastered before any succeeding topic is studied.

The remaining 19 chapters are mostly further extensions and applications of the concepts presented in the asterisked chapters. These topics can be studied at any time and in any order, provided the fundamental concepts in *preceding* asterisked chapters have already been mastered. Also, any of the topics in these 19 nonasterisked chapters may be omitted entirely from the course without jeopardizing comprehension of succeeding material.

These extensions and applications have been tied as closely as possible to the fundamental concepts. I have avoided the use of "slick" mathematical tricks and the "pyramiding" of results. Background material common to many applications has been placed in the asterisked chapters in order to minimize repetition and to allow study of any of the 19 nonasterisked topics without prior knowledge of any others.

Chapters 3 and 4 introduce the statistical techniques used in studying small systems, starting with systems of two particles and working up. I believe that studying small systems first leads to greater appreciation and understanding of the very powerful statistical techniques developed for macroscopic systems in the chapters that follow. Although I recommend that they be studied first, they are not essential for comprehension of succeeding material, and therefore are not asterisked.

Although the topical content of the book is fairly conventional, the format is not. I think that learning efficiency is increased when the student is an *active* participant in the course. Without denigrating the importance of clear presentations in lectures and texts, I feel it is at least equally important for the student to apply new knowledge to specific problems as soon as possible. The student should be discouraged from letting this active involvement lapse into a few long problem sessions coming occasionally, or just before appropriate midterms. Instead, this activity should preferably become part of a daily routine, with application to specific problems being attempted while the knowledge is still fresh. A routine of frequent, timely, and short problem-solving sessions is far superior to a few infrequent problem-solving marathons.

For this reason, the text includes a very large number of suggested homework problems for the student to work on. Rather than being found in large groups at the ends of chapters, they are placed in several smaller groups within the chapters, immediately following the development of the appropriate concepts (and at approximately one-lecture intervals). I hope that this format encourages the student to try out newly acquired knowledge immediately.

Not only should active participation be encouraged but also the more passive components of the learning process should be streamlined as much as possible. The sooner the student understands the text material, the sooner he/she can try applying it to specific problems. Therefore, I have tried to develop the concepts as simply and clearly as possible. Summaries are frequent, and are included wherever I think would be helpful to a first-time student wrestling with fundamental concepts, rather than being relegated to the end of the chapter as is customary. The summaries are also boxed for easy identification by a student reviewing previously covered material. Hopefully, this streamlining of the passive aspects of student involvement might allow more time for active problem solving.

I wish to express my appreciation to those students in my first thermodynamics course for their ideas and encouragement, to my colleagues Joe Boone and Rich Saenz for their careful scrutiny and thoughtful suggestions, to Francesca Fairbrother for her superb typing, to Linda Indig and the staff at John Wiley, and to my wife and two sons for their indulgence of my interest in this project.

KEITH STOWE

CONTENTS

Core material indicated by *

PART 1. INTRODUCTION   1

1. The Translation Between Microscopic and Macroscopic Behavior   2

2. Microscopic Behaviors and Quantum States   8

   A. Charge Quantization   8
   B. Wave Nature of Particles   9
   C. Uncertainty Principle   11
   D. Quantum States and Phase Space   13
   E. Quantization of Angular Momentum   18
   F. Magnetic Moments   22
   G. Harmonic Oscillator   24
   H. Coulomb Potential   26
   I. Description of a State   26

   Appendix 2A.   Magnetic Moment and Angular Momentum   28

PART 2. SMALL SYSTEMS   31

3. Statistics for Small Systems   32

   A. Mean Values   32
   B. The Binomial Distribution—Probabilities for Systems
       of More than One Element   35
   C. Statistically Independent Behaviors   41

   Appendix 3A.   Stirling's Formula, Eq. 3.13   45

4. Statistics for Systems of Many Elements   47

   A. Fluctuations   47
   B. The Gaussian Distribution   53
   C. The Random Walk   59

   Appendix 4A.   The Standard Deviation in Eq. 4.3   64
   Appendix 4B.   Taylor Series Expansion, Eq. 4.5   65
   Appendix 4C. $\partial^2/\partial n^2 \ln P(n)$ at $n = \bar{n}$, Eq. 4.6   66

PART 3. INTERNAL ENERGY   69

*5. Internal Energy and Equipartition   70

   A. Internal Energy   70
   B. Degrees of Freedom   71
   C. The Equipartition Theorem   73

*6. Internal Energy and the First Law   78

    A. Heat Transfer 78
    B. Work 80
    C. Particle Transfer 83
    D. The First Law 92
    E. Exact and Inexact Differentials 94
    F. Dependent and Independent Variables 99

PART 4. ENTROPY   103

*7. The States of a System   104

    A. Equilibrium 104
    B. The Fundamental Postulate 106
    C. The Spacing of States 109
    D. Density of States and the Internal Energy 111

    Appendix 7A. $\Omega_o \propto E^{\Re/2}$ 116

*8. Entropy and the Second Law   118

    A. Microscopic Examples 119
    B. Macroscopic Examples 124
    C. The Second Law 126
    D. Entropy 128

PART 5. INTERACTIONS   133

*9. The Thermal Interaction   134

    A. Temperature and the Zeroth Law 135
    B. Temperature and the Internal Energy 137
    C. Temperature and Heat Flow 140
    D. Temperature Scales and Boltzmann's Constant 141
    E. Thermometers 143
    F. Phase Transitions 146
    G. Heat Reservoirs 148
    H. System in Equilibrium with a Heat Reservoir 150
    I. Measures of Heat Capacity 155
    J. Absolute Zero and the Third Law 157

*10. The Mechanical Interaction   160

    A. Change in Volume 160
    B. Work and the Number of Accessible States 162
    C. Other Kinds of Work 165
    D. Thermal Expansion and Compressibilities 169

*11.   The Diffusive Interaction   174

   A.   Chemical Potential   174
   B.   Influence of Temperature, Pressure, and Concentration   177
   C.   Equilibrium Conditions   179
   D.   The Approach to Equilibrium   180
   E.   Chemical Potential and the Number of Accessible States   182
   F.   Salty Water and Snowflakes   183

PART 6.   CONSTRAINTS   189

*12.   Models   190

   A.   Ideal Gas   191
   B.   Real Gases   195
   C.   Liquids   198
   D.   Solids   199
   E.   Other Common Systems   202
   F.   Sample Applications   202
      F.1)   Heat Capacities   202
      F.2)   Isothermal Compressibilities   206

13.   Natural Constraints   209

   A.   Second Law Constraints   210
   B.   Maxwell's Relations   214
   C.   Sample Applications: Nondiffusive Interactions   222
   D.   Third Law Constraints   228

Appendix 13A.   Maxwell's Relations   229

14.   Imposed Constraints   232

   A.   The Reduction of Independent Variables   232
   B.   Isobaric Processes   237
   C.   Isothermal Processes   239
   D.   Adiabatic Processes   241
      D.1)   Adiabatic Processes in Gases   242
      D.2)   Adiabatic Processes in Liquids and Solids   245
   E.   Reversibility   249
   F.   Nonequilibrium Processes   253
      F.1)   Joule-Thompson (Throttling) Process   253
      F.2)   Free Expansion of a Gas   256
      F.3)   Removal of Barrier Constraints   261

15.   Engines and Refrigerators   263

   A.   The Carnot Cycle   263
   B.   The Refrigeration Cycle   267

C. *p-V* Diagrams 268
D. Efficiencies 273

PART 7. CLASSICAL STATISTICS   279

*16. Probabilities and Microscopic Behaviors   280

A. Ensembles 280
B. Probability of Being in a Certain State 284
C. Classical and Quantum Statistics 286
D. Applications of Classical Statistics 290
  D.1) Excitation Temperature 290
  D.2) Degeneracy 291
  D.3) Examples 292
  D.4) Energy Bands 294
E. Microscopic Whim and Macroscopic Fact 297

17. Equipartition   300

A. Average Energy per Degree of Freedom 300
B. Sample Applications 302
  B.1) Heat Capacity of a Gas 302
  B.2) Heat Capacity of a Solid 305
  B.3) Brownian Motion 306

18. Maxwell Distribution for Gases   309

A. Probability Distributions 309
B. Mean Values 314
C. Particle Distribution and Flux 316
D. Collision Frequency and Mean Free Path 319

Appendix 18A. Standard Integrals   322
Appendix 18B. Collision Frequency and Relative Motion   322

19. Transport Processes in Gases   324

A. Molecular Diffusion 326
B. Thermal Conductivity 327
C. Viscosity 330

20. Magnetic Properties of Materials   336

A. The Nature of the Atomic Magnets 336
B. Diamagnetism, Paramagnetism, and Ferromagnetism 338
C. Paramagnetism 342

21. The Partition Function   349

　　A. Definition   349
　　B. Calculation of Mean Values   350
　　C. Entropy and Helmholtz Free Energy   354
　　D. Many Subsystems and Identical Subsystems   357
　　E. The Partition Function for a Gas   362

Appendix 21A.   The Series $\sum\limits_{n} e^{-nx}$ 371

22. Chemical Equilibrium   373

　　A. The Thermodynamical Potential   374
　　B. Changes in Chemical Potential   376
　　C. The Law of Mass Action   378

Appendix 22A. $\mu = -kT \ln \dfrac{\zeta}{N}$ 383

23. Equilibrium Between Phases   384

　　A. Phase Diagrams   386
　　B. A Model   389
　　C. Pressure and Temperature Dependence of Phase Transitions   396

PART 8.   QUANTUM STATISTICS   401

*24. The Occupants of Quantum States   403

　　A. The Occupation Number, $\bar{n}$ 404
　　B. Comparison with Classical Statistics   409
　　C. The Limits of Classical Statistics   419

25. Survey of Applications   422

　　A. The Spectra of Accessible States   422
　　B. Quantum Gases   429
　　　　B.1   The Particle Distribution   429
　　　　B.2   Internal Energy and the Gas Laws   435
　　C. Other Quantum Systems   438

Appendix 25A.   The Gas Law for Quantum Gases   439
Appendix 25B.   Pressure Exerted by a Gas   441

26. Blackbody Radiation   443

　　A. Photons in an Oven   443

B. Principle of Detailed Balance 448
C. Energy Flux 450

27. The Thermal Properties of Solids   455

A. Lattice Vibrations 459
   A.1) The Einstein Model 459
   A.2) The Debye Model 461
B. Conduction Electrons 470
C. Heat Capacities 476

Appendix 27A. The Integral (27.16)   480
Appendix 27B. Debye Model Notation   481

28. Semiconductors and Insulators   482

A. Band Structure 482
B. Electrical Properties 487
C. Impurities 492

29. Low Temperatures   498

A. Attaining and Measuring Low Temperatures 500
B. Superfluidity 508
   B.1) The Phenomenon 508
   B.2) A Model 511
C. Superconductivity 518
   C.1) The Phenomenon 518
   C.2) A Model 519
   C.3) Stability of the Superconducting State 520

INDEX   525

INTRODUCTION TO
STATISTICAL MECHANICS AND
THERMODYNAMICS

PART 1

INTRODUCTION

This is an introductory college text, written for students who are studying statistical mechanics and thermodynamics for the first time. It assumes no prior acquaintance with the subject.

Chapters 1 and 2 present preliminary material, telling us the directions we will be going, and establishing the base from which we start. The actual study of statistical mechanics and thermodynamics begins in Chapter 3 for small systems, and in Chapter 5 for large ones.

chapter 1

THE TRANSLATION BETWEEN MICROSCOPIC AND MACROSCOPIC BEHAVIOR

If you were a very tiny bug, somehow capable of watching individual atoms, you would find them to be extremely exciting, dancing wildly about with a variety of electronic motions and configurations, and dazzling displays of interactions with neighboring atoms. By contrast, the behavior of a very large number of atoms, such as a baseball or planet, is quite sedate. Their positions, motions, and properties change continuously and predictably. How can the behavior of macroscopic systems be so predictable if their microscopic constituents are so unruly? Shouldn't there be some connection between the two?

Imagine that you are flipping a coin and that you have to predict whether it will land "heads" or "tails." Whichever way you call it, your chances of being wrong are rather high; 50% of the time you will be wrong. However, if a large number of identical coins are being flipped, you are quite safe in predicting that roughly half will be "heads" and half "tails." The larger the number of coins, the more reliable will be your prediction.

Similarly, if you are to call the outcome of rolling a single die (Figure 1.1), you will find your chances of being wrong are quite high. No matter what you predict, $\frac{5}{6}$ of the time you will be wrong. But if a large number of identical dice are rolled, then you can quite correctly predict that roughly $\frac{1}{6}$ of them will show one dot up, $\frac{1}{6}$ will show two dots up, etc. Again, the larger the number of dice, the more reliable your prediction will be.

Your predictions could go the other way, too. In the foregoing example, we used the known probabilities for one single die to predict the outcome of rolling large numbers of them. You could have observed the outcome first, and used this to infer

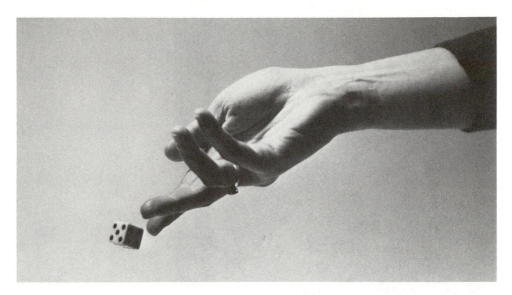

Figure 1.1 What is the probability that a rolled dice will land with two dots up? If a large number of dice were rolled, roughly what fraction of them would land with two dots up? (Elyse Rieder)

the probabilities involving a single die. For example, if you observed that in a large number of rolls of identical dice, significantly more than $\frac{1}{6}$ of them landed with one dot up, you could correctly infer that the dice were weighted somehow, giving a larger probability of any one roll landing with one dot up.

In no field of science will the system under study be a large number of flipped coins or rolled dice, although many systems will be similar to these in some ways. Nor is it the intent of this book to improve your skills as a Mississippi River boat gambler, although this may be one by-product. Rather the foregoing examples show that when a system is composed of a large number of identical elements, then you can use the observed behavior of an individual element to predict the properties of the whole system, or, conversely, you can use the observed properties of the entire system to deduce the probable behaviors of the individual elements. (See Figure 1.2.)

The study of this two-way translation between the behavior of the elements of a physical system and the properties of the system as a whole is called "statistical mechanics." One of the goals of this book is to give you the tools for making this translation, either direction, for whatever systems you wish.

PROBLEMS

1-1. Suppose you flip one million identical coins and find that six of them ended up standing on edge. If you should flip one more such coin, about what is the probability that it will end up standing on its edge? (*Answer.* 6×10^{-6}.)

1-2. Imagine that you were to deal one card from a well-shuffled deck of 52 playing cards. What is the probability that the card will be an ace? (*Answer.* $\frac{4}{52}$, since there are 4 aces in a deck.)

Figure 1.2 If you know the probabilities for one single coin flip, then you can predict the heads-tails distribution for a large number of them. Similarly, if you observe the heads-tails distribution of a large number of them, then you can infer the probabilities for any one single flip. (Fred Lyon/Photo Researchers)

1-3.   Suppose you were to deal one card from each of one million well-shuffled decks of 52 playing cards. How many of the dealt cards would be aces? (*Answer.* 7.7×10^4.)

1-4.   The outer edge of a roulette wheel is divided into 38 slots. When the wheel is spun, a ball bounces around until it ends up in one of these slots. Under normal conditions, there are equal probabilities for the ball to end up in each slot. As an observer of roulette games, how could you tell if the wheel had somehow been tampered with, improperly balanced, etc.?

In the above examples the translation between the behavior of the individual elements and the properties of the system as a whole was quite trivial, and you may wonder why a text this size is required to explain something so simple. The justification for the remainder of this text is that we would like to equip you to handle problems more sophisticated than the foregoing examples. This increased sophistication stems from two sources as explained in the following paragraphs.

First, the information we wish to know may be considerably more quantitative than that in the foregoing examples. For instance, we may wish to know not only that *roughly* half the flipped coins will land heads up, but we may wish to know something about the kinds of fluctuations about this value we might expect. After all, if we flip 100 coins, it would be quite unlikely that the heads-tails distribution would be *exactly* 50–50. Could we expect that normally somewhere between 45 and 55 would be "heads"? If not, what would be the range in values over which we could normally expect the number of "heads" to be? Once the coins have been flipped and their heads-tails configuration observed, what kinds of changes might we expect to see if we perturb the system, by jiggling the table, for example? Again, we have used the coins for illustration, but similar quantitative questions may be asked and answered for any physical system.

Second, the individual elements of a system may be considerably more complex than coins or dice, each of which had only a limited number of equally probable configurations. The element may be a molecule, for example, with a very large number of possible positions, orientations, and kinds of motion. Furthermore, the various kinds of motion, positions, or orientations may not all be equally probable, and these probabilities may change with changing parameters such as temperature and pressure.

Changes in the behaviors of the individual elements will be reflected in changes in the properties of the system as a whole (Figure 1.3). The study of how the properties of systems vary with such things as changes in temperature and pressure is called "thermodynamics." We frequently study the thermodynamics of macroscopic systems in order to gain a better understanding of the individual microscopic elements, using the tools of statistical mechanics to translate the one into the other.

Figure 1.3 In comparing solid and liquid forms, we see that differences in the probable behaviors of the individual water molecules are reflected in differences in the thermodynamic properties of the system as a whole. (K. Bendo)

PROBLEMS

1-5. Flip a coin twice. What percentage of the time did it land heads? Repeat this 4 times, each time recording the percentage of the two flips that were heads. Now flip the coin 20 times, and record what percentage of the 20 flips were heads. Repeat. For which case (2 flips, or 20 flips) is the outcome generally closer to a 50–50 heads-tails distribution?

1-6. If you flip 20 coins, why would it be unwise to bet on exactly 10 landing heads up?

1-7. A certain puddle of water has 10^{25} identical water molecules. As the temperature of this puddle falls to 0°C and below, the puddle freezes—a considerable change in the thermodynamic properties of this system. What do you suppose happens to the individual molecules to cause this remarkable change in the properties of the macroscopic system? (See Figure 1.3.)

It is not possible in one book to adequately illustrate the broad spectrum of areas of study where statistical tools are needed. Clearly, the fields of biology, geology, chemistry, meteorology, oceanography, physics, electronics, metallurgy, astronomy, etc., all have their own particular systems of interest, and each of these fields has many subfields, each with its own particular types of systems to study. These tools are also being used in fields where the individual elements are neither microscopic nor quite identical, such as in agriculture and forestry, and in "soft" sciences such as sociology and politics.

Furthermore, even the systems under study in any one discipline change over the years. The last 30 years in particular have seen a rapid expansion in the areas where the tools of statistical mechanics have been applied, with a corresponding expansion in our resulting understanding of nature. Undoubtedly, this proliferation of studies will continue.

Consequently this book is not intended to cover all the systems under study today, nor to predict what systems will be studied in the future. Instead, this book equips you with the necessary tools to handle the translation between microscopic and macroscopic properties of systems you may encounter in your career, whatever these systems may be. As long as the system is composed of large numbers of identical elements (or is composed of subsystems, each having large numbers of identical elements), then the tools presented in this text will be appropriate.

PROBLEMS

1-8. List eight systems composed either (1) of large numbers of identical elements or (2) of subsystems, each having large numbers of identical elements that you think might be of interest in any field of study.

There are certain general types of systems that are of interest over a broad spectrum of fields of study, and so we discuss these in this book. We use them to illustrate the use of statistical techniques and to help you understand the thermo-dynamical properties of these types of systems, and how these properties are related to the behavior of the individual elements. Many of these examples will be similar to systems you encounter in your careers, and therefore, can serve as a guide for you.

SUMMARY

If a system is composed of a large number of identical elements, then the probable behaviors of an individual element may be used to predict the properties of the system as a whole, or the properties of the system as a whole may be used to infer the probable behaviors of an individual element. The study of the statistical techniques used to make this two-way translation between microscopic and macroscopic behaviors of physical systems is called "statistical mechanics." The study of how changes in temperature and pressure (or similar parameters) affect the properties of systems is called "thermodynamics." From the thermodynamics of macroscopic systems we gain a better understanding of their microscopic elements.

chapter 2

MICROSCOPIC BEHAVIORS AND QUANTUM STATES

There are many properties of the elementary constituents of matter that we do not understand. We do not know fundamentally what "charge" or "mass" is, nor do we understand why one of these is quantized and the other isn't. Similarly, we do not know why particles behave like waves. This behavior results in many curious properties, including the quantization of angular momentum and in uncertainties in the values of positions and momenta. Nonetheless, whether we understand them or not, these are important properties of the microscopic constituents, and we must be prepared to use them in our translation between microscopic and macroscopic behaviors.

The basic elements of many interesting physical systems are very tiny, and their behaviors are influenced by quantum effects. Because some readers of this book may not yet have had a course in modern physics, we outline in this section some quantum effects that are important in common physical systems.

A. CHARGE QUANTIZATION

For reasons we do not yet understand, nature has provided electrical charge in fundamental units of 1.6×10^{-19} coulombs, a unit that we identify by "e."

$$e = 1.6 \times 10^{-19} \text{ C} \tag{2.1}$$

Most ordinary matter is composed of tiny atoms of one type or another, each atom having dimensions of roughly 10^{-8} cm. Each atom, in turn, is composed of a very tiny, dense, positively charged nucleus surrounded by a negatively charged cloud of electrons. The nuclear dimensions are roughly 10^{-4} times the size of the electron cloud.

The nucleus is composed of individual "nucleons," each of which is nearly 2000 times more massive than an electron. There are two types of nucleons. Some are electrically neutral and are called "neutrons," and others carry a charge of $+e$, and

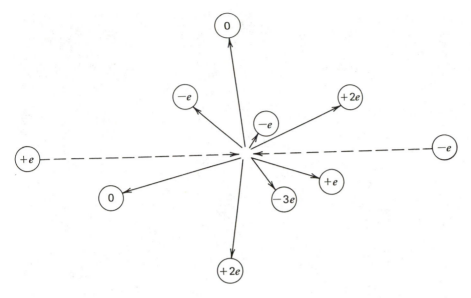

Figure 2.1 At very high energies, collisions between elementary particles often produce large numbers of "fragments." The amount of charge on any one of these fragments is always some integral number of units of e, and the total amount of charge on all particles together does not change during the collision.

are called "protons." The electrons are negatively charged, each carrying a charge of $-e$, so in an electrically neutral atom, there must be as many electrons in the electron cloud as there are protons in the nucleus.

We can shoot very high energy probes into electrons or nucleons in order to test their structure (Figure 2.1). When we do this, we find we can produce a large number of different kinds of particles as "fragments" from the collision. These fragments may consist of a wide variety of types, but the charge of each is always found to be an integral number of units of e (eg., $+2e$, 0, $-e$, etc.).

PROBLEMS

2-1. Roughly how thick is an atomic nucleus, measured in centimeters (cm)? In angstroms (Å)? In meters (m)?

2-2. What is the charge of a proton? An electron? A neutron?

2-3. After combing your hair, you find your comb has a net charge of -1.92×10^{-18} C. How many extra electrons are on your comb?

B. WAVE NATURE OF PARTICLES

At the turn of the century, it was thought that energy could be transported from one point to another through either of two distinct processes: the propagation of matter itself, or the propagation of waves. Until the 1860s, waves were thought to exist only

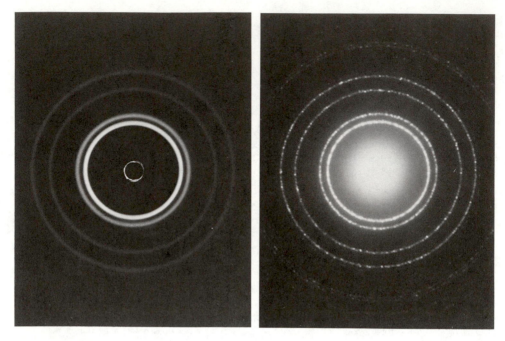

Figure 2.2   Diffraction patterns for X rays (left) and electrons (right) scattering from a periodic crystal lattice. This is a clear demonstration of the wavelike nature of electrons, because only waves could produce such a diffraction patterns. (Educational Development Center, Newton, MA.)

as the propagation of vibrations or oscillations through matter, but then the work of James Clerk Maxwell (1831–1879) demonstrated that electromagnetic radiation was also a type of wave, with the oscillations being in electric and magnetic fields rather than in matter. These waves could travel through space, without needing a material medium at all. Experiments with appropriate slits and diffraction gratings showed electromagnetic radiation to display the same diffractive behavior as mechanical waves in material media (e.g., sound waves, or ocean waves).

Then in the early twentieth century, some experiments were performed that began to blur the distinction between the two forms of energy transport. The photoelectric effect demonstrated that electromagnetic "waves" could behave like particles, and other experiments showed that "particles" could behave like waves. Shooting beams of "particles," such as electrons or protons, onto appropriate diffraction gratings yielded diffraction patterns for the reflected beam, just like waves (Figure 2.2). The wavelength (λ) for these "particle-waves" was found to be inversely proportional to the particle's momentum (p), and was governed by the same equation used for electromagnetic waves in the photoelectric effect.

$$\lambda = \frac{h}{p}$$

$$h = 6.63 \times 10^{-34} \text{ J} \cdot \text{s} \tag{2.2}$$

An alternative way of expressing this is to write the momentum of a particle in terms of its wave number, $k = 2\pi/\lambda$.

$$p = \hbar k$$

$$\hbar = \frac{h}{2\pi} = 1.05 \times 10^{-34} \, \text{J} \cdot \text{s}$$

(2.3)

We do not know *why* particles behave as waves any more than we know *why* electrical charge comes in fundamental units of "*e*." Hopefully, some of you will someday help us unravel fundamental mysteries such as these. But the experiments have been sufficiently thorough as to offer us no choice but to accept that "particles" do behave as waves with the above relationship between their wavelengths and momenta. Given this, we can set up wave equations to describe any system of particles we like.

The machinery for setting up these equations, solving them, and using these solutions to describe the behavior of the system of interest is called "quantum mechanics." In the next few pages we describe some of the results of these calculations applied to particles and particle systems that are of general interest. The validity of these results have been verified by appropriate experiments.

PROBLEMS

2-4. What is the wavelength associated with an electron moving at a speed of 10^6 m/s? What is the wavelength associated with a proton moving at this speed? What is the wavelength of a 70-kg sprinter running at 10 m/s?

C. UNCERTAINTY PRINCIPLE

Any single-valued function of the variable x can be written as a linear combination (or "superposition") of sine-wave components, having various values for the wave numbers, k (Figure 2.3). These sine-wave components may be of the form $\sin kx$ and $\cos kx$, or equivalently, e^{ikx} and e^{-ikx}, and the technique used to determine the contributions of each component to any function, $f(x)$, is called "Fourier analysis." Written mathematically, for any single-valued function $f(x)$, we can write

$$f(x) = \sum_{k=0}^{\infty} (a_k \sin kx + b_k \cos kx)$$

(2.4)

or equivalently,

$$f(x) = \sum_{k=-\infty}^{\infty} c_k e^{ikx}$$

(2.5)

where a_k, b_k, and c_k are the amplitudes of the kth component of the function $f(x)$.

Now let us consider the "wave function" that describes some particle, and let us zero in on the behavior in the variable x. (For the moment, we ignore the dependence on y, z, or t, holding them constant.) A particle is localized in space, and therefore its wave function also is localized. However, the sine-wave components each extend

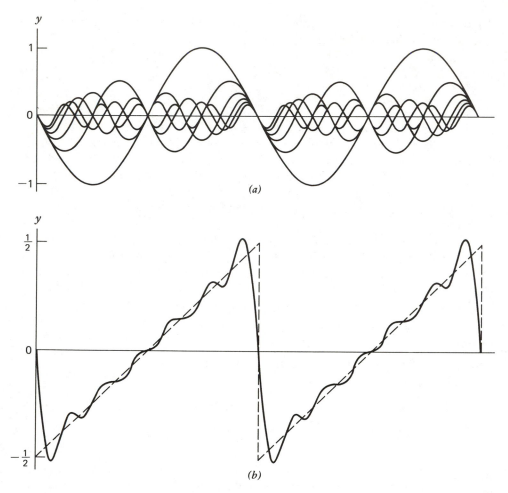

Figure 2.3   The superposition of the sine waves (a) yields the sawtooth wave (b). (After Halliday and Resnick, *Physics,* 3rd ed., John Wiley and Sons, 1978.)

forever, from $x = -\infty$ to $x = +\infty$. Consequently, if we are to construct a localized wave function from the superposition of infinitely long sine-wave components, then the superposition must be such that the components all cancel each other out everywhere except for the appropriate small region of space (Figure 2.4).

To accomplish this cancellation requires an infinite number of sine-wave components, but the bulk of the contributions will generally come from components whose wave numbers, k, lie within some finite region, Δk. As we do the Fourier analysis of various wave functions, we find a very interesting result. The more localized the wave function is in x, the broader will be the characteristic spread in k of the individual components. In fact, for most reasonable wave functions, the two can be shown to be inversely related. If Δx represents the characteristic width of the localized wave function, and Δk the characteristic spread in wave numbers of

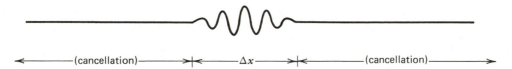

◄————(cancellation)————►◄————Δx————►◄————(cancellation)————►

Figure 2.4 In a localized wave function, the sine wave components must cancel each other out everywhere except for the appropriate localized region of space, Δx.

the sine-wave components, then

$$\Delta k = \frac{2\pi}{\Delta x}$$

or

$$\Delta x \, \Delta k = 2\pi \tag{2.6}$$

Using the relationship between wave number and momentum for a particle, (2.3), this becomes

$$\Delta x \, \Delta p_x = h \tag{2.7}$$

A wave function is a "probability amplitude," which means that the probability of finding the particle at point x is proportional to the square of the wave function amplitude there.* Also, the probability that the particle will have wave number k (or corresponding momentum $\hbar k$) is proportional to the square of the amplitude of the kth sine-wave component. So, the particle might be found anywhere within the range Δx, and might have any momentum in the range Δp. If we could make the most perfect and accurate measurements possible, then the uncertainty in our measured values of momentum and position would be related according to the foregoing expression. Of course, in practice our measurements will be less than perfect, and our resulting inaccuracies greater. So the "uncertainty principle" is more commonly written as

$$\Delta x \, \Delta p_x \geq h \tag{2.8}$$

PROBLEM

2-5. Suppose we know that a certain electron is somewhere in an atom, so our uncertainty in the position of this electron is the width of the atom, $\Delta x = 10^{-10}$ m. What is our minimum uncertainty in the x-component of its momentum? Of its velocity?

D. QUANTUM STATES AND PHASE SPACE

The fact that neither the particle's position nor its momentum can be precisely determined is not a reflection of our sloppy experimental techniques. Instead, it is a characteristic of the particle itself, resulting from its wavelike behavior. Wave functions were forced on us by our observations of nature; they were not something

---

* This may make more sense if you recall that the energy of waves in matter is proportional to the amplitude squared.

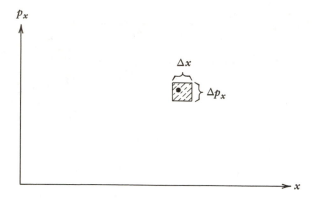

Figure 2.5 According to classical physics, a particle could be located as a point in $p_x - x$ space. That is, both its position and momentum could be specified exactly. In modern physics, however, the best we can do is to identify a particle as being somewhere within a box of area $\Delta x \, \Delta p_x = h$.

dreamed up by us for our own convenience. In fact, most of us would probably have voted against the idea, had nature given us our choice.

The uncertainty relationship (2.7) says that if we try to locate the coordinates of a particle in the two-dimensional space of (x, p_x), we will not find it at a certain point. Instead, the best we can hope to do is to say it is somewhere within a rectangle of area $\Delta x \, \Delta p_x = h$. (see Figure 2.5.) If we try to specify its position in x better, for example, then our uncertainty in p_x will increase, so the area of the rectangle $\Delta x \, \Delta p_x$ will not change (Figure 2.6).

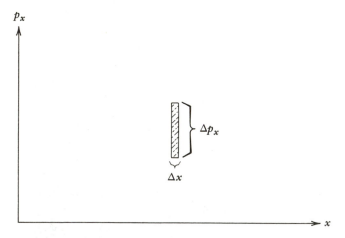

Figure 2.6 Because of the wave nature of particles, if we try to specify better the location of a particle in x-space, we lose accuracy in the determination of its momentum, p_x. The area of the quantum box (Δx Δp_x) does not change.

The position (x, y, z) and momentum (p_x, p_y, p_z) are the coordinates of a particle in a kind of six-dimensional space, called "phase space." In phase space, all coordinates are treated on an equal footing. The uncertainty relation (2.7) involves the uncertainties in the position and momentum coordinates in the two-dimensional phase space of (x, p_x). Identical relationships can be derived for uncertainties in the y and z coordinates.

$$\Delta y \, \Delta p_y = h \tag{2.7'}$$

$$\Delta z \, \Delta p_z = h \tag{2.7''}$$

We can multiply the three relationships (2.7, 2.7', 2.7'') together to get

$$\Delta x \, \Delta y \, \Delta z \, \Delta p_x \, \Delta p_y \, \Delta p_z = h^3 \tag{2.9}$$

This means that if we want to specify the position and momentum of any particle, the best we can do is to say that these coordinates are located somewhere within a six-dimensional quantum "box" or quantum "state" of volume $\Delta x \, \Delta y \, \Delta z \, \Delta p_x \, \Delta p_y \, \Delta p_z = h^3$.

As an example, consider a particle moving in the x-direction, confined to be located somewhere between 0 and x_0, and to have momentum somewhere between 0 and p_0 (Figure 2.7).

$$0 < x < x_0$$

$$0 < p < p_0$$

How many different quantum states will be available to this particle? In two-dimensional phase space, the number of different available states will equal the total

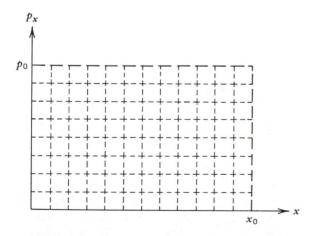

Figure 2.7 The total number of quantum states accessible to a particle whose momentum and position are confined to the region $0 \le p_x < p_0$, $0 \le x < x_0$, is equal to the total accessible area divided by the area of a single quantum state, h. The space that includes both momentum and position is called "phase space."

two-dimensional area available $(x_0 p_0)$ divided by the area of a single quantum state $(\Delta x \, \Delta p = h)$.

$$\text{(Number of states)} = \frac{\text{(total area)}}{\text{(area of one state)}} = \frac{x_0 p_0}{h} \tag{2.10}$$

We can quickly extend this reasoning to a particle in six-dimensional "phase space," whose position is confined to some finite region in x, y, and z, and whose momentum is confined to some finite region in p_x, p_y, p_z.

$$\begin{aligned} \text{(Number of states)} &= \frac{\text{(total six-dimensional volume)}}{\text{(volume of one quantum state)}} \\[2mm] &= \frac{\text{(total six-dimensional volume)}}{h^3} \\[2mm] &= \frac{V_r V_p}{h^3} \end{aligned} \tag{2.11}$$

where V_r and V_p are volumes in coordinate and momentum space, respectively. In particular, the number of quantum states available in the six-dimensional volume element $d^3 r \, d^3 p$ is given by $d^3 r \, d^3 p / h^3$.

$$\text{(Number of quantum states)} = \frac{1}{h^3} \, dx \, dy \, dz \, dp_x \, dp_y \, dp_z \tag{2.12}$$

This tells us that the number of quantum states included in any interval of any of the coordinates is directly proportional to the length of that interval. If q represents any of the phase space coordinates, then

$$\text{(Number of quantum states in interval } dq) \propto dq \tag{2.12'}$$

In many applications of statistical mechanics, we wish to calculate some property that requires us to do a summation over all states accessible to a particle. Since quantum states are normally very small and close together, it is often convenient and correct to replace the discrete summation by continuous integration, using the result (2.12).

$$\sum_{\text{states}} \rightarrow \int \frac{d^3 r \, d^3 p}{h^3}$$

Unfortunately, it is sometimes difficult to determine the precise regions of phase space over which these integrations should be carried out, because interactions between particles may constrain the regions of phase space accessible to them. For example, an atom in a solid is not allowed to roam throughout the entire solid, because interactions with its neighbors hold it to some small local region. Interactions with neighbors may similarly constrain the momenta accessible to particles in a system.

When interactions with neighbors are negligible, such as is true in a gas, the particles have access to all regions of phase space. If the system is confined to some volume V, then integration over all volume and all momentum directions (i.e., over the angles in

$d^3 p = p^2 \, dp \sin \theta \, d\theta \, d\phi$) gives

$$\sum_{\text{states}} \rightarrow \int \left(\frac{4\pi V}{h^3} \right) p^2 \, dp$$

If the particles are nonrelativistic,* then we can use $\varepsilon = p^2/2m$ to write this as a distribution of states in particle energies,

$$\sum_{\text{states}} \rightarrow \int \left[\frac{2\pi V (2m)^{3/2}}{h^3} \right] \varepsilon^{1/2} \, d\varepsilon$$

However, when the particles of a system interact appreciably with each other, these interactions may severely constrain the regions of phase space accessible to them, and the spectrum of accessible states may be quite different than that just given.

In general, we write

$$\sum_{\text{states}} \rightarrow \int g(\varepsilon) \, d\varepsilon \tag{2.13}$$

where $g(\varepsilon)$ is the number of accessible states per unit energy, and is therefore called the "density of states." As we saw above, the density of states for a system of nonrelativistic, noninteracting particles is given by

$$g(\varepsilon) = \left[\frac{2\pi V (2m)^{3/2}}{h^3} \right] \varepsilon^{1/2}$$

but for other systems where mutual interactions are not negligible, the density of states may be quite different from this. A good deal of experimental and theoretical research effort goes into the examination and illumination of the density of states accessible to the particles of various systems, because knowledge of the density of states allows us to analyze many properties of the systems, including the kinds of interactions among the individual particles.

PROBLEMS

2-6.   Consider a particle moving in one-dimension, which is confined to 10^{-4} m of space, and whose momentum is constrained to lie between -10^{-24} and $+10^{-24}$ kg m/s. Roughly how many different quantum states are available to it?

2-7.   Consider an electron moving in one dimension, which is constrained to be in a small region, 10^{-9} m long, and whose speed must be less than 10^7 m/s (i.e., the velocity is between $+10^7$ and -10^7 m/s). Roughly how many different quantum states are available to this electron?

2-8.   Consider a proton moving in three-dimensions whose motion is confined to be within a nucleus (a sphere of radius 10^{-14} m) and whose momentum must have magnitude less than $p_0 = 10^{-19}$ kg m/s. Roughly how many quantum states are available to this proton? (*Hint.* The volume of a sphere of radius p_0 is $\frac{4}{3}\pi p_0^3$.)

---

* That is, if the particles are moving at speeds well below the speed of light.

2-9.   A particle is confined to within a rectangular box with dimensions 1 cm by 1 cm by 2 cm. In addition, it is known that the magnitude of its momentum is less than 3 g-cm/s. How many states are available to it? (*Hint*. In this problem, the available volume in momentum space is a sphere of radius 3 g-cm/s.)

E.   QUANTIZATION OF ANGULAR MOMENTUM

Another surprising result of quantum mechanics is that the angular momentum of a particle (or system of particles) cannot be arbitrary. It must be expressed as

$$J = \sqrt{j(j+1)}\hbar \tag{2.14}$$

where j is either integer or half integer;

$$j = 0, 1, 2, \ldots \text{(“boson”)}$$

or

$$j = \frac{1}{2}, \frac{3}{2}, \frac{5}{2}, \ldots \text{(“fermion”)}$$

Furthermore, the component of angular momentum in any direction (e.g., the z-direction) can only have values of

$$J_z = j_z \hbar \tag{2.15}$$

where j_z is either integer or half integer, and can take on the following values only. (See Figure 2.8.)

$$j_z = j, j-1, j-2, \ldots -j \tag{2.16}$$

Since angular momentum always comes in these units of $\hbar$, we frequently identify the angular momentum of a particle by stating the value of j. A "spin $\frac{1}{2}$" particle, for example, is one for which $j = \frac{1}{2}$. Electrons, protons, and neutrons are common "spin $\frac{1}{2}$ particles." The magnitude of their angular momentum is given by

$$J = \sqrt{\frac{1}{2}\left(\frac{3}{2}\right)}\hbar = \frac{\sqrt{3}}{2}\hbar$$

and the z-component of this may have either of the values

$$J_z = +\frac{1}{2}\hbar \qquad \text{or} \qquad -\frac{1}{2}\hbar$$

which are commonly referred to as "spin-up" or "spin-down" states.

A "spin 1" particle, such as some mesons, would have total angular momentum given by

$$J = \sqrt{1(2)}\hbar = \sqrt{2}\hbar$$

and $J_z = \hbar, 0, -\hbar$.

The quantum mechanical origin of these strange restrictions on angular momentum, lies in the observation that if either the particle or the laboratory is rotated by a

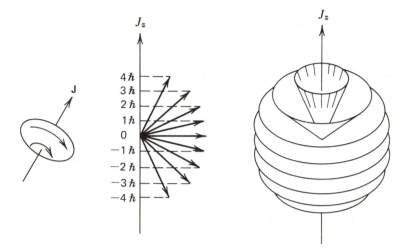

Figure 2.8 The total angular momentum of a particle or system of particles may take on any of the following values: $J = \sqrt{j(j+1)}\hbar$, where $j = 0, \frac{1}{2}, 1, 1\frac{1}{2}$, etc. The component of this along any one specified axis (e.g., the z-axis) may take on only the following values: $J_z = j_z\hbar$, where $j_z = j, j-1, j-2, \ldots, -j$. That is, both the total angular momentum and its orientation with respect to any one axis are quantized, being allowed only certain values. (What is the magnitude of the total angular momentum for the above particle?)

complete 360° around any axis, the observed situation will be the same as before the rotation. That is, for example, if you should turn completely around, both you and the world around you should be back to the same state you were before you turned. Nothing has changed, turned upside down, etc.

Since observations are correlated with the square of the wave function, this restriction on the wave function is that upon rotation by 360° around any axis, the wave function must remain unchanged, except for a possible change in sign. This provides a "boundary condition" on the wave function, and if you look carefully when studying this in your quantum mechanics course, you will see that this boundary condition provides the constraints on the possible values of angular momentum.

It turns out that particles with half-integer spin all have wave functions that change sign under rotation by 360°, and those with integer spin all have wave functions that do not change sign under 360° rotation. Particles with half-integer spin are called "fermions," and those with integer spin are called "bosons." Fermions and bosons each obey a different type of quantum statistics as we will see in a later chapter.

The angular momentum of a system of particles is due both to the spin of the individual particles, and due to their orbital angular momentum (Figure 2.9). In such cases, the symbol J is reserved for the total angular momentum. The symbols S and L are then used to designate spin and orbital angular momentum, respectively, being

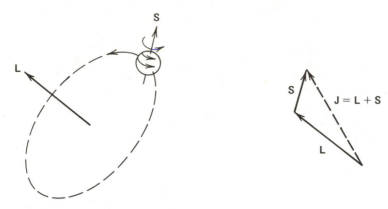

Figure 2.9   The total angular momentum of an electron in orbit about an atom (**J**) is the sum of its orbital angular momentum (**L**) plus its spin angular momentum (**S**). The total angular momentum of any system of particles is the sum of their total orbital angular momentum plus their total spin angular momentum. Because of quantum effects, the angular momenta of any single particle or system of particles are allowed only the following values:

$$J = \sqrt{j(j+1)}\hbar \qquad j = 0, \tfrac{1}{2}, 1, 1\tfrac{1}{2}, 2, \ldots$$

$$L = \sqrt{l(l+1)}\hbar \qquad l = 0, 1, 2, 3, \ldots$$

$$S = \sqrt{s(s+1)}\hbar \qquad s = 0, \tfrac{1}{2}, 1, 1\tfrac{1}{2}, 2, \ldots$$

with the condition that when s is integer (half integer) j is integer (half integer).

expressable by

$$S = n\sqrt{s(s+1)}\hbar$$

$s =$ integer or half integer, depending on the particle   (2.17)

and

$$L = \sqrt{l(l+1)}\hbar$$

$l =$ integer only   (2.18)

For example, consider a hydrogen atom with the electron in an $l = 0$ (no orbital angular momentum) orbit. Then the total angular momentum of the atom is due entirely to the spins of the electron and proton only. Since they are both spin $\tfrac{1}{2}$ particles, the z-component of the total angular momentum of this atom could take on the following values.

$$S_z = +1\hbar \qquad \text{(both spin up)}$$

$$S_z = 0 \qquad \text{(one spin up, the other spin down)}$$

$$S_z = -1\hbar \qquad \text{(both spin down)}$$

and the total spin angular momentum ($\sqrt{s(s+1)}\hbar$) could be characterized by $s = 1$ or 0, depending on the relative orientations of the two spins. Notice that two fermions combine to make a boson, no matter how the spins are aligned.

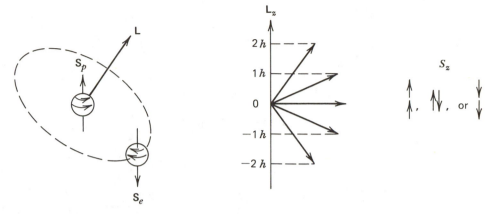

Figure 2.10 Illustration of the possible orientations of the orbital angular momentum and spin angular momentum for a hydrogen atom with the electron in an $l = 2$ orbit. The proton and electron are both spin $\frac{1}{2}$ particles ($S_z = \pm\frac{1}{2}\hbar$ for each). The possible values for the magnitude of the total angular momentum are $J = \sqrt{j(j + 1)}\,\hbar$, where $j = 3$, 2, or 1. Why is $j = 0$ not possible in this case?

Now let us extend the above example to the case where the electron is in an orbit with angular momentum given by $l = 2$. The z-component of this orbital angular momentum could possibly take on the values

$$L_z = (2, 1, 0, -1, -2)\hbar$$

depending on the orientation. Combining this with the considerations in the previous paragraph regarding the spin orientations of the two spin $\frac{1}{2}$ particles, we see that the z-component of the atom's total angular momentum (Figure 2.10) can range from $J_z = +3\hbar$ ($L_z = 2\hbar$, and both particles are spin up) to $J_z = -3\hbar$ ($L_z = -2\hbar$, and both particles are spin down).

$$J_z = (3, 2, 1, 0, -1, -2, -3)\hbar$$

Furthermore, the total angular momentum of the atom, $\sqrt{j(j + 1)}\,\hbar$, could be characterized by $j = 3, 2,$ or 1, depending on how the particles' spins are aligned relative to the orbital angular momentum. Notice again, that a system of two fermions is a boson, no matter what the orbital angular momentum or the orientation of their spins. In general, a system of an even number of fermions is a boson, and a system of an odd number of fermions is a fermion.

PROBLEMS

2-10. A hydrogen atom is sometimes found in a state where the spins of the proton and the electron parallel each other. Yet the total angular momentum of the atom is zero. How is this possible?

2-11. What are the possible values of the total angular momentum quantum number, j, for a hydrogen atom, if the electron is in a $l = 1$ orbit? What orientations of electron spin, proton spin, and electron orbital angular momenta could give each value for j?

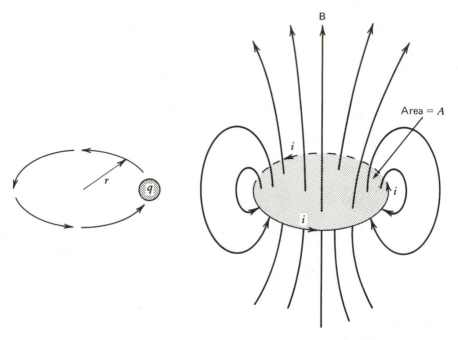

Figure 2.11   An orbiting electrical charge is a current loop. The magnetic moment of a current loop is given by $\mu = iA$, where *i* is the electrical current in the loop, and *A* is the area of the loop.

F.   MAGNETIC MOMENTS

Moving charges cause magnetic fields. A current flowing in a loop is an electromagnet, with magnetic moment (μ) equal to the product of the current intensity and the area of the loop (Figure 2.11).

If a charged particle is in an orbit, then the magnetic moment of this circulating charge will be proportional to its charge, *q*, its angular momentum, *L*, and inversely proportional to its mass, *m*.

$$\mu \propto \frac{q}{m} L$$

The reason for this relationship is that current intensity is proportional to the particle's *charge* and speed, whereas angular momentum is proportional to its *mass* and speed. So to convert from angular momentum to magnetic moment, we must divide out the mass and insert the charge in its place. We derive this result rigorously in Appendix 1A at the end of this chapter, where we show

$$\mu = \frac{q}{2m} \mathbf{L} \tag{2.19}$$

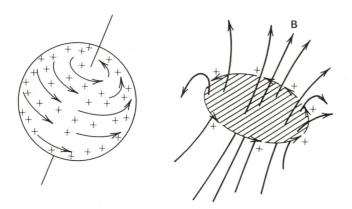

Figure 2.12 A spinning charged particle is a current loop, generating its own magnetic field, and having its own characteristic magnetic moment.

Because the angular momentum of a particle is quantized, its magnetic moment is quantized too. The energy of interaction of a magnetic moment with a magnetic field, **B**, is given by $-\boldsymbol{\mu}\cdot\mathbf{B}$. If we define the z-direction to be that of the magnetic field, then we have the interaction energy, U, given by

$$U = -\mu_z B \tag{2.20}$$

But since μ_z can have only certain discrete values, the energy U also can have only certain discrete values.

So far we have talked about the magnetic moment of a charged particle caused by its orbit. But a spinning charged particle is also a current loop (Figure 2.12), and therefore also has a magnetic moment. So the total magnetic moment of a particle may be due both to its spin and its orbit, and both are quantized.

That portion of a particle's magnetic moment caused by its spin, is not exactly equal to $(q/2m)S$, but rather is proportional to it.

$$\boldsymbol{\mu}_{\text{spin}} = g\,\frac{e}{2m}\,S \tag{2.21}$$

The constant of proportionality, g, is called the "gyromagnetic ratio." Each particle has its own characteristic gyromagnetic ratio, with that for the electron being $g = -2$, for the proton being $g = +5.58$, and that for the neutron being $g = -3.82$. (e is the fundamental unit of charge, 1.6×10^{-19} C.)

A final observation we should make concerning the result (2.19) is that the magnetic moment is inversely related to the mass. Therefore, if two particles have the same charges and the same spin or orbital angular momenta, the one with the smaller mass will have the greater magnetic moment. When we study magnetism in materials, we frequently consider electrons only. The reason is that protons are nearly 2000 times more massive, and therefore magnetic contributions from the nuclei will normally be nearly 2000 times smaller than those from the electrons.

PROBLEMS

2-12. Use the relationship (2.21) to estimate the magnetic moment of a spinning electron, given that an electron is a spin $\frac{1}{2}$ particle. If this electron were placed in an external magnetic field of 1 tesla (T), what would be the two possible values of its magnetic interaction energy? (1 tesla = 1 weber/m^2 = 1 J·s/C·m^2.)

2-13. Repeat Problem 2-12 for a proton.

G. HARMONIC OSCILLATOR

Consider any object in stable equilibrium, such as a marble in the bottom of a bowl, a pendulum hanging straight down, or an atom inside some solid. For any such object, displacement from its equilibrium position will require a force proportional to the displacement, for sufficiently small displacements. The determination of how small is "sufficiently" small and in what coordinate the displacement should be measured are important considerations, but we will not consider them here, since they depend on the specific system of interest. This relationship is frequently expressed as

$$F = -kx \tag{2.22}$$

where F is the restoring force (equal and opposite to the applied displacing force), x is the displacement from equilibrium, and k is the constant of proportionality.

When the displaced object is released, it will oscillate across its equilibrium position in what is called "simple harmonic motion." If the object has mass, m, the angular frequency of its oscillation is given by

$$\omega = \sqrt{\frac{k}{m}} \tag{2.23}$$

The application of quantum mechanics to this problem yields the interesting result that only certain energies are possible, given by

$$E = \left(n + \frac{1}{2}\right)\hbar\omega \qquad n = 0, 1, 2, \ldots \tag{2.24}$$

for a one-dimensional harmonic oscillator, and

$$E = \left(n + \frac{3}{2}\right)\hbar\omega \qquad n = 0, 1, 2, \ldots \tag{2.25}$$

for a three-dimensional harmonic oscillator. (See Figure 2.13.) The basic cause is that if a particle oscillating back and forth is treated as a wave, then a wave going equally in both directions yields a standing wave, and only standing waves of certain wavelengths (hence, certain momenta and energies) can fit into a restricted region of space, such as that defined by the harmonic oscillator potential.

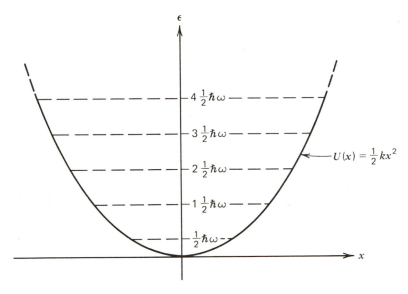

Figure 2.13 Particles within a potential well are allowed only to be in quantum states of certain energies, with finite spacings between them. The energies allowed a harmonic oscillator (i.e., a particle whose potential energy is given by $\frac{1}{2}kx^2$) are $(n + \frac{1}{2})\hbar\omega$, where n is an integer and ω is the angular frequency determined in the normal way from the force constant and the particle's mass $(\omega = \sqrt{k/m})$.

Frequently, students see the harmonic oscillator as a very specialized case and, therefore, not very useful. On the contrary, the elements of many systems in nature, including the atoms in most solids, are found at or near their equilibrium positions, so their behaviors under displacement (e.g., thermal agitation) are often very well described by this model.

The harmonic oscillator is a special case of a particle in a potential well, where the shape of the potential well is described by

$$U_{\text{h.o.}} = \frac{1}{2}kx^2 \qquad (2.26)$$

Although this one special case has a wide range of applications in nature, other kinds of potential wells are possible. However, they all show the same general characteristic—that the particle's energy may take on only certain discrete values, although the particular spectrum of energies allowed may change from one kind of potential well to another. Furthermore, this quantization of allowed energies always arises from the same fundamental consideration: if particles behave as waves, and these move back and forth in a confined region of space, then they create standing waves. Standing waves of only certain wavelengths (therefore, only certain momenta or energies) may fit in any confined region of space.

H.   COULOMB POTENTIAL

In addition to the harmonic oscillator, another type of potential well with fairly wide range of usefulness is the Coulomb potential

$$U = -k_C \frac{Ze^2}{r}$$

(2.27)

where

$$k_C = \frac{1}{4\pi\varepsilon_0} = 9 \times 10^9 \ \text{N·m}^2/\text{C}^2$$

This would, for example, approximate the potential "seen" by an outer electron on any atom or ion, where $+Ze$ was the net central charge due to the nucleus and other electrons.

The result of quantum mechanical calculations for this Coulomb potential is that the allowed energies are given by

$$E = -\frac{1}{n^2} (Z^2 E_0)$$

$$E_0 = 13.6 \ \text{eV}$$

$$n = 1, 2, 3, \ldots$$

(2.28)

PROBLEMS

2-14.   What is the Coulombic force between an electron and a proton separated by 0.5×10^{-10} m? If this same amount of force were due to a spring stretched by $x = 0.5 \times 10^{-10}$ m, what would be the force constant, k, of this spring?

2-15.   Suppose an electron was connected to a proton by a spring with spring constant equal to that which you calculated in Problem 2-14. What would be the angular frequency, ω, for the electron's oscillations? (Use Eq. 2.23.) What would be the lowest possible energy of oscillation for this electron? (*Hint*. Use Eq. 2.24 with $n = 0$.] How does this compare with the magnitude of the energy of an electron in the lowest level in a hydrogen atom (Eq. 2.28)?

I.   DESCRIPTION OF A STATE

We began Chapter 1 with a pair of examples, involving coins and dice. Each of these things could have only a limited number of configurations; a coin had only two, and a die had six.

Then in this chapter, we see that some characteristics of microscopic components of real physical systems also come only in discrete values. The charge of a particle cannot be arbitrary, but must be some finite integral number of units of "e." If the phase space available to a particle is finite, then it will be in one of only a finite number of discrete quantum states.

The angular momentum of a particle can only take on certain discrete values, as can its magnetic moment. Their orientations are further restricted so that the z-component of angular momentum or magnetic moment can only be one of a certain number of discrete values. Among other things, this means that the energy of magnetic interaction between a charged particle and an external magnetic field can only be one of a finite number of discrete values.

We also saw that the energies of particles confined in space by some potential, could only take on certain discrete values, and we gave explicit expressions for particles in harmonic oscillator and Coulombic potentials.

In each of these cases, we zeroed in on one aspect of the behavior of a particle and saw that it could be described rather simply. But the individual elements of a system may have many such properties. Although each individual property may be simple, the sheer number of different properties may make complete description of the element impossible.

For example, if the element of some system is an atom, then we may describe the state of the outer electron by giving the appropriate number, n, in Eq. 2.28. But the atom may have many inner electrons also, each being in a different state, and these each being more complicated than that of the outer electron. Then we must consider the nucleus. We would have to give the orbital and spin angular momenta of each particle and its magnetic moment. We would have to include a description of all nuclear, electromagnetic, weak, and gravitational interactions among these particles. When we finally described the internal components of any one atom, then we would have to begin describing their interactions with neighboring atoms and other external fields. Clearly, the complete description of the one atomic element could grow until it filled several volumes, and still not be complete.

Fortunately, in any particular problem there will be only one or two properties of the element that would be relevant, so we can ignore all others. For example, in considering the magnetic properties of a material, we may wish to know the magnetic moment of the outer electron only, and nothing else. Or, if we were interested in the thermal properties of a material, we may wish to know the vibrational state (the number "n" in Eq. 2.25) of the atom as a whole, and nothing else. As another example, at the start of the previous chapter, we needed only consider the heads-tails properties of the coins. Their colors, compositions, designs, interactions with the table, etc., were irrelevant for the problem we were concerned with.

Consequently, when we describe the "state" of an element, we will only give the property that is relevant for the problem we are considering. To do more would be needlessly confusing.

The state of a system is determined by the state of each of the elements. For example, the state of a system of three coins would be determined by specifying the heads-tails configuration of each, and the spin state of three electrons may be specified by identifying which electrons are spin up and which are spin down.

You may see that when the system becomes large (10^{24} electrons, for example) the description of the system becomes hopelessly long. One statement for each of 10^{24} electrons is a lot of statements! Fortunately, we use statistical methods to describe the states of large systems, and the larger the systems, the simpler and more useful these

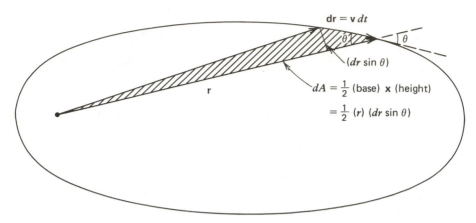

Figure 2.14   An elliptical orbit. The area of the shaded triangle is given by $dA = \frac{1}{2}(\text{base}) \times (\text{height}) = \frac{1}{2}(r) \times (dr \sin \theta) = \frac{1}{2}|\mathbf{r} \times d\mathbf{r}|$.

descriptions will be. In Chapter 3, we begin with small systems, and proceed to larger systems to illustrate the development and utility of some of these statistical techniques.

SUMMARY

Many important properties of the microscopic elements of a system are quantized. These include electrical charge, position and momentum coordinates, angular momentum, magnetic moment, and magnetic interaction energies. Also quantized are the energies of any particles confined to a restricted region of space by any kind of potential well or barrier.

The complete description of the state of an element of a system would normally be impossibly complicated. However, we will generally be interested in only one or two aspects of its behavior, and so we only mention these in our description. The state of a system is determined by specifying the state of each of its elements. This will be done statistically for large systems.

APPENDIX 2A   MAGNETIC MOMENT AND ANGULAR MOMENTUM

Consider a charge e moving in an elliptical orbit as in Figure 2.14. The area, dA, of the shaded triangle is

$$dA = \frac{1}{2}\,\text{base} \times \text{height}$$

$$= \frac{1}{2}\,|\mathbf{r}|\,|d\mathbf{r}\sin\theta|$$

$$= \frac{1}{2}\,|\mathbf{r} \times d\mathbf{r}|$$

If we use the normal right-hand rule for defining the direction vector for this current loop, we can see

$$dA = \frac{1}{2}\mathbf{r} \times d\mathbf{r}$$

Using

$$d\mathbf{r} = \mathbf{v}\, dt$$

and

$$\mathbf{L} = m\mathbf{r} \times \mathbf{v}$$

we have

$$dA = \tfrac{1}{2}\mathbf{r} \times \mathbf{v}\, dt = \frac{1}{2m}\mathbf{L}\, dt$$

The area of the entire loop is obtained by integrating the above. Since angular momentum is constant, we can take it out of the integral and get

$$A = \oint dA = \frac{1}{2m}\mathbf{L}\oint dt = \frac{1}{2m}\mathbf{L}T \tag{2A.1}$$

where T is the time in which the charge goes once completely around the loop. The current intensity, i, is the rate at which charge passes a given point on the loop. Since charge e passes any given point in time T, we have the current intensity given by

$$i = \frac{e}{T}$$

Combining this with the result (2A.1) for the area of the loop, we have for the magnetic moment

$$\mu = i A$$

$$= \frac{e}{T}\left(\frac{1}{2m}\mathbf{L}T\right) = \frac{e}{2m}\mathbf{L} \tag{2A.2}$$

PART 2
SMALL SYSTEMS

Most common systems we deal with have extremely large numbers of identical elements. For example, a glass of water has more than 10^{24} identical water molecules, and a gold ring has roughly 10^{23} identical gold atoms. A rock that you can hold in your hand contains more than 10^{23} identical molecules of each of several different kinds of minerals, and the room you are now in probably contains over 10^{27} identical nitrogen molecules and nearly as many identical oxygen molecules.

There are many smaller systems that are also quite important to us. Examples include many aspects of microelectronics, thin films and surface coatings, and many aspects of low temperature studies, where vibrational excitations in solids, vortices in liquids, charge carriers in many materials, etc., may be quite limited in number. The study of conduction in semiconductors and insulators, or switches in a computer may also involve systems of relatively small numbers of elementary components. We may wish to study some behavioral characteristic of a small population of plants, animals, or people, or we may wish to analyze the results of a small number of identical experiments. There are also many times in our everyday lives when we wish to predict the outcome of a relatively small number of similar endeavors.

From a statistical point of view, the "size" of a system is determined by the number of individual elements, rather than by the physical dimensions. For example, a herd of elephants is a "small" system and the conduction electrons in a pin is a "large" one, statistically speaking.

The thermodynamic properties of large systems are extremely predictable, even though the behavior of one individual element is not. This predictability allows us to use rather elegant and streamlined statistical tools in analyzing large systems. By contrast, the behaviors of smaller systems are more erratic and unpredictable (Figure 3.1). For these it will be necessary to analyze fluctuations as well as average values for various properties of interest. For this reason, the statistical tools used for small systems are quite detailed, and they become quite cumbersome when the number of elements in the system is large. Fortunately, this is the point where the more elegant methods for large systems become useful.

Aside from the fact that many small systems are important, and that students completing the course should feel comfortable handling any system, large or small, there are pedagogical reasons for including the study of small systems, and these suggest an advantage to studying them first. By starting with small systems, and working toward larger systems, we can better understand both the need and the underlying justification for the statistical tools used for larger systems, as developed in the succeeding chapters. Our minds can only comprehend small numbers. By starting with small, comprehensible systems, we can develop an intuition for the basic causes of the thermodynamic behaviors of larger systems, and an understanding of why the statistical methods work.

chapter 3

STATISTICS FOR SMALL SYSTEMS

In most systems, the behavior of the elementary constituents is quite unpredictable. Each could be doing several different things, and we cannot say with certainty which behavior will be displayed by any constituent at any time. At best we can only give the probabilities for the different possible behaviors. Assuming we know these probabilities, we should be able to predict the average values of various properties, even if we can't predict what they will be at any instant. Also, if we know the probabilities for the individual behaviors, we should be able to calculate the probabilities for the various possible behaviors of small systems of these constituents.

We begin our study of small systems by the straightforward application of conventional statistical techniques to systems of two elements, then three, then more. We will see that as the number of elements increases, the behavior of the system becomes more and more predictable (Figure 3.1).

A. MEAN VALUES

In this book we develop the machinery to relate the properties of a system of many elements to the probable behaviors of a single element. Since the individual elements may have many possible distinct behaviors or configurations, this will require us to do some "averaging" over possible individual behaviors. For example, when flipping a

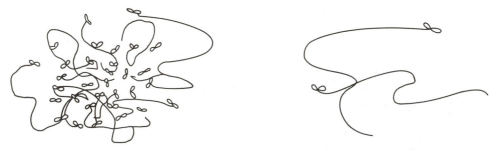

Figure 3.1   The behavior of a swarm of gnats is much more predictable than the behavior of just one or two. The larger the system, the more predictable is its behavior.

large number of coins, we know that roughly half of them will land heads up, since on the average a single coin will land heads up half of the time. In this case, the problem was simple enough for us to do in our heads, but we wish in this section to develop the method to take mean values of functions that may be more complex than the heads-tails configurations, and for systems more complex than coins.

To start our development, we imagine we have many identically prepared systems. We let P_s indicate the probability that a system is in the state s. Suppose we are interested in the average or "mean" value of some function f, which has the value f_s when the system is in state s. Then the mean value of f is determined as follows:

$$\bar{f} = \sum_s P_s f_s \tag{3.1}$$

SUMMARY

If f is some function that takes on the value f_s when the system is in state s, and if P_s is the probability of the system being in state s, then the mean value of f is given by

$$\bar{f} = \sum_s P_s f_s \tag{3.1}$$

where the sum is over all states s accessible to the system.

EXAMPLE

Suppose each system is a single coin, and f is the number of heads showing. That is, $f = 1$ when the coin is heads up, and $f = 0$ when the coin is tails up. What is the mean value of f?

Since the probability of landing heads is $\frac{1}{2}$ and that of landing tails is $\frac{1}{2}$, we have $P_{heads} = \frac{1}{2}$, $P_{tails} = \frac{1}{2}$. The mean value of f is then

$$\bar{f} = P_{heads} f_{heads} + P_{tails} f_{tails} = \frac{1}{2}(1) + \frac{1}{2}(0) = \frac{1}{2}$$

This tells us that the average number of heads per coin showing is $\frac{1}{2}$, or equivalently, heads will show up half the time.

EXAMPLE

Suppose each system is a single rolled die, and n indicates the number of dots showing upward. Suppose f is the number of dots showing upward, $f_n = n$. What is the mean value of f for this system?

For correctly balanced dice, $P_1 = P_2 = P_3 = P_4 = P_5 = P_6 = \frac{1}{6}$. The average value of f, then, is

$$\bar{f} = \sum_{n=1}^{6} P_n(n) = \frac{1}{6}(1) + \frac{1}{6}(2) + \frac{1}{6}(3) + \frac{1}{6}(4) + \frac{1}{6}(5) + \frac{1}{6}(6) = 3\frac{1}{2}$$

On the average, $3\frac{1}{2}$ dots will show up. If you rolled a million dice, you would expect your total score to be about $3\frac{1}{2}$ million.

EXAMPLE

Suppose that again the system is a single rolled dice, with n indicating the number of dots showing upward. But this time suppose $f_n = (n - 1)^2$. What is the mean value of f?

In this case the mean value of f would be given by

$$\bar{f} = \sum_{n=1}^{6} P_n(n - 1)^2 = \frac{1}{6}(0)^2 + \frac{1}{6}(1)^2 + \frac{1}{6}(2)^2 + \frac{1}{6}(3)^2 + \frac{1}{4}(4)^2 + \frac{1}{5}(5)^2 = 9\frac{1}{6}$$

There are two relationships that we may sometimes use to simplify the calculation of mean values: (1) the mean value of the sum of two functions equals the sum of their mean values and (2) the mean value of a constant times a function equals the constant times the function's mean value. These statements summarized mathematically are as follows.

SUMMARY

If f and g are functions of the state of a system and c is some constant, then

$$\overline{(f + g)} = \bar{f} + \bar{g} \tag{3.2}$$

and

$$\overline{cf} = c\bar{f} \tag{3.3}$$

These two relationships can be easily proven from the definition of mean values. (See Eq. 3.1.)

$$\overline{(f + g)} = \sum_{n} P_n(f_n + g_n) = \sum_{n}(P_n f_n + P_n g_n)$$

$$= \sum_{n} P_n f_n + \sum_{n} P_n g_n = \bar{f} + \bar{g} \tag{3.4}$$

$$\overline{cf} = \sum_{n} P_n(cf_n) = c\sum_{n} P_n f_n = c\bar{f} \tag{3.5}$$

PROBLEMS

3-1.   Suppose you roll a weighted dice, so that the probability of six dots showing upward is $\frac{1}{2}$, and the probability of it landing in each of the other five states is $\frac{1}{10}$.

(a)   What would be the average number of dots showing upward per roll? ($f_n = n$)

(b)   What would be the average of the square of the number of dots showing upward per roll? ($f_n = n^2$)

3-2.   The energy of a spin $\frac{1}{2}$ particle in an external magnetic field along the z-axis is $E = -\mu B$ if it is spin up, and $E = +\mu B$ if it is spin down. Suppose the probability of the particle being in the lower-energy state is $\frac{3}{4}$ and that of being in the higher-energy state is $\frac{1}{4}$. That is, $P_{up} = \frac{3}{4}$,

$P_{down} = \frac{1}{4}$, $E_{up} = -\mu B$, $E_{down} = +\mu B$. What would be the average value of the energy of such a particle, expressed in terms of μB?

3-3.   Suppose you flip a coin. If $f_{heads} = 5$ and $f_{tails} = 27$, what is the mean value of f per coin flip?

3-4.   Suppose you are to roll two properly balanced dice at a time. Die 1 has 6 different possible states and die 2 also has 6 different possible states, making a total of 36 different ways the two could end up. That is, for each state of die 1, there are 6 different states for die 2, making $6 \times 6 = 36$ altogether. Suppose f is the sum of the number of dots showing upward on the two dice. What is the mean value for f per roll?

B.   THE BINOMIAL DISTRIBUTION — PROBABILITIES FOR SYSTEMS OF MORE THAN ONE ELEMENT

In this section we learn how to calculate the probability for a system to be in each of its various possible states, starting with the known probabilities for the behavior of a single element. For example, starting with the known probability of a single coin to land heads up, we will be able to calculate the probability that a system of 3 coins has 2 heads and one tail, or a system of 12 coins has 5 heads and 7 tails, etc.

As is often the case with problems encountered in the sciences, the solution to this kind of problem is simple if the problem is set up correctly, and impossible if it is not. Therefore, it pays to take care in setting up the problem.

We begin with a criterion for the behavior of a single element, which must be either satisfied or not satisfied. We use p to indicate the probability that the criterion is satisfied, and q to indicate the probability that it is not. For example:

1.   Criterion: A given air molecule is in the front third of an otherwise empty room. The probability that this criterion is true is $\frac{1}{3}$, and the probability that it is not true (i.e., the air molecule is elsewhere in the room) is $\frac{2}{3}$. Therefore, we have $p = \frac{1}{3}$, $q = \frac{2}{3}$.
2.   Criterion: A flipped coin lands heads up. $p = \frac{1}{2}$, $q = \frac{1}{2}$.
3.   Criterion: A spin $\frac{1}{2}$ elementary particle in no external fields has spin up. $p = \frac{1}{2}$, $q = \frac{1}{2}$.
4.   Criterion: A rolled die lands with two dots up. $p = \frac{1}{6}$, $q = \frac{5}{6}$.
5.   Criterion: A swaggering drunk, whose next step is equally probable in all directions, takes his next step westwardly. That is, his next step is somewhere in the westward half of the area around him. $p = \frac{1}{2}$, $q = \frac{1}{2}$.

A correctly formulated criterion is either satisfied or not satisfied, so we can state with certainty that it must be one or the other.

$$p + q = 1 \tag{3.6}$$

This means, of course, that q and p are not independent variables,

$$q = 1 - p \tag{3.7}$$

but for simplicity, we will use q rather that $(1 - p)$ in most of this section.

Now suppose the system has two identical elements, which we label 1 and 2. For each element, the criterion is or is not satisfied, so we have

$$(p_1 + q_1)(p_2 + q_2) = 1 \times 1 = 1$$
$$= p_1p_2 + p_1q_2 + q_1p_2 + q_1q_2 \qquad (3.8)$$

Here we see that p_1p_2 is the probability that both elements satisfy the criterion, p_1q_2 is the probability that 1 does and 2 doesn't, q_1p_2 is the probability that 1 doesn't and 2 does, and q_1q_2 is the probability that both don't. All these respective probabilities add up to unity, because this exhausts all possible configurations with respect to the criterion of interest.

EXAMPLE

Consider two air molecules, 1 and 2, in an otherwise empty room, and suppose we are interested in the probabilities of their various possible distributions between the front one-third and the rear two-thirds of the room. What would the probabilities of the various possible configurations be?

The probability that either molecule is in the front third of the room is $\frac{1}{3}$.

$$p_1 = p_2 = \frac{1}{3}$$

Similarly, the probability that each is *not* in the front third of the room is $\frac{2}{3}$.

$$q_1 = q_2 = \frac{2}{3}$$

The probabilities for all possible configurations are given by

$$(p_1 + q_1)(p_2 + q_2) = 1$$
$$= p_1p_2 + p_1q_2 + q_1p_2 + q_1q_2$$

We see that the respective probabilities for the various configurations are

Both in front: $p_1p_2 = \frac{1}{9}$.
1 in front, 2 in back: $p_1q_2 = \frac{2}{9}$.
1 in back, 2 in front: $q_1p_2 = \frac{2}{9}$.
Both in back: $q_1q_2 = \frac{4}{9}$.

EXAMPLE

Consider two rolled dice, labeled 1 and 2, and suppose we are interested in whether or not they land each with one dot showing upward. What are the probabilities for the various possible configurations of the two?

The probability that either die lands with one dot up is equal to $\frac{1}{6}$.

$$p_1 = p_2 = \frac{1}{6}$$

The probability that either die does *not* land with one dot up is $q_1 = q_2 = \frac{5}{6}$. The probabilities for the various possible configurations of the two is then given by:

Both land with one dot up: $p_1 p_2 = \frac{1}{36}$.

1 lands with one dot up but 2 doesn't: $p_1 q_2 = \frac{5}{36}$.

1 doesn't land with one dot up, but 2 does: $q_1 p_2 = \frac{5}{36}$.

Both do *not* land with one dot up: $q_1 q_2 = \frac{25}{36}$.

For two identical elements whose respective probabilities are the same, we don't really have to keep tract of which is which.

$$p_1 = p_2 = p \qquad q_1 = q_2 = q$$

The various possible configurations of the two with regard to the criterion of interest is given by

$$(p_1 + q_1)(p_2 + q_2) = (p + q)^2$$
$$= p^2 + 2pq + q^2 \tag{3.9}$$

As we have seen, p^2 is the probability that both satisfy the criterion, q^2 is the probability that both don't, and $2pq$ is the probability that one does and the other doesn't.

Notice that in the expression $2pq$, the coefficient "2" indicates there are two different possible configurations for which one element satisfies the criterion and the other doesn't.

$$p_1 q_2 + q_1 p_2 = 2pq$$

That is, it could be that 1 satisfies the criterion and 2 doesn't, or it could be that 1 doesn't and 2 does.

If we extend our analysis to systems of three elements, we find the probabilities of the various possible configurations of the system with respect to the criterion of interest to be given by

$$(p_1 + q_1)(p_2 + q_2)(p_3 + q_3) = (p + q)^3 = 1^3 = 1$$
$$= p^3 + 3p^2 q + 3pq^2 + q^3 \tag{3.10}$$

We can see that the probabilities of the various possible states are as follows.

All satisfy the criterion: p^3.
Two satisfy the criterion and one doesn't: $3p^2 q$.
One satisfies the criterion and two don't: $3pq^2$.
None satisfy the criterion: q^3.

Notice again that in the term $3p^2 q$, for example, the coefficient "3" indicates there are three different possible configurations for which two elements satisfy the criterion and one doesn't. (See Figure 3.2.) If we label the elements by 1, 2, and 3, then these configurations would be as follows.

1. 1 and 2 satisfy the criterion and 3 doesn't.
2. 1 and 3 satisfy the criterion and 2 doesn't.
3. 2 and 3 satisfy the criterion and 1 doesn't.

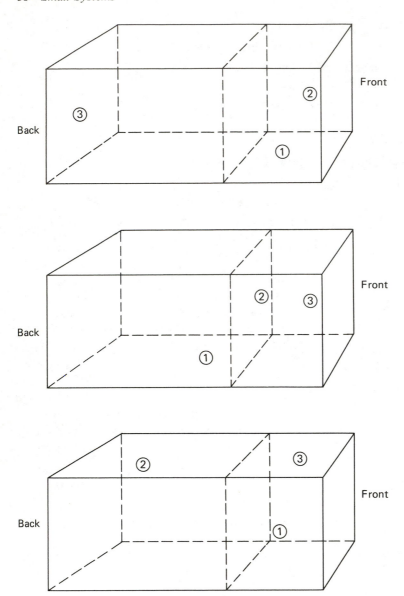

Figure 3.2 For three distinguishable particles in a room, there are three different ways of having two in front and one in back.

Equivalently, we could write

$$3p^2q = p_1p_2q_3 + p_1q_2p_3 + q_1p_2p_3$$

EXAMPLE

If we flip three coins, what is the probability that two land heads and one doesn't?

The probability of any one landing heads is given by $p = \frac{1}{2}$, and the probability of not landing heads is therefore $q = \frac{1}{2}$. The probability of any two landing heads and the other not, is given by the $3p^2q$ term in the expansion.

$$3p^2q = 3\left(\frac{1}{2}\right)^2\left(\frac{1}{2}\right) = \frac{3}{8}$$

We can continue to expand the above development to systems of four elements, or five, or any arbitrary number N. For a system of N elements, the probabilities for all the possible configurations is given by the "binomial expansion."

$$(p + q)^N = \sum_{n=0}^{N} \frac{N!}{n!(N-n)!} p^n q^{N-n} = 1^* \tag{3.11}$$

The nth term in this expansion represents the probability that n elements satisfy the criterion and the remaining $(N - n)$ elements do not. The number of *different* possible configurations of the individual elements, for which n elements satisfy the criterion and $(N - n)$ do not, is given by the "binomial coefficient," $[N!/n!(N - n)!]$, appearing in the above expression.

SUMMARY

If we have some criterion for the behavior of a single element of a system, and if $p = $ the probability that the criterion is satisfied, and $q = $ the probability that it is not satisfied $(q = 1 - p)$, then for a system of N elements, the probability of this system being in a state where n elements satisfy the criterion and the remaining $(N - n)$ elements do not satisfy it is given by

$$P_N(n) = \frac{N!}{n!(N-n)!} p^n q^{N-n} \tag{3.12}$$

The "binomial coefficient," $[N!/n!(N - n)!]$, is the number of different configurations of the individual elements, for which n satisfy the criterion and $(N - n)$ do not.

EXAMPLE

Suppose there were five air molecules in an otherwise empty room. What is the probability of two of them being in the front third of the room, and three in back?

The probability of any one being in front is $p = \frac{1}{3}$. Therefore, the probability of it not being in front is $q = \frac{2}{3}$. The state we are interested in has five air molecules $(N = 5)$, two of which are in front $(n = 2)$. The appropriate probability is

$$P_5(2) = \frac{5!}{2!3!}\left(\frac{1}{3}\right)^2\left(\frac{2}{3}\right)^3 = \frac{80}{243}$$

---

* The factorial is defined by $n! \equiv n \cdot (n - 1) \cdot (n - 2) \cdot \ldots \cdot 2 \cdot 1$, and $0! \equiv 1$.

EXAMPLE

In the above example, what is the number of different possible arrangements of the five molecules that leave two in front and three in back?

The number of different arrangements is given by the coefficient,

$$\frac{N!}{n!(N-n)!} = \frac{5!}{2!3!} = 10$$

Although the probability calculations using the binomial expansion are correct for systems of any size, these calculations become quite cumbersome for systems of more than a few elements. Suppose, for example, we wanted to calculate the probability that 40 out of 100 flipped coins land heads up. Our calculation would then be

$$P_{100}(40) = \frac{100!}{40!60!} \left(\frac{1}{2}\right)^{40} \left(\frac{1}{2}\right)^{60}$$

Although tedious, factors like $(1/2)^{40}$ can be calculated using logarithms, but the factorials (e.g., $100! = 100 \cdot 99 \cdot 98 \cdot 97 \cdot \ldots$) become overwhelming when the numbers are large.

Fortunately, there is an approximation known as Stirling's formula, which allows us to calculate factorials fairly accurately for numbers larger than the few that we would want to calculate directly. (See Appendix 3A.)

$$\ln (m!) \approx m \ln m - m + \frac{1}{2}\ln (2\pi m) \tag{3.13}$$

If we use this formula to calculate the value of $m!$ for $m = 10$, we find it accurate to within 1%, and it increases in accuracy for larger values of m.

PROBLEMS

3-5. Consider a system of four flipped coins.

(a) What is the probability of two landing heads and the other two tails?
(b) How many different configurations of the individual coins are possible that have two heads and two tails?
(c) Label the four coins 1, 2, 3, and 4. Make a chart that lists the various possible configurations of these that have two heads and two tails. Is the number of configurations on your chart the same as that predicted by the binomial coefficient in Eq. 3.12?

3-6. Consider a system of five molecules. The probability of any one being in an excited state is $\frac{1}{10}$.

(a) What is the probability of none being in an excited state?
(b) What is the probability of one and only one being in an excited state?

3-7. If you throw two dice, what is the probability of throwing "snake eyes" (each die having one dot up)?

3-8. If you throw eight dice:

   (a) What is the probability that five and only five have four dots up?
   (b) How many different configurations of the dice are there that give five and only five with four dots up?
   (c) What is the probability that five *or more* have four dots up?

3-9. Consider five spin $\frac{1}{2}$ elementary particles: (independent, and no external fields).

   (a) What is the probability that four have spin up and the other has spin down?
   (b) If these elementary particles were distinguishable (i.e., you could tell one from the other), how many different configurations of the five would give this result?

3-10. Using Stirling's formula, calculate the probability of getting exactly 500 heads and 500 tails when flipping 1000 coins.

C. STATISTICALLY INDEPENDENT BEHAVIORS

Suppose we were interested in the states of a system relative to *two* different criteria. The probability that a system of N elements is in a state where n_1 elements satisfy the first criterion and n_2 elements satisfy the second, would then be represented by $P_N(n_1, n_2)$.

Of particular interest are cases where an element's behavior with respect to one of the criteria does not affect its behavior with respect to the other. For these cases, the criteria are said to be "statistically independent." For example, consider an air molecule in an empty room (Figure 3.3). The two criteria might be as follows.

1. Whether it is in the front third of the room ($p_1 = \frac{1}{3}$, $q_1 = \frac{2}{3}$).
2. Whether it is in the top half of the room. (If we ignore gravity, $p_2 = \frac{1}{2}$, $q_2 = \frac{1}{2}$.)

Whether or not the molecule is in the front third of the room has no bearing on whether it is in the top half. The two behaviors are statistically independent.

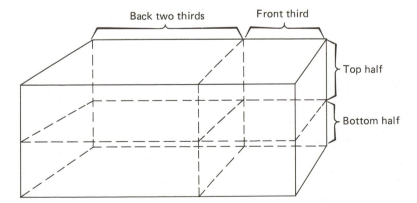

Figure 3.3 If a room is completely vacant, then whether a molecule happens to be in the top or bottom portion of the room has no bearing on whether it is in the front or back. The behaviors are statistically independent.

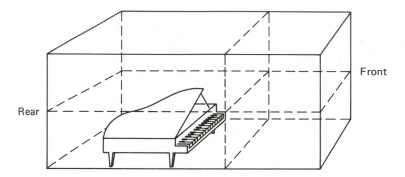

Figure 3.4 If a grand piano is in the rear, bottom portion of the room, then a molecule in the rear of the room is more likely to be in the top portion, because the piano takes up some of the space below. In this case, then, the two behaviors are not statistically independent.

The situation would be different if, for example, there was a grand piano in the rear of the room, as illustrated in Figure 3.4. In this case a molecule in the rear of the room is more likely to be in the top half where there is more room for it. Consequently, whether it is in the front or back of the room would affect the probabilities of being in the top or bottom halves, and so the two behaviors would not be statistically independent.

As another example, consider the spin orientations of two close, spin $\frac{1}{2}$ particles, such as two nucleons in the same nucleus. Because of the coupling of the magnetic moment of one with the field produced by the other, the spin orientation of one may have an effect on the probable spin orientation of the other, so the two phenomena would not be statistically independent. When we apply the mathematics of statistically independent elements to actual physical systems, we must take care to insure that the members of the system are indeed statistically independent. This often means that the elementary members that we consider may consist of more than one elementary particle—perhaps a nucleus, or an atom, or a molecule, or a group of molecules.

For statistically independent criteria, the probability that the system is in state i with respect to the first criterion and also in state j with respect to the second is given by the product of the respective individual probabilities.

$$P_{ij} = P_i P_j \tag{3.14}$$

This can be extended to probability calculations for any number of statistically independent behaviors,

$$P_{ijkl\ldots} = P_i P_j P_k P_l \cdots$$

where k is the state with respect to the third criterion, l is the state with respect to the fourth, etc.

EXAMPLE

Consider the system consisting of a single air molecule in an otherwise empty room. What are the probabilities for the various positions of that molecule with respect to the two criteria listed on the previous page?

According to Eq. 3.14, the respective probabilities are given as follows.

Front third, top half: $p_1 p_2 = \frac{1}{3} \times \frac{1}{2} = \frac{1}{6}$.

Front third, bottom half: $p_1 q_2 = \frac{1}{3} \times \frac{1}{2} = \frac{1}{6}$.

Rear two-thirds, top half: $q_1 p_2 = \frac{2}{3} \times \frac{1}{2} = \frac{2}{6}$.

Rear two-thirds, bottom half: $q_1 q_2 = \frac{2}{3} \times \frac{1}{2} = \frac{2}{6}$.

EXAMPLE

Consider a system of five air molecules in an otherwise empty room. What is the probability that two of them are in the front third of the room, and four of them are in the top half?

According to Eq. 3.14 the requested probability is the product of the two individual probabilities.

$$P_5(2, 4) = P_5(2)P'_5(4) = \left(\frac{5!}{2!3!} p_1^2 q_1^3 \right)\left(\frac{5!}{4!1!} p_2^4 q_2^1 \right)$$

$$= \left(\frac{80}{243} \right)\left(\frac{5}{32} \right) = .051$$

If f is a function of the configuration of the system with respect to only one of the criteria, then the probabilities with respect to the other criterion need not be considered in calculating the mean value of f. That is, if $f = f(i)$ is a function only of the state i with respect to the first criterion and is independent of the state j with respect to the second, then

$$\bar{f} = \sum_{i,j} P_{ij} f(i) = \sum_{j} \sum_{i} P_i P_j f(i)$$

$$= \sum_{j} P_j \sum_{i} P_i f(i) = (1) \sum_{i} P_i f(i) \tag{3.15}$$

Another result for statistically independent behaviors, that will be of interest to us in Chapter 4, is that if $f = f(i)$ is a function of the state of the system with respect to the first criterion only, and $g = g(j)$ is a function of the state of the system with respect to the second criterion only, then the mean value of the product is the product of the mean values.

$$\bar{fg} = \sum_{i,j} P_{ij} f(i)g(j) = \sum_{i} \sum_{j} P_i P_j f(i)g(j)$$

$$= \sum_{i} P_i f(i) \sum_{j} P_j g(j) = \bar{f}\bar{g} \tag{3.16}$$

PROBLEMS

3-11. You are dealing cards from a full 52-card, freshly shuffled deck. You are interested in whether the first two cards dealt will be aces.

> Criterion 1: The first card dealt is an ace.
>
> Criterion 2: The second card dealt is an ace.

Are these two criteria statistically independent?

3-12. Again you are interested in whether the first two cards dealt are aces, as in the previous problem. However, this time you return the first card to the deck and reshuffle before dealing the second card. Are the two criteria statistically independent in this case?

3-13. Answer the questions under Figure 3.5.

Figure 3.5   What is the probability that the very next card dealt will be a queen? Does it depend on what has already been dealt? How? (K. Bendo)

3-14. Suppose you have two freshly shuffled full decks of cards, and you deal one card from one and one card from the other.

   (a) What is the probability that the first card dealt will be an ace?
   (b) What is the probability that the second card dealt will be a club?
   (c) Are the two things statistically independent?
   (d) What is the probability that the first card dealt will be an ace *and* the second card dealt will be a club?

3-15 Suppose you were involved in a game where two cards were dealt in the manner of the previous problem. Suppose further that the dealer would pay you $4 if the second card dealt was a club, regardless of the first card, and you would pay him $1 if the second card was not a club. Use Eq. 3.15 to compute the mean value of the money you would win per game if you played it many times.

APPENDIX 3A   STIRLING'S FORMULA, EQ. 3-13

The derivation of Stirling's formula is based on the observation that the logarithm of a product of numbers equals the sum of the logarithms of the individual numbers.

$$\ln(m!) = \ln(1) + \ln(2) + \cdots + \ln(m)$$

$$= \sum_{k=1}^{m} \ln(k)$$

This sum can be approximated by an integral, as illustrated in Figure 3.6. If we let Δ be an interval of unit length extending from $(k - \tfrac{1}{2})$ to $(k + \tfrac{1}{2})$, then

$$\sum_{k=1}^{m} \ln(k) = \sum_{k=1}^{m} \ln(k)\Delta$$

$$\approx \int_{1/2}^{m+1/2} \ln(k)\,dk = \left(m + \frac{1}{2}\right)\ln\left(m + \frac{1}{2}\right) - m - \frac{1}{2}\ln\left(\frac{1}{2}\right).$$

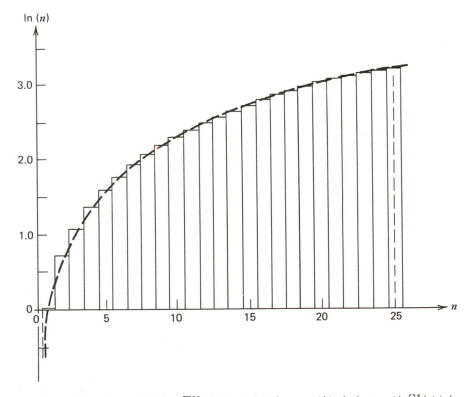

Figure 3.6   Comparison of $\ln(25!) = \sum_{n=1}^{25} \ln(n)$, which is the area within the boxes, with $\int_{1}^{25} \ln(n)\,dn$, which is the area under the curve. Since the area under the curve is slightly smaller than the area in the boxes, the value of the integration will be a bit too small. More accuracy can be obtained by integrating from $\tfrac{1}{2}$ to $25\tfrac{1}{2}$. In general, we have the following forms of Stirling's approximation: $\sum_{k=1}^{m} \ln k \approx \int_{1}^{m} \ln k\,dk = m \ln m - m$, or more accurately, $\sum_{k=1}^{m} \ln k \approx \int_{1/2}^{m+1/2} \ln k\,dk = m \ln m - m + \tfrac{1}{2}\ln(2em)$.

If we ignore terms of order $1/m$ or smaller, this can be juggled into the following form.

$$\ln (m!) = \sum_{k=1}^{m} \ln (k)$$

$$\approx m \ln m - m + \frac{1}{2} \ln (2em)$$

Using this formula to calculate values of $m!$, we find the results to be consistently low by nearly 7% for values of m greater than 10, which isn't bad considering the crude nature of our derivation. A more careful derivation gives a similar result with only a slight modification in the third term: e being replaced by π.

$$\ln (m!) \approx m \ln m - m + \frac{1}{2} \ln (2\pi m) \tag{3A.1}$$

Since the last term is much smaller than the other two for large values of m, it is sometimes omitted for simplicity.

chapter 4

STATISTICS FOR SYSTEMS OF MANY ELEMENTS

The techniques developed in Chapter 3 for translating between the behaviors of individual constituents and the behaviors of small systems is correct for systems of any size. Unfortunately, it becomes extremely cumbersome when applied to any but the smallest systems. Fortunately, there is a way of streamlining our calculations for systems of more than just a few constituents.

In the preceding chapter we saw how we could use the binomial distribution of Eq. 3.11 to calculate probabilities of all the various possible configurations of the elements of a system. Although this method is correct for systems of any size, it is quite slow and cumbersome when applied to any but the very smallest systems, and it frequently gives us more information than we care to know.

A.   FLUCTUATIONS

Suppose we were interested in the outcomes of flipping 1000 coins. We could go meticulously through the calculations of Eq. 3.12, calculating the probability for each of the 1001 possible outcomes. We would get

$$P_{1000}(0) = 9.3 \times 10^{-302}$$

$$P_{1000}(1) = 9.3 \times 10^{-299}$$

$$\vdots$$

$$P_{1000}(495) = .0240$$

$$\vdots$$

$$P_{1000}(500) = .0253$$

$$\vdots$$

$$P_{1000}(1000) = 9.3 \times 10^{-302}$$

This is a great deal of work, and is more information then normally would be useful.

It would probably be more helpful to us if we could just keep two numbers in mind.

1. The mean value of the number of coins that would land heads up if the coin-flip experiment were repeated many times.
2. Some measure of the fluctuations we could expect around this value.

In the above example of 1000 coins, for instance, we know that very nearly half would land heads. Nonetheless, if we were wagering on the outcome, it would not be wise to bet that exactly 500 land heads, because the number may well turn out to be 499, or 502, etc. In fact, from the above list of probabilities we can see that we would win less than 3% of the time if we bet on exactly 500 landing heads. Instead, we would be wiser to bet that the number of heads will be within some range near 500, and it would be nice to know the size of that range of characteristic fluctuations before making our bet.

To analyze mean values and characteristic fluctuations, we imagine that we create a large number of identically prepared systems.* (For example, imagine we have a large number of systems of 1000 flipped coins each.) For these, the average number of elements per system that satisfy the criterion is given by the product of the probability that a single element does (p) times the number of elements per system (N).

$$\bar{n} = pN \tag{4.1}$$

This can either be thought of as the definition of the probability p, or it can be derived from the binomial distribution of Eq. 3.11 and the definition of mean values, Eq. 3.1.

EXAMPLE

If a large number of coins were flipped many times, how many would land heads up, on the average?

Since the probability of any one landing heads is $p = \frac{1}{2}$, then on the average $\frac{1}{2}$ of the total ($\frac{1}{2}N$) would land heads.

EXAMPLE

If a large number of dice are rolled many times, how many would land with one dot up, on the average?

Since the probability of any one landing with one dot up is $p = \frac{1}{6}$, we would expect one-sixth of the dice ($\frac{1}{6}N$) to land with one dot up, on the average.

The average fluctuation of n about its mean value, $\bar{n}$, must be zero, due to the definition of mean values.

$$\overline{\Delta n} = \overline{(n - \bar{n})} = \bar{n} - \bar{\bar{n}} = \bar{n} - \bar{n} = 0$$

This is because the positive fluctuations cancel the negative ones.

* A large number of identically prepared systems is called an "ensemble."

A more meaningful determination of the characteristic fluctuation is to take the average of the fluctuations squared (thus giving only positive numbers), and then take the square root of that. This is called the "standard deviation," and is usually given the symbol σ.

$$(\text{standard deviation})^2 = \sigma^2 = \overline{(n - \bar{n})^2}$$

or

$$(\text{standard deviation}) = \sigma = [\overline{(n - \bar{n})^2}]^{1/2} \tag{4.2}$$

Notice that the square of the standard deviation is simply a mean value, so it can be calculated using the definition of mean values, Eq. 3.1, with the probabilities P_n given by Eq. 3.12.

$$\sigma^2 = \overline{(n - \bar{n})^2} = \sum_n P_n (n - \bar{n})^2$$

$$= \sum_{n=0}^{N} \left(\frac{N!}{n!(N - n)!} p^n q^{N-n} \right) (n - \bar{n})^2$$

This calculation is a little bit tedious, so we relegate it to the Appendix 4A. But it gives a simple result,

$$\sigma^2 = Npq$$

or equivalently,

$$\sigma = \sqrt{Npq} \tag{4.3}$$

In the next section we will show that for large numbers of events, N, the distribution of n about the mean is of a certain form, called a "Gaussian distribution." We will see that for a Gaussian distribution, the probability that n is within one standard deviation of $\bar{n}$ is .68. So when wagering on the outcome of flipping 1000 coins, bet that the number of heads will be within one standard deviation of 500. You'll win 68% of the time.

According to the result (4.3), the standard deviation increases only as the square root of the number of elements per system, N. This is in contrast to the mean value of n, which increases linearly with the number of elements per system. This means that although both the mean value, $\bar{n}$, and the fluctuations about this mean value, σ, increase as the systems get larger, the *relative* fluctuations, $\sigma/\bar{n}$, *decrease*. This is illustrated in Figure 4.1.

$$\frac{\sigma}{\bar{n}} = \frac{\sqrt{Npq}}{Np} = \sqrt{\frac{q}{Np}} \propto \frac{1}{\sqrt{N}} \tag{4.4}$$

EXAMPLE

Consider systems of 100 flipped coins and systems of 10,000 flipped coins. What are the average number of heads, the standard deviation, and the relative fluctuation in each case?

For the systems of 100 flipped coins, the average number of heads would be 50.

$$\bar{n}_{100} = pN = \left(\frac{1}{2}\right)(100) = 50$$

Frequency of occurrence

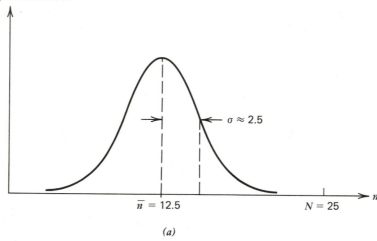

$$\bar{n} = 12.5 \qquad\qquad N = 25$$

(a)

Frequency of occurrence

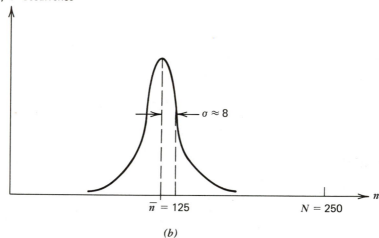

$$\bar{n} = 125 \qquad\qquad N = 250$$

(b)

Figure 4.1 Plot of frequency of occurrence, $P_N(n)$, as a smooth function of n, for $N = 25$ (a) and $N = 250$ (b). For this particular criterion, $p = q = \frac{1}{2}$. Notice that as N increases, the absolute width of the peak increases, but its relative width decreases.

The standard deviation would be

$$\sigma_{100} = \sqrt{Npq} = \sqrt{(100)\left(\frac{1}{2}\right)\left(\frac{1}{2}\right)} = \sqrt{25} = 5$$

The relative fluctuation would be

$$\frac{\sigma_{100}}{\bar{n}_{100}} = \frac{5}{50} = 10\%$$

For the system of 10,000 flipped coins we have, correspondingly,

$$\bar{n}_{10,000} = pN = \left(\frac{1}{2}\right)(10,000) = 5,000$$

$$\sigma_{10,000} = \sqrt{Npq} = \sqrt{(10,000)\left(\frac{1}{2}\right)\left(\frac{1}{2}\right)} = \sqrt{2500} = 50$$

and

$$\frac{\sigma_{10,000}}{\bar{n}_{10,000}} = \frac{50}{5000} = 1\%$$

Notice that the relative fluctuation for systems of 100 flipped coins is 10% whereas that for 10,000 flipped coins is only 1%. This means that for 100 flipped coins, the number of heads will usually (i.e., 68% of the time) be between 40 and 60% of the total, whereas for the larger systems of 10,000 flipped coins, the number of heads will usually be between 49 and 51% of the total. The larger the system, the more predictable will be its behavior.

SUMMARY

Consider many identically prepared systems of N element each. If p is the probability that any one element satisfies the criterion of interest, and q is the probability that it does not, the average number of elements that satisfy the criterion is given by

$$\bar{n} = pN \qquad (4.1)$$

the standard deviation for the fluctuations of n about this value is given by

$$\sigma = \sqrt{Npq} \qquad (4.3)$$

and the relative fluctuations are given by

$$\frac{\sigma}{\bar{n}} = \sqrt{\frac{q}{Np}} \qquad (4.4)$$

For sufficiently large systems, the value of n will be within one standard deviation of $\bar{n}$ 68% of the time.

In the sciences, the fluctuation of a variable is sometimes more interesting than its mean value. For example, the average current from an alternating current source is zero, as the current goes in each direction half of the time. Similarly, the average velocity of a molecule of a material in a container is also zero, although individual molecules may each be traveling with very high speeds. In these and many other examples, the mean value of the variable may be misleading, suggesting no motion at all. In these cases, the standard deviation of the variable may be much more illuminating.

When the mean value of a variable is zero, the standard deviation of that variable is sometimes referred to as its "root mean square" value. This term is, of course, a natural consequence of the definition of the standard deviation, being the square root of the mean of the squares.

EXAMPLE

Consider systems of 100 molecules in otherwise empty rooms. What would be the average number of molecules in the front third of the rooms, the standard deviation about this value, and the relative fluctuation about this value?

For this case, $N = 100$, $p = \frac{1}{3}$, and $q = \frac{2}{3}$. Therefore we have

$$\bar{n} = pN = 33\tfrac{1}{3}$$

$$\sigma = \sqrt{Npq} = \frac{10\sqrt{2}}{3} \approx 4.7$$

$$\frac{\sigma}{\bar{n}} = .14$$

EXAMPLE

Repeat the above for systems of 10^{28} molecules in otherwise empty rooms.

For this case, $N = 10^{28}$, and p and q remain the same as above.

$$\bar{n} = pN = 3.3 \times 10^{27}$$

$$\sigma = \sqrt{Npq} = 4.7 \times 10^{13}$$

$$\frac{\sigma}{\bar{n}} = 1.4 \times 10^{-14}$$

PROBLEMS

4-1.   Consider a set of identically prepared systems of 100 rolled dice apiece. Suppose we were interested in the number of dice per system with one dot showing upward. For these systems calculate the following.

(a)   The mean number of dice per system with one dot up ($\bar{n}$).
(b)   The standard deviation about this value.
(c)   The relative fluctuation about $\bar{n}$.

4-2.   Repeat the above for a set of systems of 10^8 rolled dice apiece.

4-3.   Using the theorem that the mean value of a constant times a function equals the constant times the mean value of the function $(\overline{cf} = c\bar{f})$, prove that $\overline{(n - \bar{n})^2} = \overline{n^2} - \bar{n}^2$. (*Hint. $\bar{n}$ is a constant.*)

4-4.   For air at room temperature, the probability of one molecule being in an excited electronic state is about 10^{-10} (i.e., $p = 10^{-10}$, $q = 1$). In a typical room there are about 10^{28} air molecules. Using these numbers, calculate the following.

(a) The mean number of electronically excited air molecules in a room at any time.
(b) The standard deviation about this value.
(c) The relative fluctuation about this value.

B. THE GAUSSIAN DISTRIBUTION

With the binomial formula, (3.11) or (3.12), we can calculate the probabilities $P_N(n)$ for all possible states of the system—that is, for all possible numbers of elements, n, that satisfy the criterion. For systems of many elements, each probability calculation can be simplified by using logarithms and Stirling's formula. Nonetheless, for these larger systems there are so many different possible states ($N + 1$ of them) that to calculate the entire spectrum of probabilities for all possible states could be an extremely tedious task.

We should suspect there might be an easier way of doing things, since for larger systems we should find the distribution of probabilities to be reasonably smooth (Figure 4.2), with states that are nearly the same having nearly the same probabilities. For example, we would expect $P_{1000}(491)$ to be close to $P_{1000}(490)$ and $P_{1000}(492)$, and probably would even lie somewhere between these two values.

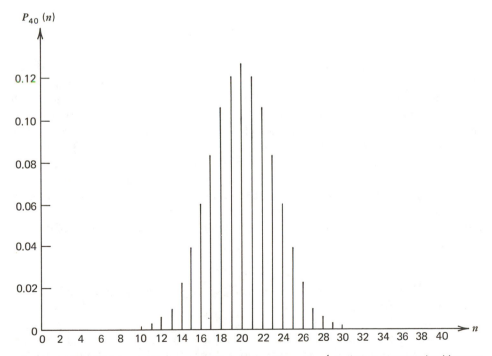

Figure 4.2 Plot of $P(n)$ versus n for a system of 40 elements with $p = q = \frac{1}{2}$. For large systems we should expect the distribution of probabilities to be rather smooth, with neighboring values of n having neighboring *probabilities, P(n)*.

Indeed, there is a simpler way for calculating the probabilities for systems of many elements. The entire distribution of probabilities over all possible states can be given in terms of the two parameters $\bar{n}$ and σ discussed in the previous section. Clearly, this will be a great deal simpler than doing the many individual probability calculations using the binomial formula.

The simplified distribution is of a certain form, called a "Gaussian distribution." Its derivation involves making approximations, but these approximations become increasingly reliable as the number of elements in the system gets larger. Therefore, this Gaussian distribution is useful in those cases where the binomial formula is not. Conversely, in cases where the binomial calculations are useful and simple, this Gaussian distribution cannot be used because the approximations lose validity for small systems. In this way, the two types of probability calculations are complementary to each other.

Our derivation of this simplified formula will involve a mathematical tool called a "Taylor series expansion." Suppose we know the value of some function $f(x)$ and all its derivatives at some point $x = x_0$. Then we can calculate the value of this function at some other point, $x = x_1$, through the following formula. (See Appendix 4B for derivation.)

$$f(x_1) = f(x_0) + \sum_{m=1}^{\infty} \frac{1}{m!} \frac{\partial^m f}{\partial x^m}\bigg|_{x=x_0} (x_1 - x_0)^m$$

$$= f(x_0) + \frac{\partial f}{\partial x}\bigg|_{x=x_0} (x_1 - x_0) + \frac{1}{2} \frac{\partial^2 f}{\partial x^2}\bigg|_{x=x_0} (x_1 - x_0)^2 + \cdots \tag{4.5}$$

Notice that if the function is a constant, then only the first term is nonzero. If the function is linear in x, only the first two terms are nonzero, and so on. The smoother the function, and the closer x_1 is to x_0, the more accurately the first few terms will be in approximating the function at $x = x_1$. Therefore, in using the Taylor series expansion, it is advantageous to apply it to functions which are as smooth as possible, and to choose x_0 to be as close as possible to the values of x_1 that are of interest to us.

The function we are concerned with is $P_N(n)$. We drop the subscript "N" for simplicity, and we consider $P(n)$ to be a continuous function of n. We are going to develop a formula for $P(n)$ using a Taylor series expansion, and to maximize the accuracy of our result, we will try to satisfy the two criteria which increase the accuracy of the first few terms of the expansion:

1.   Choose as smooth a function as possible.
2.   Choose the reference point, n_0, to be as close as possible to the values of n that we are interested in.

To satisfy the first criterion, we choose to expand the logarithm of $P(n)$, because the logarithm of a function varies much more slowly than does the function itself.

Concerning the second criterion we observe that systems are most likely to be found in states of highest probability. If $n_{\max}$ represents the state for which $P(n)$ is a

Frequency of occurrence

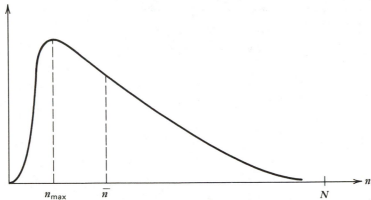

Figure 4.3   Example of a skewed distribution for which the maximum does not occur at the mean value of *n*.

maximum, then we are most likely to be interested in values of $P(n)$ for n near n_{max}. States with n further from n_{max} will occur less frequently and therefore be of less interest to us. We assume that

$$n_{max} = \bar{n}$$

which may not always be true. Figure 4.3 shows a case where $\bar{n}$ is significantly different from n_{max}. Fortunately, as we saw in the preceding section, the distribution gets more and more peaked as the system gets larger. Any difference between n_{max} and $\bar{n}$ becomes smaller as the distribution becomes more peaked. (See Figure 4.1.) So for large systems (large values of N) the above approximation is justified.

We now expand the logarithm of $P(n)$ about the point $n = \bar{n}$ according to the Taylor series formula.

$$\ln P(n) = \ln P(\bar{n}) + \frac{\partial}{\partial n} \ln P(n)\bigg|_{n=\bar{n}} (n - \bar{n})$$

$$+ \frac{1}{2} \frac{\partial^2}{\partial n^2} \ln P(n)\bigg|_{n=\bar{n}} (n - \bar{n})^2 + \cdots \tag{4.6}$$

The first derivative of a function at its maximum is zero.

$$\frac{\partial}{\partial n} \ln P(n)\bigg|_{n=\bar{n}} = 0$$

The second derivative of a function at its maximum must be negative. We can calculate this second derivative using the definition of differentials and the values for $P(n)$ given by the binomial formula (3.12). For example, the first derivative is

$$\frac{\partial P(n)}{\partial n} \approx \frac{\Delta P(n)}{\Delta n} = \frac{P(n + 1) - P(n)}{(n + 1) - n} = \cdots$$

Completing this and then extending it to second derivatives is straightforward, and is done in Appendix 4C, yielding the result

$$\frac{\partial^2}{\partial n^2} \ln P(n)\bigg|_{n=\bar{n}} = -\frac{1}{Npq} = -\frac{1}{\sigma^2} \tag{4.7}$$

We can use the same differential technique to evaluate the third-order term, $\partial^3/\partial n^3 \ln P(n)|_{n=n}$, and so on, but we find that as

$$|n - \bar{n}| \ll Npq$$

or equivalently,

$$|n - \bar{n}| \ll \sigma^2 \tag{4.8}$$

we can ignore third- and higher-order terms in comparison to the first three terms in the expansion. Our expansion is now

$$\ln P(n) = \ln P(\bar{n}) - \frac{1}{2\sigma^2}(n - \bar{n})^2$$

or

$$P(n) = P(\bar{n})e^{-(n-\bar{n})^2/2\sigma^2} \tag{4.9}$$

To calculate the value of $P(\bar{n})$ in the above expression, we ensure that the sum of the probabilities of all possible values of n must be equal to 1. Equivalently, we can say with certainty that the system must be in one of the possible states.

$$\sum_n P(n) = 1$$

$$= \sum_n P(n)\,\Delta n \qquad (\text{since } \Delta n = 1)$$

$$\approx \int_{n=-\infty}^{\infty} P(n)\,dn$$

$$= P(\bar{n})\int_{n=-\infty}^{\infty} e^{-[(n-\bar{n})^2/2\sigma^2]}\,dn$$

$$= P(\bar{n})(\sqrt{2\pi}\,\sigma) \tag{4.10}$$

Consequently, we have

$$P(\bar{n}) = \frac{1}{\sqrt{2\pi}\,\sigma}$$

and

$$P(n) = \frac{1}{\sqrt{2\pi}\,\sigma}\,e^{-[(n-\bar{n})^2/2\sigma^2]} \tag{4.11}$$

To find the probability that n is within one standard deviation of $\bar{n}$, we must integrate Eq. 4.11 between the limits of $\bar{n} - \sigma$, and $\bar{n} + \sigma$. This turns out to be very difficult to do, being done best numerically, and yielding the value .68, which has been referred to earlier. (See Figure 4.4.)

P(n)

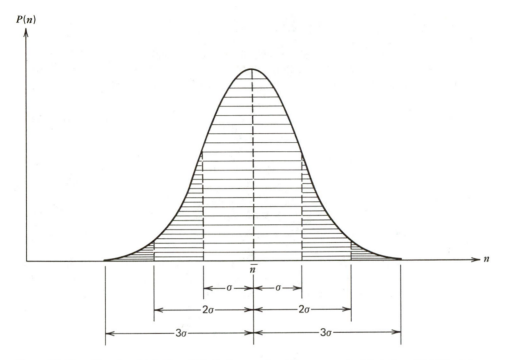

Figure 4.4 For a Gaussian distribution, 68% of all events lie within one standard deviation of $\bar{n}$, 68% of the remainder lie between one and two standard deviations of $\bar{n}$, 68% of the remainder between two and three standard deviations of $\bar{n}$, etc.

Let's pause for a paragraph to reflect on how our expectations for the accuracy of the Taylor series expansion were borne out in the result. We expected the results to be most accurate for values of n near the expansion point, $\bar{n}$. Indeed, we see from Eq. 4.9, that the prediction for $n = \bar{n}$ is

$$P(n = \bar{n}) = P(\bar{n})e^{-0} = P(\bar{n})$$

which is exactly correct. At values of n far away from this, we expect that the expansion is not quite as valid. Indeed, if we consider a system for which $N = 1000$, $p = \frac{1}{2}, q = \frac{1}{2}$ (hence, $\bar{n} = 500$, $\sigma = 15.8$), we see that the formula (4.11) gives probabilities for the system having $n = -1$ or $n = 1001$ elements satisfying the criterion, for example. Both of these cases are impossible! The respective calculations give very small values for these probabilities.

$$P_{1000}(-1) = P_{1000}(1001) = 10^{-216}$$

Although extremely small, this is not exactly zero as it should be, so the formula is not exactly correct for values for n far from $\bar{n}$. Of course, in the case of $n = -1$ or $n = 1001$ we have $|n - \bar{n}| = 501$ and $\sigma^2 = 250$, so the criterion (4.8) is not satisfied and we shouldn't expect the results to be exactly correct in this case.

SUMMARY

For a system of N elements, each of which has the probability p of satisfying the criterion of interest, and probability q of not satisfying it, then the probability that n elements satisfy the criterion and the remaining $(N - n)$ do not, is given by

$$P_N(n) = \frac{N!}{n!(N - n)!} p^n q^{(N - n)} \tag{3.12}$$

or

$$P_N(n) = \frac{1}{\sqrt{2\pi}\sigma} e^{-[(n - \bar{n})^2/2\sigma^2]} \quad \text{(if } n - \bar{n} \ll \sigma^2) \tag{4.11}$$

where

$$\bar{n} = pN \tag{4.1}$$

$$\sigma = \sqrt{Npq} \tag{4.3}$$

The first form is correct for all systems, but it is most useful for systems of only a few elements. The second form is more useful for systems of many elements where the constraint, $|n - \bar{n}| \ll \sigma^2$, is usually satisfied.

EXAMPLE

Suppose there are 3000 air molecules in an otherwise empty room. What is the probability that exactly 1000 of them will be in the front third of the room at any time?

For this problem $N = 3000$, $n = 1000$, $p = \frac{1}{3}$, $q = \frac{2}{3}$. Therefore, $\bar{n} = pN = 1000$, and $\sigma = \sqrt{Npq} = 25.8$. For this case, $n = \bar{n}$, so the exponent in the Gaussian formula (4.11) is

$$\frac{(n - \bar{n})^2}{2\sigma^2} = 0$$

Consequently, we have

$$P_{3000}(1000) = \frac{1}{\sqrt{2\pi}\,\sigma} e^{-0} = 1.6 \times 10^{-2}$$

EXAMPLE

For the above case, what is the probability of exactly 1100 air molecules being in the front third of the room at any time?

In this case, everything is the same as above except that $n = 1100$. So the exponent in the Gaussian formula (4.11) is

$$\frac{(n - \bar{n})^2}{2\sigma^2} = \frac{(100)^2}{2(25.8)^2} = 7.5$$

Consequently, we have

$$P_{3000}(1100) = \frac{1}{\sqrt{2\pi}\,\sigma} e^{-7.5} = 8.8 \times 10^{-6}$$

EXAMPLE

In the above example, how many different combinations of the air molecules are there, such that 1100 are in the front third of the the room, and the rest in back? The number of different combinations is given by the binomial coefficient

$$\frac{N!}{n!(N - n!)}$$

with $N = 3000$ and $n = 1100$. This may be calculated using Stirling's formula, (3.13), and gives the result

$$\frac{3000!}{1100!1900!} = 10^{858}$$

That is a lot of different possible combinations!

PROBLEMS

4-5. Consider identically prepared systems, each having 600 rolled dice. Suppose you are interested in the number of dice per system, n, with six dots showing upward

   (a)   What is the average value of n?
   (b)   What is the standard deviation, σ?
   (c)   In the distribution of probabilities over the various possible states,

$$P_{600}(n) = Ae^{-B(n-n)^2}$$

   what are the values of the coefficients A and B?
   (d)   If you rolled 600 dice, what is the probability of exactly 100 landing with six dots up?
   (e)   What is the probability of exactly 96 landing with six dots up?

4-6. In the above problem, what is the number of different possible combinations of the dice such that 100 show six dots up and the remaining 500 do not?

4-7. If you flip 100 coins, what is the probability of exactly:

   (a)   50 landing heads?
   (b)   49 landing heads?
   (c)   48 landing heads?
   (d)   42 landing heads?

4-8. From the Gaussian formula (4.11) calculate the ratio of $P_N(n = \bar{n} \pm \sigma)$ to $P_N(n = \bar{n})$.

C. THE RANDOM WALK

One important further application of the tools developed in this chapter is the study of the motion of objects where that motion occurs in individual, discrete "steps." If each step is random in direction, being independent of the direction of the preceding or succeeding steps, then the study of the net motion is referred to as the "random walk problem."

    A very commonly used example for this type of problem is the study of the net motion of drunks, who begin their strolls from a single lightpost, but who are so intoxicated that each step may be in any random direction, and of a range of different lengths. The direction of a step might be influenced by such things as wind, a slope to the ground, etc., and the length might be influenced by their height, or external factors as well, but given the probabilities of the various directions and lengths for a single step, we can use the tools of this chapter to answer the following two questions:

1. After each drunk has taken N steps, what will be their average position relative to the starting point?
2. How spread out will they be? That is, what will be the standard deviation of their positions relative to their average position?

For those of you who may wonder what this problem has to do with physical systems, notice, for example, that the diffusion of an ammonia molecule through air (Figure 4.5) is analogous. After each collision with an air molecule it may go in any direction, and it may travel various distances between collisions. But given the probabilities of directions and distances, we can calculate (once ammonia molecules are sprung loose somewhere in a room) where on the average they will be after N collisions, and what the standard deviation about this value would be. Similarly, the travel of electrons through a metal and the diffusion of a "hole" through a semiconductor are random walk problems.

In many of these problems, we may question if the motion in any one step is completely random. When an ammonia molecule collides with an air molecule, for

Figure 4.5   The diffusion of ammonia molecules through a room can be treated as a random walk problem. The distance traveled between two successive collisions with air molecules can be thought of as one random "step." (Keith Stowe)

example, it is more likely to scatter forward than backward, so its motion after the collision is not completely independent of its motion before the collision. But clearly, after some number of collisions, any trace of its previous motion will be lost, so we could fit this type of problem into the random walk framework simply by having a single "step" encompass the appropriate number of collisions. These considerations usually make little difference in the result, so for our purposes in this section, we ignore them.

Motion in more than one dimension can be broken up into its individual components, so we need develop the formalism for motion in only one dimension. We consider motion in the x direction, letting s be the displacement during a single step.

If there is no wind, hill, etc., so that each drunk is equally likely to take a step in either direction, then the average distance gone per step, $\bar{s}$, is zero. But because of external factors, one direction might be favored over another.

Let the function $P(s)\,ds$ describe the probability of a single step having length in the range between s and $s + ds$. According to the definition of mean values, then, the average length of a step would be given by

$$\bar{s} = \int_{-\infty}^{\infty} sP(s)\,ds \tag{4.12}$$

The standard deviation for the lengths of single steps is by definition

$$\overline{(\Delta s)^2} = \overline{(s - \bar{s})^2} = \int_{-\infty}^{\infty} (s - \bar{s})^2 P(s)\,ds \tag{4.13}$$

This can be reduced to the following form: (See Problem 4-3.)

$$\overline{(\Delta s)^2} = \overline{(s - \bar{s})^2} = \overline{(s^2 - 2\bar{s}s + \bar{s}^2)}$$
$$= \overline{s^2} - 2\bar{s}\bar{s} + \bar{s}^2$$
$$= \overline{s^2} - \bar{s}^2 \tag{4.14}$$

The value of $\bar{s}^2$ in this expression is obtained by squaring the value of $\bar{s}$ calculated according to Eq. 4.12, and $\overline{s^2}$ also can be calculated from the definition of mean values.

$$\overline{s^2} = \int_{-\infty}^{\infty} s^2 P(s)\,ds \tag{4.15}$$

The equations (4.12) and (4.14) tell us the average net motion and standard deviations in their net motion after each drunk has taken a single step. We wish to extend this to calculating the average net motion, S, and standard deviation, $\sigma^2 = \overline{(S - \bar{S})^2}$, after each drunk has taken N steps.

The total distance gone for a single drunk is the sum of the distances gone during each step.

$$S = \sum_{i=1}^{N} s_i$$

The average distance gone after N steps is just the product of the number of steps, N, times the average distance gone per step, $\bar{s}$.

$$\bar{S} = N\bar{s} \tag{4.16}$$

The deviation in any one drunk's position from the mean position of all the drunks is

$$S - \bar{S} = \sum_{i=1}^{N} s_i - N\bar{s} = \sum_{i=1}^{N} s_i - \sum_{i=1}^{N} \bar{s} = \sum_{i=1}^{N} (s_i - \bar{s}) = \sum_{i=1}^{N} \Delta s_i$$

The average of the square of this is the square of the standard deviation.

$$\sigma^2 = \overline{(S - \bar{S})^2} = \overline{\left(\sum_{i=1}^{N} \Delta s_i \right)^2} = \overline{\left(\sum_{i=1}^{N} \Delta s_i \right) \left(\sum_{j=1}^{N} \Delta s_j \right)}$$

$$= \overline{\sum_{i=1}^{N} \sum_{j=1}^{N} \Delta s_i \Delta s_j} = \sum_{i=1}^{N} \sum_{j=1}^{N} \overline{\Delta s_i \Delta s_j}$$

For the random walk, the individual steps are independent of each other. We take advantage of this by separating out the terms in this last product for $i \neq j$, and use the result (3.16) for statistically independent behaviors.

$$\sigma^2 = \sum_{i=1}^{N} \overline{(\Delta s_i)^2} + \sum_{i=1}^{N} \sum_{j \neq i} \overline{\Delta s_i \, \Delta s_j}$$

$$= N\overline{(\Delta s)^2} + \sum_{i=1}^{N} \sum_{j \neq i} \overline{\Delta s_i \, \Delta s_j}$$

Since

$$\overline{\Delta s_i} = \overline{(s_i - \bar{s})} = \bar{s} - \bar{s} = 0$$

we see that the last term is zero, and

$$\sigma^2 = N\overline{(\Delta s)^2} \tag{4.17}$$

SUMMARY

For the random walk problem in any one dimension, if $P(s)\,ds$ is the probability that a single "step" falls in the range between s and $s + ds$, then after N steps the average position of the objects, $\bar{S}$, and the square of the standard deviations of their positions, σ^2, are given by

$$\bar{S} = N\bar{s} \tag{4.16}$$

$$\sigma^2 = N\overline{(\Delta s)^2} \tag{4.17}$$

where

$$\bar{s} = \int_{-\infty}^{\infty} sP(s)\,ds \tag{4.12}$$

and

$$\overline{(\Delta s)^2} = \overline{(s - \bar{s})^2} = \int_{-\infty}^{\infty} (s - \bar{s})^2 P(s)\,ds \tag{4.13}$$

or equivalently,

$$\overline{(\Delta s)^2} = \overline{s^2} - \bar{s}^2 \tag{4.14}$$

with

$$\overline{s^2} = \int_{-\infty}^{\infty} s^2 P(s)\,ds \tag{4.15}$$

Notice that the average distance gone increases linearly with N, whereas the standard deviation increases only as the square root of N. In comparison to the distance gone, the position of a random walker in one dimension gets *relatively* more predictable as N increases (provided that $\bar{s} \neq 0$), but *absolutely* less predictable.

$$\frac{\sigma}{\bar{S}} \propto \frac{1}{\sqrt{N}} \tag{4.16}$$

This interesting result is similar to that for the binomial probability distribution studied earlier in this chapter, where we saw that the value of the standard deviation, σ, increased as $N^{1/2}$, although its relative value, $\sigma/\bar{n}$, decreased as $N^{-1/2}$.

EXAMPLE

Consider the migration of a conduction electron in a metal. The random thermal motion of the electrons causes each to undergo about 10^{12} collisions per second with atoms and other electrons, going a root mean square distance of about 1 Å between collisions $[\overline{(\Delta s)^2} = (1 \text{ Å})^2]$. Normally, this motion is completely random. With all directions of travel equally likely, the average length of a "step" in any single direction is zero.

However, in the presence of an electric field, one direction may be slightly favored over another. In a typical case, the average net distance traveled between collisions is 10^{-4} Å in the direction favored by the field. (Notice that this is extremely small compared to the root mean square length of a step, so the influence of the electric field is small compared to the random thermal motion.) For this case, calculate the average distance gone by an electron per second, and the standard deviation about that value.

The number of steps taken in one second is $N = 10^{12}$, so the average distance gone is

$$\bar{S} = N\bar{s} = (10^{12})(10^{-4} \text{ Å}) = 10^{-2} \text{ m}$$

The standard deviation is given by

$$\sigma^2 = N\overline{(\Delta s)^2} = (10^{12})(1 \text{ Å})^2$$

or

$$\sigma = 10^6 \text{ Å} = 10^{-4} \text{ m}$$

Notice that after 10^{12} steps, the standard deviation is small compared to the average displacement, although for any one step, the reverse is true. The larger the number of steps, the more predictable is the behavior.

PROBLEMS

4-9. An ammonia bottle is opened very briefly in the center of a large room releasing many ammonia molecules into the air. These ammonia molecules go an average of $s_0 = 10^{-5}$ m in any direction between collisions with air (and other ammonia) molecules, and collide on the average of 10^7 times per second. After each collision they are equally likely to go in any direction.

(a)   What is the average displacement in one dimension (say the z-dimension) for any one "step"?

(*Hint.* $s_z = s_0 \cos \theta$ in spherical coordinates. Taking the average over all solid angles gives

$$\bar{s}_z = \frac{1}{4\pi} \iint (s_0 \cos \theta) \sin \theta \, d\theta \, d\phi \,)$$

(b)   What is the square of the standard deviation for any one step?

(*Hint.* $\overline{(\Delta s_z)^2} = \overline{s_z^2} - \bar{s}_z^2$. $\bar{s}_z^2$ is obtained by squaring the answer to part (a), and $\overline{s_z^2}$ is obtained by averaging over all solid angles,

$$\overline{s_z^2} = \frac{1}{4\pi} \iint (s_0 \cos \theta)^2 \sin \theta \, d\theta \, d\phi \,)$$

(c)   What is the average displacement in the z-direction of the escaped NH_3 molecules after 2 s?

(d)   What is the standard deviation about the value obtained in part (c)?

(e)   If you were standing 6 m from the bottle, how long would it take before more than 32% of the ammonia molecules were farther from the bottle than you?

4-10.   A hole is migrating through a semiconductor that has no external field, so the hole is equally likely to go in any direction after collision with a lattice site. Such a collision occurs roughly 10^{13} times per second, and the hole goes an average of 10^{-8} cm between collisions. Using the hints in the previous problem to help you, calculate:

(a)   What is the average displacement in one dimension for any one step?

(b)   What is the standard deviation about this value?

(c)   What would be the average displacement in this dimension after 10 μs?

(d)   What would be the standard deviation about the value in part (c)?

4-11.   Energy produced in the center of the sun has a hard time finding its way out. You are going to calculate about how long it takes an average photon to get out. On the average in the sun, a photon will go about 1 cm between collisions with hydrogen atoms, and on the average it is held about 10^{-8} s by the hydrogen atom before being reemitted in a completely random direction. Hence, the photon takes about 10^8 "steps" per second.

(a)   What is the average distance traveled in any one dimension per step?

(b)   What is the standard deviation about the value obtained in part (a)?

(c)   The radius of the sun is about 7.0×10^{10} cm. About how many "steps" must be a photon take before having a 32% chance of being outside the sun?

(d)   At the rate of 10^8 steps per second, how many years does the number of steps in part (c) correspond to? (1 yr $\approx \pi \times 10^7$ s.)

APPENDIX 4A   THE STANDARD DEVIATION IN EQ. 4.3

In this appendix we derive the formula for the square of the standard deviation, using the binomial distribution for identically prepared systems of N elements each. It can be written as

$$\sigma^2 = \overline{(\Delta n)^2} = \overline{(n - \bar{n})^2} = \overline{n^2 - 2\bar{n}n + \bar{n}^2} = \overline{n^2} - \overline{2\bar{n}n} + \overline{\bar{n}^2}$$

Since $2\bar{n}$ is a constant, we have from Eq. 3.3,

$$\overline{2\bar{n}n} = (2\bar{n})\bar{n} = 2\bar{n}^2$$

and

$$\overline{\bar{n}^2} = \bar{n}^2$$

Therefore,

$$\overline{(\Delta\bar{n})^2} = \overline{n^2} - \bar{n}^2$$

To calculate $\overline{n^2}$ we use the definition of mean values, Eq. 3.1, and the trick that

$$n^2 p^n = \left(p \frac{\partial}{\partial p}\right)^2 p^n$$

This gives us the following result:

$$\overline{n^2} = \sum_n n^2 P_N(n) = \sum_n n^2 \left[\frac{N!}{n!(N - n)!} p^n q^{N-n}\right]$$

$$= \left(p \frac{\partial}{\partial p}\right)^2 \sum_n \left[\frac{N!}{n!(N - n)!} p^n q^{N-n}\right] = \left(p \frac{\partial}{\partial p}\right)^2 (p + q)^N$$

$$= \left(p \frac{\partial}{\partial p}\right) [pN(p + q)^{N-1}] = [p^2 N(N - 1)\underbrace{(p + q)^{N-2}}_{1} + pN\underbrace{(p + q)^{N-1}}_{1}]$$

$$= p^2 N(N - 1) + pN = Np[pN - p + 1]$$

Noticing that $-p + 1 = q$, and that $Np = \bar{n}$, we have

$$\overline{n^2} = Np(Np + q) = (Np)^2 + Npq = \bar{n}^2 + Npq$$

Subtracting $\bar{n}^2$ from both sides we get

$$\sigma^2 = \overline{n^2} - \bar{n}^2 = Npq$$

APPENDIX 4B   TAYLOR SERIES EXPANSION, EQ. 4.5

Any continuous, single-valued function of x, $f(x)$, can be represented as a polynomial in x of the form

$$f(x) = c_0 + c_1(x - a) + c_2(x - a)^2 + c_3(x - a)^3 + \cdots + c_n(x - a)^n + \cdots \qquad \text{(B.1)}$$

where a is some constant, as are the coefficients $c_0, c_1, c_2, \ldots$. Taking the differentials of Eq. B.1 with respect to x gives,

$$f'(x) = c_1 + 2c_2(x - a) + 3c_3(x - a)^2 + \cdots + nc_n(x - a)^{n-1} + \cdots$$

$$f''(x) = 2c_2 + 6c_3(x - a) + \cdots + n(n - 1)c_n(x - a)^{n-2} + \cdots$$

$$f^{(n)}(x) = n!c_n + \frac{(n + 1)!}{1!} c_{n+1}(x - a) + \cdots$$

By evaluating each of these at the point $x = a$, we see that

$$f(a) = c_0$$
$$f'(a) = c_1$$
$$f''(a) = 2!c_2$$
$$\vdots$$
$$f^{(n)}(a) = n!c_n$$
$$\vdots$$

Hence, in the polynomial expression (B.1), all the coefficients are given by

$$c_n = \frac{1}{n!} f^{(n)}(a)$$

Putting this result into the polynomial (B.1), we have

$$f(x) = f(a) + f'(a)(x - a) + \frac{1}{2!} f''(a)(x - a)^2 + \cdots + \frac{1}{n!} f^{(n)}(a)(x - a)^n + \cdots$$

$$= \sum_{n=0}^{\infty} \frac{1}{n!} f^{(n)}(a)(x - a)^n$$

This is called the "Taylor series expansion" of the function $f(x)$ around the point $x = a$.

APPENDIX 4C $\quad \partial^2/\partial n^2 \ln P(n)$ AT $n = \bar{n}$, EQ. 4.6

The expression for the first derivative of $\ln P(n)$ is

$$\frac{\partial \ln P(n)}{\partial n} \approx \frac{\Delta \ln P(n)}{\Delta n} = \frac{\ln P(n + 1) - \ln P(n)}{(n + 1) - n} = \ln P(n + 1) - \ln P(n)$$

$$= \ln \frac{N!}{(n + 1)!(N - n - 1)!} p^{n+1}q^{N-n-1} - \ln \frac{N!}{n!(N - n)!} p^n q^{N-n}$$

$$= \ln \frac{n!(N - n)!}{(n + 1)!(N - n - 1)!} \frac{p^{n+1}q^{N-n-1}}{p^n q^{N-n}} = \ln \frac{N - n}{n + 1} \frac{p}{q}$$

This expression for the first derivative is used in calculating the second derivative as follows:

$$\frac{\partial^2 \ln P(n)}{\partial n^2} \approx \frac{\Delta \dfrac{\partial \ln P(n)}{\partial n}}{\Delta n} = \frac{\dfrac{\partial \ln P(n + 1)}{\partial n} - \dfrac{\partial \ln P(n)}{\partial n}}{1}$$

$$= \ln \frac{N - n - 1}{n + 2} \frac{p}{q} - \ln \frac{N - n}{n + 1} \frac{p}{q}$$

$$= \ln \frac{N - n - 1}{N - n} \frac{n + 1}{n + 2} = \ln \left(1 - \frac{1}{N - n}\right) + \ln \left(1 - \frac{1}{n + 2}\right)$$

Using the expansion $\ln (1 - x) \approx -x$, for small x, and ignoring 2 in comparison to n, this becomes

$$\left. \frac{\partial^2 \ln P(n)}{\partial n^2} \right|_{n=\bar{n}} \approx -\frac{1}{N-n} - \left. \frac{1}{n+2} \right|_{n=\bar{n}} \approx -\frac{1}{N-n} - \left. \frac{1}{n} \right|_{n=\bar{n}} = -\frac{1}{N-\bar{n}} - \frac{1}{\bar{n}}$$

$$= -\left(\frac{1}{N-Np} + \frac{1}{Np} \right) = -\frac{1}{N}\left(\frac{1}{1-p} + \frac{1}{p} \right) = -\frac{1}{N}\left(\frac{1}{q} + \frac{1}{p} \right)$$

$$= -\frac{1}{Npq}(p+q) = -\frac{1}{Npq}$$

This result is Eq. 4.7.

PART 3
INTERNAL ENERGY

In Chapter 4, you learned that the statistical tools used in the study of large systems would be quite simple and elegant in comparison to those used for small systems. These tools are also extremely powerful, being useful in the studies of a wide variety of properties for a wide variety of systems. Indeed, they have a far wider range of usefulness than could possibly be covered in a single book such as this. Most of the remainder of this book develops these tools, illustrating some of this wide range of areas over which they can be used.

In this and the following chapter, we begin our study of large systems by giving a qualitative picture of some of the physical properties that the more formal developments in the succeeding chapters will deal with. We are particularly interested in the internal energy of a system, and how it can be altered. We will pretend we are extremely tiny microscopic observers, who can see individual atoms dancing about and interacting with each other. In this way, we can develop some intuitive feeling for internal energy and how it can be changed through interactions.

✻ chapter 5

INTERNAL ENERGY AND EQUIPARTITION

The elements of most systems have many different ways of storing energy, such as through their own motions or through interactions with neighbors. What are these ways that energy can be stored, and how is the energy distributed among them?

A. INTERNAL ENERGY

The internal energy of a system is the sum of the energies of the individual elements of the system, which includes both their energy of motion, and their potential energies arising from interactions with each other. The internal energy of an iron bar, for example, would include the kinetic energies of the conduction electrons, the potential energies of the iron atoms due to interactions with neighboring atoms, and their kinetic energies of vibration about their equilibrium positions (Figure 5.1).

Examples of what would *not* be internal energy of the iron bar would be its gravitational potential energy if it were sitting on the edge of a cliff, ready to fall off, or its kinetic energy if it were flying through the air. These forms of energy include more than just the elements of the iron bar alone. They include their gravitational interactions with the earth, and the relative motion of the two bodies. If the "system" were enlarged to include both the iron bar and the earth, then these would contribute to the internal energy of this larger system, but they are not part of the internal energy of the iron bar itself.

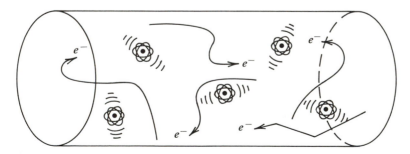

Figure 5.1 The internal energy of an iron bar would include kinetic energies of conduction electrons, and the vibrational energies of the iron atoms.

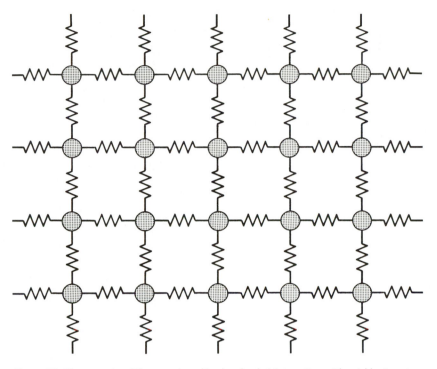

Figure 5.2 The atoms in solids are anchored in place by their interactions with neighboring atoms. These interactions make the system behave as if they were connected together by tiny springs.

Usually, a system is composed of more than one kind of element, and each kind of element may have many ways in which it can store energy. For example, the individual atoms in solids are anchored in place by their electrostatic interactions with neighboring atoms, which act like tiny springs (Figure 5.2). They vibrate across their equilibrium positions like tiny harmonic oscillators, with the energy changing back and forth from potential energy at the endpoint of the oscillation, to kinetic energy at the midpoint of the oscillation. On the average, half of the energy will be potential energy ($\frac{1}{2}kr^2$) and half will be kinetic ($\frac{1}{2}mv^2$). Energy might also be stored in tranlational kinetic energy, such as for conduction electrons in metals or molecules of a gas, or perhaps in rotational kinetic energy of polyatomic molecules.

B. DEGREES OF FREEDOM

Each way that each element of a system can store energy is called a "degree of freedom." Counting the degrees of freedom for a system takes some care and is best learned through examples, such as those that follow.

Consider the atoms of a solid, each of which vibrates about its equilibrium position as a tiny harmonic oscillator. Each atom can store potential energy ($\frac{1}{2}kr^2$) and kinetic

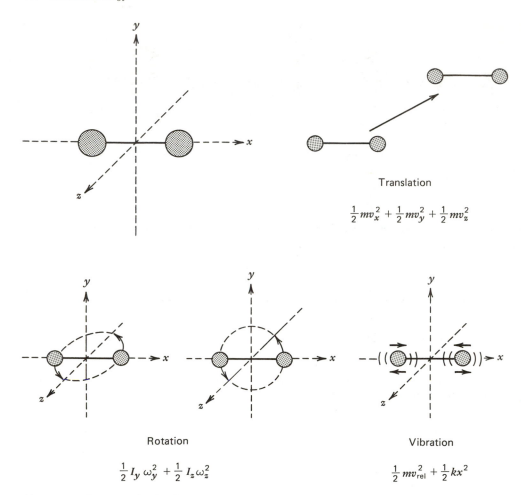

Figure 5.3 A diatomic molecule is like a little dumbbell. It may have 3 translational, 2 rotational, and 2 vibrational degrees of freedom.

energy ($\frac{1}{2}mv^2$), and on the average it will store equal amounts of each. But since it can vibrate in each of the three dimensions, there are three ways it can store potential energy

$$U = \frac{1}{2}kx^2 + \frac{1}{2}ky^2 + \frac{1}{2}kz^2 \tag{5.1}$$

and three ways it can store kinetic energy

$$E_k = \frac{1}{2}mv_x^2 + \frac{1}{2}mv_y^2 + \frac{1}{2}mv_z^2 \tag{5.2}$$

Each atom, then, has six degrees of freedom, and N such atoms in the solid would give the system $6N$ degrees of freedom altogether.

As another example, consider a gas of diatomic molecules (Figure 5.3). Each molecule may move freely in each of the three dimensions, so it has 3 "translational degrees of freedom." If P represents its center of mass momentum, then

$$E_{\text{translational}} = \frac{P_x^2}{2m} + \frac{P_y^2}{2m} + \frac{P_z^2}{2m}$$

In addition, the molecule may rotate. If I_y and I_z represent the moments of inertia about the axes perpendicular to the molecular axis, and L_y and L_z the angular momentum components about these axes, then

$$E_{\text{rotational}} = \frac{L_y^2}{2I_y} + \frac{L_z^2}{2I_z}$$

The molecule may also be able to store energy in its vibrations along the molecular axis. If p_x represents the relative momentum along this axis, and x the relative separation, then

$$E_{\text{vibrational}} = \frac{p_x^2}{2M} + \frac{1}{2}kx^2$$

This adds up to 7 degrees of freedom,[*] altogether, for a diatomic molecule. A gas of N such molecules would have $7N$ degrees of freedom.

We will see in later chapters that quantum effects may sometimes restrict the number of degrees of freedom available. For example, at room temperature a diatomic nitrogen molecule may translate and rotate, but not vibrate, so that quantum effects restrict the number of degrees of freedom of a nitrogen molecule in a room to 5 rather than 7.

C. THE EQUIPARTITION THEOREM

In the above examples, the energy stored in any one degree of freedom can be expressed in the form

$$E = bq^2$$

where b is a constant and q is a position or momentum coordinate. Most forms of energy we commonly encounter are expressed in this manner.[†]

The similarity in form for these various ways that energy can be stored, has a very useful and interesting consequence. Using probability theory and the formal similarity of the various energy expressions (bq^2), we can show that, on the average, equal amounts of energy are stored in all degrees of freedom. This is called the "equipartition theorem."

---

[*] Three translational, 2 rotational, and 2 vibrational degrees of freedom.
[†] Some, such as rotational kinetic energy, or energy stored in a particle's electromagnetic field, require appropriate generalization of position and momentum coordinates.

A general proof of this theorem could be given here, but it would be rather clumsy, due to our lack of familiarity with appropriate statistical techniques. For this reason, the general proof will be given later in the book, after we have developed more powerful and elegant statistical tools. At that point, we will also show that the average energy per degree of freedom ($\bar{\varepsilon}$) is directly proportional to the temperature (T),

$$\bar{\varepsilon} = \frac{1}{2} kT$$

where k is Boltzmann's constant,

$$k = 1.381 \times 10^{-23} \text{ J/K}$$

Even in the absence of the general proof, it is fairly easy to demonstrate the plausibility of equipartition. During collisions, kinetic energy tends to be transferred from particles initially having greater kinetic energy to those of lower initial kinetic energy, rather than vice versa. As an example, when a fast-moving ball collides with a slow one (or a stationary one) on a pool table, usually the slow-moving ball gains speed at the expense of the fast one as a result of the collision. This transfer of kinetic energy through collisions means that the kinetic energies will tend to even out, and all the particles of a system will tend to have the same kinetic energy on the average.

The same is true for potential energy. If the potential energy of a particle can be written in the form bq^2, (e.g., $\frac{1}{2}kx^2$), then it is a simple harmonic oscillator. For simple harmonic oscillators, the energy is being continually transferred back and forth between kinetic and potential forms, with half the total energy being stored in each, on the average.*

Thus we see that for familiar examples, the average energy stored in all kinetic degrees of freedom would be the same, and that would also be the same as the average energy stored in all potential degrees of freedom.

EQUIPARTITION THEOREM

On the average, the internal energy will be distributed equally among all those degrees of freedom whose energies are expressible in the form bq^2. This applies to all degrees of freedom of most common systems.

In previous courses you may have learned that temperature is a measure of the kinetic energies of the individual elements of a system (Figure 5.4). But since energy is stored equally in all degrees of freedom (on the average) temperature is a measure of energy stored in *any* degree of freedom, kinetic or potential, as long as potential

---

* You can show this by writing $x = A \cos \omega t$, where the angular frequency for a harmonic oscillator is given by $\omega = \sqrt{k/m}$. If you then average the potential energy, $\frac{1}{2}kx^2$, and the kinetic energy, $\frac{1}{2}mv^2$, over any number of complete cycles, you find that the two averages are equal.

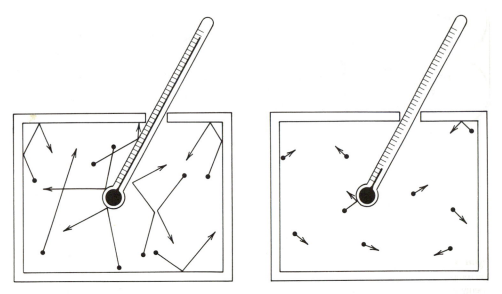

Figure 5.4 If the particles have large kinetic energies, then the system is hot. If the particles have small kinetic energies, the system is cold.

energies are measured relative to the *bottom* of the potential well* and are expressible in the form bq^2. This is the case for most common physical systems.

This fact enables us to use intuition gained from our everyday experiences to understand processes that can alter the internal energy of a system, even though we cannot directly observe the behaviors of the individual atoms, electrons, molecules, etc. Anything that increases the temperature increases the internal energy, and anything that decreases the temperature decreases the internal energy.

Sometimes we add energy to a system without changing its temperature at all. Sometimes, the reason for this is that new degrees of freedom are opening up in which the added energy is stored. In other cases the added energy goes into removing the individual elements from the potential wells in which they were bound, and so the added energy goes into changing the zero-energy reference point from which potential energies are measured. The new degrees of freedom, or the liberation of the particles from their former entrapment, give the system new physical characteristics, and we say it has undergone a "phase change."

As an example, consider the melting of an ice cube at 0°C. The added energy goes into liberating water molecules from their lattice positions in the crystal. Formerly, we measured their potential energies ($\frac{1}{2}kr^2$) relative to the bottom of the potential well, but now the liberated molecules have a different zero-energy reference point. But the average kinetic energies—and therefore the temperature—remain unchanged during this transition (Figure 5.5).

---

* The position of the bottom of the potential well depends on the particle's interactions with its neighbors. In general, this reference level changes when particles go from one system to another.

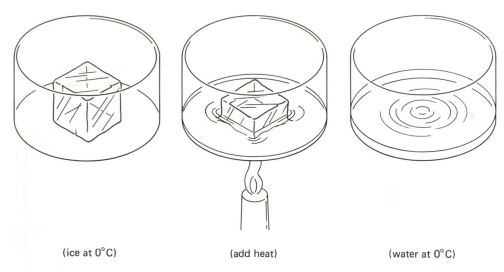

(ice at 0°C)   (add heat)   (water at 0°C)

Figure 5.5   In normal phase transitions, energy can be added to the system without changing the temperature. If the temperature doesn't change, the kinetic energy of the particles doesn't change. Where, then, does the added energy go?

SUMMARY

Internal energy is the energy stored in the motions and interactions of the elements (e.g., atoms, electrons, molecules, etc.) of a system. Often, systems are composed of more than one kind of element, and each element may store energy in many different ways. Each way that each element can store energy is called a "degree of freedom."

On the average, the internal energy of a system is distributed equally among all degrees of freedom. Temperature is a measure of the average kinetic energy per element, and is therefore proportional to the average energy per degree of freedom. When energy is added to a system, it may either go into increasing the energy per degree of freedom, in which case the temperature rises, or the system may undergo a phase change, in which case either the number of degrees of freedom is increased, or the level of the reference point for potential energies is raised (or some combination of the two).

These ideas will be developed rigorously in later chapters. They are presented here only in order to help you develop an intuition for what internal energy is, and so you can use your own experiences with temperature changes and phase changes to help you understand the processes which change the internal energy of a system.

There are three kinds of ways through which the internal energy of a system may be changed.

1.   Adding or removing heat energy.
2.   Doing work on the system, or letting the system do work on something else.

3. Adding or removing particles that undergo reactions (chemical, electrostatic, nuclear, etc.) with the particles of the system.

These are called "thermal," "mechanical," and "diffusive" interactions, respectively, and we will study these in the next chapter to see qualitatively how each affects the internal energy of a system, from a microscopic point of view.

PROBLEMS

5-1. How many degrees of freedom has a carbon atom in a diamond? (*Hint*. Is it bound, and how can it move?)

5-2. Why do you suppose a molecule of water vapor (H_2O) has 3 rotational degrees of freedom, and a molecule of nitrogen gas (N_2) has only two?

5-3. Assuming that a conduction electron in a metal is free to roam anywhere within the metal (not being constrained to any small region by any potential well), how many degrees of freedom does it have?

5-4. Consider the average potential energy of a water molecule bound in an ice crystal, and the average potential energy of one in the liquid state. Which is lower? How do you know?

5-5. Consider the phase change for iron from solid to liquid forms.

   (a) How many degrees of freedom does each iron atom have before it is melted?
   (b) How many degrees of freedom does each iron atom have after it has melted?
   (c) Did the number of degrees of freedom of the conduction electrons change?
   (d) Did the number of degrees of freedom of the whole system increase or decrease?
   (e) What happens to the energy put into the iron to melt it (on a microscopic scale)?

✳ chapter 6

INTERNAL ENERGY AND THE FIRST LAW

Since energy is conserved, any energy gained by one system must be lost by another. What are the ways in which energy can be transferred from one system to the next, and what is happening to the individual microscopic components during each of these processes?

Since energy is conserved, energy can be gained by one system only at the expense of another. Consequently, to change the internal energy of one system requires some sort of interaction with another. These interactions fall into three general categories according to the method in which the energy is transferred. These are (1) heat being transferred to or from a system, (2) work being done on or by a system, and (3) particles being transferred to or from a system.

A. HEAT TRANSFER

One way of changing the internal energy of a system is to add or remove heat energy. For example, if the system is an iron bar, we could add heat energy to it by holding it over a flame, or remove heat from it by putting it in ice water. In many cases, heat transfer will involve more than two systems. For example, when heating a pot of water on the stove, the heat energy must go from the stove burner to the pot first, and then from the pot to the water inside it.

To keep things simple, in the study of thermodynamics we usually consider the interactions between systems two at a time. Once we understand the interactions of just two systems, it is usually easy to extend this understanding to three or more interacting systems.

The thermal interaction between two systems is represented pictorially by a sketch similar to that of Figure 6.1, where we have two systems, A_1 and A_2, separated by some thin, thermally conducting, rigid membrane. The two systems together are then surrounded by a rigid insulating surface, to indicate no interactions of any kind with any other outside systems. The purist may point out that the membrane itself constitutes a third system, so for the benefit of these purists, we might consider the thin membrane to actually be the imaginary line of contact between two systems, such as between the flame and the iron bar.

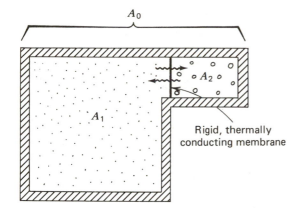

Figure 6.1 We can think of two systems interacting thermally as being separated by a rigid, thermally conducting membrane. For simplicity we will usually consider interacting systems only two at a time. Normally, we will call these two systems A_1 and A_2, and we consider the combined system, A_0, to be completely isolated from the rest of the universe.

Now let's take a submicroscopic look at what happens when a hotter body is placed in contact with a cooler one. Since temperature is a measure of the average kinetic energy per particle (i.e., atom, molecule, conduction electron, or whatever), the individual particles of the hotter body have higher kinetic energies on the average than do those of the cooler body. Along the surface of contact between the two systems, collisions will occur between particles of one and the particles of the other.

When a fast-moving particle, having large kinetic energy, collides with a slow-moving particle, having small kinetic energy, usually the slow one will speed up and the fast one will slow down as a result of the collision. For example, when a fast-moving cue ball collides with another ball sitting still on the table, both will usually end up moving, but some energy gets transferred from the one initially in motion to the one initially sitting still. You can think of possible cases where the slow ball might end up moving slower and the fast one moving faster after the collision, but these cases will be relatively rare. Usually, energy is transferred from the particle of higher kinetic energy to the one of lower kinetic energy when they collide. Consequently, when a large number of collisions have occurred among the particles of the two systems, energy will be transferred from the higher-energy particles to the lower-energy particles on the average, so heat flows from the hotter system toward the cooler system (Figure 6.2). This transfer of heat through the collisions of the individual particles is referred to as "conduction."

Heat may also be transferred by electromagnetic radiation. More energetic particles tend to radiate more energy than do less energetic ones, due to their more violent interactions and greater accelerations. Therefore, radiative interactions also result in the net transfer of energy from the hotter system to the cooler one.

As we have seen, whether the thermal interaction be radiative or conductive, the energy transfer is from hot to cold, rather than vice versa. This is just another reflection of the fact that energy tends to be shared equally by all degrees of freedom. When the two systems are interacting with each other, energy tends to be transferred

(hot)   (cold)

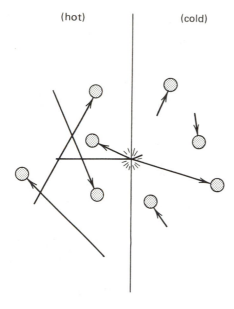

Figure 6.2 In the collision between particles, energy is usually transferred from the particle of greater kinetic energy to the one of lower kinetic energy. Consequently, through the collisions of particles along the boundary between two systems, energy tends to flow from the hotter system to the cooler one, rather than vice versa.

from the system whose particles have higher kinetic energies to the system whose particles have lower kinetic energies, until the average energy stored in each degree of freedom for both systems is the same.

SUMMARY

When two systems are in thermal contact, heat is transferred from one to the other through the collisions of their constituent particles along their boundaries. In a collision, energy tends to be transferred from the particle initially having greater kinetic energy to the particle having lower kinetic energy, and therefore heat tends to flow from hotter bodies to cooler ones.

PROBLEM

6-1.   In a later chapter, we will give an alternative statistical proof that when two systems are in thermal contact the heat energy must flow from the hotter one towards the cooler one. Why do you suppose that it is necessary to use statistical methods in this proof? (*Hint.* In any single collision, is the energy necessarily transferred from the hotter material to the cooler material, or must we average over many collisions?)

B.   WORK

Another way to increase the internal energy of a system is to do work on it. From our introductory physics course, we may remember that work is the product of force times distance.

There are many types of forces that can do work on a system. For example, electrostatic forces may cause a slight shift in the positions of the negatively charged electrons, or the positively charged nuclei in some material, such as a pencil or a gold ring. Not very much work would be done on any one charged particle, since the average position of each would be moved only a very tiny amount. But these objects contain some 10^{24} charged particles, so altogether a large amount of work could be done on the system as a whole. Similarly, magnetic forces could shift motion or the orientations of some of the magnetic moments of the individual electrons or atoms.

These forces that do the work on a microscopic or atomic scale are difficult to represent on a diagram where atomic dimensions are far too small to be seen. Consequently, work done on a system is customarily represented by some mechanical force exerted on a system, which results in a change in its volume.

The type of force is irrelevant. As long as it causes some displacement of matter within the system, work will have been done. With electric and magnetic forces, the energy is given immediately and directly to the particles throughout the system. Work done by some mechanical forces first transfers energy to the particles at the boundary where the force is applied, and this is then transferred by these particles to their neighbors, and so on, until it gets distributed throughout the system. But the end result, the effect of the work on changing the internal energy of the system, is not dependent on the type of force, or on whether the work is done immediately throughout the system, or on the boundary first and then subsequently shared among the other particles of the system.

For these reasons, we represent work done by one system on another as a mechanical interaction, with two systems being separated by a movable piston, as in Figure 6.3. Again, we surround both systems by a rigid, thermally insulating, nonporous wall, to indicate no interactions with other outside systems. If we wish to consider the mechanical interaction only, then we will use a thermally insulating piston as in Figure 6.3*a*, and if we wish to consider them interacting both thermally and mechanically, then

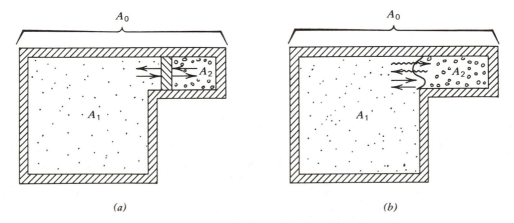

(a)　　　　　　　　　　　　(b)

Figure 6.3　(a) Two systems that are interacting mechanically can be thought of as being separated by a thermally insulating movable piston. (b) If two systems are interacting both thermally and mechanically, they can be thought of as being separated by a flexible thermally conducting membrane.

we consider the two systems to be separated by a flexible, thermally conducting membrane, as in Figure 6.3*b*.

Let's examine how work changes the internal energy of a system from a microscopic viewpoint. We consider the collisions between the particles and the walls of the container, and since we don't wish to consider heat transfer to the walls, we require these collisions to be elastic. (Alternatively, we could consider the system and the container to be of the same temperature, so that on the average the collisions result in no net transfer of energy between the two.)

When a particle undergoes an elastic collision with a stationary wall, it neither gains nor loses energy in the process. This is *not* true, however, if the wall is moving. To see this, consider throwing a perfectly elastic ball against the back of a truck (Figure 6.4). If the truck is stationary, the ball will bounce back going as fast as you threw it. If the truck is moving away from you, however, the ball will come back more slowly than you threw it, and if the truck is moving toward you, the ball will come back faster than you threw it. Similarly, when particles of a system collide elastically with the boundary of their container, they will gain energy if the boundary is moving in, compressing the system, and they will lose energy if the boundary is moving out, causing expansion of the system (Figure 6.5). Therefore, for systems that exert pressure on the walls of their container through particle collisions (e.g., gases, or solids under pressure), their internal

(a)   (b)

Figure 6.4   (a) If you throw a Ping-Pong ball against a stationary truck, the ball bounces back with as much energy as it had before the collision (assuming elastic collision and ignoring air friction). (b) But if the truck is moving away, the Ping-Pong ball bounces back with less energy than it had initially.

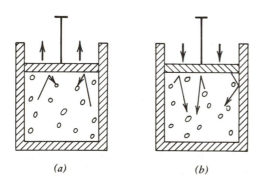

(a)   (b)

Figure 6.5   (a) As a gas expands, its molecules strike a receding piston, and therefore lose kinetic energy. As a gas does work, it loses internal energy. (b) When a gas is being compressed, its molecules collide with an incoming piston, and therefore they gain kinetic energy. When work is done on a gas, its internal energy increases.

energy increases as they are further compressed, and it decreases if they are allowed to expand.

SUMMARY

The internal energy of a system will increase or decrease when work is done on or by the system. There are a wide variety of forces that can do work on systems, but they all give the same result. So for illustrative purposes, we generally consider mechanical forces which change the volume of a system.

For systems under pressure (i.e., whose particles collide with the container walls), further compression increases the internal energy, and expansion reduces it. This is because particles colliding with inward-moving walls tend to gain energy, and those colliding with outward-moving walls tend to lose energy.

PROBLEM

6-2.   Consider a system that is *not* under pressure whose volume is reduced (i.e., container wall moved inward).

(a)   Is any work done on the system? Why or why not?

(b)   Explain from a microscopic (e.g., atomic or molecular) point of view, why the internal energy is not increased in this case. (*Hint.* If it is *not* under pressure, are there any molecules colliding with the container walls?)

C.   PARTICLE TRANSFER

Each element of a system can store energy in two possible ways: as kinetic energy and as potential energy. The kinetic energy of an element can be due to the motion of its center of mass, or to the relative motions of its subcomponents, such as rotation or vibration of a polyatomic molecule. The potential energy can be due to the interactions of an element with its neighbors, or to the interactions of the element's subcomponents among themselves, such as the interaction between hydrogen and oxygen atoms in one water molecule.

Potential energies must be measured relative to some reference level. Each system has its own peculiar zero-energy reference level from which potential energies are most conveniently measured, and as long as particles are not being transferred from one system to another, we need not be concerned with how the zero-energy reference level in one system compares to that in another.

For example, if the system of interest is a group of balls rolling and bouncing about on a tabletop, it is most convenient to measure the potential energy of each relative to the tabletop, so that the potential energy of one resting on the tabletop is zero, and the potential energy of one that has bounced to a height h above the tabletop is equal to mgh. Similarly, for a system of balls on the floor, it is most convenient to measure potential energies (mgh) relative to the floor level, with the height, h, being the height above the floor. As long as each ball remains in its particular system, all energy calculations for any one system may be measured relative to the zero-energy reference level which is most convenient.

However, if the balls of the two systems should intermix, with balls from the table top rolling off and joining those on the floor, and/or those on the floor bouncing onto the table, then we must measure all potential energies for members of both systems relative to the same reference level. It will not do to make energy calculations that sometimes use the tabletop, and other times use the floor, as zero-energy reference levels. When particles are going from one system to another, we must be sure to establish a common zero-energy reference point from which all potential energies are measured.

The "particles" of most physical systems are atoms or molecules, and we normally establish the universal zero-energy reference level as the potential energy of an isolated particle (i.e., somewhere in outer space) that has no interactions with neighbors of any type.

With this definition of the zero-energy reference level, the total internal energy of most systems must be negative. In all solids and liquids, for example, the particles are bound together, which means that their potential energies are negative. They are bound together by mutually attractive forces and their kinetic energies are insufficient to overcome this entrapment.

For example, consider a particle trapped in a harmonic oscillator potential well. If μ measures the energy of the lowest point in that potential well, then its potential energy is given by

$$u = \mu + \frac{1}{2} kr^2$$

where μ is negative. The total energy of that particle is the sum of the potential and kinetic energies

$$\varepsilon = u + \varepsilon_k = \mu + \frac{1}{2} kr^2 + \frac{1}{2} mv^2$$

If the particle is bound, this energy is negative.

It is often useful to work with the "thermal energy" of a particle, which is defined as the energy measured relative to the lowest point in the potential well. With this definition we can write for the particle in the above example

$$\varepsilon = \varepsilon_{\text{thermal}} + \mu$$

where

$$\varepsilon_{\text{thermal}} = \frac{1}{2} kr^2 + \frac{1}{2} mv^2$$

The total internal energy of an entire system of N particles may be written as

$$E = E_{\text{thermal}} + N\mu$$

where E_{thermal} is the sum of the kinetic and potential energies of all particles, *where their potential energies are measured relative to the bottom of the potential well*, μ.

Notice that the equipartition theorem really applies to only the *thermal* energies of systems. It applies to forms of energy expressed as bq^2, such as $\frac{1}{2}kr^2$, which is a potential energy *measured relative to the bottom of the potential well*, and is not the absolute potential energy of a particle.

SUMMARY

The particles of many systems have potential energy due to interactions with neighbors. We can define the "thermal energy" of a particle as the sum of its kinetic energy and its potential energy *measured relative to the bottom of its potential well*. If μ is the energy of the bottom of the potential well for each particle, then

$$\varepsilon = \varepsilon_{\text{thermal}} + \mu$$

The total internal energy of a system is given by

$$E = E_{\text{thermal}} + N\mu$$

where E_{thermal} is the total kinetic energy and potential energy of all particles, where their potential energies are measured relative to the lowest point in their potential wells (μ). If the system is bound (e.g., solids and liquids), the thermal energy is insufficient to free the particles from their potential wells.

$$E = E_{\text{thermal}} + N\mu < 0, \quad \text{(if bound)}$$

The equipartition theorem actually applies to the distribution of the thermal energy of the system only.

When systems are not exchanging particles, then we will usually not be concerned with changes in zero-energy reference levels. In this case we can choose any reference level that is convenient for measuring the internal energies of systems. This most convenient reference level is clearly the bottom of the potential wells, so we make the shift

$$E \rightarrow E - N\mu$$

and then the internal energy is identical to the "thermal energy."

$$E = E_{\text{thermal}}$$

This will be the case for most systems covered in this book.

If a particle initially all by itself and having zero energy is inserted into some system, interactions with its neighbors in this new environment will cause some shift in its potential energy. In many cases, these forces are attractive, so the particle falls into a potential well and its potential energy is negative. But in some cases, the interactions will be repulsive, and the particle's potential energy will be positive. An example would be the insertion of a particle into a gas under heavy pressure.

When the interactions among the particles are attractive, then the potential energy released when a new particle is added is converted into increased thermal energy of the system as a whole (Figure 6.6). On a microscopic level, what is happening is that the mutual attraction between the newly added particle and its neighbors causes them initially to move toward each other. This increased energy of motion, or kinetic energy, is then converted into random thermal motion through subsequent collisions.

The attraction among elements of a system is evident in many familiar processes. In freshman chemistry, many of us have had the opportunity to add concentrated sulfuric

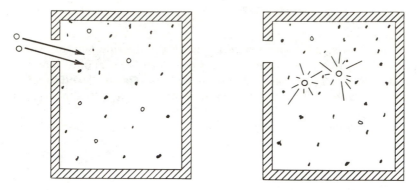

Figure 6.6 Thermal energy may be released as new particles join the system and interact with their new neighbors.

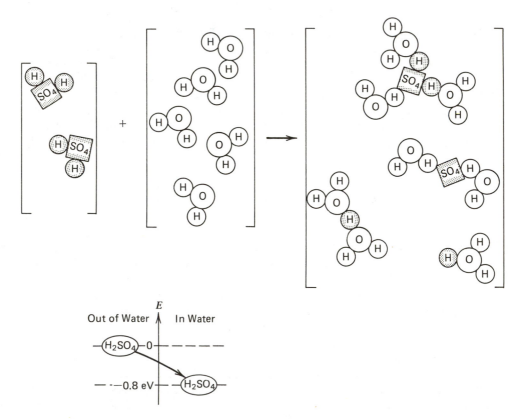

Figure 6.7 The molecules of sulfuric acid and those of water are both electrically polarized. Electrostatic attraction between oppositely charged ends of the two kinds of molecules causes the sulfuric acid molecules to fall into an electrostatic potential well when they enter the water. For dilute solutions, the depth of this well is about 0.8 eV. This is released as thermal energy.

acid to water and witness the dramatic rise in temperature of the solution. We can do it quite carefully, not exerting much force, and we can keep it away from external heat sources, so the amount of heat added to the system and the work done on it are both quite negligible. Yet the quick rise in temperature tells us the thermal motion of the particles has increased dramatically. Clearly there is a great deal of electrostatic attraction between the highly polarized water and sulfuric acid molecules, causing the release of potential energy as the latter are added to the system (Figure 6.7).

The added molecules need not be different from the major components of the system itself. Water molecules in water vapor join water in droplets, releasing a large amount of thermal energy—on the order of 540 calories for each gram of water condensed. Similarly, thermal energy is released when molecules in the liquid phase join those in the solid phase, as water freezes.

When we burn wood in our fireplace, or coal in our barbecues, we start with a system of mostly hydrocarbons and oxygen with relatively small thermal energy, and transform it into a system of mostly H_2O and CO_2 gases with a great deal of thermal energy, as evidenced by the high temperatures and glow of the escaping gases (Figure 6.8). Clearly, the components of the H_2O and CO_2 gas molecules attract each other more strongly than they did when in hydrocarbon and oxygen molecules. Equivalently, the electronic configurations in the H_2O and CO_2 molecules are of lower potential energy than those of the hydrocarbon and O_2 molecules.

The interaction need not be electrostatic or chemical. We may begin with a system of deuterium nuclei in heavy-hydrogen compounds in a thermonuclear warhead. By adding some heat or doing some work on this system, we may ignite a thermonuclear explosion, during which helium nuclei, free neutrons, and other fragments are

Figure 6.8 In burning hydrocarbons, we transfer particles from one system to another, which is accompanied by a release of thermal energy due to the interactions among the particles. We can think of the particles of both systems to be hydrogen, carbon, and oxygen atoms. They are being transferred from a system in which they are arranged as hydrocarbon and oxygen molecules to a system where they are arranged as carbon dioxide and water molecules. Thermal energy is released because their mutual interactions in the second system are stronger than in the first.

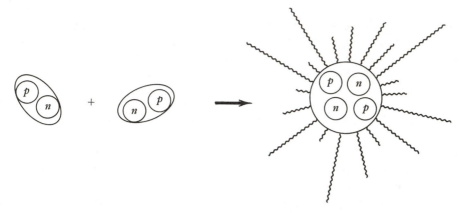

Figure 6.9 As the elements of the system of deuterium nuclei are transferred to the system of helium nuclei, a great deal of energy is released through their interactions.

produced at the expense of some of the deuterium (Figure 6.9). Clearly, the thermal energy of the system increases enormously during the explosion—far more than can be accounted for by the heat added or work done to ignite it.

In each of these examples, the thermal energy of one system is increased as particles are added to it from another. In the first case, the thermal energy of the dilute solution increased as particles were transferred to it from the concentrated solution. In the condensation of water vapor, thermal energy is released to the water droplets as molecules are transferred to it from the vapor phase. When hydrocarbons burn, the thermal energy of the H_2O and CO_2 gases is increased as particles are added to this system from the system of hydrocarbons and oxygen. Finally, in the warhead example, the thermal energy of the system of fusion products increases as particles are transferred to it from the system of deuterium nuclei.

We call this exchange of particles between systems the "diffusive interaction," and we usually consider the diffusive interactions between only two systems at a time, as we do with thermal and mechanical interactions. The extension to three or more systems is straightforward. We represent the diffusive interaction schematically as illustrated in Figure 6.10.

The potential energy of a particle in a system will depend on its position relative to its neighbors. If we were to remove all the thermal energy we could from that particle, we would necessarily leave it standing motionless* at that point where its potential energy is a minimum. (See Figure 6.11.) It follows that if we wished to add a particle to the system in such a way as to not bring any thermal energy at all into the system with it, then that particle would have to be added with an energy exactly equal to the energy of the lowest point in its potential well, and no more (Figure 6.11).

This energy, which is the energy of the very lowest point in the particle's potential well, is called the "chemical potential," and it is given the symbol, μ. If ΔN particles are

---

* Ignore quantum effects for the moment.

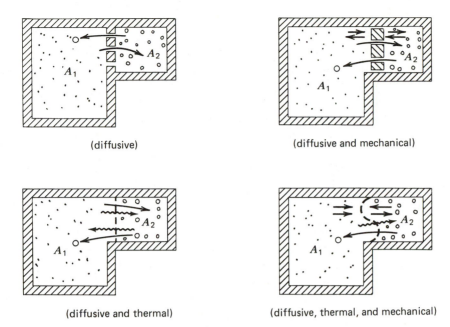

(diffusive)   (diffusive and mechanical)

(diffusive and thermal)   (diffusive, thermal, and mechanical)

Figure 6.10 We can think of two systems interacting diffusively as being separated by a porous, rigid, thermally insulating wall. If they are interacting both diffusively and mechanically, we can think of them as being separated by a porous, thermally insulating, movable piston. If they are interacting both diffusively and thermally, we can think of them as being separated by a porous, rigid, thermally conducting membrane. And if they are engaging in all three types of interactions, we can think of them as being separated by a porous, flexible, thermally conducting membrane.

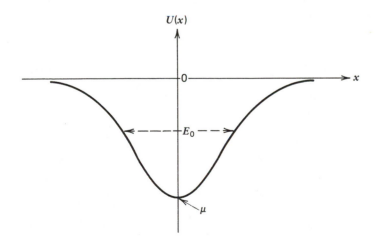

Figure 6.11 Plot of potential energy of some particle as a function of its position. If a particle has energy E_0, it will oscillate back and forth across the potential well, with its energy transferring back and forth between kinetic and potential forms. The energy measured relative to the bottom of the potential well ($U = \mu$) is called the "thermal energy." If we wish to add a particle to the system in such a way that it has no thermal energy at all, then its energy must be $E = \mu$.

added to the system without adding any thermal energy at all ($\Delta Q = \Delta W = 0$), then the internal energy of the system changes by

$$\Delta E = \mu \Delta N$$

Conversely, if we wish to add energyless particles to the system, so that the total internal energy remains unchanged, then the thermal energy released to the system by each particle will be equal to the depth of the potential well it falls into, which is caused by its interactions with its neighbors. The thermal energy gained will be equal and opposite to the maximum potential energy released.

$$\Delta E = 0 = \Delta E_{\text{thermal}} + \mu \Delta N$$

As we have seen, if particles are going from one system to another, the common zero-energy reference level for measuring potential energies is the energy of a particle that is undergoing no interactions of any kind with any neighbors. Therefore, when the interactions between particles of a system are mostly attractive, the chemical potential is negative, and when the interactions are mostly repulsive, the chemical potential is positive (Figure 6.12).

The random thermal motions of a particle's neighbors causes continual fluctuations in its own potential well. Therefore, the chemical potential must be defined as the *average* depth of the potential well. Roughly half the time the actual depth of the potential well will be lower than this, and half the time it will be higher.

The chemical potential is generally *not* the same as a particle's average potential energy. Only when all the thermal energy is extracted from a system will the particles be found at the bottom of their potential wells.* As thermal energy is added, some goes into kinetic degrees of freedom and some into potential degrees of freedom, in which case a particle's average potential energy will be somewhat above the level of the bottom of the potential well.

We expect that the interactions between neighboring molecules would be influenced by how fast they are moving and by how close they are together. Therefore, we expect the chemical potential, μ, to be dependent on the system's temperature, T, and its volume, V. But it will also depend on how many molecules of this type are present, N. For example, as the number of sulfuric acid molecules in solution becomes very large, then the available water molecules must be distributed among large numbers of H_2SO_4 molecules (Figure 6.13). This will mean fewer water molecules interacting with any one H_2SO_4 molecule, on the average, so the total interaction per H_2SO_4 molecule (alternatively, the "depth" of the potential well) will lessen. Therefore, the chemical potential should depend on all three of these variables, T, V, and N.

$$\mu = \mu(T, V, N)$$

The depth of the potential well depends on the strength of the interactions. Nuclear interactions are extremely strong, and chemical potentials of the reactants are measured in megaelectron volts (MeV). This implies that particles engaging in nuclear reactions each release thermal energies of several megaelectron volts as they fall into

---

* Ignore quantum effects.

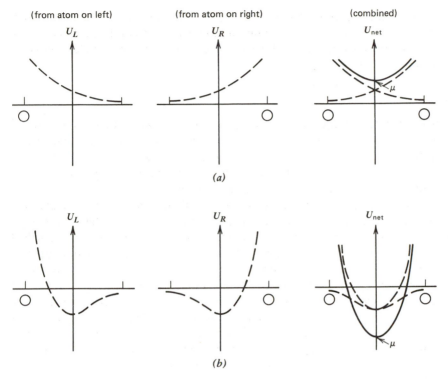

Figure 6.12 Plots of the potential energy of an atom versus its position, due to interaction with atoms on both sides. (a) If the interatomic forces are repulsive, the chemical potential will be positive. (b) If the interatomic forces are attractive, the chemical potential will be negative.

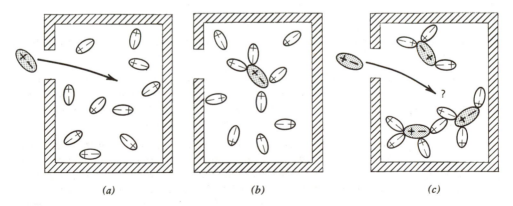

Figure 6.13 Consider the dissolution of a system of one kind of molecule (e.g., sugar or acid) in another system (e.g., water). (a) As the first molecules enter the new system, they fall into an electrostatic potential well (b) due to the attraction between oppositely charged portions of neighboring molecules. Their chemical potential is negative, and they release thermal energy to the system as they dissolve. (c) As more molecules are added to the system, the depth of the potential well decreases (i.e., the chemical potential rises) because there are fewer remaining free solvent molecules with which the newly added molecules can interact.

potential wells of these depths. The chemical potentials involved in chemical and electrostatic interactions are typically a million times smaller than this, with chemical potentials, or thermal energies released per particle, being measured in electron volts (eV) or fractions thereof.

SUMMARY

The particles of systems frequently have potential energies due to their interactions with each other. Potential energies must be measured relative to some reference level, and this level is generally the bottom of the potential well in which a particle sits. As long as the system is *not* interacting diffusively, all energies are most conveniently measured relative to this particular reference level.

However, when systems are exchanging particles, then some universal reference level must be established from which the energies of particles of all the interacting systems are measured. This universal zero-energy reference level is customarily taken as the potential energy of an isolated particle undergoing no interactions at all.

The chemical potential, μ, is defined to be the energy of the lowest point in a particle's potential well, measured relative to this universal reference level. When a particle goes from one system to another, new interactions with different neighbors in the new environment cause a shift in its chemical potential. The chemical potential is negative if interactions are predominantly attractive, and positive if interactions are predominantly repulsive. A shift in chemical potential corresponds in a shift in the depth of the bottom of the potential well (or a shift in the zero-energy reference level peculiar to a system).

Interactions among neighbors, and therefore the chemical potential, are affected by the temperature, volume, and number of particles in a system.

PROBLEMS

6-3. Is the chemical potential for a H_2SO_4 molecule entering fresh water positive or negative? How do you know?

6-4. Suppose the chemical potential for a certain salt in solution A is -0.6 eV, and its chemical potential in solution B is -0.8 eV. If these two solutions are separated by a semipermeable membrane, which allows the salt ions through, but not the solvents, will the salts tend to diffuse from side A to side B, from side B to side A, or neither? Why?

6-5. The boiling point of water is considerably higher than the boiling point for other light molecules, such as NH_3 and CH_4. It is even much higher than the boiling point for molecules nearly twice as heavy, such as N_2 and O_2. Why do you suppose this is? How is your answer related to the chemical potential of the molecules of these materials?

D.   THE FIRST LAW

In the preceding three sections we examined each of the three ways that the internal energy of a system may be changed. (See Figure 6.14.)

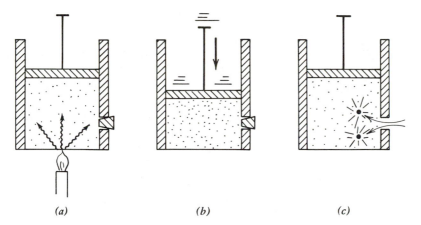

(a) *(b)* *(c)*

Figure 6.14 Illustration of the three kinds of processes through which the internal energy of a system may be increased. (a) Adding heat energy (dQ). (b) Doing work on the system ($-dW$). (c) Adding particles ($\mu\,dN$). We can write the change in internal energy of a system as $\Delta E = \Delta Q - \Delta W + \mu\,\Delta N$. The ΔW has a negative sign because ΔW represents work done by the sytem, which is the negative of work done on the system.

1. By adding or removing heat energy.
2. By doing work on the system, or letting the system do work on something else.
3. By adding or removing particles from the system, which undergo interactions with their neighbors.

> If we let dE represent the incremental change in internal energy of the system, dQ the increment of heat added, dW the increment of work done *by* the system (note that this is the negative of the work done *on* the system), and dN the incremental change in the number of particles in the system, then we write the first law of thermodynamics as
>
> $$dE = dQ - dW + \mu\,dN$$
>
> where μ is the chemical potential.

It is customary to let dW denote the work done *by* the system rather than the work done *on* the system, and that is why the dW term is preceded by the minus sign.

If there are several different kinds of particles in the system, then the last term should become $\sum_i \mu_i\,dN_i$, where the sum is over all different types of particles. Similarly, if more than one kind of work is being done by the system (e.g., change in volume, magnetic work, etc.), then the dW should be replaced by a sum over the various different kinds of work, $\sum_j dW_j$.

E. EXACT AND INEXACT DIFFERENTIALS

Consider some system undergoing thermal and mechanical interactions, but whose number of particles remain fixed. In this case, the first law is

$$dE = dQ - dW$$

Although we can talk of the "total internal energy" of the system, it is improper to refer to the "total heat content," or the "total work done by it." The total internal energy of a system is measurable (in principle), but whether that energy entered the system as heat added, or as work done on it, or as some combination of the two, cannot be determined *a posteriori*.

For example, suppose we measured the internal energy of an iron bar and then left the room while a friend changed its internal energy. When we return, we can measure the increase in internal energy, ΔE, but we will not be able to tell whether that change was made by adding heat to the bar, or by doing work on it. If the bar is hotter, we may think that heat was added. However, the temperature could also have arisen by doing work on the bar, such as by hitting it with a hammer. Similarly, if we return to find the bar squeezed in a clamp as in Figure 6.15, we may think that work had been done on the bar. However, it could be that our friend first cooled the bar until it contracted enough to slip into the clamp. Then he or she may have heated the bar up again so that it expanded against the clamp, without the clamp having moved at all. So again, all we can determine is that the internal energy has changed, but we have no way of knowing by what route that change was made. In summary, one thing that makes the internal energy different from these other two variables is that the total internal energy of a system can be measured at any time. However, in performing this measurement, one has no way of knowing how much of this internal energy entered the system as heat, and how much entered as work done on the system.

In mathematical terms, the change in the internal energy of a system is an "exact differential," because any change in internal energy is a function of the end points only (e.g., initial and final temperatures and volumes), and not a function of the path taken between initial and final states.

On the other hand, dQ and dW are "inexact differentials," because the total heat added, ΔQ, and the total work done, ΔW, in going from initial to final states does depend on the path taken. In the above examples involving the iron bar, we saw how the same final state could be attained by doing work ($-\Delta W$) without adding any heat ($\Delta Q = 0$), by adding heat (ΔQ) without doing any work ($-\Delta W = 0$), or an infinite number of possible combinations of the two. Therefore, the amount of heat added (ΔQ) or work done (ΔW) *cannot* be determined from the end points alone.

In many textbooks on thermodynamics, the distinction between exact and inexact differentials is indicated by placing a bar on the differential symbol d for inexact differentials. The first law for nondiffusive interactions would then be written as

$$dE = đQ - đW \tag{6.1}$$

and that for diffusive interactions as well would be

$$dE = đQ - đW + \mu\,dN \tag{6.2}$$

$\Delta Q = 0;\ \Delta W \neq 0$:

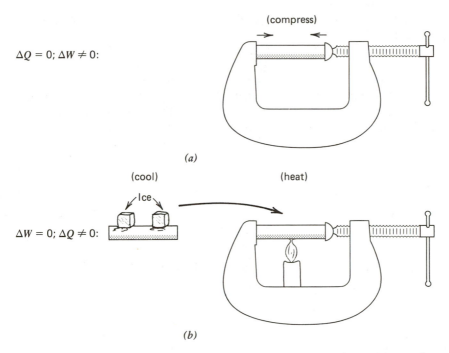

$\Delta W = 0;\ \Delta Q \neq 0$:

Figure 6.15 Illustration of two quite different paths for getting from the same initial state to the same final state. (a) The iron bar is put into the clamp and then compressed. In this process, work is done on the iron bar, but no heat energy is added or removed. ($\Delta W \neq 0;\ \Delta Q = 0$). (b) The iron bar may first be cooled so that it contracts and can be slipped into the clamp that has been already adjusted for the desired final length. Then heat can be added to bring the temperature up to the desired final temperature. In doing this, the bar expands and becomes stuck in the clamp. In this process, no work is done on the bar, but there is some net heat transfer. ($\Delta W = 0;\ \Delta Q \neq 0$) In both processes the iron bar ends up at the same final temperature and pressure.

Notice that the differential symbol d in the last term has no bar on it, because the number of particles in a system is determinable, and therefore any change in the number of particles, ΔN, can be determined from analysis of initial and final states alone. That is, if the system starts with 100 particles and ends up with 103, we know that $\Delta N = 3$, and we need not worry about which particular thermodynamic route the system took in going from initial to final state.

Above, we have considered the physical implications of exact and inexact differentials involved in the first law. We now look into the more general mathematical implications of these. The particular variables that characterize a system may differ from one system to the next, but the general mathematical formalism applies to all.

Consider some single-valued function of two variables. (We choose two variables only for simplicity. The extension to more variables is straightforward.)

$$F = F(x, y)$$

If the initial and final values of the two variables are (x_i, y_i) and (x_f, y_f), respectively, then the change in F is given by

$$\Delta F = F(x_f, y_f) - F(x_i, y_i) \tag{6.3}$$

Notice that the change in F is uniquely determined by the two endpoints, and not by the route taken. It does not matter, for example, if x changed first and y later, if y changed first and x later, if they both changed together, or which of the infinite variety of possible routes was taken. Since F is a single-valued function of x and y, we only need to know the initial and final values of these two variables to know the change in F.

The differential form of F is given by

$$dF = \frac{\partial F}{\partial x}\, dx + \frac{\partial F}{\partial y}\, dy \tag{6.4}$$

Therefore, the way to determine if some differential

$$g(x, y)\, dx + h(x, y)\, dy \tag{6.5}$$

is exact or not, we need to see if we can find some function, $F(x, y)$, such that

$$\frac{\partial F}{\partial x} = g \qquad \text{and} \qquad \frac{\partial F}{\partial y} = h \tag{6.6}$$

If we can, the differential is exact, and if we can't, the differential is inexact. Alternatively, we can use the identity

$$\frac{\partial^2 F}{\partial y\, \partial x} = \frac{\partial^2 F}{\partial x\, \partial y} \tag{6.7}$$

By comparing the differential form (6.5) to Eq. 6.6, we can see that for exact differentials,

$$\frac{\partial g}{\partial y} = \frac{\partial h}{\partial x} \tag{6.8}$$

EXAMPLE

Is $(2xy)\, dx + (x^2)\, dy$ an exact differential?

In this example, $g = 2xy$ and $h = x^2$. The function $F = x^2 y + (\text{constant})$ is such that

$$\frac{\partial F}{\partial x} = g \qquad \text{and} \qquad \frac{\partial F}{\partial y} = h$$

Alternatively we can try the second method. For this case

$$\frac{\partial g}{\partial y} = 2x$$

and

$$\frac{\partial h}{\partial x} = 2x$$

The two are the same, so the differential is exact.

EXAMPLE

Is $3xy\,dx + x^2\,dy$ an exact differential?

In this example, $g = 3xy$ and $h = x^2$. We can find no function F such that both

$$\frac{\partial F}{\partial x} = g \quad \text{and} \quad \frac{\partial F}{\partial y} = h$$

Alternatively, we can try the second method. For this example

$$\frac{\partial g}{\partial y} = 3x$$

and

$$\frac{\partial h}{\partial x} = 2x$$

These are not the same. Through either method we reach the conclusion that $3xy\,dx + x^2\,dy$ is *not* an exact differential.

SUMMARY

A differential $g(x, y)\,dx + h(x, y)\,dy$ is exact if there is a function $F(x, y)$ such that

$$\frac{\partial F}{\partial x} = g \qquad \frac{\partial F}{\partial y} = h \tag{6.6}$$

Alternatively, it is exact if

$$\frac{\partial g}{\partial y} = \frac{\partial h}{\partial x} \tag{6.8}$$

The integral of an exact differential depends only on the endpoints of the integral, and not on the path taken.

$$\int_i^f dF(x, y) = F(x_f, y_f) - F(x_i, y_i) \tag{6.9}$$

In the first law, the change in internal energy, dE, is an exact differential, as is the change in number of particles, dN. However, the heat added, dQ, and work done by the system, dW, are not exact differentials. Consequently, when the internal energy of a system is changed by a measured amount, ΔE, there is no way of knowing how much of that energy entered as heat, ΔQ, and how much as work done, $-\Delta W$, without knowing the particular thermodynamic path followed in going from initial to final states.

PROBLEMS

6-6. Test each of the following differentials to see if they are exact. Try using both methods.

(a) $-y \sin x\,dx + \cos x\,dy$

(b) $y\,dx + x\,dy$

(c) $yx^3e^x\,dx + x^3e^x\,dy$
(d) $(1 + x)ye^x\,dx + xe^x\,dy$
(e) $4x^3y^{-2}\,dx - 2x^4y^{-3}\,dy$

6-7.   Consider the path integral of the exact differential

$$dF = 2xy\,dx + x^2\,dy$$

Integrate this from point (1, 1) to (4, 3) along paths 1 and 3 in Figure 6.16. Are the two results the same?

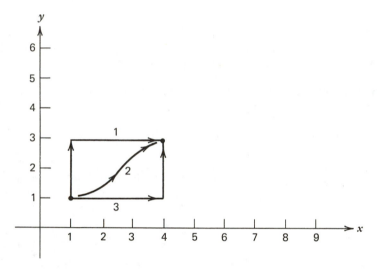

Figure 6.16   The intergral of an exact differential between initial and final points is independent of the path taken.

6-8.   Consider the path integral of the inexact differential

$$dG = 3xy\,dx + x^2\,dy$$

Integrate this from point (1, 1) to (4. 3) along paths 1 and 3 in Figure 6.16. Are the two results the same?

6-9.   In Chapter 7 we will see that the work done by a system can be written as $dW = p\,dV$. Compute the work done by a system as its pressure and volume change from the initial values (p_i, V_i) to the final values (p_f, V_f), by evaluating the integral

$$\Delta W_{if} = \int_{p_i, V_i}^{p_f, V_f} p\,dV$$

(a)   Along path 1 in Figure 6.17. [*Answer.* $\Delta W = p_i(V_f - V_i)$.]
(b)   Along path 2 in Figure 6.17.
(c)   Is the amount of work done by the system dependent on the path taken?

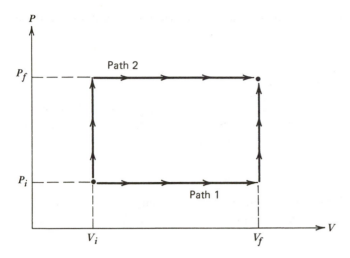

Figure 6.17 Illustration of two of the possible paths for a system to follow in going from some initial pressure and volume (p_i, V_i) to some final pressure and volume (p_f, V_f).

F. DEPENDENT AND INDEPENDENT VARIABLES

In thermodynamics, we often have many different system variables, such as internal energy, entropy, temperature, pressure, volume, chemical potential, number of particles, and many others, of which only two or three are independent (depending on the number of types of interactions the system engages in). This could cause much confusion, particularly when the partial derivatives of functions are involved.

For example, consider some function, F, which is a function of many different variables ($q, r, s, t, u, v, w, x, y, \ldots$), only two of which are independent. If we are to take the partial derivative with respect to one of the variables, for example, $\partial F/\partial u$, we must hold the "other" independent variable constant. But how will we let people know which of the various possible "other" variables was the one we held constant?

There are two customary ways of doing this. The notations

$$\frac{\partial F(u, x)}{\partial u} \tag{6.10}$$

and

$$\left.\frac{\partial F}{\partial u}\right)_x \tag{6.11}$$

are the normal ways of indicating that there were only two independent variables involved and that x was the one held constant while the derivative with respect to u was taken. Of these two notations, the first is more commonly used in mathematics and the second more common in thermodynamics, so we will use the second notation more in this book.

Similarly, the notation

$$\frac{\partial F}{\partial w}\bigg)_{x,y}$$

would indicate there are three independent variables (three types of interactions which the system is undergoing), and that the variables x and y are the ones held constant while the partial derivative with respect to w is taken.

EXAMPLE

Consider a function of four variables, $F(w, x, y, z)$, only two of which are independent. For example, suppose we can write y and z in terms of w and x as

$$y = w^2 x$$

$$z = e^w$$

If we are given that

$$F(w, x, y, z) = w + xyz$$

find

$$\frac{\partial F}{\partial y}\bigg)_z$$

First we must convert everything into terms of y and z, as these are the two to be considered independent in doing the partial derivative. Solving for w and x in terms of y and z, we have

$$w = \ln z$$

$$x = yw^{-2} = \frac{y}{(\ln z)^2}$$

Now we can write F in terms of the two independent variables (x, y) alone,

$$F = w + xyz = \ln z + \frac{y^2 z}{(\ln z)^2}$$

and take the appropriate partial derivative

$$\frac{\partial F}{\partial y}\bigg)_z = \frac{2yz}{(\ln z)^2}$$

PROBLEMS

6-10.   Suppose $w = xy$, $z = x^2/y$. Express:

(a) z as a function of w and y, $z(w, y)$
(b) z as a function of w and x, $z(w, x)$
(c) w as a function of z and y, $w(z, y)$

(d) w as a function of z and x, $w(z, x)$

(e) x as a function of y and z, $x(y, z)$

6-11.   For Problem 6-10, evaluate the following partial differentials.

(a) $\partial z/\partial w)_y$

(b) $\partial w/\partial y)_z$

(c) $\partial x/\partial y)_z$

(d) $\partial y/\partial w)_z$

6-12.   Consider the variables x, y, z, u, v, where $x(u, v) = u^2v$; $y(u, v) = u^2 + 2v^2$, $z(x, y) = xy$. Any one of these five variables can be expressed as a function of any other two.

(a)   What is y as a function of x and v, $y(x, v)$?

(b)   What is z as a function of u and v, $z(u, v)$?

(c)   What is v as a function of y and u, $v(y, u)$?

(d)   What is $\partial y/\partial x)_v$?

(e)   What is $\partial x/\partial u)_v$?

(f)   What is $\partial x/\partial u)_y$?

PART 4

ENTROPY

In Chapter 2 we studied the statistics of small systems. By "small" we meant systems that had a comprehensible number of elements, and we handled the statistics in such a way that each element had only two possible states—either it satisfied our criterion or it didn't. The statistical tools introduced were a bit cumbersome, but they were correct nonetheless.

Often, systems of interest to us contain incomprehensible numbers of elements and each element may have an incomprehensible number of states available to it. There are roughly 10^{28} nitrogen molecules in a typical room full of air, for example, and each molecule has more than 10^{33} quantum states available to it considering its center of mass motion alone. If we also considered rotational states, the number would be much larger. With the number of different states for just one molecule being so large, the number of different states for the system as a whole is clearly overwhelming.

To help us handle such enormous numbers, we must develop a new statistical tool. In Chapter 7 we develop some of the underlying concepts needed for understanding this tool, and in the following chapter we present the tool itself—the second law of thermodynamics.

✳ chapter 7

THE STATES OF A SYSTEM

When energy is added to a system, it increases the number of states accessible to the individual constituents, because it gives them the possibility of occupying higher-energy states that were previously inaccessible. Since a macroscopic system is made up of myriads of particles, each of which may be in any of a number of states, the number of configurations or "states" for the macroscopic system must be truly overwhelming. Furthermore, a small increase in the number of states accessible to the individual particles would result in a very large increase in the number of possible configurations of the macroscopic system. In this chapter we examine the dependence of the number of accessible states on the internal energy of the system and the relative probabilities of being in the various possible configurations.

A consequence of the wave nature of particles is that we cannot ever determine the position and momentum coordinates of a particle exactly. Instead, the best we can possibly do is to determine them to within some very small but finite range (see Eq. 2.9). This minimum determinable range of values constitutes a small region of six-dimensional phase space* called a "quantum state." Although the dimensions of a quantum state are very small, they are finite. This means that a finite (although perhaps very large) number of quantum states span all the possible values of a particle's position and momentum coordinates.

If a macroscopic system is composed of a large number of particles, each of which could be in a large number of different possible quantum states, then the total number of different possible configurations (or "states") for that system will be extremely large. However, no matter how large this is, it is still finite. This is of crucial importance because we know how to deal with large numbers. Infinity would be a different matter.

A. EQUILIBRIUM

An isolated system is said to be in "equilibrium" when the probabilities of it being in the various possible states do not vary with time. For example, consider the system

---

* Recall that the range of possible values of position and momentum define a six-dimensional space called "phase space."

consisting of ammonia molecules that we release in the front of a room. When first released, the quantum states located near the point of release have relatively high probabilities of containing the molecules, while those at the rear of the room have no chance at all. But, as time goes on, the molecules migrate, and the probabilities change until finally they are equally likely to be found anywhere in the room. At that point, the system is said to be in "equilibrium."

As another example, consider changing the internal energy of a material by compressing it or by adding heat, as in Figure 7.1. Initially, the energy is given to the molecules at the boundary, where the energy transfer took place. Initially, then, the system is not in equilibrium because the system is more likely to be found in the states for which molecules near the boundary have higher energies. But in time, interactions between the molecules distribute this energy throughout the system until all possible states again are equally probable, and the system is in "equilibrium."

The characteristic time that is required for a system to reach equilibrium, after being perturbed, is called its "relaxation time." The relaxation time depends both on the type of perturbation and on the system. If the system is air in a gallon jug, for example, the relaxation time for a pressure differential introduced at a surface (i.e., work done on the system) is about a millisecond, whereas the relaxation time for a temperature differential introduced on one surface (i.e., heat added to the system) may be several minutes. The relaxation time for spin states of conduction electrons in metals at room temperatures would be measured in nanoseconds, whereas the relaxation time for the heads-tails configurations of coins placed on the ground may be measured in geological time scales, requiring earthquakes, volcanoes, etc., to scramble them sufficiently to be considered in equilibrium.

When applying statistical tools to processes involving macroscopic systems, it is often helpful if the systems can be considered to always be near equilibrium. This requires that when systems interact, the transfer of heat, work, or particles must proceed at a rate which is slow compared to the relaxation time of the systems for that particular process. When this happens, we say the process is "quasi-static."

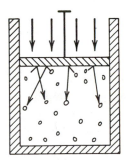 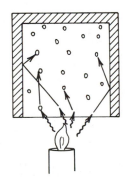

Figure 7.1 When mechanical work is done on a system, or heat is added to a system, the additional energy is first given to particles near the boundary and is subsequently shared with the rest of the system through the mutual interactions of the particles.

SUMMARY

An isolated system is said to be in "equilibrium" when the probabilities of it being in the various accessible states do not vary with time. If a system is perturbed, or some constraint on it released, the characteristic time required for it to come into equilibrium is called its "relaxation time" for that process. When interactions between systems proceed at a rate that is sufficiently slow that the systems can always be considered near equilibrium, the process is called "quasi-static."

PROBLEM

7-1. Consider the addition of a small amount of heat energy at the boundary of some system.

  (a) For this process would you expect the relaxation time in metal or in a gas to be shorter? Why?
  (b) How would you expect the relaxation for this process in glass to compare to that for gases? To that for metals? Why? (*Hint*. Consider the strength of the interaction between elements of the systems, and the mobility of conduction electrons.)

B. THE FUNDAMENTAL POSTULATE

The tools for the statistical analysis of the behaviors of large systems are based on one single, very important postulate.

FUNDAMENTAL POSTULATE

An isolated system in equilibrium is equally likely to be in any of its accessible states.

This postulate seems to be quite reasonable, but this in itself does not justify its adoption and use. After all, Nature has not always chosen the path that at first seemed most reasonable in the eyes of mankind. Instead, we must justify its use by comparing the results of experiments, with predictions based on this postulate. This has been done again and again for a wide variety of systems with amazing success, and by now the validity of this postulate is accepted without question in the scientific community. Although most of the remainder of this book is devoted to illustrating the applications of this postulate, the new student will just have to accept this postulate for now and trust that the author would not bother to develop a book full of applications based on a false premise.

If the number of states accessible to the entire system is given by Ω_o, then the probability that it will be found in any one of them is (Figure 7.2)

$$P = \frac{1}{\Omega_o}$$

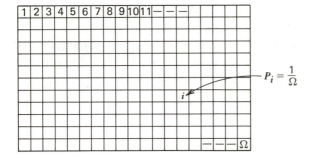

Figure 7.2 If there are Ω states available to the system, and all are equally probable, then the probability of the system being in any one state is $1/\Omega$.

For example, if there are four accessible states altogether, the probability that it is in any one of them is $\frac{1}{4}$.

We can extend this reasoning to find the probability of an isolated system being found in a certain subset of all the accessible states. If Ω_i is the number of states in the subset of interest, then the probability, P_i, that the system will be found in this subset is

$$P_i = \frac{\text{number of states in the subset}}{\text{number of accessible states altogether}} = \frac{\Omega_i}{\Omega_o} \qquad (7.1)$$

EXAMPLE

Consider a system of three distinguishable spin $\frac{1}{2}$ particles (Figure 7.3), for which the z-component of each particle's magnetic moment is either $+\mu$ or $-\mu$, depending on its spin orientation. Suppose there is an external magnetic field, B,

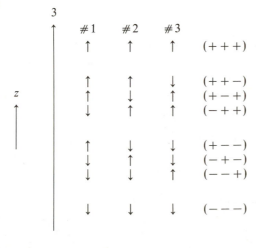

Figure 7.3 There are eight different possible arrangements of spin orientations for a system of three distinguishable spin $\frac{1}{2}$ particles. We give these different arrangements in four groups, according to the net angular momentum of the system.

oriented along the z-direction. If the energy of the system is $-\mu B$, what is the probability that the orientation of the first particle's magnetic moment is $+$?

There are three different spin states for the three particles all giving total energy of $-\mu B$. In terms of their orientations, these are $(+ + -)$, $(+ - +)$, and $(- + +)$. Of these, two have the orientation of the first particle being $+$, so

$$P(\mu_z \text{ of state 1 is } +) = \frac{(\text{number of states with } \mu_z \text{ of 1 being } +)}{(\text{total number of states})}$$

$$= \frac{2}{3}$$

EXAMPLE

The electrons in a plasma are confined to stay within volume V, and to have momenta less than $p_o = 10^{-22}$ kg-m/s. What is the probability that the momentum of one certain electron is less than $p = 10^{-23}$ kg-m/s?

According to Eq. 2.11, the total number of quantum states available to it is

$$\Omega_o = \frac{V_r V_p}{h^3} = \frac{V_r}{h^3}\left(\frac{4}{3}\pi p_o^3\right)$$

(V_r and V_p are volumes in coordinate and momentum space, respectively) and the number of states having momenta less than $p = 10^{-23}$ kg-m/s is

$$\Omega_i = \frac{V_r V_p}{h^3} = \frac{V_r}{h^3}\left(\frac{4}{3}\pi p^3\right)$$

So the probability of it being in subspace Ω_i is

$$P_i = \frac{\Omega_i}{\Omega_o} = \frac{p^3}{p_o^3} = \left(\frac{10^{-23}}{10^{-22}}\right)^3 = 10^{-3}$$

PROBLEMS

7-2. Consider a system of three spin $\frac{1}{2}$ particles, each having magnetic moment μ. If there is no external magnetic field ($B = 0$), then all spin states are of the same energy and are equally accessible.

   (a) Write down all the possible spin configurations of the three (see Figure 7.3).
   (b) What is the probability that the z-component of the magnetic moment of the system is $-\mu$?
   (c) What is the probability that the spin of particle 1 is up ($+$) *and* the spin of particle 2 is down ($-$)?

7-3. Consider a system of four flipped pennies.

   (a) What is the probability that two are heads and two are tails? (*Hint.* Write down all the possible heads-tails configurations and count.)
   (b) What is the probability that all four are tails?

(c) Suppose someone told you that two of them were heads and two were tails, but not which were which. Then what is the probability that penny 1 is heads *and* penny 3 is tails?

7-4. A particle is confined to be within a rectangular box of dimensions 1 cm by 1 cm by 2 cm. Furthermore, the magnitude of its momentum is constrained to be less than 3×10^{-19} g cm/s.

  (a) What is the probability of it being in any one single quantum state?
  (b) What is the probability that the magnitude of its momentum is less than 2×10^{-19} g cm/s?

7-5. An otherwise empty room contains 6×10^{26} air molecules. How many times more probable is it to have them split exactly 50–50 between front and back halves of the room than to have a 49–51 split? (*Hint.* Recall from Chapter 2 that the number of different states for which n of N elements satisfy the criteria and the remaining $N - n$ do not is given by

$$\frac{N!}{n!(N-n)!}$$

Use this plus Stirling's approximation (p. 40) to solve this problem.)

C. THE SPACING OF STATES

It is often convenient to identify the state of a system by its internal energy, E. Frequently, this does not uniquely identify the state, because several different states may have the same energy. For example, if the system is three spin $\frac{1}{2}$ particles, each having magnetic moment μ and sitting in an external magnetic field B, we may refer to the "state" of energy μB. This does not uniquely identify the state, however, because there are three different spin configurations all having this same energy $(+--, -+-,$ and $--+)$. Similarly, we may identify the "state" of a hydrogen atom by giving its energy, or perhaps just by giving the value of the principle quantum number, n, in the formula

$$E_n = \frac{-13.6 \text{ eV}}{n^2} \tag{7.2}$$

This does not identify the angular momentum of the atom, and in general there are many different angular momentum states for each energy level.

When several different states have the same energy, we say that the state (or "energy level") is "degenerate." An equivalent way of saying that there are n different states of energy E_0 is to say that the state of energy E_0 is "n-times degenerate." In the above examples of three spin $\frac{1}{2}$ particles we could say that the $E = \mu B$ state is "three-times degenerate," or in the case of the hydrogen atom we could say that the $E = -3.4$ eV ($n = 2$) state is "eight-times degenerate," for example.

The concept of degenerate states is only used in reference either to very small systems or to the individual elements of larger systems. The reason for this is that for large systems, the degeneracy of each possible energy level is incomprehensibly large. For example, there are about $10^{10^{28}}$ different ways of arranging the air molecules in the room in which you are sitting so that half of them are in the front half, and the remainder in back. Clearly, all these different arrangements have the same energy and clearly this number is infinitesimal in comparison to the number of different arrangements among quantum states (as opposed to front and back halves of the room) having the same energy.

Not only is the degeneracy of any one level huge for large systems, but the spacing between different energy states is nearly infinitesimal in comparison to the total internal energy of the system. The spacing between neighboring quantum energy levels for the system of air in your room is about 10^{-24} eV, or about 10^{-51} of the total internal energy of the air.

When dealing with large systems, then, we cannot even hope to be able to control the particular quantum energy level for the system as a whole. The best we can hope to do is hold the system to within some energy range of finite width, ΔE, which is very large in comparison to the spacing of the energy levels of the system.

We should expect the number of states available to the system to be proportional to the size of the allowed energy range, ΔE. That is, if we double the allowed range of energies, we double the number of states accessible to the system. If $\Omega(E, \Delta E)$ is the number of accessible states in the range between E and $E + \Delta E$, then we can write

$$\Omega(E, \Delta E) = g(E)\Delta E \tag{7.3}$$

where $g(E)$ is called the "density of states," and it expresses the number of accessible states per unit energy increment. (See Figure 7.4.)

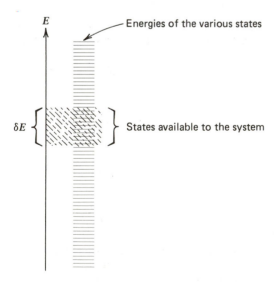

Figure 7.4 If the energy of a system is confined to the range δE, the number of states accessible to the system depends on both the density of states (i.e., the number of states per unit energy interval) and the size of the range δE.

SUMMARY

For small systems, energy levels are either nondegenerate, or the degree of degeneracy is a small number. Furthermore, the spacings between different energy levels are generally appreciable fractions of the total internal energy.

For large systems, however, the degree of degeneracy of any one energy level is huge, and the spacings between neighboring energy levels are nearly infinitesimal in comparison to the total internal energy. For large systems, the number of accessible states in the energy range between E and $E + \Delta E$ is proportional to the product of the density of states, $g(E)$, and the size of the allowed range, ΔE

$$\Omega(E, \Delta E) = g(E)\Delta E \tag{7.3}$$

PROBLEM

7-6.   Consider an air molecule of mass 4.8×10^{-26} kg in a standard-sized classroom (3 m $\times$ 10 m $\times$ 12 m).

   (a)   If the average speed of an air molecule at room temperature is 300 m/s, what is the average magnitude of its momentum?

   (b)   Double your answer to part (a), and call this the maximum allowed value for the magnitude of the molecule's momentum, p_o. What is the number of different quantum states accessible to this molecule? (*Hint.* Use Eq. 2.12. Available volume in momentum space is $V_p = \frac{4}{3}\pi p_o^3$.)

   (c)   What is the kinetic energy of the air molecule, using $v = 300$ m/s?

   (d)   What is the total kinetic energy of all the air molecules (about 10^{28} of them) in the room?

   (e)   If $V_p = \frac{4}{3}\pi p_o^3$, what is the minimum value of p_o for a molecule such that there is just one quantum state available to it in the room? (*Hint.* $V_r V_p/h^3 = 1$.)

   (f)   How much kinetic energy does the answer to part (e) correspond to?

   (g)   Assuming that the answer to part (f) is roughly the spacing between quantum energy levels allowed the room full of air, what fraction is this of the total internal energy of the system? [See part (d).]

D.   DENSITY OF STATES AND THE INTERNAL ENERGY

We will now demonstrate that the density of states, $g(E)$, is an extremely rapidly increasing function of the internal energy of the system, E. We will show in particular that for a system having $\mathfrak{N}$ degrees of freedom, the density of states typically goes as

$$g(E) \propto E^{\mathfrak{N}/2}$$

We start by noting that the number of states available to the entire system is the product of the number of states available to the individual components of the system. More formally we can write

$$\Omega_o = \prod_{i=1}^{\mathfrak{N}} \Omega_i \tag{7.4}$$

where Ω_o is the number of states available to the entire system, and Ω_i is the number of states available in the ith degree of freedom.

For example, consider the number of heads-tails states available to a system of coins. For a single coin, there are just two possible states. For a system of two coins, there are $2 \times 2 = 4$ possible states, because for each configuration of the first, there are two possible configurations for the second. For a system of three coins, there are $2 \times 2 \times 2 = 8$ different possible states, and so on. The total number of states available to any system is the product of the number of states available to the individual components. More examples of this are presented in homework Problems 7-7 through 7-10.

An immediate consequence of Eq. 7.4 is that for most macroscopic physical systems, the number of accessible states must be an extremely rapidly increasing function of the energy of the system. Consider the number of states available in the ith degree of freedom only. As the internal energy of the entire system, E, is increased, we should expect the number of states available in the ith degree of freedom to increase, because all the previous states are still available, plus additional ones allowed by the additional energy. We should expect, then

$$\Omega_i(E) \propto E^\chi \tag{7.5}$$

where χ is some finite positive number. Putting this into Eq. 7.4 we see that for a system of $\mathfrak{N}$ degrees of freedom, the number of states available to the entire system should go as

$$\Omega_o(E) = \Omega_1 \cdot \Omega_2 \cdot \Omega_3 \cdots \propto (E^\chi)^\mathfrak{N} = E^{\chi\mathfrak{N}}. \tag{7.6}$$

Since $\mathfrak{N}$ is around 10^{22} to 10^{28} for most common macroscopic systems, we can see that the number of accessible states increase extremely rapidly with the internal energy of the system.

The exponent χ is equal to $\frac{1}{2}$ for most ordinary systems, as is proven in Appendix 7A, and as can be seen intuitively as follows. The number of quantum states accessible to any degree of freedom is directly proportional to the size of the allowed range in the appropriate position or momentum coordinate, q. (See Eq. 2.12.) For most systems, the energy stored on any one degree of freedom is proportional to q^2 (e.g., $(\frac{1}{2m})p^2$, $\frac{1}{2}kx^2$, etc.), so the maximum energy stored in one degree of freedom is proportional to the square of the allowed range in the coordinate q.

$$\varepsilon_{max} = bq_{max}^2 \qquad (b = \text{constant})$$

Therefore, the number of accessible quantum states is proportional to $\varepsilon_{max}^{1/2}$. The maximum energy per degree of freedom increases as total internal energy increases, and so we have

(number of quantum states accessible to any degree of freedom)

$$\propto q_{max} \propto \varepsilon_{max}^{1/2} \propto E^{1/2}$$

or

$$\Omega_i \propto E^{1/2} \qquad \left(\chi = \frac{1}{2}\right). \tag{7.7}$$

An example of a special case where the value of the exponent (χ) is *not* equal to $\frac{1}{2}$ is that of relativistic motion. In this case, the energy ε is proportional to the *first* power of the momentum, rather than the second power:

$$\varepsilon = cp$$

where c is the speed of light. For this case, we have

(number of quantum states accessible to this degree of freedom)

$$\propto p_{max} \propto \varepsilon_{max} \propto E$$

So for this special case, the exponent (χ) is equal to 1, rather than $\frac{1}{2}$.

In any case, we see that the number of quantum states accessible to any degree of freedom is given by

$$\Omega_i = E^\chi$$

where E is the internal energy of the system, and χ is a positive constant, normally equal to $\frac{1}{2}$. For simplicity of presentation, we will henceforth use $\chi = \frac{1}{2}$ in our development.

With this result, we can return to Eq. 7.4 and see that for a large system having $\mathfrak{N}$ degrees of freedom and internal energy E, the number of states accessible to it is given by

$$\Omega_o(E) = \prod_{i=1}^{\mathfrak{N}} \Omega_i = (\text{const})E^{\mathfrak{N}/2} \tag{7.8}$$

In practice, we are not able to control the internal energy of any system exactly. Rather, the best we can do is to keep it in some range between E and $E + \Delta E$. The wider the range of allowed energies, the larger will be the number of states accessible to the system. Since the number of accessible states will be proportional to both $E^{\mathfrak{N}/2}$ and to ΔE, we can write

$$\Omega_o(E, \Delta E) = (\text{const})E^{\mathfrak{N}/2} \Delta E \tag{7.9}$$

where $\Omega_o(E, \Delta E)$ represents the number of accessible states whose total internal energies lie in the range between E and $E + \Delta E$. We can see immediately that the density of state, $g(E)$, is given by

$$g(E) = (\text{const})E^{\mathfrak{N}/2} \tag{7.10}$$

SUMMARY

The number of states available to a system is the product of the numbers of states accessible to the individual degrees of freedom.

$$\Omega_o = \prod_{i=1}^{\mathfrak{N}} \Omega_i \tag{7.4}$$

The number of states available to the individual degrees of freedom must increase as the internal energy of the system increases.

$$\Omega_i \propto E^\chi \qquad \chi > 0 \tag{7.5}$$

This means that the number of states accessible to the entire system is an extremely rapidly increasing function of the internal energy.

$$\Omega_o \propto E^{\chi\mathfrak{N}} \tag{7.6}$$

The number of states accessible in any degree of freedom is proportional to the allowed range in the appropriate momentum or position coordinate, q. Since the energy in that degree of freedom is normally proportional to q^2, the number of accessible states is proportional to the square root of the allowed range of energies.

$$\Omega_i \propto q_i \propto \varepsilon_i^{1/2}$$

Since the range of allowed energies in any degree of freedom is directly proportional to the internal energy of the entire system, we have

$$\Omega_i \propto E^{1/2} \tag{7.7}$$

and

$$\Omega_o(E) \propto E^{\mathfrak{R}/2} \tag{7.8}$$

The number of accessible states having energies in the range between E and $E + \Delta E$ is given by

$$\Omega_o(E, \Delta E) = (\text{const})E^{\mathfrak{R}/2}\,\Delta E \tag{7.9}$$

where the density of states, $g(E)$, is given by

$$g(E) = (\text{const})E^{\mathfrak{R}/2} \tag{7.10}$$

Take a moment to consider just how rapidly the density of states increases with the system's internal energy. For a small, but macroscopic, system of 2×10^{24} degrees of freedom, the density of states increases as

$$g(E) \propto E^{10^{24}}$$

This is a very rapidly increasing function of the energy!

EXAMPLE

Consider a room full of air which has typically 10^{28} molecules, and 5×10^{28} degrees of freedom. (Each molecule has 3 translational and 2 rotational degrees of freedom.) If room temperature is increased by 1°C, by what factor does the number of states available to the system increase?

As we have seen, temperature is related to internal energy because it is a measure of average energy per degree of freedom. Room temperature is around 300 K, so an increase from 300 K to 301 K is an increase of about 0.33% in internal energy. Consequently, the number of available states would increase by a factor of about

$$\left(\frac{E_2}{E_1}\right)^{\mathfrak{R}/2} = (1.0033)^{2.5 \times 10^{28}}$$

$$= 10^{3.6 \times 10^{26}}$$

So an increase of just 1°C in room temperature is a *phenomenal* increase in the number of available states for the system.

For those purists among us who insist that χ will not always be equal to $\frac{1}{2}$ for any conceivable system, we should point out that it is not necessary that $\chi = \frac{1}{2}$ for the

Figure 7.5 How many different heads-tails states are there available to this system of six coins? Is it equal to the product of the number of states available to each? (Kathy Bendo).

statistical tools developed in this book to apply; it is only necessary that the number of available states be a very rapidly increasing function of the internal energy of the system. It can be seen from Eq. 7.6

$$\Omega \propto E^{\chi \mathfrak{N}} \tag{7.6}$$

that for macroscopic systems (e.g., $\mathfrak{N} \sim$ Avogadro's number), Ω will be a very rapidly increasing function of E for any positive value of χ.*

PROBLEMS

7-7. Make a table showing the number of different possible heads-tails configurations for a system of:

    (a)   Two coins.
    (b)   Three coins.
    (c)   Four coins.
    (d)   Is the number of different configurations equal to the product of the numbers of states available to the individual coins in each case?

7-8. Answer the question in the caption to Figure 7.5.

---

* That χ must be positive can be shown using the *fundamental postulate*. If a system is equally likely to be found in all its accessible states, then an increase in internal energy must mean that additional higher-energy states must have opened up that the system could go into. Otherwise, the internal energy wouldn't have changed.

7-9. How many different states are available to a system of five dice?

7-10. How many different states are available to a system of 10^{24} molecules, each of which could be in any one of five different states?

7-11. A certain system has 4×10^{24} degrees of freedom. By what factor does the number of available states increase if the internal energy is doubled? (Assume that the energy range, ΔE, is the same for both cases.)

7-12. Consider a cup of water.

  (a) Roughly how many water molecules are in this system?
  (b) If the molecules are unable to vibrate, but can translate, and rotate around all three orthogonal axes, how many degrees of freedom has each molecule?
  (c) If you raise the temperature of this water from room temperature to boiling, the internal energy increases by what factor (roughly)?
  (d) Estimate the factor by which the density of states increases when the temperature of a cup of water is raised from room temperature to the boiling point.

7-13. In a certain system, the density of states doubles when the internal energy is doubled.

  (a) Is this a macroscopic or a microscopic system?
  (b) How many degrees of freedom does this system have?

APPENDIX 7A $\Omega_o \propto E^{\mathfrak{N}/2}$

Consider a system having $\mathfrak{N}$ degrees of freedom. According to Eq 2.12, the number of states accessible to the ith degree of freedom is given by

$$\Omega_i \propto \int dq_i$$

where the range of integration is over all allowed values of q_i. The number of states accessible to the entire system is the product of the numbers of states available to the individual degrees of freedom (Eq. 7.4).

$$\Omega_o = \prod_{i=1}^{\mathfrak{N}} \Omega_i \propto \prod_{i=1}^{\mathfrak{N}} \int dq_i \tag{7A.1}$$

If the energy in the ith degree of freedom can be expressed as

$$\varepsilon_i = bq_i^2$$

then we can make the substitution of variables

$$r_i = \sqrt{b}\, q_i \qquad (\varepsilon_i = r_i^2) \tag{7A.2}$$

and the above expression for the number of states for the entire system becomes

$$\Omega_o = (\text{const}) \prod_{i=1}^{\mathfrak{N}} \int dr_i \tag{7A.3}$$

where we have absorbed all the factors of $b^{-1/2}$ into the constant of proportionality.

The integration is subject to the constraint that the sum of the internal energies in all degrees of freedom equals the total internal energy E.

$$\sum_{i=1}^{\mathfrak{N}} \varepsilon_i = \sum_{i=1}^{\mathfrak{N}} r_i^2 = E \tag{7A.4}$$

According to the Pythagorean theorem, this constraint makes the integration of (7A.3) a surface integral over an $\mathfrak{N}$-dimensional sphere of radius $E^{1/2}$.

$$\Omega_o = (\text{const}) \prod_{i=1}^{\mathfrak{N}} \int_s dr_i, \qquad \sum_{i=1}^{\mathfrak{N}} r_i^2 = E$$

The surface area of an $\mathfrak{N}$-dimensional sphere is proportional to the radius raised to the $(\mathfrak{N} - 1)$th power.*

$$\Omega_o(E) = (\text{const})(\sqrt{E})^{\mathfrak{N} - 1}$$

If we ignore 1 in comparison to $\mathfrak{N}$, this becomes

$$\Omega_o(E) = (\text{const})E^{\mathfrak{N}/2} \tag{7A.5}$$

---

* The surface of a two-dimensional sphere (i.e., a circle) is the circumference, and is proportional to the first power of the radius ($2\pi r$). The surface of a three-dimensional sphere is proportional to the second power of the radius ($4\pi r^2$), and so on.

* chapter 8

ENTROPY AND THE SECOND LAW

For macroscopic systems, the number of accessible states is extremely sensitive to the internal energy. Consequently, when two systems are interacting in any way, the number of states accessible to the combined system is extremely sensitive to the distribution of the energy between the two. Virtually all accessible states correspond to an energy distribution very close to some optimum value. But what does this have to do with entropy and the fact that the entropy never decreases?

In Chapter 7 we showed that the density of states for macroscopic systems is an extremely rapidly increasing function of the internal energy. An important consequence of this result is that when any two systems are interacting in any manner that permits the exchange of energy between them, the number of states available to the combined system will be *extremely* sensitive to the distribution of the available energy between the two. Very small shifts in the energy distribution will cause very large changes in the number of accessible states.

Consider a plot of the number of states accessible to the combined system as a function of the distribution of total energy E_0 between subsystems A_1 and A_2, such as Figure 8.1, which peaks at some "optimum" value of the energy distribution. Because the number of accessible states is so very sensitive to the distribution of energy, any deviation of the energy distribution from the optimum value will cause *dramatic* reduction in the number of available states.

Since the combined system is equally likely to be found in any of its accessible states, the combined system is almost surely to be found with the energy distribution very near this optimum value, because that is where virtually all the available states are located. Any other distribution of energies between the two subsystems is extremely unlikely, as the number of available states is so greatly reduced.

Consequently, when two systems are allowed to interact with each other, such as through the exchange of some form of work, heat, or particles (Figure 8.2), the result of the interaction will be to redistribute the internal energies in such a way that the number of states available to the combined system is maximized. This is the idea that provides the basis of the most powerful tools of statistical mechanics. You can see that it is a statement of probabilities, but we elevate this statement of probabilities to the stature of a "law." We do this, because even though there is some finite calculable probability that the "law" will be broken, this probability is so minute that we can rest assured that we'll never see it violated by any macroscopic system.

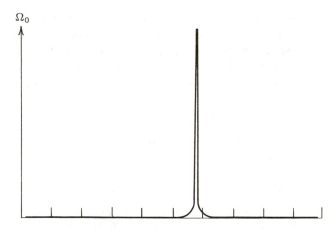

Figure 8.1 Plot of the number of states accessible to the combined system, A_0, as a function of the distribution of the energy between systems A_1 and A_2. For macroscopic systems, the number of accessible states is *extremely* sensitive to the energy distribution, with an extremely narrow, sharp peak at the optimum value.

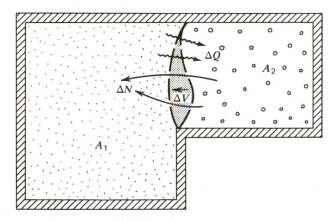

Figure 8.2 In general, there are many different ways that two systems may interact, including the performance of various types of work (e.g., the exchange of volume), the exchange of various kinds of particles, and the exchange of heat energy.

This chapter is devoted to developing, illustrating, and quantifying this powerful statement of probabilities.

A. MICROSCOPIC EXAMPLES

Consider some system A_0, as in Figure 8.2, which is composed of two interacting subsystems, A_1 and A_2. For simplicity, we wish to consider the interactions of systems only two at a time, so the combined system, A_0, must be completely isolated from the rest of the universe. The two subsystems A_1 and A_2, however, may be mutually interacting in any manner: thermally, mechanically, diffusively, or any combination of these. Assume that we have allowed sufficient time for them to come into equilibrium.

The internal energies of the two subsystems may change as a result of their interaction, but they are subject to the constraint that the energy of the combined system is constant.

$$E_0 = E_1 + E_2 = \text{constant} \tag{8.1}$$

The number of states available to the combined system is the product of the number of states available to the subsystems. (See Eq. 7.4.)

$$\Omega_0 = \Omega_1 \Omega_2 \tag{8.2}$$

Suppose our systems are quite small, with A_1 having 6 degrees of freedom and A_2 having 10. From Eq. 7.8 we have for any system of $\mathfrak{N}$ degrees of freedom

$$\Omega \propto E^{\mathfrak{N}/2} \tag{8.3}$$

Therefore, the number of states available to A_1 and A_2 would be

$$\Omega_1 \propto (E_1)^3$$
$$\Omega_2 \propto (E_2)^5 \tag{8.4}$$

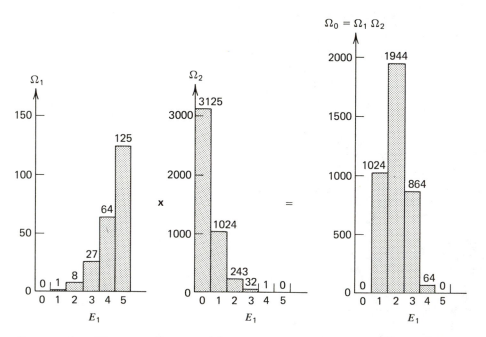

Figure 8.3 Plots of Ω_1 versus E_1, Ω_2 versus E_1 (where $E_2 = E_0 - E_1$), and Ω_0 versus E_1 for $\mathfrak{N}_1 = 6$, $\mathfrak{N}_2 = 10$. For simplicity, the energy is assumed to come in units of 1, with the total energy of the system being $E_0 = 5$.

For simplicity, we will take the constant of proportionality in the above equations to be 1, although the results of our following calculations would be the same for any other value of this constant.

Suppose the energy comes in units of 1, and that the energy of the combined system is 5.

$$E_0 = E_1 + E_2 = 5 \tag{8.5}$$

There are six different possible ways of distributing this energy between A_1 and A_2 as given below. We also give the number of states available to subsystems, A_1 and A_2, according to Eq. 8.4 and to the combined system, A_0, according to Eq. 8.2. Figure 8.3 also displays the number of states available to the combined system as a function of the energy distribution.

Table 8.1   Number of Accessible States for
Two Small Interacting Systems
with $\mathfrak{N}_1 = 6$ and $\mathfrak{N}_2 = 10$

E_1	E_2	Ω_1	Ω_2	$\Omega_0 = \Omega_1\Omega_2$
0	5	0	3125	0
1	4	1	1024	1024
2	3	8	243	1944
3	2	27	32	864
4	1	64	1	64
5	0	125	0	0
			Total	3896

Since this microscopic system is equally likely to be in any of these accessible states, the most probable energy distribution is $E_1 = 2$, $E_2 = 3$. Half of the available states ($1944/3896 = .50$) have this energy distribution, so the system will have this energy distribution half of the time.

For slightly larger systems, the distribution will be considerably more peaked. Consider a microscopic system that is twice as large as the previous one, $\mathfrak{N}_1 = 12$ and $\mathfrak{N}_2 = 20$. According to Eq. 8.3 the number of states available to each is given by

$$\Omega_1 \propto (E_1)^6$$
$$\Omega_2 \propto (E_2)^{10}$$

Again, we take the constant of proportionality to be 1, for simplicity, and again we imagine the energy coming in units of one and the energy of the combined system to be

$$E_0 = E_1 + E_2 = 5$$

For this case we make another table indicating the number of states available to the combined system, A_0, for each of the six possible distributions of energy between subsystems A_1 and A_2. The results are also illustrated in Figure 8.4.

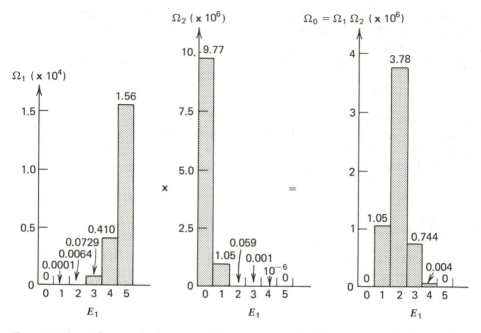

Figure 8.4 Plots of Ω_1 versus E_1, Ω_2 versus E_1 (where $E_2 = E_0 - E_1$), and Ω_0 versus E_1 for $\mathfrak{N}_1 = 12$ and $\mathfrak{N}_2 = 20$. For simplicity, the energy is assumed to come in units of 1, with the total energy of the system being $E_0 = 5$.

Table 8.2   Number of Accessible States for Two Small
Interacting Systems with $\mathfrak{N}_1 = 12$ and
$\mathfrak{N}_2 = 20$

E_1	E_2	Ω_1	Ω_2	$\Omega_0 = \Omega_1\Omega_2$
0	5	0	9.77×10^6	0
1	4	1	1.05×10^6	1.05×10^6
2	3	64	5.90×10^4	3.78×10^6
3	2	729	1.02×10^3	0.744×10^6
4	1	4.10×10^3	1	0.004×10^6
5	0	1.56×10^4	0	0
			Total	5.57×10^6

For this system, over two-thirds $(3.78/5.57 = .68)$ of all accessible states have the $E_1 = 2$, $E_2 = 3$ energy distribution, so the system will have this energy distribution more than two-thirds of the time.

As a final example, consider another microscopic system 10 times as big as the last one, with $\mathfrak{N}_1 = 120$ and $\mathfrak{N}_2 = 200$. For this case, we have

$$\Omega_1 \propto (E_1)^{60}$$

$$\Omega_2 \propto (E_2)^{100}$$

and making the same simplifying assumptions as before, the number of states available to the combined system for each of the six possible distributions of energy is as follows

Table 8.3   Number of Accessible States for Two
Small Interacting Systems with
$\mathfrak{N}_1 = 120$ and $\mathfrak{N}_2 = 200$

E_1	E_2	Ω_1	Ω_2	Ω_0
0	5	0	7.9×10^{69}	0
1	4	1	1.6×10^{60}	1.6×10^{60}
2	3	1.2×10^{18}	5.2×10^{47}	6.2×10^{65}
3	2	4.2×10^{28}	1.3×10^{30}	5.5×10^{58}
4	1	1.3×10^{36}	1	1.3×10^{36}
5	0	8.7×10^{41}	0	0
			Total	6.2×10^{65}

For this system, the $E_1 = 2$, $E_2 = 3$ energy distribution is by far the most probable, because 99.9997% of all accessible states have this energy distribution. Consequently, we could be fairly certain that the system has this distribution of energies at any instant in time. Nonetheless, there is some finite probability (.000003) that the system will *not* have this distribution of energies.

In the next section we will see that when the system is *macroscopic*, with $\mathfrak{N}_1$ and $\mathfrak{N}_2$ being numbers like Avogadro's number, this probability for some other distribution of energies becomes impossibly small.

PROBLEMS

8-1.   Consider a system A_0, composed of subsystems A_1 and A_2, for which the number of degrees of freedom is given by $\mathfrak{N}_1 = 3$, $\mathfrak{N}_2 = 4$. If energy comes in units of 1 and the energy of the combined system is 4:
   (a)   What is the probability that the system will be in a state with $E_1 = 1$, $E_2 = 3$?
   (b)   Which distribution of energies is most probable, and what is the probability of the system having this distribution?

8-2.   Repeat Problem 8-1 for a system with $\mathfrak{N}_1 = 5$, $\mathfrak{N}_2 = 6$.

8-3.   Consider a system A_0 composed of three interacting subsystems, A_1, A_2, and A_3. Suppose that $\mathfrak{N}_1 = 4$, $\mathfrak{N}_2 = 5$, $\mathfrak{N}_3 = 6$, and that the energy comes in units of 1 with the total energy being given by $E_0 = E_1 + E_2 + E_3 = 4$. Use Eqs. 7.4 and 7.8, with the constant of proportionality equal to 1, to do the following.
   (a)   Make a table showing $\Omega_1, \Omega_2, \Omega_3$, and Ω_0 for each of all the possible distributions of the energy among these three systems (15 different distributions in all).
   (b)   Which energy distribution is most probable, and what is its probability?

8-4.   Review the definition of "equilibrium." Explain why it is necessary for the combined system to be in equilibrium for the probability calculations in the above problems and examples to be valid. (*Hint*. You will need to use the *fundamental postulate*.)

B.   MACROSCOPIC EXAMPLES

In the preceding section, we examined hypothetical microscopic systems. We saw that as the number of degrees of freedom increased, the probability of finding one particular distribution of energies between the subsystems also increased.

Now we wish to consider one further example, similar to the first example in the previous section, except that the number of degrees of freedom for each subsystem is larger by a factor of 10^{24}. Again, we assume that energy comes in units of 1, and that the energy of the combined system be constrained by

$$E_0 = E_1 + E_2 = 5$$

For this case, however, the number of degrees of freedom for each of the subsystems is given by

$$\mathfrak{N}_1 = 6 \times 10^{24} \qquad \mathfrak{N}_2 = 10 \times 10^{24}$$

Again for simplicity, we choose the constant of proportionality in Eq. 8.3 to be 1 (i.e., $\Omega = 1E^{\mathfrak{N}/2}$). Our table of values for Ω_1, Ω_2, and Ω_0 for each of the six possible distributions of energy between the subsystems is now as follows.

Table 8.4   Number of Accessible States for Two
Large Interacting Systems with
$\mathfrak{N}_1 = 6 \times 10^{24}$ and $\mathfrak{N}_2 = 10 \times 10^{24}$

E_1	E_2	Ω_1	Ω_2	$\Omega_0 = \Omega_1\Omega_2$
0	5	0	$10^{6.99 \times 10^{24}}$	0
1	4	1	$10^{6.02 \times 10^{24}}$	$10^{6.02 \times 10^{24}}$
2	3	$10^{1.81 \times 10^{24}}$	$10^{4.77 \times 10^{24}}$	$10^{6.58 \times 10^{24}}$
3	2	$10^{2.86 \times 10^{24}}$	$10^{3.01 \times 10^{24}}$	$10^{5.87 \times 10^{24}}$
4	1	$10^{3.61 \times 10^{24}}$	1	$10^{3.61 \times 10^{24}}$
5	0	$10^{4.19 \times 10^{24}}$	0	0
			Total	$10^{6.58 \times 10^{24}}$

A word of warning is in order here; the similarity in appearance in some of the above exponents is deceiving. Figures like $10^{6.02 \times 10^{24}}$ and $10^{6.58 \times 10^{24}}$ in Table 8.4 may look similar, but in fact the difference is enormous. The one distribution ($E_1 = 2$, $E_2 = 3$) is by far the most probable. In fact, it is $10^{0.56 \times 10^{24}}$ times more probable than all the other distributions combined! This means that the probability that the system does *not* have this optimum distribution of energies is only 1 chance in $10^{5.6 \times 10^{23}}$. It is perhaps more impressive if we write this as 1 chance in $10^{560,000,000,000,000,000,000,000}$. To give this number some perspective, we might note that the length of one second compared to the age of the universe is only 1 part in 10^{18}, and the volume of a grain of sand compared to the volume of the entire observable universe is just 1 part in 10^{90}. Clearly, these odds are overwhelming!

Hopefully, we should all be impressed by these kinds of probabilities for macroscopic systems. But we should also be concerned that the result may be an artifact of some of our simplifying assumptions.

Anything that simply multiplies the number of states, Ω_0, by a constant, would not effect the relative probabilities, as the ratios of numbers of states would be the same. For example, we assumed the constant of proportionality in Eq. 8.3 was 1, but the results would be the same no matter what it was.

The assumption that energy comes in units of one-fifth the total internal energy of the combined system, is not very realistic, however, and using more realistic values does have some effect on the probability calculations. We should have instruments capable of detecting energy changes of a millionth of the system's internal energy, and in some cases we could detect changes as small as a billionth of the total. It would be impossible to reconstruct a table similar to the previous ones with energies extending from 0 to 5 in units of 5×10^{-9}, because this would involve 10^9 separate calculations! Rather, we will just compare two neighboring energy distributions, such as $(E_1 = 2, E_2 = 3)$ and $(E_1 = 2 + \Delta,\ E_2 = 3 - \Delta)$, where $\Delta = 5 \times 10^{-9}$. Using Eq. 8.3 we can find for this case

$$\Omega_0(E_1 = 2, E_2 = 3) = 10^{3.6 \times 10^{14}}[\Omega_0(E_1 = 2 + \Delta, E_2 = 3 - \Delta)]$$

This shows that it is about $10^{360,000,000,000,000}$ times more probable to find this particular macroscopic system in a state with the $(E_1 = 2, E_2 = 3)$ distribution in energies, than in a state with energies shifted by only one-billionth of the total internal energy.

These odds are still overwhelming! For two interacting macroscopic systems in equilibrium, there will be one optimum distribution of energies between the two, and the chance of the system being in any state with any measurably different distribution of energies is infinitesimal. If we had an extremely quick and accurate instrument that could measure energies in increments of one-*billionth* of the total internal energy of the system, and this instrument could give us readings at *microsecond* intervals, we would have to wait roughly $10^{360,000,000,000,000}$ times longer than the age of the universe to see the distribution of energies change just once!

PROBLEMS

8-5. In constructing Table 8.4 we assumed that the numbers of degrees of freedom for each of the two subsystems were given by $\mathfrak{N}_1 = 6 \times 10^{24}$ and $\mathfrak{N}_2 = 10 \times 10^{24}$. Using the values given in Table 8.4, construct a similar table for subsystems which are twice as large. [*Hint.* $E^{2\mathfrak{N}_1} = (E^{\mathfrak{N}_1})^2$.]

8-6. Consider a system consisting of two interacting subsystems, for which $\mathfrak{N}_1 = 24 \times 10^{24}$ and $\mathfrak{N}_2 = 20 \times 10^{24}$. Construct a table similar to Table 8.4 listing Ω_1, Ω_2, and Ω_0. (*Hint.* Notice $\mathfrak{N}_1 = 4\mathfrak{N}_1'$, $\mathfrak{N}_2 = 2\mathfrak{N}_2'$, where $\mathfrak{N}_1'$ and $\mathfrak{N}_2'$ are the degrees of freedom used in constructing Table 8.4. Then see the hint in Problem 8-5.)

8-7.   Consider the two numbers 10^{11} and 10^{12}. Which is larger? How many times larger?

8-8.   Consider the two numbers $10^{6.01 \times 10^{24}}$ and $10^{6.02 \times 10^{24}}$. Which is larger? How many times larger?

8-9.   What number is a billion times larger than $10^{2 \times 10^{20}}$?

8-10.   Suppose that $\Omega_1 = E_1^{10^{24}}$ and $\Omega_2 = E_2^{2 \times 10^{24}}$. Compute the ratio $\Omega_0(E_1 = 2, E_2 = 3)/\Omega_0(E_1 = 2.01, E_2 = 2.99)$. (*Hint.* Use logarithms.)

8-11.   Make estimates of the numbers of degrees of freedom for several macroscopic systems that you see around you.

C.   THE SECOND LAW

In the previous section we saw that when two interacting macroscopic systems are in equilibrium, the distribution of energies between them will be such that the number of states available to the combined system is a maximum. This result was independent of the type of interaction, whether it be thermal, mechanical, diffusive, or any combination of these.

The energies of the systems will be functions of some variable parameters describing the interactions. For example, if the interaction is diffusive (e.g., chemical), then the energies will depend on the numbers of particles of the various types, N_i. If the interaction is mechanical, then the energies will depend on the volumes, magnetic fields, electric fields, etc. And if the interaction is thermal, then the energies will be dependent on parameters related to the heat exchange between the two.

Since the energies are functions of these "system variables," we can reword our previous result as follows.

When two interacting macroscopic systems are in equilibrium, the values of the various system variables will be such that the number of states available to the combined system is a maximum.

There is a very important corollary to this result, which applies to interacting systems that have not yet reached equilibrium. It is called the "second law of thermodynamics."

SECOND LAW

As two interacting macroscopic systems approach equilibrium, the changes in the system variables will be such that the number of states available to the combined system *increases*. Or, in the approach to equilibrium,

$$\Delta\Omega_0 > 0$$

There are equivalent alternative ways of expressing the second law, as we will see later in this section.

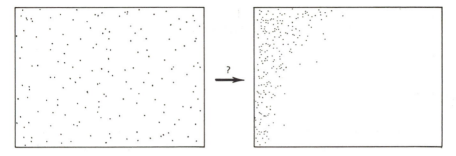

Figure 8.5 Air molecules in a room. There is a finite probability that all of the air molecules in a room will arrive at one end simultaneously. The probability is extremely small, however.

Notice that the second law is based on probabilities, whereas the first law seems to reflect inviolable fact. The second law does not work for small systems, whereas the first law does. Even for large, macroscopic systems, there is some infinitesimal probability that the second law could be violated. However, the chance that any living human will ever see it violated is extremely small. If we had some instrument capable of measuring any system variable with an accuracy of parts per billion, and if this machine repeated these measurements 10^6 times per second, then we would have to wait roughly $10^{360,000,000,000,000}$ times longer than the age of the universe to see even the slightest violation even once, for typical macroscopic systems.

For example, there is some small probability that the air of the room in which you are presently sitting will all rush over to one corner, leaving you to suffocate (Figure 8.5). There is some small probability that water will flow uphill, or begin to boil as ice cubes are added. There is some small probability that your blood could start transporting carbon dioxide *to* your cells and oxygen *away*, rather than vice versa. But all these things are extremely unlikely. Even if we had extraordinarily accurate machines that could detect even the inception of such anomalous behavior, lasting only a microsecond, we would still have to wait longer than $10^{10^{14}}$ times the age of the universe to record even one such case.

Clearly, we can base our studies on a law whose chance of violation is so minuscule, and we can rest assured that we will never witness a violation. In fact, every moment of our existence we bet our very own lives on these odds, and needless to say, we always win.

PROBLEMS

8-12. List some processes that you can think of that would violate the second law without violating the first law.

8-13. If water were to flow uphill, what would have to happen to its temperature if the first law is not to be violated? (*Hint.* If potential energy increases but total energy doesn't, what must happen?)

8-14. Write out the number 10^{100} longhand (i.e., don't use exponential notation) and time how long it takes you to do it. How long would it take you to write out $10^{360,000,000,000,000}$ longhand? (1 year $= 3.17 \times 10^7$ s).

D. ENTROPY

The number of states available to a macroscopic system is an extremely large and unwieldy number. Numbers on the order of $10^{10^{24}}$ are typical. The logarithm of the number of accessible states is a much smaller and more manageable number. Consequently, when using the second law to do quantitative calculations, we will find it much more convenient to work with the logarithm of Ω rather than Ω itself.

For this reason, we generally do not work with the number of accessible states, Ω, directly, but rather we work with a quantity called "entropy," which is given the symbol S and is defined as follows:

$$\text{Entropy} = S \equiv k \ln \Omega \qquad (5.6)$$

where k is a constant of proportionality, called Boltzmann's constant, which is inserted to give entropy units that are convenient for working with typical macroscopic systems.

Entropy has another property that makes it convenient to use in practical applications. Whereas the number of states available to the combined system is multiplicative,

$$\Omega_0 = \Omega_1 \Omega_2$$

the entropy of the combined system is additive,

$$k \ln \Omega_0 = k \ln (\Omega_1 \Omega_2) = k \ln \Omega_1 + k \ln \Omega_2$$

or

$$S_0 = S_1 + S_2$$

This makes it algebraically similar to other system variables such as the internal energy, volume, or number of particles.

$$E_0 = E_1 + E_2$$
$$V_0 = V_1 + V_2$$
$$N_0 = N_1 + N_2$$

EXAMPLE

Consider a system A_0 consisting of interacting subsystems A_1 and A_2, for which $\Omega_1 = 10^{20}$ and $\Omega_2 = 2 \times 10^{20}$. What is the number of states available to the combined system, Ω_0. Also, compute the entropies S_1, S_2, and S_0 in terms of Boltzmann's constant.

The number of states accessible to the combined system is

$$\Omega_0 = \Omega_1\Omega_2 = 2 \times 10^{40}$$

The entropies S_1 and S_2 are given by

$$S_1 = k \ln \Omega_1 = k \ln (10^{20}) = k \ln (e^{2.30})^{20} = 46.0k$$

$$S_2 = k \ln \Omega_2 = k \ln (2 \times 10^{20}) = k \ln (2) + k \ln (e^{2.30})^{20} = 46.7k$$

The entropy, S_0, can be calculated either from Ω_0 or by adding S_1 and S_2.

$$S_0 = k \ln \Omega_0 = k \ln (2 \times 10^{40}) = k \ln (2) + k \ln (e^{2.3})^{40} = 92.7k$$

or

$$S_0 = S_1 + S_2 = 92.7k$$

The word "entropy" is not a part of our everyday vocabulary, as is the word "energy," and students frequently tend to think there is something inherently magical or mystical about entropy. Furthermore, there are enough simplistic and incorrect definitions of entropy to be found in introductory texts, that some confusion among the students is justified. Please notice that entropy is nothing inherently magical. It is simply a convenient measure of the number of states accessible to a system. It is more convenient than using the number of states directly for two reasons.

1. It is smaller and more manageable.
2. Entropies are additive, whereas numbers of states are multiplicative.

Because the entropy is the logarithm of the number of accessible states, when Ω increases, so does S, and when Ω is a maximum, so is S. (See Figure 8.6.)

$$\Delta\Omega > 0 \text{ implies } \Delta S > 0$$

$$\Omega = \Omega_{max} \text{ implies } S = S_{max}$$

Consequently, the second law can be stated in terms of the entropy rather than the number of accessible states.

SECOND LAW (ALTERNATIVE)

As two interacting macroscopic systems approach equilibrium, the changes in the system variables will be such that the entropy of the combined system increases.

$$\Delta S_0 > 0$$

Once the interacting systems have reached equilibrium, then the number of accessible states will be maximized, and there will be no further changes.

$$\Delta S_0^{\text{equilibrium}} = 0$$

Consequently, the second law is sometimes written as follows.

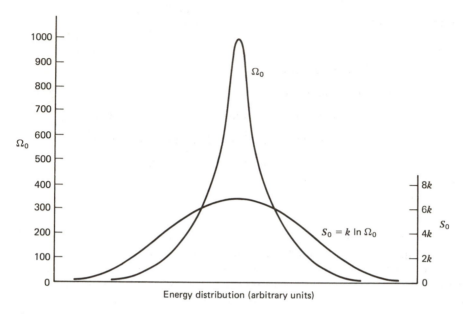

Figure 8.6 Plot of number of accessible states (Ω_0) and entropy (S_0) versus energy distribution for some small system. The entropy is the more slowly varying of the two even for this small system. The difference between the two becomes more pronounced as the size of the system increases.

SECOND LAW (ALTERNATIVE ALTERNATIVE)

For any two interacting systems (whether or not they are yet in equilibrium) the entropy of the combined systems cannot decrease.

$$\Delta S_0 \geqslant 0$$

In the above development, we considered just two interacting subsystems. But each of these two subsystems could itself be composed of several component systems. For example, A_1 could have subcomponents A_a, A_b, A_c, etc., and A_2 could have subcomponents A_s, A_t, A_u, etc.

$$A_1 = A_a + A_b + A_c + \cdots$$

$$A_2 = A_s + A_t + A_u + \cdots$$

$$A_0 = A_1 + A_2 = A_a + A_b + A_c + \cdots + A_s + A_t + A_u + \cdots$$

The number of states accessible to the combined system would be given by

$$\Omega_0 = \Omega_1 \Omega_2$$
$$= \Omega_a \Omega_b \Omega_c \cdots \Omega_s \Omega_t \Omega_u \cdots$$

and its entropy would be given by

$$S_0 = S_1 + S_2$$
$$= S_a + S_b + S_c + \cdots + S_s + S_t + S_u + \cdots$$

In our derivation we placed no restriction on the nature of the subsystems A_1 and A_2, nor on the numbers and kinds of ways in which they interact, so the result for two interacting subsystems is equally valid for any number of interacting subsystems.

EXAMPLE

Consider three interacting systems for which $\Omega_1 = 10^{10^{24}}$, $\Omega_2 = 2 \times 10^{10^{24}}$, and $\Omega_3 = 3 \times 10^{2 \times 10^{24}}$. What is Ω_0, and what are S_1, S_2, S_3, and S_0 in terms of k?
  The number of states, Ω_0 is given by

$$\Omega_0 = \Omega_1\Omega_2\Omega_3 = 1 \cdot 2 \cdot 3 \cdot 10^{(1+1+2) \times 10^{24}} = 6 \times 10^{4 \times 10^{24}}$$

The entropies are given by

$$S_1 = k \ln (10^{10^{24}}) = k \ln (e^{2.3})^{10^{24}} = 2.3 \times 10^{24}k.$$

$$S_2 = k \ln (2 \times 10^{10^{24}}) = k \ln (2) + k \ln (e^{2.3})^{10^{24}} = 2.3 \times 10^{24}k$$

$$S_3 = k \ln (3 \times 10^{2 \times 10^{24}}) = k \ln (3) + k \ln (e^{2.3})^{2 \times 10^{24}} = 4.6 \times 10^{24}k$$

$$S_0 = S_1 + S_2 + S_3 = 9.2 \times 10^{24}k$$

or equivalently,

$$S_0 = k \ln (6 \times 10^{4 \times 10^{24}}) = k \ln (6) + k \ln (e^{2.3})^{4 \times 10^{24}}$$
$$= 9.2 \times 10^{24}k.$$

PROBLEMS

8-15. Consider two small interacting systems, A_1 and A_2, for which $\Omega_1 = 2$ and $\Omega_2 = 4$.

(a) What is Ω_0?
(b) What are S_1 and S_2 in terms of Boltzmann's constant, k?
(c) What is S_0 in terms of k?

8-16. Repeat Problem 8-15 for $\Omega_1 = 200$ and $\Omega_2 = 400$.

8-17. Repeat Problem 8-15 for $\Omega_1 = 2 \times 10^{10^{24}}$ and $\Omega_2 = 4 \times 10^{10^{24}}$. [Be careful in handling the exponents for part (a)!]

8-18. Consider five small interacting systems for which $\Omega_1 = 1$, $\Omega_2 = 2$, $\Omega_3 = 3$, $\Omega_4 = 4$, and $\Omega_5 = 5$.

(a) What is the number of states accessible to the combined system (Ω_0)?
(b) What are S_1, S_2, S_3, S_4, and S_5 in terms of Boltzmann's constant, k?
(c) Compute the entropy of the combined system (S_0) two ways (in units of Boltzmann's constant):
(1) By using the answer to part (a).
(2) By using the answer to part (b).

This chapter on entropy and the second law is the most important chapter in the whole book, and it is perhaps worthwhile to review what was covered.

SUMMARY

In Chapter 4 it was promised that the *fundamental postulate* would form the basis of all statistical tools used in the study of large systems. This postulate states that a system in equilibrium is equally likely to be found in any of its accessible states.

When two or more systems are interacting, the number of states for the combined system is an extremely sensitive function of the distribution of energy among them, so that virtually all the accessible states correspond to just one optimum distribution of energies. Since all states are equally probable, but virtually all correspond to one distribution of energies, then the system will always be found in a state that has this optimum energy distribution. For typical macroscopic systems, the probability of its being found in some state having some other distribution of energies is so small that even with the fastest and most accurate detection devices imaginable, we would have to wait roughly $10^{360,000,000,000,000}$ times the age of the universe to see this happen even once.

The distribution of energies is a function of the "system variables." These could be volumes, number of particles, magnetic fields, etc., depending on the types of interactions between the subsystems. The above statement regarding the infinitesimal probability of the combined system having other than some "optimum" distribution of energies among the subsystems, is also a statement of the infinitesimal probability of these system variables having other than some "optimum" values.

We define a quantity called "entropy" (given the symbol S) which is proportional to the logarithm of the number of accessible states, making it much smaller and easier to work with. The fact that Ω_0 will be at a maximum when the system is in equilibrium means that S_0 will also be a maximum.

$$S_0 = k \ln \Omega_0$$

$$\Omega_0 = \text{maximum implies } S_0 = \text{maximum}$$

We will rarely refer directly to the fundamental postulate itself again, but the needed information is carried in the law that in equilibrium, S_0 will be a maximum. This is what we will use henceforth.

PART 5

INTERACTIONS

In Part 4 of this book we showed that the number of accessible states is extremely sensitive to the internal energy of the system. Consequently, anything that changes the internal energy should have an appreciable impact on the system's properties.

Since energy is conserved, energy gained by one system must be lost by another. There are many different processes by which internal energy can be transferred between interacting systems, but they can be divided up into three general classes.

1. The transfer of heat energy from one system to another.
2. Work done by one system on another.
3. The transfer of particles from one system to another.

In the following three chapters we study each of these types of interactions separately, to see what effect they bear on the properties of systems.

chapter 9

THE THERMAL INTERACTION

As children, we quickly learn that heat flows from hot objects to cold ones, but we do not know why. If it flowed the other direction, energy would still be conserved. To understand the reason, we need to know what "hot" and "cold" mean—that is, what temperature really measures. Perhaps then we can also understand why adding the same amount of heat to different objects causes different changes in temperature, and why adding heat to objects undergoing phase transition may cause no change in temperature at all.

We have seen that for any macroscopic system, the number of accessible states is an extremely rapidly increasing function of the internal energy. If Ω is the number of accessible states, then

$$\Omega = (\text{const})E^{\mathfrak{R}/2} \tag{7.14}$$

where $\mathfrak{R}$ is on the order of Avogadro's number for macroscopic systems, and where the internal energy, E, is measured relative to the appropriate zero-energy reference level, $N\mu$. The entropy (S) is proportional to the logarithm of Ω, so it varies less rapidly with E, and it is more convenient for use in practical calculations.

$$S = k \ln \Omega = k \left(\frac{\mathfrak{R}}{2}\right) \ln E + (\text{const}) \tag{9.1}$$

In this chapter, we examine in greater detail the dependence of the entropy, S, on the internal energy, E, for two systems interacting thermally, but not mechanically or diffusively (Figure 9.1). In general, there will be as many system variables as there are types of interactions between the two systems, but for simplicity we will let a change in volume represent mechanical interactions of any kind, and a change in N represent diffusive interactions of any kind. The variation of S with E for thermal interactions only, then, can be represented by

$$\left.\frac{\partial S}{\partial E}\right)_{V,N}$$

where holding V constant implies no mechanical interactions of any kind and constant N implies no exchange of particle of any kind.

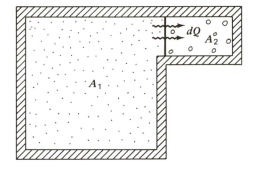

Figure 9.1 In this chapter we consider two systems interacting thermally, but not mechanically or diffusively.

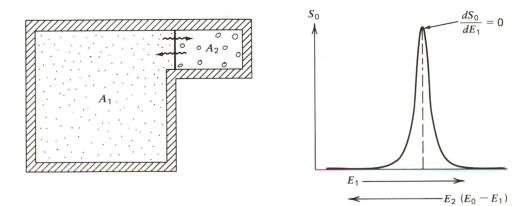

Figure 9.2 When two interacting systems are in equilibrium, the distribution of the total energy E_0 between them will be such that S_0 is a maximum.

A. TEMPERATURE AND THE ZEROTH LAW

Consider two systems, A_1 and A_2, interacting thermally as in Figure 9.1. When the two are in equilibrium,* the energy distribution between them is such that the entropy, S_0, of the combined system is at its maximum value. That is, if we plot S_0 as a function of E_1 (or of $E_2 = E_0 - E_1$) as in Figure 9.2, then at equilibrium

$$\left. \frac{\partial S_0}{\partial E_1} \right)_{V,N} = 0$$

---

* When two systems interact thermally and are in equilibrium, we say they are in "thermal equilibrium."

By using the relationships

$$S_0 = S_1 + S_2$$

and

$$dE_1 = -dE_2$$

we can rewrite this as

$$\left.\frac{\partial S_0}{\partial E_1}\right)_{V,N} = \left.\frac{\partial S_1}{\partial E_1}\right)_{V,N} + \left.\frac{\partial S_2}{\partial E_1}\right)_{V,N}$$

$$= \left.\frac{\partial S_1}{\partial E_1}\right)_{V,N} - \left.\frac{\partial S_2}{\partial E_2}\right)_{V,N} = 0$$

or

$$\left.\frac{\partial S_1}{\partial E_1}\right)_{V,N} = \left.\frac{\partial S_2}{\partial E_2}\right)_{V,N} \tag{9.2}$$

This tells us that there is some property of each system, $\partial S/\partial E)_{V,N}$, which is the same for two systems in thermal equilibrium. It is convenient to give this property a name.

We define a parameter T, called "temperature," as follows.

$$\left.\frac{\partial S}{\partial E}\right)_{V,N} \equiv \frac{1}{T} \tag{9.3}$$

Equation 9.2 tells us

$$\frac{1}{T_1} = \frac{1}{T_2}$$

so we arrive at the following observation.

When two systems are in thermal equilibrium, their temperatures are equal.

It is easily seen that the converse to this statement is also true. That is, when two systems have the same temperature, they are in thermal equilibrium. For two systems of the same temperature in thermal contact, we have

$$0 = \frac{1}{T_1} - \frac{1}{T_2} = \left.\frac{\partial S_1}{\partial E_1}\right)_{V,N} - \left.\frac{\partial S_2}{\partial E_2}\right)_{V,N} = \left.\frac{\partial (S_1 + S_2)}{\partial E_1}\right)_{V,N} = \left.\frac{\partial S_0}{\partial E_1}\right)_{V,N}$$

so the entropy of the combined system is a maximum, and we have equilibrium.

Now consider two systems, A_1 and A_2, each in thermal equilibrium with a third system, A_3, so that

$$T_1 = T_3 \quad \text{and} \quad T_2 = T_3$$

There is an algebraic theorem, which states that if $T_1 = T_3$ and $T_2 = T_3$, then $T_1 = T_2$. We translate this into the "zeroth law of thermodynamics."

> ### ZEROTH LAW
>
> If two systems are each in thermal equilibrium with a third system, then they are in thermal equilibrium with each other.

B.   TEMPERATURE AND INTERNAL ENERGY

With the information we now have at our disposal, we should be able to develop an intuitive feeling for what this parameter, T, measures. The entropy is a measure of the number of accessible states, and $1/T$ measures how this varies with the system's internal energy.

$$\frac{1}{T} = \frac{\partial S}{\partial E}\bigg)_{V,N} = \frac{\partial}{\partial E}\left(k \ln \Omega(E)\right)_{V,N} \tag{9.4}$$

The number of accessible states, $\Omega(E)$, is an increasing function of the system's internal energy. If the system has $\mathfrak{N}$ degrees of freedom, then we can write

$$\Omega(E) = (\text{const})E^{\chi \mathfrak{N}} \qquad \chi > 0 \tag{9.5}$$

For the specific (but common) case where the energy stored in each degree of freedom can be written in the form $\varepsilon = bq^2$, where b is a constant and q a position or momentum coordinate, we saw in Eq. 7.8 that χ in the above equation has the value of $\frac{1}{2}$.

$$\Omega(E) = (\text{const})E^{\mathfrak{N}/2} \qquad (\varepsilon = bq^2) \tag{7.14}$$

Putting this expression for $\Omega(E)$ into Eq. 9.4 yields

$$\frac{1}{T} = \frac{\frac{1}{2}\mathfrak{N}k}{E}$$

or

$$E = \frac{1}{2}\mathfrak{N}kT \qquad (\varepsilon = bq^2) \tag{9.5'}$$

For the more general case represented by Eq. 9.5 we have

$$E = \chi\mathfrak{N}kT$$

In either case, it appears that the temperature, T, is a rather direct measure of the internal energy of the system.

There is a slight swindle in the above development that is frequently ignored in introductory texts, because the correction turns out to be of minor importance in many common systems. Nonetheless, you can see that something appears to be wrong in the above result, because it seems to say that the internal energy depends on only one parameter, T. What if the system is interacting both thermally *and* mechanically, for example? We know there are as many independent variables as there are types of interactions, so the energy in general would depend on *two* things instead of just one.

What is disguised in the above expressions is that the energy E is measured relative to the appropriate zero-energy reference level, $N\mu$. This measure of the energy is commonly referred to as the "thermal energy" of the system, and is given by

$$E_{thermal} = E_{total} - N\mu$$

or, equivalently,

$$E_{total} = E_{thermal} + N\mu$$
$$= \tfrac{1}{2}\mathfrak{N}kT + N\mu$$

If each of the particles has v degrees of freedom, then the total number of degrees of freedom for the system is

$$\mathfrak{N} = vN$$

and the above result becomes

$$E_{total} = \frac{v}{2} NkT + N\mu$$

$$= N\left(\frac{v}{2} kT + \mu\right) \tag{9.6}$$

The chemical potential, or zero-energy reference level, is a result of interactions between particles of the system, which depends on their spacing as well as their motion. Therefore, even though the "thermal energy" of a system may depend only on its temperature, the total internal energy depends (in general) on as many variables as there are types of interactions between systems, because the zero-energy reference point may shift as a result of these interactions even if the temperature does not.

SUMMARY

Temperature is a measure of the "thermal energy" of the particles of a system, which is the energy measured relative to the zero-energy reference point. If each particle has v degrees of freedom, and the energy in each degree of freedom can be expressed as $\varepsilon = bq^2$, then the average energy per particle can be written as

$$\bar{\varepsilon} = \bar{\varepsilon}_{thermal} + \bar{\varepsilon}_{reference\ level}$$

$$= \frac{v}{2} kT + \mu \tag{9.7}$$

The total internal energy of a system of N particles is

$$E_{total} = N\bar{\varepsilon} = N\left(\frac{v}{2} kT + \mu\right)$$

$$= \frac{v}{2} NkT + N\mu \tag{9.6}$$

and the "thermal" portion of this energy is

$$E_{thermal} = \frac{v}{2} NkT \tag{9.8}$$

There are many common processes for which the chemical potential does not change noticeably. For these, the change in temperature is a fairly direct measure of the change in internal energy of the system.

$$\Delta E \approx \frac{v}{2} N k \Delta T$$

However, there are also many common processes (e.g., change of phase) during which the chemical potential does vary considerably, and for these we cannot use temperature changes as a measure of changes in internal energy.

PROBLEMS

9-1. An "ideal gas" is a gas whose molecules have only kinetic degrees of freedom and no energy of mutual interaction.

   (a) For an ideal gas of N molecules, $\Omega(E)$ is proportional to E raised to what power?
   (b) Show that for an ideal gas of N molecules, $E = \frac{3}{2} NkT$.

9-2. For a system of N linear harmonic oscillators, show that $E = NkT$.

9-3. Suppose there are 10^{28} diatomic air molecules in a room, each of which can translate in three directions, and rotate around two orthogonal rotational axes.

   (a) How many degrees of freedom does this system have?
   (b) If Boltzmann's constant is given by $k = 1.38 \times 10^{-23}$ J/K, about what is the internal energy of this system at room temperature?

9-4. Starting with $\Omega(E) = (\text{const.})E^{x\Re}$, where E is the thermal energy of the system, use the definitions of entropy and temperature to derive the relationships between E_{thermal} and T and between $\bar{\varepsilon}_{\text{thermal}}$ and T. (*Hint.* Parallel the development from Eqs. 9.4 to 9.8.)

9-5. Two small systems, A_1 and A_2, are in thermal equilibrium. The number of states accessible to each increases with its energy according to $\Omega_1 = C_1(E_1)^{10}$, and $\Omega_2 = C_2(E_2)^8$, where C_1 and C_2 are constants. The total energy of the combined system is fixed at $E_0 = E_1 + E_2 = 10^{-18}$ J.

   (a) How many degrees of freedom have systems A_1 and A_2?
   (b) What is the value of E_1 and E_2 when these systems are in equilibrium? (*Hint.* Use $\partial \Omega_0 / \partial E_1 = 0$, $\Omega_0 = \Omega_1 \Omega_2$, and $E_2 = E_0 - E_1$.)
   (c) What is the entropy of the combined system in equilibrium? (*Hint.* $S_0 = k \ln \Omega_0 = k \ln (\Omega_1 \Omega_2)$, and $k = 1.38 \times 10^{-23}$ J/K.)
   (d) What is the temperature of this system? (*Hint.* $1/T = \partial S_1 / \partial E_1$, and $S_1 = k \ln \Omega_1$.)

9-6. If the entropy of a certain system changes by 1 J/K when the internal energy is doubled, how many degrees of freedom does the system have? (*Hint.* $S_2 - S_1 = k \ln \Omega_2 - k \ln \Omega_1 = k \ln (\Omega_2/\Omega_1)$, where $\Omega \propto E^{\Re/2}$.)

9-7. If a system has 10^{24} degrees of freedom, by how much does the entropy increase when the internal energy is increased by 10%? (*Hint.* See hint in Problem 9-6.)

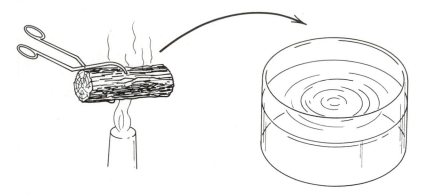

Figure 9.3 When a hot iron bar is put into cold water, will heat flow from the bar to the water, or from the water to the bar? Why?

C. TEMPERATURE AND HEAT FLOW

We have shown that when two systems are in equilibrium, their temperatures are equal, and vice versa. But what happens when two systems of different temperatures are placed in thermal contact? From our practical experience we know that heat flows from the hotter one to the cooler one until thermal equilibrium is established, and in this section we show that this behavior is demanded by the second law.

The first law (energy conservation) does not tell us which direction the heat must flow. Instead, it only says that however much flows out of one object must flow into the other. That is, energy cannot be created or destroyed in the process. Consider, for example, placing a hot iron bar in a bucket of cold water (Figure 9.3). It would not violate the first law if the heat flowed from the cooler water and into the hot iron bar, so that the bar gets hotter and hotter and melts, while the water gets colder and colder and freezes. As long as the heat lost by the water is gained by the iron bar, energy is conserved and the first law is satisfied.

To prove that this process would violate the Second Law, we consider two systems, A_1 and A_2 (Figure 9.4), which are interacting thermally, but which are not yet in thermal equilibrium. The change in entropy for either system is

$$\Delta S = \frac{\partial S}{\partial E}\bigg)_{V,N} \Delta E = \frac{\Delta E}{T}$$

According to the second law, the change in entropy for the combined system must be greater than zero.

$$\Delta S_0 = \Delta S_1 + \Delta S_2 = \frac{\Delta E_1}{T_1} + \frac{\Delta E_2}{T_2} > 0$$

The first law (i.e., conservation of energy) demands that energy lost by A_1 must be gained by A_2, and vice versa.

$$\Delta E_2 = -\Delta E_1$$

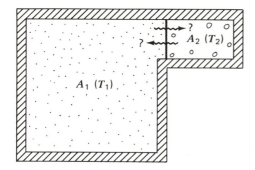

Figure 9.4   When two thermally interacting systems are not yet in thermal equilibrium, which way will the heat flow? In order for the entropy of the combined system to increase, it must flow from the hotter system to the cooler one, so this is the way it will go according to the second law.

With this, the above expression for the increase of entropy of the combined system is

$$\Delta S_0 = \Delta E_1 \left(\frac{1}{T_1} - \frac{1}{T_2} \right) > 0$$

In this form, we can see that the requirement of the second law ($\Delta S_0 > 0$) becomes:

1.   If $T_2 > T_1$, then $\Delta E_1 > 0$.
2.   If $T_1 > T_2$, then $\Delta E_1 < 0$.

That is, the energy is gained by A_1 if it is cooler, and lost by A_1 if it is hotter. In other words, the energy must flow from the hotter system to the cooler system, and not vice versa.

SUMMARY

If two thermally interacting systems are not yet in thermal equilibrium, then the second law demands that the energy must flow from the hotter system to the cooler one, and not vice versa.

D.   TEMPERATURE SCALES AND BOLTZMANN'S CONSTANT

We have learned that the average thermal energy per degree of freedom is proportional to the product kT.*

$$\bar{\varepsilon} = \frac{E}{\mathfrak{N}} = \frac{1}{2} kT \tag{9.7}$$

Since both the thermal energy, E, and the number of degrees of freedom, $\mathfrak{N}$, are

---

* Recall that the thermal energy is the internal energy measured relative to the zero-energy reference level, $N\mu$.

measurable properties of a system, the value of Boltzmann's constant, k, and the temperature, T, are interdependent. We may either choose a value for Boltzmann's constant, k, which establishes the temperature,

$$T = \frac{1}{k}\frac{2E}{\mathfrak{N}}$$

or we may choose a value for the temperature, T, which establishes Boltzmann's constant,

$$k = \frac{1}{T}\frac{2E}{\mathfrak{N}}$$

but we cannot do both.

It is wise to choose a temperature scale according to some accurate and easily reproduced standard. Water is a very common and convenient substance, and it has the property that as the pressure on it is reduced from atmospheric pressure, the boiling point becomes lower (Figure 9.5), and the freezing point rises slightly, until at a pressure

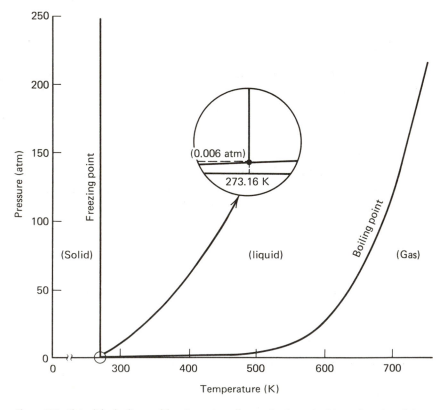

Figure 9.5 Plot of the boiling and freezing points of water (horizontal axis) as a function of the pressure (vertical axis). The area around the triple point is enlarged.

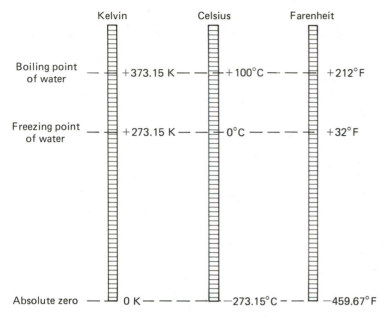

Figure 9.6   Comparison of Kelvin, Celsius, and Farenheit temperature scales.

of about 0.006 atmosphere, the boiling and freezing points are the same. The temperature and pressure at which all three phases are in equilibrium with each other is called the "triple point" of water.

Because water is so common, and the triple point is easily reached, our temperature scale is now based on this convenient standard.

The temperature of the triple point of water is defined to be

$$T_{\text{triple point}} \equiv 273.16 \text{ K, exactly} \qquad (9.9)$$

This fixes the value of Boltzmann's constant at

$$k = 1.381 \times 10^{-23} \text{ J/K} = 0.864 \times 10^{-4} \text{ eV/K} \qquad (9.10)$$

This temperature scale has the property that under atmospheric pressure, water freezes at 273.15 K and boils at a temperature 100° higher, 373.15 K. The Celsius temperature scale is derived from the Kelvin temperature scale by subtracting 273.15 (Figure 9.6).

$$T(^\circ\text{C}) = T(\text{K}) - 273.15$$

E.   THERMOMETERS

The temperature of any system can be measured directly if it has any property that varies with its temperature. Such properties are called "thermometric parameters," and could include such things as color, length, texture, smell, electrical resistance, etc., depending on the system. Some examples of systems with thermometric parameters are listed below.

System	*Thermometric Parameter*
mercury column	length of column
finger	nerve impulses to brain
gas at constant volume	pressure
gas at constant pressure	volume
steel bar	length of bar
cricket	frequency of chirp
electrical resistor	resistance
hog	curliness of tail

Once calibrated, we can determine the temperature of a system at any future date by simply measuring its thermometric parameter.

This often turns out to be more trouble than it is worth. We are interested in the temperatures of so many things that we could spend a great deal of our days calibrating their thermometric parameters. And the constraints we would have to place on them so that these calibrated parameters could remain reliable temperature indicators could cause great inconvenience.

For example, we could spend a few days determining how the volume of water in a swimming pool is related to its temperature. However, if we wished to use this as a temperature indicator in the future, we would have to make sure there was no change in volume due to other causes, such as evaporation, splashing, rainfall, etc. We would have to keep our swimming pool covered and unused at all times, and that would doubtlessly be unacceptable. Similar constraints would have to be imposed to measure temperatures of stove burners, ovens, air, human metabolism, and many other systems commonly of interest to us.

It is usually more convenient to use thermometers. Thermometers are relatively small systems with easily measured thermometric parameters. They have been precalibrated, and the necessary constraints are placed on them to ensure that their thermometric parameter remains a reliable temperature indicator.

If we place a thermometer in thermal contact with a system whose temperature we wish to know, heat energy will flow between the two until thermal equilibrium is established. We know that when in thermal equilibrium, the temperatures of the two systems are the same, so we need only read the temperature of the thermometer to know the temperature of the system of interest as well.

The choice of an appropriate thermometer generally involves some compromise over concerns of convenience, price, size, accuracy, and range of utility. Thermometers should be small, because we don't want them to significantly alter the properties of the system whose temperature we are measuring. "Smallness," therefore, is determined on a relative basis; a thermometer small enough to measure outside air temperature may not be small enough to measure the temperature of a drop of water.

A popular compromise over these concerns is a thermometer consisting of colored alcohol, or mercury, in a closed glass tube. The thermometric parameter is the length of

Figure 9.7 A familiar type of thermometer is a column of mercury or colored alcohol in a closed glass tube. Expansion of the fluid in the bulb at the base pushes more of the fluid up into the tube, causing the length of the column to increase. From the reading on the thermometer, we know the temperature of the system of interest, because the temperatures of the two are the same when they are in thermal equilibrium.

the fluid column (Figure 9.7). For more demanding scientific work, temperature is often determined from the electrical resistance of tiny resistors.

The problem of what to use as the standard thermometer from which all others are calibrated involves the requirements that its measurements are reproducible, and that it work over the widest possible range. Clearly, crickets, hogs, and fingers have limitations in both these areas. The volume of solids and liquids generally have greater reproducibility and wider ranges of utility, but still they change phases (solids to liquids and liquids to gases) at sufficiently high temperatures. Electrical resistors also suffer similar breakdown beyond a certain temperature range. The widest useful range for reproducible temperature measurement are provided by gases, so the "standard" thermometer has been chosen to be a gas at constant volume.

For an "ideal" gas held at constant volume, the pressure is directly proportional to the temperature. (See Chapter 12, Section A.)

$$T = (\text{const}) \cdot p$$

This is because at higher temperatures, the gas molecules collide more frequently and more forcefully with the walls of the container. The constant of proportionality can be determined by measuring the pressure of the gas at some standard reference point, such as the freezing point of water, and thereafter any temperature can be determined from the measured pressure.

In real gases, however, interactions between the gas molecules cause slight deviations from the linear relationship between p and T, so in making real gas thermometers, some effort must be made to reduce these intermolecular interactions. This is done as follows: (1) by choosing a gas whose intermolecular forces are weak (helium is best, but hydrogen is cheaper and not too bad) and (2) by putting the gas at very low pressures to begin with. (If the gas is sparse, then intermolecular distances are large and intermolecular interactions are negligible.)

PROBLEMS

9-8. Would a gas at constant pressure or a gas at constant volume have a larger range of utility as a thermometer? Why?

9-9. Suppose you have a constant volume gas thermometer, containing an ideal gas in a glass bulb, and you wish to measure the temperature of your bath water. For an ideal gas,

$pV = NkT$, where p is the pressure, V is the volume (held constant), and N is the number of molecules in the gas. You place it first in ice water, and measure the pressure of the gas in the bulb to be 7.3 mm Hg. Then you place the bulb in your bath water, and the pressure in the bulb rises to 8.6 mm Hg. If the temperature of the ice water (at atmospheric pressure, hence not at the triple point) is 273.15 K, what is the temperature of the bath?

9-10.   Suppose you don't like the way the temperature scale is defined, and so you wished to define one that gives a nice round number for the value of Boltzmann's constant. You measure your temperatures on this scale in °R (for "degrees round"). In units of °R, what would be the boiling point of water if:

(a) $k = 1.0 \times 10^{-16}$ erg/°R?
(b) $k = 10^{-4}$ eV/°R?
(c) $k = 1.0$ J/°R?
(d) $k = 1.0$ eV/°R?

9-11.   What would be the value of Boltzmann's constant if the temperature of the triple point were defined as $T_{\text{triple point}} = 100$ K? What would be the boiling point of water on this scale?

9-12.   For an "ideal gas" each molecule has three degrees of freedom (translational only). Suppose you calibrated your temperature scale by saying that 300 K was defined to be the temperature of a mole of an ideal gas that has a thermal energy of 3740 J. What would be the value of Boltzmann's constant, k? (A mole has Avogadro's number of molecules.)

9-13.   Consider some ice at $-1°$C in a glass of water at $+10°$C. For each calorie of heat energy that flows from the water to the ice, what is the change of entropy for the:

(a)   Ice?
(b)   Water?
(c)   Combined system?

(*Hint.* How are ΔS, T, and ΔE related?)

F.   PHASE TRANSITIONS

According to its definition, Eq. 9.3, temperature is a measure of the rate of increase of entropy with energy.

$$\frac{1}{T} \equiv \frac{\partial S}{\partial E}\bigg)_{V,N} \tag{9.3}$$

At higher temperatures, the rate of increase is smaller, as illustrated in Figure 9.8. We can see that this must be so from our definition of entropy,

$$S = k \ln \Omega(E)$$

where

$$\Omega(E) = (\text{constant})E^{\Re/2}$$

Consequently, entropy increases only logarithmically in E,

$$S = \frac{1}{2}\Re k \ln E + \text{const} \tag{9.1}$$

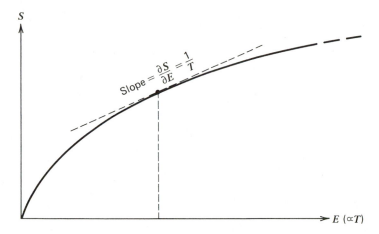

Figure 9.8 Plot of entropy versus internal energy for some system. The slope of the curve is $1/T$. Since $E \propto T$, as E increases, the slope ($1/T \propto 1/E$) decreases.

and so the *rate* of increase *decreases* as the energy increases.

$$\left.\frac{\partial S}{\partial E}\right)_{V,N} = \frac{\Re k}{2E} = \frac{1}{T}$$

However, when a system is undergoing phase transition, we add heat to it without changing its temperature. How is it that we can increase the system's internal energy without changing its temperature? According to Eq. 9.6 ($E = \frac{1}{2}\Re kT$, or $1/T = \Re k/2E$) this appears to be impossible!

The answer to this apparent paradox is that either or both of two things can keep the ratio $\Re k/2E$ constant. One possibility is that the number of degrees of freedom can increase as energy is added to the system, keeping the ratio $\Re/E$ constant.

The other possibility is that the added heat releases particles from the potential wells that bind them. In their new phase, the zero-energy reference point for the measurement of potential energy is higher than it was in the previous phase. That is, our expressions for the number of accessible states, and entropies of systems, really involve the *thermal energies* (see Section C, Chapter 6).

$$E_{\text{thermal}} = E_{\text{total}} - N\mu$$

where $N\mu$ is the zero-energy reference level. If E were to represent the *total* internal energy, rather than just the thermal portion of it, our expressions for entropy and temperature would be as follows.

$$\Omega(E) = (\text{const})(E - N\mu)^{\Re/2} \tag{9.11}$$

$$S = \frac{1}{2}\Re k \ln (E - N\mu) + \text{const} \tag{9.12}$$

$$\frac{1}{T} = \frac{\Re k}{2(E - N\mu)} \tag{9.13}$$

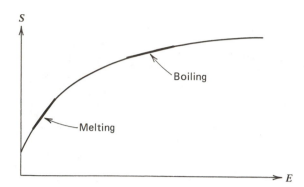

Figure 9.9 Plot of entropy versus internal energy for some system. Flat regions indicate normal phase transitions. They are regions where the slope of the curve (1/T) remains constant even as energy is added and the internal energy increases.

At many phase transitions, then, the zero-energy reference point increases as energy is added to the system, so that $(E - N\mu)$ remains unchanged. This happens, for example, when ice melts or as water evaporates. We'll continue to use the symbol E, rather than $E - N\mu$ in our dealings with systems, but we hope the student keeps in mind that energies are measured from some reference level, and that level may change from one phase to another.

Since $1/T$ measures the slope of the S versus E curve, regions of constant temperature correspond to regions of constant slope. For many systems, such plots have two flat regions, as in Figure 9.9, corresponding to solid-liquid and liquid-gas phase transitions. Additional flat regions may be present, corresponding to phase transitions of other types.

G. HEAT RESERVOIRS

In this section we wish to examine properties of systems that are sufficiently large that their temperatures do not change noticeably when a small amount of heat, ΔQ, is added to or removed from them. Such systems are called "heat reservoirs."

Notice that whether a system serves as a heat reservoir depends both on the size of the system and the size of ΔQ. For example, a glass of water could be considered a heat reservoir for $\Delta Q = 10^{-6}$ calories, but not for $\Delta Q = 10^6$ calories. The ocean, however, could be a heat reservoir for both. In the infinitesimal limit $\Delta Q \rightarrow dQ$, any system—large or small—can be a heat reservoir.

We wish to consider how the entropy of a reservoir changes when a small amount of heat is added to it. Using a Taylor series expansion,* we write the entropy as

$$S(E + \Delta Q) = S(E) + \frac{\partial S}{\partial E}\bigg)_{V,N} (\Delta Q) + \frac{1}{2}\frac{\partial^2 S}{\partial E^2}\bigg)_{V,N} (\Delta Q)^2 + \cdots$$

* See Appendix 4B.

In the infinitesimal limit we can ignore terms of second order and higher in (dQ), giving us

$$S(E + dQ) = S(E) + \left.\frac{\partial S}{\partial E}\right)_{V,N} dQ = S(E) + \frac{dQ}{T}$$

or

$$S(E + dQ) - S(E) = dS = \frac{dQ}{T}$$

Consequently, for any system, large or small,

$$dQ = T\,dS \qquad (9.14)$$

This equation tells us how the entropy changes when a small amount of heat energy is added to a system in equilibrium. It is possible to increase the entropy of a system without adding any heat, however. For example, the removal of some barrier or other constraint may increase the number of states accessible to a system (increasing its entropy) without the addition of any heat. Therefore, when applying this result, we must be careful to make sure that the increase in entropy was caused by heat added to a system in equilibrium, rather than through the removal of some constraint, or through some other nonequilibrium process.

We have seen that the heat added to a system is an inexact differential. However, the change in entropy is exact, as the number of accessible states is a measurable property of the system, so the initial and final entropies can be determined independently of the path taken.

EXAMPLE

By what factor does the number of states available to a heat reservoir increase if 1 eV of heat energy is added when the reservoir has a temperature of 300 K?

The entropy is a measure of accessible states, so we can calculate the increase in accessible states from the change in entropy.

$$\Delta S = S_2 - S_1 = k \ln \Omega_2 - k \ln \Omega_1 = k \ln \frac{\Omega_2}{\Omega_1}$$

Combining this with Eq. 9.14 we have

$$\Delta S = \frac{\Delta Q}{T} = k \ln \frac{\Omega_2}{\Omega_1}$$

or

$$\frac{\Omega_2}{\Omega_1} = e^{\Delta Q/kT} \qquad (9.15)$$

Putting in the given values of ΔQ and T, we have for this case the number of accessible states increasing by a factor of

$$\frac{\Omega_2}{\Omega_1} = e^{39} = 5.5 \times 10^{16}$$

This may seem like a terribly large increase to be accomplished by just 1 eV of energy. But recall that the number of accessible states is an *extremely* rapidly varying function of the internal energy. The above is just one example of this fact.

H.   SYSTEM IN EQUILIBRIUM WITH A HEAT RESERVOIR

We now consider a system A_1, in thermal equilibrium with a very large reservoir, A_2. The average internal energy of system A_1 is $\bar{E}_1$, and we know it will fluctuate somewhat via thermal exchange with the large reservoir. From previous probability calculations, we expect that the smaller A_1 is, the larger will be the relative fluctuations in E_1.

From the *fundamental postulate* we know that the probability that system A_1 will have a specific energy E_1 is proportional to the number of states accessible to the combined system under these conditions, $\Omega_0(E_1)$. Using the symbol P to represent probability, we have

$$P(E_1) = (\text{const})\Omega_0(E_1)$$

The entropy of the combined system, S_0, is also related to the number of accessible states,

$$S_0 = k \ln \Omega_0$$

or

$$\Omega_0 = e^{S_0/k} \tag{9.16}$$

so we can use our knowledge of the behavior of entropies to make the required probability calculation

$$P(E_1) = (\text{const})e^{S_0(E_1)/k} \tag{9.17}$$

We use a Taylor series expansion to find the entropy of the combined system when E_1 differs from its mean value by an amount ΔE_1.

$$S_0(E_1 = \bar{E}_1 + \Delta E_1) = S_0(E_1 = \bar{E}_1) + \left(\frac{\partial S_0}{\partial E_1}\right)_{\bar{E}_1} (\Delta E_1) + \frac{1}{2}\left(\frac{\partial^2 S_0}{\partial E_1^2}\right)_{\bar{E}_1} (\Delta E_1)^2 + \cdots \tag{9.18}$$

We can simplify this expression by noting that

$$\left.\frac{\partial S_0}{\partial E_1}\right)_{\bar{E}_1} = 0 \tag{9.19}$$

because in equilibrium the entropy of the combined system is a maximum. The quadratic term can be simplified by writing

$$\frac{\partial^2}{\partial E_1^2} S_0 = \frac{\partial^2}{\partial E_1^2} (S_1 + S_2) = \frac{\partial}{\partial E_1}\left(\frac{\partial S_1}{\partial E_1} + \frac{\partial S_2}{\partial E_1}\right)$$

$$= \frac{\partial}{\partial E_1}\left(\frac{\partial S}{\partial E_1} - \frac{\partial S}{\partial E_2}\right) = \frac{\partial}{\partial E_1}\left(\frac{1}{T_1} - \frac{1}{T_2}\right) \tag{9.20}$$

Since system A_2 is a large reservoir, its temperature doesn't change with small changes in E_1, so we have

$$\frac{\partial}{\partial E_1}\left(\frac{1}{T_2}\right) = 0 \tag{9.21}$$

Finally, using Eq. 9.5′ for system A, we have

$$\frac{\partial}{\partial E_1}\left(\frac{1}{T_1}\right) = \frac{\partial}{\partial E_1}\left(\frac{\mathfrak{N}k}{2E_1}\right) = -\frac{\mathfrak{N}k}{2E_1^2} \tag{9.22}$$

Combining the results (9.21) and (9.22) with Eq. 9.20 we have

$$\left.\frac{\partial^2}{\partial E_1^2} S_0\right)_{\bar{E}_1} = -\frac{\mathfrak{N}k}{2\bar{E}_1^2}$$

This along with Eq. 9.19 can be inserted into the Taylor series expansion, (9.18), to yield

$$S_0(E_1 = \bar{E}_1 + \Delta E_1) = S_0(E_1 = \bar{E}_1) - \frac{\mathfrak{N}k}{4\bar{E}_1^2}(\Delta E_1)^2$$

If we put this result into Eq. 9.17 we have

$$P(E_1) = (\text{const})e^{[S_0(\bar{E}_1) - (\mathfrak{N}k/4\bar{E}_1^2)(\Delta E_1)^2]/k}$$

Absorbing the factor $e^{S_0(\bar{E}_1)/k}$ into the constant of proportionality, we have a Gaussian distribution in energy (see Chapter 4, Section B),

$$P(E_1) = (\text{const})e^{-(\Delta E_1)^2/2\sigma^2}$$

where the standard deviation, σ, is

$$\sigma = \sqrt{\frac{2}{\mathfrak{N}}}\,\bar{E}_1$$

We know that the probability that system A, has energy in the range between E_1 and $E_1 + dE_1$ will be proportional to the size of the range dE_1, so we can write this probability as

$$P(E_1)\,dE_1 = (\text{const})e^{-(\Delta E_1)^2/\sigma^2}\,dE_1$$

Integrating this over all E_1 and setting the result equal to 1 (since the probability is certainty that the system has some value of E_1), we find the value of the constant of proportionality as we did in Eq. 4.10.

For a system in thermal equilibrium with a large reservoir, the probability that the system has energy in the range between E and $E + dE$ is given by

$$P(E)\,dE = \frac{1}{\sqrt{2\pi}\sigma}\,e^{-(\Delta E)^2/2\sigma^2}\,dE \tag{9.23}$$

where

$$\sigma = \sqrt{\frac{2}{\mathfrak{N}}}\,\bar{E} \tag{9.24}$$

and

$$\Delta E = (E - \bar{E})$$

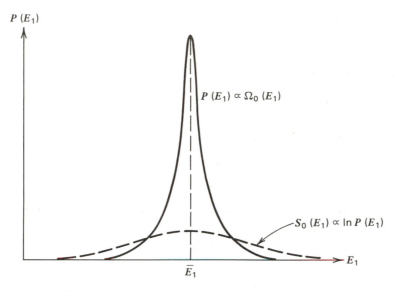

Figure 9.10   The probability distribution in the energy of the system A_1 is peaked at $E_1 = \bar{E}_1$, because the distribution corresponding to the most accessible states in the distribution that the system will have most often. Since entropy is proportional to the logarithm of the number of accessible states, we can say equivalently that the most probable distribution of energy between the two systems is that which maximizes the total entropy.

In spite of the mathematical nature of the derivation, this result should not come as any surprise. From the fundamental postulate, we know that the combined system is most likely to be found where accessible states are most plentiful. Therefore the average value of the energy of system A, corresponds to the energy distribution yielding the most available states. Any deviation from this energy distribution must cause a *reduction* in the number of accessible states, and consequently a reduction in the probability of the system having that energy distribution (Figure 9.10).

Since entropy is a measure of accessible states, we can say equivalently that we normally expect the combined system to have an energy distribution which maximizes the entropy. Any deviation from this optimum energy distribution must correspond to a reduction in entropy.

Our result (9.23) allows us to calculate characteristic fluctuations in the energy of a system in thermal equilibrium with a large reservoir. If we plot the probability distribution as a function of the energy E, we see it is a Gaussian distribution peaked at $E = \bar{E}$, with the measure of the width of the peak being given by σ (Figure 9.11). Although the absolute value of the peak width increases as the size of the system increases,

$$\text{width} = \sigma = \sqrt{\frac{2}{\mathfrak{N}}}\,(\mathfrak{N}\bar{\varepsilon}) = \sqrt{2\mathfrak{N}}\,\bar{\varepsilon} \tag{9.25}$$

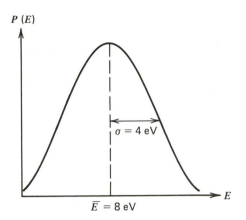
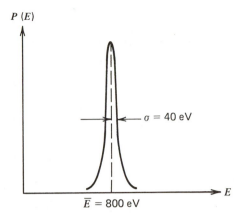

Figure 9.11 Plots of P(E) versus E for small systems with $\mathfrak{N} = 8$ and $\mathfrak{N} = 800$ degrees of freedom, respectively, where the average energy per degree of freedom is $\bar{\varepsilon} = 1$ eV in both systems. For the larger system, the width of the peak is greater (40 eV as opposed to 4 eV), but the relative width is smaller ($\frac{1}{20}$ as opposed to $\frac{1}{2}$).

(where $\bar{\varepsilon} =$ average energy per degree of freedom), the relative width of the peak decreases.

$$\text{relative width} = \frac{\sigma}{\bar{E}} = \sqrt{\frac{2}{\mathfrak{N}}} \tag{9.26}$$

EXAMPLE

Consider a small system for which $\mathfrak{N} = 16$, in equilibrium with a large reservoir at temperature 310 K. If you were to plot the probability $P(E)$ as a function of E for this system, for what value of E would this distribution peak, what would be the standard deviation for this distribution in energies, and what would be the relative width of the peak?

The distribution peaks for the average energy, $E = \bar{E}$, given by ($\mathfrak{N} = 16$, $T = 310$ K)

$$E = \bar{E} = \frac{1}{2}\mathfrak{N}kT = 3.4 \times 10^{-20} \text{ J}$$

The standard deviation is

$$\sigma = \sqrt{\frac{2}{\mathfrak{N}}}\bar{E} = 1.2 \times 10^{-20} \text{ J}$$

and the relative width of the peak is

$$\text{relative width} = \frac{\sigma}{\bar{E}} = \sqrt{\frac{2}{\mathfrak{N}}} = .35$$

EXAMPLE

Repeat the above calculations for a system of $\mathfrak{N} = 16 \times 10^{24}$ degrees of freedom. In this case we have

$$\bar{E} = \frac{1}{2} \mathfrak{N}kT = 3.4 \times 10^4 \text{ J}$$

$$\sigma = \sqrt{\frac{2}{\mathfrak{N}}} \bar{E} = 1.2 \times 10^{-8} \text{ J}$$

and

$$\text{relative width} = \frac{\sigma}{\bar{E}} = \sqrt{\frac{2}{\mathfrak{N}}} = .35 \times 10^{-12}$$

PROBLEMS

9-14. If you add 20 J of heat to a chunk of ice at $-20°C$ (assume the chunk is big enough so that the temperature doesn't change):

(a) What is the change in entropy of the ice?
(b) By what factor does the number of states available to the ice increase? [*Hint.* The factor is $\Omega_{\text{final}}/\Omega_{\text{initial}}$. $\Delta S = k \ln \Omega_f - k \ln \Omega_i = k \ln (\Omega_f/\Omega_i)$.]

9-15. (a) How many joules of heat energy would you have to add to the Atlantic Ocean to double the number of states accessible to it? Assume that $T = 4°C$, which is the water's average temperature. (*Hint.* See hint to Problem 9-14.)
(b) Would the answer be the same if you were dealing with a cup of water at $4°C$ instead?

9-16. Consider a small system having 100 degrees of freedom in thermal equilibrium with a large reservoir at room temperature. In a plot of the probability distribution $P(E)$ for the energy of the small system, as a function of its energy:

(a) At what value of E would the distribution peak?
(b) What would be the width of this peak, as indicated by the standard deviation, σ?
(c) What would be the relative width of the peak?

9-17. Repeat Problem 6-16 for a system having 10^{22} degrees of freedom.

9-18. A cannister containing Avogardo's number of N_2 molecules is in thermal equilibrium with a large heat reservoir. Each molecule has 3 translational and 2 rotational degrees of freedom.

(a) How many degrees of freedom has the whole gas?
(b) The mean energy of the gas is 3200 J. What is the temperature of the reservoir?
(c) You want to bet someone that at any instant, the energy of the gas will be within __% of its mean value (3200 J). What is the smallest percentage you could use here and still have a good bet (i.e., a 68% chance of winning)? (*Hint.* For a Gaussian distribution there is a 68% probability that the variable will be within one standard deviation of its mean value.)

I. MEASURES OF HEAT CAPACITY

If you put a pot of water on one stove burner, and a slice of bread on another, and then turn both burners on high, you will find that the bread is charred to a crisp long before the pot of water even becomes lukewarm. Similarly, a piece of toast is cool enough to eat a few seconds after it leaves the toaster, but a bowl of soup takes much longer to cool down.

Some objects have a greater capacity for holding heat then others. That is, some objects can absorb or release large amounts of heat with only small changes in temperature, and others undergo large temperature changes when only small amounts of heat energy are exchanged.

We quantify this concept by defining the "heat capacity" of an object to be a measure of how much heat energy must be added in order to raise its temperature by one standard unit. Normally the standard unit is a Celsius degree, but occasionally the Fahrenheit degree is used.

In general, a system may be interacting in many ways with its environment. Adding heat energy to it may also stimulate interactions of other types. The heat capacity is influenced by these other interactions, so we usually need to specify under what conditions the heat capacity is measured.

We seldom measure heat capacities of systems interacting diffusively with their environment, so the number of particles is assumed to be constant unless otherwise stated. On the other hand, objects whose heat capacities are to be determined are frequently interacting mechanically with their environment. If we measure their heat capacity at constant pressure (e.g., atmospheric pressure) then they may expand as heat is added, and if we measure the heat capacity at constant volume, then the pressure may increase as heat is added. The technique for measuring heat capacities at constant pressure or constant volume is illustrated schematically in Figure 9.12.

We let y represent the parameters that are to be held constant as the heat capacity is determined. There will be one such parameter for each type of interaction between the system and its environment. The heat capacity for the system is defined by

$$C_y \equiv \frac{\partial Q}{\partial T}\bigg)_y = \text{``heat capacity at constant } y.\text{''} \qquad (9.27)$$

Obviously, the heat capacity of an object depends on its size. The heat capacity of an army tank would be much greater than that of a teaspoon, for example, even though they are made of the same material.

There are two commonly used measures of heat capacity which depend on the materials of a system, but not its size. One is the heat capacity per mole,* or "molar heat capacity," and the other is the heat capacity per unit mass, or "specific heat."

---

* A mole is 6.022×10^{23} molecules, or equivalently, the amount of the material such that its mass in grams is equal to the mass of one molecule in atomic mass units.

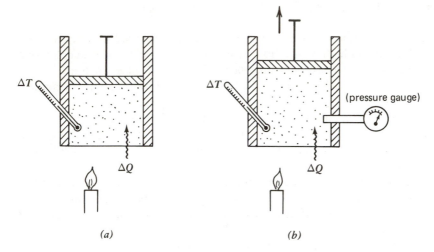

(a)   *(b)*

Figure 9.12 Measuring heat capacities by adding measured amounts of heat, ΔQ, and recording the increase in temperature, ΔT. (a) $C_V = \Delta Q/\Delta T)_V$. As the heat is added, the piston is kept fixed, so the volume remains constant. (b) $C_p = \Delta Q/\Delta T)_p$. As the heat is added, the system is allowed to expand, so that the pressure remains constant.

Table 9.1   Specific Heats of Various Common Substances, at Constant Pressure (except for the last two items, values are measured at room temperature)

Substance	Specific Heat (Cal/g/°C)
Aluminum	0.21
Copper	0.09
Gold	0.03
Acetone (liquid)	0.53
Ethyl alcohol (liquid)	0.58
Marble	0.21
Dry leather	0.36
Synthetic rubber	0.45
Sugar	0.27
Table salt	0.21
Wood	0.42
Air	0.25
Helium (gas)	1.25
Water	1.00
Ice (below freezing)	0.50
Liquid nitrogen ($-200°C$)	0.47

For a system consisting of n moles of a substance, having mass m, we define

$$\text{molar heat capacity at constant } y = c_y \equiv \frac{1}{n} C_y \qquad (9.28)$$

$$\text{specific heat at constant } y = c'_y \equiv \frac{1}{m} C_y \qquad (9.29)$$

Table 9.1 lists the specific heats of various common substances.

Heat capacities are not defined at most ordinary phase transitions, because added latent heat causes no change in temperature.

J. ABSOLUTE ZERO AND THE THIRD LAW

Consider what happens as you remove energy from a system. The number of accessible states is a rapidly increasing function of the internal energy, so as you *remove* energy, the number of accessible states rapidly *declines*.

Suppose you could continue removing energy until you reduce the number of accessible states to one. At this point the system would be in this one state of lowest energy, and it could give up no more energy because there are no lower-lying states for it to fall into. You could remove no more energy from this system. Its energy has reached the zero-energy reference level, and its temperature is therefore "absolute zero."

Since the number of accessible states is just one, the entropy is zero.

$$S = k \ln \Omega = k \ln 1 = 0 \qquad (6.30)$$

Notice that this result does not depend on the size of the box that the system is in, or what pressure it is under, or the strength of the magnetic field, etc. No matter what conditions or constraints the system is held under, there will be one state of lowest energy, and when the system is in that state its entropy is zero.

The above observation provides us the "third law of thermodynamics."

THIRD LAW

The entropy of a system goes to zero as the temperature goes to zero, no matter what the values of the external parameters (e.g., pressure, volume, magnetic field, etc.).

Notice also that a system will have just one lowest-lying state, no matter how large or small it is. That the above is valid no matter what the value of N (the number of particles) gives rise to additional interpretations of the third law.

The above formulation assumes that the lowest-lying level is not degenerate. If the lowest-lying level were five-times degenerate, for example, then at absolute zero the number of accessible states would be 5, and the entropy would be

$$S_0 = k \ln 5$$

For this reason, the third law is sometimes expressed with the words "some constant, S_0" replacing the word "zero" in the foregoing boxed statement. However, although degenerate ground states appear in theoretical models, in real physical systems there seems to always appear some form of weak interaction that breaks the degeneracy at very low energies, leaving one state lower than the others. For this reason, we will use the above boxed-in form in this book, but the student should understand why there is such a variety of forms of it appearing in the literature.

Among other things, the third law requires that heat capacities must vanish at absolute zero. To show this, imagine you were to heat up a system from absolute zero to some finite temperature, T_0, while holding parameters y constant. The entropy would be given by

$$S(T_0) = \int_0^{T_0} dS = \int_0^{T_0} \frac{dQ}{T} = \int_0^{T_0} \frac{C_y \, dT}{T}$$

The last integral diverges unless C_y vanishes at $T = 0$. Consequently,

$$C_y \xrightarrow[T \to 0]{} 0$$

The third law can be used to determine the amount of energy released or absorbed in any chemical reaction, providing the heat capacities of all reactants are known.

EXAMPLE

Consider the reaction $A + B \to C$, where the molar heat capacities are given by

$$c_A = 5T^{1/2} \text{ cal/mole-K}$$

$$c_B = 8T^{1/3} \text{ cal/mole-K}$$

$$c_C = 15T^{1/3} \text{ cal/mole-K}$$

where T is measured in degrees Kelvin. If this reaction is carried out at 300 K, how much heat is released per mole of substance C produced?

The heat released will be given by

$$\Delta Q = T \Delta S$$

where $T = 300$ K and $\Delta S = S_C - (S_A + S_B)$. The entropy, S_A at 300 K is given by

$$S_A(300 \text{ K}) = \int_0^{300} \frac{c_A \, dT}{T} = 5 \frac{\text{cal}}{\text{K}} \int_0^{300} \frac{T^{1/2} \, dT}{T} = 173 \frac{\text{cal}}{\text{K}}$$

Similar calculations of $S_B(300 \text{ K})$ and $S_C(300 \text{ K})$ give

$$S_B(300 \text{ K}) = 161 \text{ cal/K}$$

$$S_C(300 \text{ K}) = 301 \text{ cal/K}$$

Therefore,

$$\Delta S = -33 \text{ cal/K}$$

and the heat released is (The negative sign indicates that heat is leaving the system rather than entering it.)

$$\Delta Q = T \Delta S = -9900 \text{ cal}$$

PROBLEMS

9-19. Why can't heat capacities be defined at phase transitions?

9-20. How much energy is required to raise by 1°C the temperature of a kilogram of (use Table 9.1):

(a) Water?
(b) Gold?
(c) Marble?
(d) Wood?

9-21. Which of the materials in Problem 9-20 would make the most efficient reservoir for the storage of solar heat? Why do you suppose daily and seasonal temperature changes are greater inland than on the coast?

9-22. Consider a material that tends to expand as the temperature rises. Do you think c_p or c_v would be larger for this material? Why?

9-23. We wish to determine the number of degrees of freedom, v, of a single water molecule by measuring its specific heat. The number of degrees of freedom of N water molecules is given by

$$\mathfrak{N} = vN$$

(a) How many water molecules are there in a gram of water?
(b) How many calories of heat energy are required to raise the temperature of 1 g of water by $\Delta T = 1$ K?
(c) With the above knowledge, use the relationship

$$\Delta E = \frac{1}{2} \mathfrak{N} k \, \Delta T$$

to determine v, the number of degrees of freedom per molecule.

9-24. Consider the chemical reaction $2A + B \rightarrow C$. Suppose you are going to produce 1 mole of C at 500 K and atmospheric pressure. The molar heat capacities of these substances at atmospheric pressure are all zero between 0 and 10 K, and they are constant above 10 K, being given by

$$c_A = 93 \text{ cal/mole-K} \qquad (T > 10 \text{ K})$$

$$c_B = 86 \text{ cal/mole-K}$$

$$c_C = 262 \text{ cal/mole-K}$$

(a) How much heat will be released in this reaction?
(b) How can the system lose entropy and not violate the second law?

chapter 10

THE MECHANICAL INTERACTION

We can do work on a system by exerting a force over a distance. There are many different ways of doing this. Work done on a system may cause a change in temperature. If the temperature is held constant (e.g., immersion in a water bath), then the work done may be reflected in changes in some other property, such as the number of accessible states, or entropy.

Chapter 9 dealt with systems interacting thermally. In this chapter we study their mechanical interactions.

A.   CHANGE IN VOLUME

Consider some system expanding its perimeter as in Figure 10.1. If it exerts force F_1 against a small area a_1, and that area moves distance ds_1, then the work done by the system on a_1 is given by

$$dW_1 = \mathbf{F}_1 \cdot d\mathbf{s}_1 = F_1 \, ds_1 \cos \theta = \left(\frac{F_1}{a_1}\right)(a_1 \, ds_1 \cos \theta) = p \, dV_1$$

where p is the pressure, and dV_1 the small change in volume due to the movement of area a_1.

If we add up all the bits of work done on each little area a_i along the perimeter of the system, we have

$$
\begin{aligned}
dW_{\text{total}} &= dW_1 + dW_2 + \cdots \\
&= p \, dV_1 + p \, dV_2 + p \, dV_3 + \cdots \\
&= p(dV_1 + dV_2 + dV_3 + \cdots) \\
&= p \, dV_{\text{total}}
\end{aligned}
$$

> If a system having pressure p changes volume by an amount dV, then the work done by this system is given by
> $$dW = p \, dV \qquad\qquad (10.1)$$

The change in volume is an exact differential, because initial and final volumes are measurable. The inexact differential, dW, can be changed into the exact differential, dV,

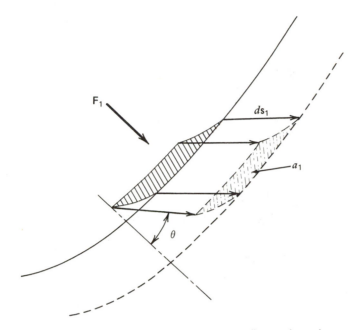

Figure 10.1 Consider some expanding system. A small piece of its surface area, a_1, is displaced a distance ds_1 by a force F_1 acting at an angle θ to the direction of the displacement.

by multiplying by $1/p$. Similarly, in Chapter 9 we saw that the inexact differential, dQ, could be changed into the exact differential, dS, by multiplying by $1/T$.

EXAMPLE

A gas under a pressure of 100 atm expands against a piston. If the cross section of the piston is 0.04 m² and it is pushed back a distance of 0.1 m, what is the work done by the gas?

The work is given by

$$\Delta W = p \, \Delta V = (10^7 \text{ N/m}^2)(4 \times 10^{-3} \text{ m}^3) = 4 \times 10^4 \text{ J}$$

EXAMPLE

Suppose the above expansion was done without any heat energy or particles being added or removed from the system. Suppose further that the gas had 10^{28} molecules, each having 5 degrees of freedom. What is the change in temperature of the gas?

If no heat or particles are added or removed, then

$$dE = -p \, dV$$

But

$$E = \frac{1}{2} \Re k T$$

where

$$\mathfrak{N} = 5 \times 10^{28}$$

Consequently, we have

$$\frac{1}{2} \mathfrak{N} k \Delta T = -p \Delta V$$

or

$$\Delta T = -\frac{p \Delta V}{\frac{1}{2} \mathfrak{N} k} = -\frac{4 \times 10^4 \text{ J}}{3.5 \times 10^5 \text{ J/K}} = -.11 \text{ K}$$

PROBLEMS

10-1.   A gas having 10^{27} degrees of freedom is under a pressure of 150 atm. It is allowed to expand by 0.01 m^3 without any heat or particles being added or removed in the process.

  (a)   What is the work done by the gas?
  (b)   What will be the change in temperature of the gas?

10-2.   As liquid water is compressed at constant temperature, its pressure increases rapidly, according to the formula

$$p = \left[1 + 2.5 \times 10^4 \left(1 - \frac{V}{V_0} \right) \right] \text{atm}$$

where V_0 is its volume under atmospheric pressure.

  (a)   If some water has a volume of 1 liter at atmospheric pressure, what will be its volume at the bottom of the ocean where the pressure is 500 atm?
  (b)   How much work is done by a liter of water that is brought to the surface from the ocean bottom? (*Hint.* The pressure isn't constant so you'll have to integrate $\int p \, dV$.)
  (c)   How many water molecules are in a liter of water? (One mole has a mass of 18 g.)
  (d)   If each water molecule has 6 degrees of freedom, and a one-liter sample of water is retrieved from the ocean bottom without heat or particles being added or removed in the process, what will be the change in the temperature of the water sample?

B.   WORK AND THE NUMBER OF ACCESSIBLE STATES

For systems interacting thermally, mechanically, and diffusively, the change in energy is given by the first law.

$$dE = T \, dS - p \, dV + \mu \, dN \tag{10.2}$$

In this equation, there are seven variables (E, T, S, p, V, μ, N) only three of which are *independent* (there are only three types of interactions). We can write any one of them in terms of any other three.

$$E = E(T, S, p)$$

$$\mu = \mu(p, V, N)$$

$$T = T(p, \mu, N)$$

$$\text{etc.}$$

It is frequently convenient to express the energy in terms of S, V, and N, because their differentials (dS, dV, dN) are exact, and because the easily measured properties of temperature, pressure, and chemical potential are given by (Figure 10.2)

$$T = \frac{\partial E}{\partial S}\bigg)_{V,N}$$

$$p = -\frac{\partial E}{\partial V}\bigg)_{S,N}$$

$$\mu = \frac{\partial E}{\partial N}\bigg)_{S,V} \tag{10.3}$$

as can be seen from Eq. 10.2.

Because the second law is so useful, we are frequently interested in dealing with the entropy of a system. From Eq.10.2 we get the following expression for change in entropy

$$dS = \frac{1}{T}\,dE + \frac{p}{T}\,dV - \frac{\mu}{T}\,dN \tag{10.4}$$

and the easily measured parameters of temperature, pressure, and chemical potential are given by

$$\frac{1}{T} = \frac{\partial S}{\partial E}\bigg)_{V,N}$$

$$\frac{p}{T} = \frac{\partial S}{\partial V}\bigg)_{E,N}$$

$$\frac{\mu}{T} = -\frac{\partial S}{\partial N}\bigg)_{E,V} \tag{10.5}$$

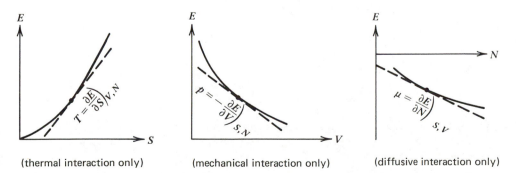

(thermal interaction only)   (mechanical interaction only)   (diffusive interaction only)

Figure 10.2 The temperature is the slope of the plot of E versus S for a system undergoing thermal interactions only (V and N are constant). The pressure is the negative of the slope of the plot of E versus V for a system undergoing mechanical interactions only. The chemical potential is the slope of E versus N for a system undergoing diffusive interactions only (plotted here for the case of attractive interactions among particles).

In this chapter we are particularly concerned with the second of these relationships,

$$p = T \frac{\partial S}{\partial V}\bigg)_{E,N}$$

In terms of the number of accessible states, we have

$$S = k \ln \Omega$$

and

$$p = kT \frac{\partial \ln \Omega}{\partial V}\bigg)_{E,N} = kT \frac{\Delta \ln \Omega}{\Delta V}\bigg)_{E,N} \qquad (10.6)$$

This relationship tells us that if a system's volume is changed with no particles being added, and the internal energy of the system remains fixed, then

$$\frac{p \Delta V}{kT} = \Delta \ln \Omega = \ln \Omega_2 - \ln \Omega_1 = \ln \frac{\Omega_2}{\Omega_1}$$

or

$$\frac{\Omega_2}{\Omega_1} = e^{p \Delta V/kT} \qquad (10.7)$$

The number of accessible states changes exponentially with changes in the volume.

The above relationship (10.7) is really the same as Eq. 9.15

$$\frac{\Omega_2}{\Omega_1} = e^{\Delta Q/kT} \qquad (9.15)$$

because, according to the first law

$$\Delta E = \Delta Q - p \Delta V + \mu \Delta N$$

we have $\Delta Q = p \Delta V$ for processes where $\Delta E = \Delta N = 0$.

SUMMARY

Rearranging the terms in the first law as follows,

$$dS = \frac{1}{T} dE + \frac{p}{T} dV - \frac{\mu}{T} dN$$

we see that

$$\frac{p}{T} = \frac{\partial S}{\partial V}\bigg)_{E,N} = \frac{\partial (k \ln \Omega)}{\partial V}\bigg|_{E,N}$$

As a result, if a system changes volume by an amount ΔV, with its internal energy and number of particles remaining constant, the number of accessible states will change by a factor by

$$\frac{\Omega_2}{\Omega_1} = e^{p \Delta V/kT} \qquad (10.7)$$

According to Eqs. 10.5 and 10.6, pressure is a measure of how the entropy—or number of accessible states—increases with volume. We see that the pressure is greatest when the relative increase in accessible states is greatest. One way of looking at pressure, then, is as a measure of how hard a system is trying to gain additional accessible states (or equivalently, to increase its entropy) through the aquisition of additional volume.

EXAMPLE

If 20 m^3 of air at room temperature and atmospheric pressure is allowed to expand by 0.001%, by what factor does the number of accessible states increase? Assume no energy or particles are added to or removed from the system.

For this case, $\Delta V = 2 \times 10^{-4}$ m^3, $p = 10^5$ N/m^2, and $T = 300$ K. From Eq. 10.6 we have

$$\Delta \ln \Omega = \frac{p \Delta V}{kT}$$

This can be written equivalently as

$$\ln \Omega_2 - \ln \Omega_1 = \ln \frac{\Omega_2}{\Omega_1} = \frac{p \Delta V}{kT}$$

or

$$\frac{\Omega_2}{\Omega_1} = e^{p \Delta V / kT} = e^{4.8 \times 10^{21}} = 10^{2.1 \times 10^{21}}$$

In previous chapters, we saw that the number of accessible states is an extremely rapidly increasing function of the internal energy for macroscopic systems. From the above example, we find that the number of accessible states is an extremely rapidly varying function of its volume as well. Since entropy is a measure of the number of states, and since it depends on the three variables (E, V, N), it should come as no surprise in the next chapter when we show that the number of accessible states varies extremely rapidly with N also.

C. OTHER KINDS OF WORK

In the previous section we analyzed the work done by a system

$$dW = \mathbf{F} \cdot d\mathbf{s}$$

For the specific case of a system expanding its volume (e.g., Figure 10.3) we found

$$dW = p \, dV$$

But there are many other kinds of forces that can cause movement of matter over some distance, resulting in work being done.

For example, in addition to mechanical forces, magnetic forces, electric forces, gravitational forces, and many others can result in the displacement of material and some net work being done. For each of these, the product

$$dW = \mathbf{F} \cdot d\mathbf{s}$$

Figure 10.3 This diver's depth gauge is an elegantly simple application of the mechanical interaction between two systems. The tube around the perimeter has an opening at the bottom. It is plugged to the right of this opening, and is open to the left. As the pressure increases, water enters the opening, moving clockwise around the gauge, and compressing the air trapped in front of it as it goes. One determines depth by how far the air in the tube has been compressed. (Keith Stowe)

can be converted into a form like

$$dW = \mathscr{F} \, dx$$

where $\mathscr{F}$ represents some generalized force, and dx is a change in some parameter resulting from displacement of matter. Some examples are given in Table 10.1.

Table 10.1   Some Examples of Changing Parameters and Their Generalized Forces That Result in Work Being Done by a System

Changing Parameter, dx	Generalized Force, $\mathscr{F}$	dW
Position	Mechanical force	$F_x \, dx$
Volume	Pressure	$p \, dV$
Electrical potential	$-$(Charge)	$-q \, dU$
Magnetic field	Magnetic moment	$\mu \, dB_0{}^a$
Separation	$-$(Gravitational force)	$-\left(G \dfrac{m_1 m_2}{r^2} \right) dr$

[a] Unfortunately, the symbol commonly used for magnetic moment (μ) is the same as that commonly used for chemical potential. Hopefully, you will be able to distinguish them according to context. We use B_0 for the applied external magnetic field.

Any of the developments given in this book involving pressure and volume would be the same for any of the generalized forces and their corresponding changing parameters. The first law would become

$$dE = T\,dS - \mathscr{F}\,dx + \mu\,dN \tag{10.2}$$

and the corresponding changes in equations (10.3) through (10.6) would be

$$p = -\frac{\partial E}{\partial V}\bigg)_{S,N} \rightarrow \mathscr{F} = -\frac{\partial E}{\partial x}\bigg)_{S,N} \tag{10.3}$$

$$dS = \frac{1}{T}\,dE + \frac{p}{T}\,dV - \frac{\mu}{T}\,dN \rightarrow dS = \frac{1}{T}\,dE + \frac{\mathscr{F}}{T}\,dx - \frac{\mu}{T}\,dN \tag{10.4}$$

$$\frac{p}{T} = \frac{\partial S}{\partial V}\bigg)_{E,N} \rightarrow \frac{\mathscr{F}}{T} = \frac{\partial S}{\partial x}\bigg)_{E,N} \tag{10.5}$$

$$p = kT\frac{\partial \ln \Omega}{\partial V}\bigg)_{E,N} \rightarrow \mathscr{F} = kT\frac{\partial \ln \Omega}{\partial x}\bigg)_{E,N} \tag{10.6}$$

For example, if the interaction were electrostatic, then we would have the charge given by

$$(-q) = -\frac{\partial E}{\partial U}\bigg)_{S,N}$$

and

$$\frac{(-q)}{T} = \frac{\partial S}{\partial U}\bigg)_{E,N}$$

from Eqs. 10.3 and 10.5, respectively.

Finally, a system could be undergoing more than one type of mechanical interaction at a time. For example, if interacting electrostatically, magnetically, and with change of volume at the same time, we would have

$$dW = p\,dV - q\,dU + \mu\,dB_0$$

Consequently, in addition to its dependence on S and N, the interaction will depend on as many additional parameters as there are types of mechanical interactions. The first law would read

$$dE = T\,dS - \sum_i \mathscr{F}_i\,dx_i + \mu\,dN \tag{10.8}$$

and the various generalized forces would be given by

$$\mathscr{F}_i = -\frac{\partial E(S, N, x_1, x_2, \ldots)}{\partial x_i}$$

or

$$\frac{\mathscr{F}_i}{T} = \frac{\partial S(E, N, x_1, x_2, \ldots)}{\partial x_i}$$

EXAMPLE

Consider a system in an impenetrable, rigid container. The system has net charge of $q = +10^{-12}$ C. By what amount does the entropy of the system change when the electrostatic potential of the system is decreased by 1 V, assuming that the temperature remains at 300 K during the process?

For this process, $dW = -q\,dU$, so $-q$ and the change in electrostatic potential, dU, replace p and dV, respectively, in our earlier results. From Eq. 10.5 we have

$$-\frac{q}{T} = \frac{\partial S}{\partial U}\bigg)_{E,N}$$

and so for this process

$$\Delta S = -\frac{q}{T}\,\Delta U = 3.3 \times 10^{-15} \text{ J/K}$$

EXAMPLE

In the above problem, by what factor does the number of accessible states increase?
The change in entropy is related to the increase in accessible states through

$$\Delta S = S_2 - S_1 = k \ln \Omega_2 - k \ln \Omega_1 = k \ln \frac{\Omega_2}{\Omega_1}$$

Consequently, we have

$$\frac{\Omega_2}{\Omega_1} = e^{\Delta S/k} = e^{2.4 \times 10^8} = 10^{1.0 \times 10^8}$$

PROBLEMS

10-3. In the last example, why was it necessary for the system to be in an "impenetrable, rigid container" for the answer to be exactly correct? What other constraint is necessary?

10-4. By how much must the entropy of a system increase, if the number of accessible states doubles?

10-5. Consider 1 m³ of steam at a temperature of 600°C and a pressure of 50 atm. By how much must its volume expand if the number of accessible states is doubled?

10-6. What is the magnetic moment of a system if the internal energy decreases by 10^{-14} J when the external magnetic field strength increases by 1 T (all other parameters held constant)? (*Hint.* Look at Table 10.1 and Eq. 10.3.)

10-7. A rubber ball is in contact with a heat reservoir that keeps its temperature constant at 300 K. (Since $E \propto T$, the internal energy remains constant as well.) If at a pressure of 1.001 atm its volume compresses by 10^{-3} cm³:

    (a) What is the change in entropy?
    (b) By what factor does the number of accessible states change?

10-8.   A ferromagnet at 300 K has a magnetic moment of $\mu_z = 10^{-3}$ J/T. It is sitting in an external field oriented along the z-axis of strength $B_z = 0.1$ T. The external field suddenly changes so that the number of accessible states doubles.

(a)   Did the field strength increase or decrease?
(b)   By how much?

10-9.   The number of states accessible to an ideal gas having energy in the range between E and $E + \delta E$ is given by

$$\Omega_0 = (\text{const})V^N E^{3N/2}\, \delta E,$$

where V is the volume of the gas and N the number of molecules. (See Chapter 7, Section D.)

(a)   Using Eq. 10.6 show that $pV = NkT$.
(b)   This is sometimes written $pV = nRT$, where n is the number of moles of the gas, and R is called the "gas constant." What is R in terms of Boltzmann's constant and Avogadro's number?

D.   THERMAL EXPANSION AND COMPRESSIBILITIES

Most materials tend to expand when heated. For gases, increased thermal motion of the molecules causes them to collide more forcefully with the walls of the container, pushing them outward. For solids and liquids, increased thermal motion of the molecules causes them to oscillate with greater amplitudes about their equilibrium positions, which usually results in greater average intermolecular spacings and the consequent increase in volume.

A measure of the relative increase in volume per unit increase in temperature is called the "coefficient of volume expansion" of the material, and is defined as follows:

$$\beta = \text{"coefficient of volume expansion"} \equiv \frac{1}{V}\frac{\partial V}{\partial T}\bigg)_p \tag{10.9}$$

In general, the coefficient of volume expansion for any material is a function of temperature and pressure.

$$\beta = \beta(T, p)$$

For sufficiently small* changes in temperature, the change in volume of a material is given by

$$\Delta V = \frac{\partial V}{\partial T}\bigg)_p \Delta T$$

$$= V\beta\,\Delta T \tag{10.10}$$

---

* The new volume can be expressed in a Taylor series expansion

$$V(T\,\Delta T) = V(T) + \left(\frac{\partial V}{\partial T}\right)\Delta T + \left(\frac{1}{2!}\frac{\partial^2 V}{\partial T^2}\right)\Delta T^2 + \cdots$$

From this, we see that the change in volume, $\Delta V = V(T + \Delta T) - V(T)$, is linear in ΔT only if ΔT is small enough that second- and higher-order terms are negligible.

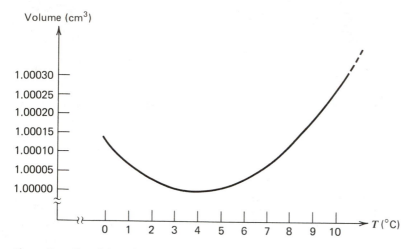

Figure 10.4 Plot of the volume of one gram of water versus temperature. Its volume decreases with increased temperature between 0 and 4°C, and then increases after that.

Usually, coefficients of volume expansion are positive. However, for pure water between 0 and 4°C it is negative (Figure 10.4). That is, the volume of water gets smaller as its temperature increases in this range. For systems undergoing phase changes, coefficients of volume expansions are often undefined. The coefficients of volume expansion for materials are usually recorded for a pressure of one atmosphere, unless otherwise stated.

The coefficients of volume expansion are much smaller for solids and liquids than for gases, being typically $4 \times 10^{-5}/°C$ for solids, $4 \times 10^{-4}/°C$ for liquids, and $4 \times 10^{-3}/°C$ for gases. There is much variation in these values for solids and liquids, but not much variation for the gases, as is seen in Table 10.2.

Table 10.2 Coefficients of Volume Expansion, $\beta = \dfrac{1}{V}\dfrac{\partial V}{\partial T}\Big)_p$, for Various Substances at Room Temperature and Atmospheric Pressure

	$\beta(/K)$		$\beta(/K)$		$\beta(/K)$
Solids		Liquids		Gases	
Aluminum	5.5×10^{-5}	Alcohol	9.0×10^{-4}	Air	3.7×10^{-3}
Brass	5.6×10^{-5}	Olive oil	6.8×10^{-4}	CO_2	3.8×10^{-3}
Marble	3.4×10^{-5}	Water	2.1×10^{-4}	N_2	3.7×10^{-3}
Steel	4.0×10^{-5}	Mercury	1.8×10^{-4}	H_2	3.7×10^{-3}
Quartz	3.4×10^{-5}			H_2O	4.2×10^{-3}
Oak (wood)	11.2×10^{-5}				

A measure of the relative increase in *length* (X) per unit increase in temperature is called the "coefficient of linear expansion" of the material, and is defined analogously as

$$\alpha = \text{"coefficient of linear expansion"} \equiv \frac{1}{X}\frac{\partial X}{\partial T}\bigg)_p \qquad (10.11)$$

It is likewise a function of the temperature and pressure,

$$\alpha = \alpha(T, p)$$

and for sufficiently small changes in temperature, the change in linear dimensions of the material is given by

$$\Delta X = \frac{\partial X}{\partial T}\bigg)_p \Delta T$$

$$= X\alpha\,\Delta T \qquad (10.12)$$

The coefficient of volume expansion can be related to the coefficient of linear expansion in the following way. If the material originally has volume $V = XYZ$, and is heated up by ΔT, then the new volume, V', is

$$V' = V + \Delta V = V(1 + \beta\,\Delta T)$$

But the new volume can also be written in terms of the new dimensions (X', Y', Z') as

$$V' = X'Y'Z' = X(1 + \alpha\,\Delta T)Y(1 + \alpha\,\Delta T)Z(1 + \alpha\,\Delta T)$$
$$= XYZ(1 + \alpha\,\Delta T)^3$$
$$= V(1 + \alpha\,\Delta T)^3$$

Keeping only first-order terms in ΔT this last expression becomes

$$V' = V(1 + 3\alpha\,\Delta T)$$

Comparing this to expression (10.10), we see that the coefficient of volume expansion is just three times the coefficient of linear expansion,

$$\beta = 3\alpha \qquad (10.13)$$

providing the coefficient of linear expansion is the same in each of the three dimensions. For some crystals, the coefficients of linear expansion ($\alpha_x, \alpha_y, \alpha_z$) are different in different dimensions, and so the above result must be appropriately modified. (See homework Problem 10-10.)

In addition to its dependence on temperature, the volume of most materials also varies with pressure. A measure of the relative change in volume per unit increase in pressure (temperature held constant) is called the "isothermal compressibility."

$$\kappa = \text{"isothermal compressibility"}$$

$$\equiv -\frac{1}{V}\frac{\partial V}{\partial p}\bigg)_T \qquad (10.14)$$

The isothermal compressibility of any material is a function of the temperature and pressure.

$$\kappa = \kappa(T, p)$$

For sufficiently small changes in pressure, Δp, the corresponding change in volume is given by

$$\Delta V = \frac{\partial V}{\partial p}\bigg)_T \Delta p$$

$$= -V\kappa\,\Delta p \qquad (10.15)$$

Sometimes people refer to the "bulk modulus" of a material, which is the reciprocal of the isothermal compressibility

$$\text{bulk modulus} = -V\frac{\partial p}{\partial V}\bigg)_T = \frac{1}{\kappa}$$

SUMMARY

The dimensions of any material tend to vary with temperature and pressure. The coefficient of linear expansion (α), coefficient of volume expansion (β), and isothermal compressibility (κ) are measures of these properties, defined by

$$\alpha = \frac{1}{X}\frac{\partial X}{\partial T}\bigg)_p \qquad (10.9)$$

$$\beta = \frac{1}{V}\frac{\partial V}{\partial T}\bigg)_p \qquad (10.11)$$

$$\kappa = -\frac{1}{V}\frac{\partial V}{\partial p}\bigg)_T. \qquad (10.14)$$

Each of these properties varies from one material to the next, and for any one material they depend on the the temperature and pressure.

PROBLEMS

10-10. A certain solid is not isotropic, having different coefficients of linear expansion in each dimension.

$$\alpha_x = \frac{0.5 \times 10^{-5}}{°C} \qquad \alpha_y = \frac{1.5 \times 10^{-5}}{°C} \qquad \alpha_z = \frac{2 \times 10^{-5}}{°C}$$

Calculate what its coefficient of volume expansion is.

10-11. Suppose you wanted to measure the coefficient of volume expansion of some solid. Suppose that in addition to the solid you have available a cylinder with a moveable piston with a temperature and a pressure gauge through the side (as in Figure 10.5), a

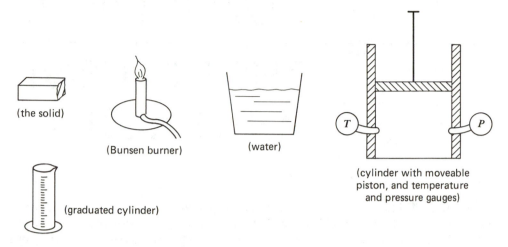

(the solid)

(Bunsen burner)

(water)

(graduated cylinder)

(cylinder with moveable piston, and temperature and pressure gauges)

Figure 10.5   How would you measure the coefficient of volume expansion of some solid, having at your disposal only the materials in this figure?

Bunsen burner, a graduated cylinder, and water. How would you go about measuring β for the solid? (*Hint.* Maybe you should measure β for the water first.)

10-12.   With the apparatus of the previous problem (see Figure 10.5), how would you measure the isothermal compressibility of the solid?

✳ chapter 11

THE DIFFUSIVE INTERACTION

By pouring gasoline on a fire, we quickly discover than the energy of a system can change when particles are added. But how does this square with conservation of energy? When one system gains energy, doesn't another have to lose it? If the energy of the fire increases, doesn't the energy of some other system necessarily decrease?

In the previous two chapters we studied the effects of thermal and mechanical interactions, respectively, on the properties of a system. In this chapter we examine the third kind of interaction—the diffusive, or "chemical" interaction.

Chemical potential (μ) enters the first law on an equal footing with temperature (T) and pressure (p).

$$dE = T\,dS - p\,dV + \mu\,dN$$

But chemical potential is not talked about in our everyday conversation as are the other two. Because lack of familiarity breeds lack of confidence in working with it, we devote the first two sections of this chapter to a qualitative review of chemical potential, so that you maintain some familiarity with the concept.

A. CHEMICAL POTENTIAL

When new particles are introduced into a system, they may release thermal energy to the system as a result of their interactions with the other particles already there. This may happen even if the particle initially has no energy of its own at all.

For example, if we could take a motionless water molecule from outer space, where it has neither kinetic energy of motion nor potential energy due to interactions with others, and if we inserted this energyless particle into a glass of liquid water, it would still be able to release some thermal energy to its new surroundings (Figure 11.1). It can do this because of electrostatic attraction to its neighbors in its new environment. Because of this attraction, it moves toward its neighbors, picking up kinetic energy as it goes. This kinetic energy gained by the new particle and its neighbors is soon

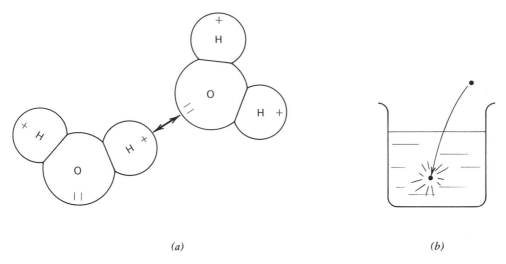

(a)   *(b)*

Figure 11.1   (a) The electrically polarized water molecule interacts attractively with its neighbors. (b) This causes it to fall into a potential well and release thermal energy when a new water molecule joins liquid water.

transferred to others through collisions, so the system as a whole gains thermal energy,* even though the new particle had none at all to begin with.

   The reason for this apparent contradiction of the Law of Conservation of Energy, is that in its new environment its potential energy is *negative*, due to the electrostatic interactions with its neighbors. In the new environment the water molecule "falls into a potential well," and it gains kinetic energy as it loses potential energy. Consequently, energy is not spontaneously created; it is simply converted from potential to kinetic form.

   In principle, then, we could determine the chemical potential of particles in a system by adding one more isolated, energyless particle to the system and measuring the system's gain in thermal energy. Since thermal energy gained (ΔQ) is equal to the depth of the potential well the particle falls into, we have

$$\mu = -\Delta Q$$

In practice, however, it is not possible to make such measurements involving either isolated or energyless particles. Instead, we add a known number of particles, ΔN, having known internal energy to a system at constant volume. If ΔQ is the increased thermal energy of the combined system, then (see Figure 11.2)

$$\Delta Q = -\mu \, \Delta N$$

or

$$\mu = -\frac{\Delta Q}{\Delta N}$$

* "Thermal energy" is a term referring to kinetic energy, plus potential energy measured relative to the bottom of each particle's potential well. It corresponds to the heat energy that would be absorbed by the system in raising its temperature from zero to the present temperature, if this could be done without changing the zero-energy reference level, $N\mu$.

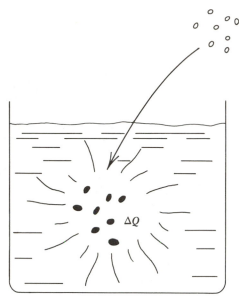

Figure 11.2 If thermal energy ΔQ is released when ΔN energyless particles are added to a system, then the chemical potential of these particles in that system is $\mu = -\Delta Q/\Delta N$.

EXAMPLE

Suppose 10^{21} molecules of some acid at room temperature are added to a liter of water at room temperature, and the temperature rises by 0.1°C. What is the chemical potential of an acid molecule in the water? Assume the self-interaction among the acid molecules is negligible compared to their interaction with water molecules.

The heat capacity of a liter of water is 10^3 cal/°C, so $\Delta Q = 10^2$ cal. A liter of water has about 3.3×10^{25} water molecules, so we can ignore the heat capacity of the acid molecules by comparison.

Each acid molecule goes from an acid environment, where its potential well was of negligible depth, to a water environment, where the depth of its potential well is μ. The total thermal energy gained is equal to the total potential energy lost, so we have

$$\Delta Q = -\mu \Delta N$$

where

$$\Delta Q = 10^2 \text{ cal} \qquad \Delta N = 10^{21}$$

This gives us the chemical potential of this acid in water as

$$\mu = -\frac{10^2}{10^{21}} \text{ cal} = -2.6 \text{ eV}$$

In Section D we will demonstrate that particles tend to flow from regions of higher chemical potential toward regions of lower chemical potential. This should come as no surprise. For the same reason that water flows downhill rather than up, all matter tends to flow toward regions of lower potential energy.

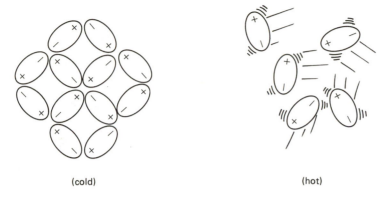

(cold)   (hot)

Figure 11.3 We expect the chemical potential to depend on the temperature, because the average strengths of the interactions between molecules may depend on their state of motion. For example, at cold temperatures, the molecules may order themselves in such a fashion that brings oppositely charged portions of neighboring molecules near each other. At higher temperatures, thermal motion may break up this ordered arrangement.

B. INFLUENCE OF TEMPERATURE, PRESSURE, AND CONCENTRATION

We will now try to get a qualitative feeling for how the chemical potential should depend on other system variables. We are now familiar with many variables that describe a system (Ω, S, E, T, p, V, μ, N), and others are yet to come. If the system is undergoing three types of interactions, then there are three independent variables, and we should be able to express any one in terms of any other three. (Often, this is more easily said than done, however.)

In particular, we should be able to see intuitively that chemical potential should vary with temperature, pressure, and particle concentration.

When the particles are moving faster, their interactions are usually more intense and vary more quickly. Therefore, we expect the chemical potential to change at higher temperatures (Figure 11.3).

At higher pressures, the particles of a system are brought closer together, which usually increases the strength of their mutual interactions. These increased interactions may be predominantly attractive or repulsive, so the chemical potential may either lessen or rise, respectively, with increased pressure.

Finally, chemical potential may either increase or decrease with particle concentration, depending on the nature of the interaction. If the particles are interacting attractively with other molecules of their own type (e.g., water with water), then increased concentrations mean more close neighbors, more interactions, and therefore more negative chemical potential (Figure 11.4). If the particles are mostly interacting with particles of another type (e.g., a solute in a solvent), then increasing their own numbers decreases the relative abundance of the other type of particle, thereby decreasing their average interaction energy. So in this case, chemical potential would become less negative with increased concentration (see Figure 6.13.)

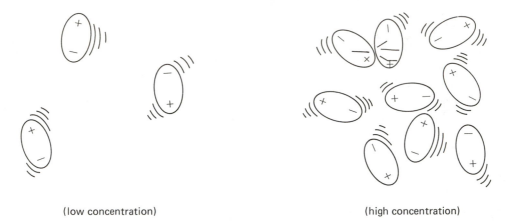

(low concentration)   (high concentration)

Figure 11.4 We also expect the chemical potential to depend on the number of particles. At higher particle concentrations the strength and frequency of interactions between particles would be different than at low concentrations.

PROBLEMS

11-1. In the vapor phase, water molecules are rather far apart, and you can assume their mutual interactions to be small on the average. That is for water in the vapor phase $\mu \approx 0$. When water vapor condenses at 100°C and atmospheric pressure, it releases 540 calories per gram of condensate. Using this information, calculate the chemical potential for water in the liquid state at 100°C.

11-2. In Problem 11-1 we ignored the fact that the volume of the liquid water increases as the water condenses ($\Delta E = 0 = \Delta Q - p\Delta V + \mu\Delta N$, so $\mu\Delta N = -\Delta Q + p\Delta V$.) Calculate $p\Delta V$ for the accumulation of 1 cm³ of liquid water at atmospheric pressure, and compare it to ΔQ or $\mu\Delta N$ for the same amount of liquid water condensate. Were we justified in ignoring $p\Delta V$ by comparison?

11-3. A certain sugar has a chemical potential of -1.6 eV in water and -0.9 eV in oil. Suppose some of this sugar was dissolved in oil and you wished to remove it. Can you suggest a way to do it, taking advantage of the difference in the two chemical potentials?

11-4. From a molecular point of view, why is the chemical potential of:

(a) Liquid water less (i.e., more negative) than that of water vapor?
(b) Ice less than that of liquid water?
(c) Salt in water less than that of salt in oil?

11-5. A certain salt has chemical potential of -0.4 eV in the crystalline form. When one mole of this crystalline salt is dissolved in a large tub of water, 2500 cal of heat energy are released. What is the chemical potential of the salt in water?

11-6. When a certain acid condenses from the vapor phase at room temperature, it releases latent heat of 4600 cal/mole.

(a) Assuming the self-interactions (hence the chemical potential) negligible in the vapor state, use the above information to calculate the chemical potential of this acid in the liquid form at room temperature.

(b)   When 10^{-2} mole of this acid in the liquid form is added to one liter of water at room temperature, the temperature of the water rises 0.1°C. What is the chemical potential for this acid in water?

C.   EQUILIBRIUM CONDITIONS

Consider two systems, A_1 and A_2, interacting thermally, mechanically, and diffusively as in Figure 11.5. We can rearrange the terms in the first law

$$\Delta E = T \Delta S - p \Delta V + \mu \Delta N$$

to find an expression for the change of entropy for either system

$$\Delta S = \frac{1}{T} \Delta E + \frac{p}{T} \Delta V - \frac{\mu}{T} \Delta N \tag{11.1}$$

Since these two systems are interacting only with each other, we have

$$\Delta E_2 = -\Delta E_1$$
$$\Delta V_2 = -\Delta V_1$$
$$\Delta N_2 = -\Delta N_1 \tag{11.2}$$

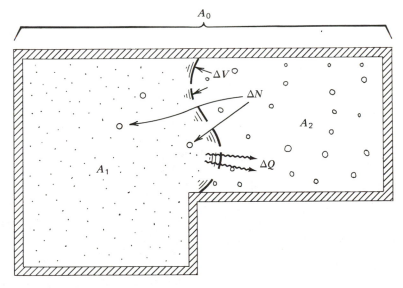

Figure 11.5   Consider two systems, A_1 and A_2, interacting thermally, mechanically, and diffusively. The combined system, A_0, is completely isolated from the rest of the universe, so that the total internal energy, volume, and number of particles does not change. Therefore, any energy, volume, or particles gained by one of the subsystems must come from the other. $\Delta E_1 = -\Delta E_2$, $\Delta V_1 = -\Delta V_2$, $\Delta N_1 = -\Delta N_2$. When they are in equilibrium, the entropy of the combined system will be a maximum.

and the change in entropy for the combined system due to small redistribution of energy, volume, or numbers of particles between the two systems is

$$\Delta S_\circ = \Delta S_1 + \Delta S_2$$

$$= \frac{1}{T_1} \Delta E_1 + \frac{p_1}{T_1} \Delta V_1 - \frac{\mu_1}{T_1} \Delta N_1 + \frac{1}{T_2} \Delta E_2 + \frac{p_2}{T_2} \Delta V_2 - \frac{\mu_2}{T_2} \Delta N_2$$

Using the relationships (11.2) this becomes

$$\Delta S_\circ = \left(\frac{1}{T_1} - \frac{1}{T_2}\right) \Delta E_1 + \left(\frac{p_1}{T_1} - \frac{p_2}{T_2}\right) \Delta V_1 - \left(\frac{\mu_1}{T_1} - \frac{\mu_2}{T_2}\right) \Delta N_1 \qquad (11.3)$$

When the two systems are in equilibrium, the entropy of the combined system is a maximum. The maximum of any function is the point where its derivatives are zero, or equivalently, the point where its value doesn't change with small changes in the variables. Therefore, when the entropy of the combined system is maximum, it won't vary with small variations in the distribution of energy, volume, or particles between the two interacting subsystems. At equilibrium, then,

$$\Delta S_\circ = 0 = \left(\frac{1}{T_1} - \frac{1}{T_2}\right) \Delta E_1 + \left(\frac{p_1}{T_1} - \frac{p_2}{T_2}\right) \Delta V_1 - \left(\frac{\mu_1}{T_1} - \frac{\mu_2}{T_2}\right) \Delta N_1 \qquad (11.4)$$

and since there are three independent variables, (E_1, V_1, N_1), each term must be zero individually.

$$\frac{1}{T_1} - \frac{1}{T_2} = 0$$

$$\frac{p_1}{T_1} - \frac{p_2}{T_2} = 0$$

$$\frac{\mu_1}{T_1} - \frac{\mu_2}{T_2} = 0$$

This gives us the following result.

When two systems interacting thermally, mechanically, and diffusively are in equilibrium,

$$T_1 = T_2$$

$$p_1 = p_2$$

$$\mu_1 = \mu_2 \qquad (11.5)$$

D.   THE APPROACH TO EQUILIBRIUM

We now consider two systems interacting thermally, mechanically, and diffusively, but that are not yet in equilibrium. Since they are not yet in equilibrium, the entropy of the combined system will not yet have reached its maximum value, but rather it will be increasing.

$$\Delta S_0 > 0 \tag{11.6}$$

We can use this fact to show that in the approach to equilibrium, heat energy must flow from the hotter system to the cooler one, volume will be gained by the one under greater pressure at the expense of the other, and particles will tend to flow from the system of higher chemical potential toward the one of lower chemical potential (Figure 11.5)

We start with equation (11.3), and make the following substitution for ΔE_1 (since we are interested in flow of heat, volume, and particles).

$$\Delta E_1 = \Delta Q_1 - p_1 \Delta V_1 + \mu_1 \Delta N_1 \tag{11.7}$$

This gives us the following form of Eq. 11.3.

$$\Delta S_o = \left(\frac{1}{T_1} - \frac{1}{T_2}\right)\Delta Q_1 + \frac{1}{T_2}(p_1 - p_2)\Delta V_1 - \frac{1}{T_2}(\mu_1 - \mu_2)\Delta N_1 \tag{11.8}$$

If we treat (Q_1, V_1, N_1) as the three independent variables (i.e., any one can vary without either of the others changing at all), then we see that each of the three terms in Eq. 11.8 must individually be greater than zero if $\Delta S_o > 0$.

$$\left(\frac{1}{T_1} - \frac{1}{T_2}\right)\Delta Q_1 > 0$$

$$\frac{1}{T_2}(p_1 - p_2)\Delta V_1 > 0$$

$$-\frac{1}{T_2}(\mu_1 - \mu_2)\Delta N_1 > 0 \tag{11.9}$$

The implications of these three conditions as follows.

1. If $T_1 > T_2$, ΔQ_1 must be negative, and if $T_2 > T_1$, ΔQ_1 must be positive.
2. If $p_1 > p_2$, ΔV_1 must be positive, and if $p_2 > p_1$, ΔV_1 must be negative.
3. If $\mu_1 > \mu_2$, ΔN_1 must be negative, and if $\mu_2 > \mu_1$, ΔN_1 must be positive.

SUMMARY

Consider two interacting systems that are not yet in equilibrium. The requirement (second law) that the entropy of the combined system increase demands that the following occur.

1. If they are interacting thermally, then heat will flow from the hotter towards the cooler, and not vice versa.
2. If they are interacting mechanically, then volume is gained by the one having higher pressure at the expense of the other, and not vice versa.
3. If they are interacting diffusively, then particles flow from the one of higher chemical potential toward the one of lower chemical potential, and not vice versa.

E.   CHEMICAL POTENTIAL AND THE NUMBER OF ACCESSIBLE STATES

For systems interacting thermally, mechanically, and diffusively, the first law is written

$$dE = T\,dS - p\,dV + \mu\,dN \tag{10.2}$$

From this we see that

$$\mu = \frac{\partial E}{\partial N}\bigg)_{S,V}$$

That is, the chemical potential is a measure of how the internal energy of the system changes per particle added, when no heat is transferred nor work is done. In other words, it is the energy that a particle must have if it is to be transferred to a system without releasing thermal energy, and with no work being done on the system in the process.

We can rearrange the first law as follows.

$$dS = \frac{1}{T}\,dE + \frac{p}{T}\,dV - \frac{\mu}{T}\,dN \tag{11.1}$$

from which we see

$$-\frac{\mu}{T} = \frac{\partial S}{\partial N}\bigg)_{E,V}$$

or

$$-\mu = T\frac{\partial S}{\partial N}\bigg)_{E,V} = kT\frac{\partial \ln \Omega}{\partial N}\bigg)_{E,V} \tag{11.10}$$

Since systems tend to go toward configurations of higher entropy, systems will tend to attract new particles if $\partial S/\partial N$ is positive, and get rid of particles if $\partial S/\partial N$ is negative. Equivalently, systems will tend to attract particles if μ is negative and expel particles if μ is positive. Again, this is just a reflection that particles tend to go toward regions of low potential energy and away from regions of higher potential energy. In most cases, one system will be competing with another for the same particles, in which case the system with most negative chemical potential (i.e., greatest $\partial S/\partial N$) will win out.

In previous chapters we have seen that the number of accessible states is an *extremely* rapidly increasing function of the internal energy and of the volume. Here we will show that it is also an extremely rapidly increasing function of the number of particles.

Suppose ΔN energyless particles are added to a system at constant volume. According to Eq. 11.10, the number of accessible states increases by a factor given by

$$\Delta \ln \Omega = -\frac{\mu}{kT}\Delta N$$

Since

$$\Delta \ln \Omega = \ln \Omega_2 - \ln \Omega_1 = \ln \frac{\Omega_2}{\Omega_1}$$

we can express the *factor* by which the number of accessible states increases as

$$\frac{\Omega_2}{\Omega_1} = e^{-(\mu\,\Delta N/kT)} \tag{11.11}$$

SUMMARY

If ΔN particles of zero energy are added to a system held at constant volume, the number of states accessible to the entire system changes by a factor of

$$\frac{\Omega_2}{\Omega_1} = e^{-(\mu\,\Delta N/kT)} \tag{11.11}$$

EXAMPLE

Consider a system at room temperature for which $\mu = -3.0$ eV. By what factor does the number of accessible states increase when a single energyless particle is added?

For this problem we have $\mu = -3$ eV, $\Delta N = 1$, and $T = 300$ K. Putting these in Eq. 11.11 gives us

$$\frac{\Omega_2}{\Omega_1} = e^{116} = 2.4 \times 10^{50}$$

which is an amazing increase for just one single added particle.

F. SALTY WATER AND SNOWFLAKES

To illustrate the ideas developed in this chapter, we consider two specific examples: the dissolution of salt in water and the formation of snowflakes.

Imagine that we place a block of salt, NaCl, in a bucket of pure water. From our experience we know that immediately the water molecules begin eating away at the salt. The water gets saltier and the salt block grows smaller (Figure 11.6).

The reason for this is that even though the individual salt molecules are attracted quite strongly by the other salt molecules in the salt block, they are attracted even more strongly by the water molecules. In a salt block, the interactions between any one NaCl molecule and its neighbors cause it to be in a potential well of a depth of about 3 eV. So the chemical potential of a NaCl molecule in a salt block is

$$\mu_{\text{NaCl in salt block}} \approx -3.0 \text{ eV}$$

But the slightly stronger electrostatic interactions with neighboring water molecules means that the chemical potential of NaCl in water is even lower.

$$\mu_{\text{NaCl in water}} \approx -3.1 \text{ eV}$$

Since their electrostatic interactions with water molecules are stronger (equivalently, their chemical potential in water is lower), the salt molecules leave the salt block and enter the water environment.

But as the water gets saltier, more and more of the water molecules are tied up around the dissolved salt ions, and fewer are left to interact with new salt molecules (Figure 11.7). This means that the chemical potential of a salt molecule in the water is rising. When a new salt molecule enters the water, there are fewer water molecules available to interact with it. Hence, the potential well that the salt molecule falls into is shallower, and it gives up less heat on dissolution.

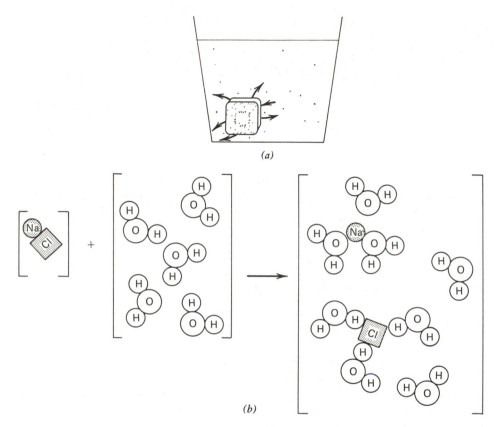

(a)

(b)

Figure 11.6   Interaction of salt block with water (a) from a molecular viewpoint (b). When the salt block is first placed in the water, the sodium and chlorine ions find themselves attracted more strongly by the water molecules than by the ions in the salt block. Consequently, their chemical potential is lower in the water, and they go into solution.

Eventually the water becomes salty enough that the chemical potential of a salt molecule in it has risen to the point where it is the same as that of a salt molecule in the salt crystal. At this point, the two systems are in "diffusive equilibrium." The salt block pulls just as hard on a nearby salt molecule as does the water, and so for every salt molecule dissolved, another will replace it on the salt block. The salt block gets no smaller, and the water no saltier.

As a second example, we consider how snowflakes are formed. Imagine that somewhere up in a cloud, a small, submicroscopic ice crystal condenses, and begins to float around through subfreezing, water-saturated air. Since the air is saturated, the chemical potential of a water molecule in it is nearly zero.

$$\mu_{\text{saturated air}} = 0$$

This just states that if you put a zero-energy water molecule into saturated air, the air molecules show little or no desire to hold onto it. (If they did, then the air wouldn't be saturated.) As a result, the water molecule falls into no potential well, and gives off no heat.

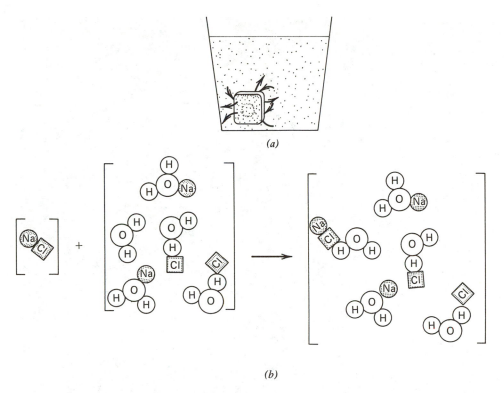

Figure 11.7 As the salt concentration increases, there are fewer water molecules available to join up with additional sodium or chlorine ions. Consequently, the depth of the electrostatic potential well experienced by these ions decreases as the salt concentration increases. When the chemical potential of the salt ions in water has risen to the point where it is equal to their chemical potential in the salt block, then the salt ions are attracted equally by both media. The rate that they go from the salt block into the water is equal to the rate that they return to the salt block, and there is no net flow.

The water molecules in an ice crystal, however, do hold onto other water molecules—that's what holds crystals together. If you put a perfectly cold water molecule on an ice crystal, the electrostatic forces of the nearby molecules allow it to fall into a potential well and give off heat. At a temperature of just below freezing, the chemical potential of a water molecule in an ice crystal is about

$$\mu_{\text{ice crystal}} - 0.6 \, \text{eV}$$

We know the water molecules will go from the higher chemical potential to the lower one. Consequently, as the small ice crystal floats around in the water-saturated air, it will grow. Because of the electrostatic configuration of a water molecule, the crystal grows faster in some directions than in others. This is responsible for the intricate designs of the large snowflakes we see (Figure 11.8).

Notice that the air doesn't have to be completely saturated for the snowflake to grow. As long as the relative humidity is high enough that the chemical potential of a water molecule in the air is still above that in the ice crystal, the snowflake will grow.

Figure 11.8 Because the chemical potential of a water molecule is lower in an ice crystal than it is in water-saturated (or nearly saturated) air, the molecules leave the air and join the crystal. (From *National Geographic Magazine*, Vol. 149, No. 3, March 1976. Photo by Robert F. Sisson.)

PROBLEMS

11-7. Consider two systems, A_1 and A_2, interacting mechanically but not thermally or diffusively. Show that:
   (a) $p_1 = p_2$ when they are in equilibrium.
   (b)   If they are not yet in equilibrium, volume is gained by the one having greater pressure.

11-8. Consider two systems interacting diffusively, but not mechanically or thermally. Show that:

   (a) $\mu_1 = \mu_2$ when they are in equilibrium.
   (b)   If they are not yet in equilibrium, then particles flow from the higher chemical potential toward the lower one.

11-9. Why does glass crystallize as it ages?

11-10. Consider a system of 10^{24} particles at room temperature for which $\mu = -1$ eV. By what factor does the number of accessible states increase if the number of particles is increased by 0.01% without adding energy to, or doing work on the system?

Figure 11.9 Some large crystals of NaCl ("table salt"). (about half size in this photo) What would happen if this were put in a bucket of fresh water? Why? Why doesn't the same thing happen in air? (American Museum of Natural History)

Figure 11.10 In a glass of ice water at $0°C$, there is no net flow of water molecules between the ice and the liquid water. Using this observation, what can you say about the relative values of the chemical potentials for water molecules in the two phases at $0°C$? (Keith Stowe)

11-11. Answer the question in the caption to Figure 11.9.

11-12. Answer the question in the caption to Figure 11.10.

PART 6

CONSTRAINTS

By this point in our study, we are familiar with the fundamental concepts underlying the statistical translation between the microscopic and macroscopic behaviors of large systems. We have been introduced to the three basic types of interactions between systems, and we have seen how the exchange of heat, work, or particles will influence some of the fundamental properties of systems.

These concepts, which have been presented in the preceding chapters, form the statistical basis of thermodynamics. Since these concepts can be applied to an infinite variety of problems, we cannot hope to anticipate all the types of problems that will be encountered during the careers of the students. The best we can do is to acquaint you with the fundamental concepts, which will be valid no matter what your particular problem of interest may be. We provide specific applications to types of systems of fairly wide general interest, in an attempt to illustrate the application of fundamental concepts, but these illustrative examples cannot possibly be complete. For most of the specific problems you encounter in your lives, you will have to go back to fundamentals, using specific examples such as those presented in this book only as guidelines for your particular work.

Fortunately, most of your various specific studies will have some things in common. Usually you will be interested in reducing the number of independent variables involved, in order to simplify your study. This can be done in many ways.

One way is through the use of specific models that can be used to find relationships among some of the independent parameters. Such models are as varied as is the range of systems to be studied, and we present only some of the more common ones to illustrate how they may be applied.

In addition, there are some natural constraints placed on the behavior of all processes. Some of these result from nature's desire to maximize entropy. Also, since there are more possible variables (e.g., T, S, p, V, μ, N, E, and others) than there are types of interactions, there are interrelationships among these parameters that enable us to write our expressions in terms of whatever set of variables we choose.

Finally, because of the nature of the system we are studying, our experimental setup, or our area of theoretical interest, we may wish to impose our own constraints on the process under study, such as doing it with a system of fixed size, fixed volume, at a fixed temperature, etc. Such imposed constraints simplify both the theoretical expressions and the experimental study.

An example of the application of constraints (some natural and some imposed) to a process of general interest is the study of engines and refrigerators, which is carried out in the final chapter of this section.

✳ chapter 12

MODELS

Each kind of system has certain characteristics. For example, gases tend to be more compressible than liquids, and liquids flow better than do solids. These characteristics can be expressed as some interrelationship among variables, such as how volume varies with pressure or number of particles, or how internal energy varies with entropy and volume. These interrelationships among variables are called "equations of state," and these are quite useful, because they allow us to transfer information from one set of variables to any other set that might be more appropriate for our particular problem or experimental setup. Since the properties of the system are due to the behaviors of the microscopic constituents, models for these microscopic constituents can be translated into equations of state for the macroscopic system.

So far we have learned that any thermodynamic property of a system depends on as many different parameters as there are kinds of interactions (Figure 12.1). For simplicity, we have represented all the various possible types of mechanical interactions as $p\,dV$, and all the various possible diffusive interactions by $\mu\,dN$,* and for this case, any one property will depend on three others. For example,

$$dE = T\,dS - p\,dV + \mu\,dN$$

$$dS = \frac{1}{T}(dE + p\,dV - \mu\,dN)$$

etc.

Frequently, we deal with systems of fixed size, such as an iron bar or a bucket of water, for which the number of particles does not vary. For these nondiffusive systems, any one property depends on two others, rather than three.

For each system, then, there must be some definite interrelationship among the various parameters. An equation expressing such an interrelationship is called an "equation of state." We are frequently quite interested in knowing an equation of state for whatever system we are studying because it facilitates our analysis and understanding.

An equation of state could express the interrelationship among any set of variables. Frequently, equations of state involve easily measured quantities such as temperature, pressure, and volume, but in principle there is no reason why they couldn't interrelate entropy, temperature, and internal energy, or any other such set of parameters.

---

* These should more properly be given by $\sum_i \mathcal{F}_i\,dx_i$ and $\sum_i \mu_i\,dN_i$, respectively.

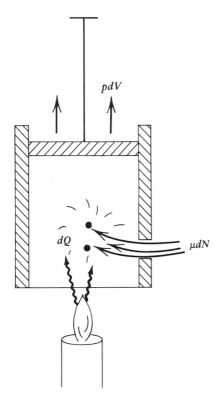

Figure 12.1 There are three different types of interactions, which means that any property of any system engaging in these interactions depends on three independent variables. If more than one type of particle is entering or leaving the system, or if more than one type of work is being done on the system, then there are correspondingly more independent variables.

Through the use of idealized models, we can frequently get a qualitative understanding of how real systems work, and from these models we can derive the interdependence of various parameters. As a check on the validity of our models, we can go into the laboratory and see if the interdependence among the parameters is indeed what the model predicts. That is, the ultimate tests of our understanding are made in the laboratory.

It is important to realize that equations of state must exist, whether or not we are clever enough to discover what they are. There are a great deal more parameters used to describe any system than there are types of interactions that it engages in, so there must be a great deal of interdependence among them. From our point of view, the discovery of an equation of state places one more constraint on the behavior of a system, but in reality that constraint exists whether we discover it or not.

A. IDEAL GAS

The "ideal gas" is one whose molecules behave as infinitesimally small, perfectly elastic "billiard balls." They do not interact with each other at a distance so their energies are entirely kinetic, with the energy of one molecule of mass m and momentum p being given by

$$\varepsilon = \frac{p^2}{2m} = \frac{1}{2m} p_x^2 + \frac{1}{2m} p_y^2 + \frac{1}{2m} p_z^2 \tag{12.1}$$

Each molecule, then, has 3 degrees of freedom, and a gas of N molecules has $3N$ degrees of freedom altogether.

As we saw in Chapter 2, the number of quantum states available to a particle confined to some region in six-dimensional space is proportional to the size of that region.

$$\Omega_{1 \text{ particle}} = \frac{\Delta x \, \Delta y \, \Delta z \, \Delta p_x \, \Delta p_y \, \Delta p_z}{h^3}$$

$$= (\text{const}) \Delta x \, \Delta y \, \Delta z \, \Delta p_x \, \Delta p_y \, \Delta p_z \tag{2.12}$$

Since the number of different states available to a system of N elements is the product of the number of states available to each (Eq. 7.4), we have

$$\Omega_{N \text{ particles}} = \prod_{i=1}^{N} \Omega_i = \prod_{i=1}^{N} \left(\frac{1}{h^3} \right) \int dx_i \, dy_i \, dz_i \, dp_{i_x} \, dp_{i_y} \, dp_{i_z} \tag{12.2}$$

where the limits on the integrals are determined by the gas being confined to some volume V, and the total energy of the gas is given by E. Doing the volume integral in Eq. 12.2 we have

$$\Omega_{N \text{ particles}} = \left(\frac{1}{h^{3N}} \right) V^N \int \prod_{i=1}^{N} dp_{i_x} \, dp_{i_y} \, dp_{i_z} \tag{12.3}$$

Since

$$p_{1_x}^2 + p_{1_y}^2 + p_{1_z}^2 + p_{2_x}^2 + \cdots + p_{N_z}^2 = 2mE$$

the integral over the momenta in Eq. 12.3 is equivalent to that over the surface of a $3N$-dimensional sphere* of radius $\sqrt{2mE}$. The surface area of a $3N$-dimensional sphere is proportional to the radius raised to the $(3N - 1)$th power, so the number of states available to an ideal gas of N particles is given by (ignoring 1 in comparison to $3N$)

$$\Omega_{N \text{ particles}} \propto V^N (\sqrt{2mE})^{3N}$$

If we absorb the factor of $(2m)^{3N/2}$ into the constant of proportionality, we have for the number of states available to an ideal gas of N particles,

$$\Omega_{\text{ideal gas}} = (\text{const}) V^N E^{3N/2} \tag{12.4}$$

Since the entropy of a system is defined as $k \ln \Omega$, the entropy of an ideal gas is

$$S_{\text{ideal gas}} = k \ln \Omega_{\text{ideal gas}} = (\text{const}) + Nk \ln V + \frac{3}{2} Nk \ln E \tag{12.5}$$

---

* The surface area of a two-dimensional sphere is given by $\int_S dx \, dy = (2\pi)r$ where $x^2 + y^2 = r^2$. That of a three-dimensional sphere is given by $\int_S dx \, dy \, dz = (4\pi)r^2$, where $x^2 + y^2 + z^2 = r^2$, etc.

Now we can apply the relationships*

$$\left.\frac{\partial S}{\partial E}\right)_V = \frac{1}{T} \tag{9.3}$$

$$\left.\frac{\partial S}{\partial V}\right)_E = \frac{p}{T} \tag{10.5}$$

to the above result (12.5) to get

$$\frac{1}{T} = \frac{\frac{3}{2}Nk}{E}$$

and

$$\frac{p}{T} = \frac{Nk}{V}$$

These are more commonly written as follows.

For an ideal gas, we have the following equations of state:

$$E = \tfrac{3}{2}NkT \tag{12.6}$$

$$pV = NkT \tag{12.7}$$

The second of these is called the "ideal gas law."

The first of these relationships we already know to be true, because we know from the equipartition theorem that an average energy of $\tfrac{1}{2}kT$ is associated with each of the $3N$ degrees of freedom. The internal energy does not depend on the volume at all, because the particles don't interact at a distance and so their average separation doesn't matter.

The second of the above relationships, the ideal gas law, agrees with our physical intuition. We can consider the pressure exerted on the walls of the container by the gas molecules colliding with it, as illustrated in Figure 12.2. According to the ideal gas law, this pressure varies with temperature and volume according to

$$p = Nk\frac{T}{V}$$

This indicates that if the temperature is held constant, then the pressure increases as the volume decreases, reflecting the increased frequency of collision between each molecule and the container walls. Furthermore, if the volume is held constant, the pressure increases with temperature, reflecting both greater collisional frequency and greater impulse delivered per collision when the molecules move at these higher velocities.

---

* These were derived from the first law, $T\,dS = dE + p\,dV$, or $dS = (1/T)\,dE + (p/T)\,dV$, and the observation that $dS = \partial S/\partial E)_V\,dE + \partial S/\partial V)_E\,dV$. The number of particles (N) is fixed.

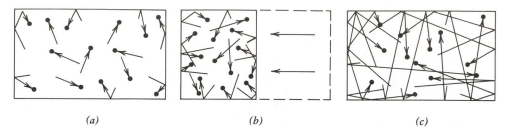

Figure 12.2   (a) The pressure in a gas is caused by the collisions of the molecules with the walls of the container. (b) If the volume of the container is smaller, the density of the molecules is larger, and collisions with the walls are therefore more frequent. (c) If the temperature is increased, the molecules move faster, thereby increasing both the frequency and the forcefulness of the collisions.

PROBLEMS

12-1.   Consider a two-dimensional ideal gas, confined to an area A in the x-y plane. The number of quantum states available to a single molecule confined to region $\Delta x \, \Delta y$ and having momentum in the range defined by $\Delta p_x \, \Delta p_y$ is given by $\Delta x \, \Delta y \, \Delta p_x \, \Delta p_y / h^2$.

  (a)   How many degrees of freedom does each molecule have?
  (b)   How many degrees of freedom does a system of N molecules have?
  (c)   What is the total internal energy of a system of 10^{22} molecules at a temperature of 300 K?
  (d)   Following a development similar to that in Eqs. 12.2 through 12.4, find the number of quantum states available to a system of N particles, constrained to be within area A and to have total internal energy E. (That is, write Ω as a function of A and E.)

12-2.   Consider a one-dimensional ideal gas, confined to a length L along the x-axis.

  (a)   How many degrees of freedom does a system of N molecules have?
  (b)   Following a development similar to that in Eqs. 12.2 through 12.4 find the number of quantum states available to a system of N molecules, constrained to be within region L, and to have total internal energy of E.
  (c)   From the answer to part (b), write the entropy of the system in terms of L and E. (See Eq. 12.5 for comparison.)
  (d)   For a one-dimensional system, the work done by it in changing its length by dL is $dW = F \, dL$. (For three-dimensional systems we had $dW = p \, dV$.) Write dE as a function of dS and dL, and write dS as a function of dE and dL.
  (e)   Derive an "ideal gas law" for this one-dimensional system from $F/T = \partial S/\partial L)_E$, using the expression for S you derived in part (c).

12-3.   The number of quantum states accessible to a certain system of N particles is given by

$$\Omega = (\text{const}) e^{\alpha V^{4/5}} E^{2N}$$

where α is a constant.

  (a)   Write an expression for the entropy in terms of V and E.
  (b)   Using $1/T = \partial S/\partial E)_V$, find how the internal energy depends on the temperature.
  (c)   How many degrees of freedom does this system have?
  (d)   Using $p/T = \partial S/\partial V)_E$, find the interrelationship between p, V, and T.

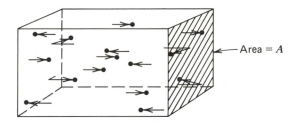

Figure 12.3 The ideal gas law can be derived by considering the pressure exerted on an end wall of a container that holds N gas molecules each moving with the single speed v_x in either the $+x$ *or* $-x$ direction. Once the answer is obtained for this hypothetical case, we can average over all speeds, using $\frac{1}{2}mv_x^2 = \frac{1}{2}kT$. (Area of wall $= A$, volume of container $= V$.)

12-4. Calculate the isothermal compressibility of 47 liters of CO_2 gas at room temperature (300 K) and atmospheric pressure (10^6 dynes/cm^2), assuming it obeys the ideal gas law reasonably well.

12-5. Suppose a container of volume V has N gas molecules of mass m, half of which are moving in the $+x$-direction with speed v_x. Consider the elastic collisions of these molecules with the wall of the container in the y-z plane, having area A (Figure 12.3).

   (a) In terms of N and V, what is the density of particles moving in the $+x$-direction? (*Ans. N/2V.*)

   (b) How many molecules collide with the wall per second? [*Ans.* $(N/2V)Av_x$.]

   (c) How much impulse is given the wall per collision? (*Ans.* $2mv_x$.)

   (d) What is the average force exerted on the wall altogether?

   (e) What is the average pressure exerted on the wall?

   (f) Now suppose there were a distribution of molecular speeds so that we had to average over v_x^2. Rewrite the answer to (e), using the equipartition theorem to express v_x^2 in terms of T. (If your answer isn't the ideal gas law, you made a mistake somewhere.)

B. REAL GASES

When dealing with real gases, we may find it necessary to modify one or both of the relationships (12.6) and (12.7). For example, the molecules of real gases may have more than 3 degrees of freedom each. In addition to translational degrees of freedom, they may also be able to store energy in rotational, vibrational or other motions (Figure 12.4). Since an average energy of $\frac{1}{2}kT$ is stored in each degree of freedom, the relationship (12.6) would have to be modified to read

$$E = \frac{v}{2}NkT \tag{12.8}$$

where v is the number of degrees of freedom per molecule.

    In addition, our model for ideal gases assumed that the molecules were infinitesimal, and did not interact with each other at a distance. This may be an adequate

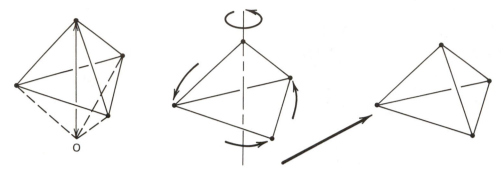

Figure 12.4 Many molecules may have several vibrational and rotational degrees of freedom in addition to their translational degrees of freedom.

model for real gases when their densities are low enough that their molecular sizes are small compared to their spacings, and their mutual interactions can be ignored. But at higher concentrations, where their mutual interactions are *not* negligible, and where their molecular sizes are *not* small in comparison to their separations, we would expect the ideal gas model to be inadequate.

We begin our modification of Eq. 12.7 by rewriting it in terms of the molar volume,* v, and the gas constant, $R = N_A k$.

$$pv = RT \qquad (12.7')$$

According to this relationship, it is possible to squash a gas into an arbitrarily small volume, simply by making the pressure large enough!

For real gases, this cannot be true. Once we have compressed the gas to the point where the molecules are "touching" each other, as in a liquid, we cannot hope to compress it further. If b is the molar volume of the material in its liquid phase, then the molar volume cannot be less than b. So the necessary limit on the molar volume of a real gas can be incorporated into the above relationship (12.7') through the replacement of v by $v - b$.

$$v \rightarrow v - b$$

An alternative way of understanding this change is to see that the actual volume available to any one molecule is the total volume (v) minus the volume occupied by all the other molecules (b).

Another deviation in the behavior of the molecules of a real gas from the perfectly elastic billiard ball idea, is that they mutually interact at a distance with a slightly attractive force. This is the cause of their condensation at sufficiently low temperatures. This attraction means that there is a force tending to hold the molecules together, in addition to that exerted by the walls of the container. So the "effective" pressure, which confines these molecules, is greater than that which would be determined from a pressure gauge. This means that the factor of p in Eq. 12.7' would have to be modified.

---

* A "mole" is Avogadro's number of molecules. The "molar volume" is the volume occupied by one mole.

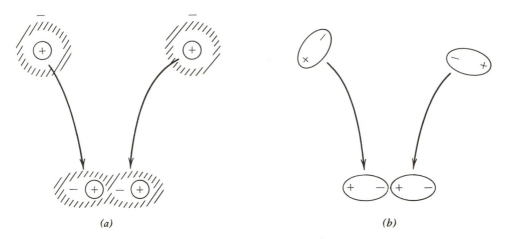

Figure 12.5 Electrostatic attraction between neighboring electrically neutral gas molecules can be accomplished in two basic ways. (a) If the charge distribution on the molecules is originally symmetric, then as they approach each other, electrostatic forces can cause slight distortions of their charge distributions, with the shift being in such a way as to bring unlike charges closer together and like charges farther apart. (b) If the gas molecules are already polarized, then as they approach each other they can rotate in such a way as to bring unlike charges closer together.

$$p \rightarrow (p + \text{mutual attraction})$$

This mutual attraction is usually electrostatic, and is caused by induced charge polarizations as two molecules approach each other* (Figure 12.5). It can be shown to be inversely proportional to the sixth power of the average molecular separation, or equivalently the square of the molar volume ($1/r^6 \propto 1/v^2$).

These two effects are incorporated into a modification of the ideal gas law, called the "van der Waals equation of state" for a real gas. If v is the molar volume of the gas, then

$$\left(p + \frac{a}{v^2}\right)(v - b) = RT \qquad (12.9)$$

where the parameters a and b depend on the particular gas, b being the molar volume of the liquid state and a being related to the polarizability of the gas molecules.

We see that the two modifications are the most significant when the molar volume is small. For more rarified real gases, $v \gg b$ and $p \gg a/v^2$, so the modifications can be ignored and the ideal gas law is good enough.

---

* In many cases the molecules are inherently polarized, but have random orientations. The "induced polarization" is then a matter of two dipoles tending to line up as they approach each other.

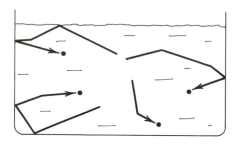

Figure 12.6 In some applications, the molecules in liquids can be treated as molecules in a gas, confined by the boundaries of the container and the surface of the liquid.

C. LIQUIDS

A good model for liquids does not yet exist. For some calculations, the liquid can be thought of as a gas, since molecules are free to roam throughout the volume defined by the liquid's surface and the boundaries of the container (Figure 12.6). Because of interactions with neighbors, a molecule moves considerably more sluggishly in a liquid than it would in the gas phase, and so it is given an "effective mass," m^*, which is much larger than its real mass. Even with this modification, the gas model is quite inadequate for most purposes. For example, we know the model leads to the equation of state $pv = RT$, which implies that at any temperature the volume can be halved simply by doubling the applied pressure. This is absurd.

For some purposes, the empirical van der Waals equation of state is adequate.

$$\left(p + \frac{a}{v^2}\right)(v - b) = RT$$

If the parameter b is thought of as some minimum volume for the liquid state, then for molar volumes near b, very large increases in the pressure would be required to cause very small changes in v. We know this to be the case for liquids.

The pressure term $(p + a/v^2)$ cannot be exactly correct, because at the small intermolecular distances characteristic of liquids, many other types of interactions are important besides those due to electrostatic charge polarizations. On the other hand, the molar volume of a liquid changes only very little, always being near the value $v \approx b$, so whatever the dependence of these other interactions have on molecular separations, it will be nearly constant in the liquid phase. That is, we should make the replacement

$$\left(p + \frac{a}{v^2}\right) \rightarrow \left(p + \frac{a}{v^2} + f(v)\right)$$

where $f(v)$ is due to other interactions. But since $v \approx b =$ constant for liquids, we have

$$\frac{a}{v^2} + f(v) \approx \frac{a}{b^2} + f(b) = \text{constant}$$

so the only effect these other interactions would have is to alter the constant in the pressure term, making it somewhat different from that which would result from the electrostatic polarization interaction alone.

Consequently, we would expect the van der Waals equation might be useful in acquiring some qualitative insight into the behavior of liquids. We find it especially useful in understanding gas-liquid phase transitions, since it can be applied to both phases. For quantitative calculations, however, it is only approximate at best, and the value of the parameter a would differ from that used for the gaseous phase.

Although our studies of liquids could be greatly improved if someone could come up with a good model, we are quite fortunate that, with the exception of water, most common systems we encounter on earth are solids, gases, or other systems for which we do have models. In addition, we have the good fortune that many of the results obtainable from statistical mechanics are very basic and model independent. Indeed, all of the results we've derived in previous chapters are true for all systems, no matter what they are.

D.   Solids

The distinguishing feature of solids is that the individual atoms are not free to roam about within the boundaries of the material, but are rather confined to very small regions of space by electromagnetic interactions with neighbors. An atom would normally be found near the bottom of the potential well caused by these interactions. But with thermal agitation we would expect some oscillations in the atom's position across this equilibrium position.

When an atom is displaced a small distance (x) from its equilibrium position, the restoring force is directly proportional to the amount of displacement $(F = -kx)$, which means that the potential energy is given by (See Section G, Chapter 2 or Figure 27.1.)

$$U = \frac{1}{2} kx^2 \tag{12.10}$$

where the force constant k is given by

$$k = \frac{\partial^2 U}{\partial x^2}\bigg|_{x=0} \tag{12.11}$$

Because the stacking of atoms may be different in some directions than others, the force constant, k, may be different in different directions, but with an appropriate transformation of variables (See homework Problem 12-7) this can always be made into the form

$$V(x, y, z) = \frac{1}{2} k(x^2 + y^2 + z^2) = \frac{1}{2}kr^2 \tag{12.12}$$

Each atom is a three-dimensional harmonic oscillator, having 6 degrees of freedom.

Because of the coupling between atoms, it is sometimes convenient to think of a solid as being a lattice of atomic masses coupled by springs (Figure 5.2). The spring constants are determined by the strength of the electromagnetic interactions among neighboring atoms according to Eq. 12.11. This model turns out to be particularly convenient when energies are small and quantum effects important.

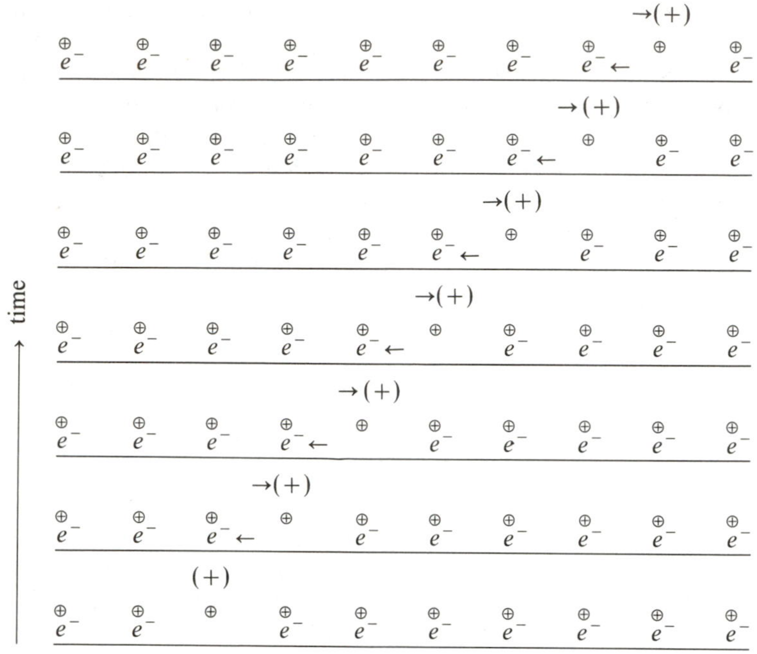

→(+)

⊕ ⊕ ⊕ ⊕ ⊕ ⊕ ⊕ ⊕ ⊕ ⊕
e⁻ e⁻ e⁻ e⁻ e⁻ e⁻ e⁻ e⁻←   e⁻

→(+)

⊕ ⊕ ⊕ ⊕ ⊕ ⊕ ⊕ ⊕ ⊕ ⊕
e⁻ e⁻ e⁻ e⁻ e⁻ e⁻ e⁻←   e⁻ e⁻

→(+)

⊕ ⊕ ⊕ ⊕ ⊕ ⊕ ⊕ ⊕ ⊕ ⊕
e⁻ e⁻ e⁻ e⁻ e⁻ e⁻←   e⁻ e⁻ e⁻

→(+)

⊕ ⊕ ⊕ ⊕ ⊕ ⊕ ⊕ ⊕ ⊕ ⊕
e⁻ e⁻ e⁻ e⁻ e⁻←   e⁻ e⁻ e⁻ e⁻

→(+)

⊕ ⊕ ⊕ ⊕ ⊕ ⊕ ⊕ ⊕ ⊕ ⊕
e⁻ e⁻ e⁻ e⁻←   e⁻ e⁻ e⁻ e⁻ e⁻

→(+)

⊕ ⊕ ⊕ ⊕ ⊕ ⊕ ⊕ ⊕ ⊕ ⊕
e⁻ e⁻ e⁻←   e⁻ e⁻ e⁻ e⁻ e⁻ e⁻

(+)

⊕ ⊕ ⊕ ⊕ ⊕ ⊕ ⊕ ⊕ ⊕ ⊕
e⁻ e⁻   e⁻ e⁻ e⁻ e⁻ e⁻ e⁻ e⁻

time →

Figure 12.7 When an electron moves in one direction to fill up a vacancy, it leaves a vacancy behind. The sequential motion of electrons in one direction to fill a vacancy results in the vacancy, with its net positive charge, moving sequentially in the opposite direction.

When one atom starts vibrating, it can send waves down the lattice. This vibrational wave, then, is not fixed to one lattice point, but rather may travel freely throughout the solid, being constrained only by the boundaries and possible quantum effects. The quantum of energy in such an elastic wave is called a "phonon," and because these phonons travel throughout the solid's volume, they can often be treated as a gas—a "phonon gas."*

A particular subgroup of solids are the metals, and metals contain another important component. In addition to the lattice of harmonic oscillators, or the phonon gas, a metal also has conduction electrons. The range of energies of these electrons (called the "conduction band") is sufficiently high that these electrons are not tied to individual atoms, but rather the conduction electrons are mutually shared by all atoms. Because of their mobility, these "free" electrons are responsible for the high electrical conductivities of metals, and they can often be thought of as an "electron gas," which is confined only by the boundaries of the metals. In a typical metal, each atom contributes one or two electrons to the conduction band, so the number of conduction electrons in this electron gas is typically one or two times the number of atoms. Electrons that

* Do not confuse this with a "photon gas," which we will study later. "Photons" refer to electromagnetic waves propagating through space, whereas "phonons" refer to vibrational waves propagating through solids.

remain bound to the atoms are called "valence electrons." They occupy a lower range of energies called the "valence band."

Semiconductors may also have conduction electrons, but they are typically many orders of magnitude fewer than those found in metals of comparable sizes. Conduction in semiconductors may also be accomplished by vacancies or "holes" in the valence band. An electron migrating from one atom to fill up the vacancy in the next, leaves a vacancy behind, and an atom which is missing one electron has a positive charge, $+e$. Consequently, the migration of valence electrons in one direction to fill up vacancies on neighboring atoms, is equivalent to the backwards migration of positively charged "holes." (See Figure 12.7.)

Since models for solids frequently have more than one component, it is often helpful to simplify our studies by concentrating on one property of one component at a time. For example, we may wish to study the heat capacity of the lattice alone, or the magnetic moment of the electron gas alone, etc.

PROBLEMS

12-6. At standard temperature and pressure, a mole of liquid water occupies a volume of $18 \ cm^3$, and a mole of water vapor occupies 22.4 liters.

    (a) Find the characteristic dimension of a single water molecule from the volume of a mole of the liquid.

    (b) Roughly what is the characteristic separation of molecules in the water vapor?

    (c) In the gas phase, how many times larger is the characteristic separation than the molecular size?

12-7. Consider an atom in some solid that is anchored in place by electrostatic interactions with its neighbors. For small displacements from its equilibrium position, the restoring force is directly proportional to its displacement,

$$F_x = -k_1 x \qquad F_y = -k_2 y \qquad F_z = -k_3 z$$

where the force constants in different dimensions are different. Suppose you wanted to make a coordinate transformation so the potential energy could be written as

$$V = \frac{1}{2} k_1 (x'^2 + y'^2 + z'^2)$$

What are (x', y', z') in terms of (x, y, z)?

12-8. If we could consider the conduction electrons in a metal to be an ideal gas, what would be the root mean square speed of such an electron at room temperature? (*Hint.* In an ideal gas the average energy per particle is given by E/N, where $E = \frac{3}{2} NkT$.)

12-9. The equipartition theorem tells us that the average energy per degree of freedom is $\frac{1}{2}kT$.

    (a) With this, calculate the root mean square speed of a gold atom vibrating around its equilibrium position in a gold wire at room temperature.

    (b) How does this compare to the root mean square speed of a conduction electron in this wire? (Assume the conduction electrons behave like an ideal gas.)

(c)   How does the average kinetic energy of a gold atom compare with the average kinetic energy of a conduction electron in this wire?

E.   OTHER COMMON SYSTEMS

In addition to the solid, liquid, and gas phases of substances, there are many other kinds of systems to which we may wish to apply the powerful tools of statistical mechanics and thermodynamics. Each field of study has its own set of systems of interest, some of which are common to several different fields.

   In the succeeding chapters we will frequently use models to illustrate the applications of various techniques or tools, or to help better understand certain kinds of systems. Sometimes we will use models already discussed, but in other cases we will wish to develop models for other common or interesting kinds of systems. These include the following models.

1.   Engines and refrigerators (Chapter 15).
2.   Magnetic properties of materials (Chapter 20).
3.   Chemical interactions (Chapter 22).
4.   Phase transitions (Chapter 23).
5.   Photon gases (Chapter 26).
6.   Thermal properties of solids (Chapter 27).
7.   Electrical properties of semiconductors and insulators (Chapter 28).
8.   Superfluidity and superconductivity (Chapter 29).

F.   SAMPLE APPLICATIONS

We now examine a few specific examples of how models help us in our thermodynamical studies.

F.1   Heat Capacities

Consider a system of a fixed number of particles (N = constant). According to the first law, the heat energy transferred to the system is given by

$$dQ = dE + p\,dV \tag{12.13}$$

If each particle has v degrees of freedom, then the system as a whole has vN degrees of freedom, and its internal energy is given by

$$E = \frac{v}{2} NkT + N\mu$$

according to the equipartition theorem. With this, the expression for the heat transfer (12.14) becomes

$$dQ = \frac{v}{2} Nk\,dT + N\,d\mu + p\,dV$$

Unless the system is undergoing a phase change, the change in the zero-energy reference level ($N\,d\mu$) is usually very small compared to the other two terms when heat is added to a system. Ignoring this term, we can write an expression for the heat added per mole (dq) as

$$dq = \frac{v}{2} R\, dT + p\, dv \tag{12.14}$$

where dv is the change in molar volume, and the gas constant R is given by

$$R = N_A k$$

The molar heat capacity at constant volume is given by

$$c_V = \left(\frac{\partial q}{\partial T}\right)_V = \frac{v}{2} R \tag{12.15}$$

and that at constant pressure is given by*

$$c_p = \left(\frac{\partial q}{\partial T}\right)_p = \frac{v}{2} R + p\left(\frac{\partial v}{\partial T}\right)_p \tag{12.16}$$

or

$$c_p = c_V + p\left(\frac{\partial v}{\partial T}\right)_p \tag{12.17}$$

> For a system of particles having v degrees of freedom each, the molar heat capacities are given by
>
> $$c_V = \frac{v}{2} R \tag{12.15}$$
>
> $$c_p = c_V + p\left(\frac{\partial v}{\partial T}\right)_p \tag{12.17}$$

The interesting thing to notice in the result (12.15) is that the molar heat capacity depends on the number of degrees of freedom per molecule. That is, the study of a *macroscopic* property of a system will reveal its *microscopic* structure!

In Eq. 12.17 we see that the difference in the two molar heat capacities depends on how the volume of a system at constant pressure changes as its temperature increases.

$$c_p - c_V = p\left(\frac{\partial v}{\partial T}\right)_p \tag{12.17}$$

For gases, we expect this difference to be large, because large changes in volume accompany large changes in temperature.

For liquids and solids, the changes in volume are much less dramatic. In fact very large increases in temperature result only in very minor changes in volume. So for liquids and solids the factor $p(\partial v/\partial T)_p$ is almost negligible and the molar heat capacities c_p and c_V are very nearly the same. (See Table 12.1.) In fact, in most cases we don't even bother to distinguish between the two unless the system we are studying is a gas.

* To make (T, p) the two independent variables in Eq. 12.14, you must first write $dv = (\partial v/\partial T)_p\, dT + (\partial v/\partial p)_T\, dp$.

Table 12.1   Difference Between Molar Heat Capacities at Constant Volume and at Constant Pressure for Various Materials at 0°C (in units of the gas constant, R)

Material	$(c_p - c_V)/R$
Gases	
Air	1.00
Ammonia	1.05
Carbon dioxide	1.02
Water vapor	1.14
Liquids	
Ethyl alcohol	0.70
Water (at 20°C)	0.045
Water (at 0°C)	−0.014
Solids	
Aluminum	0.0068
Brass	0.0050
Diamond	0.00015
Ice	0.027
Gold	0.0052

EXAMPLE

Calculate the difference between c_p and c_V for an ideal gas.
   For an ideal gas

$$pv = RT$$

so

$$p\,dv + v\,dp = R\,dT$$

or

$$p\,dv = R\,dT - v\,dp$$

Consequently,

$$p\left.\frac{\partial v}{\partial T}\right)_p = R$$

and

$$c_p - c_V = R \tag{12.18}$$

EXAMPLE

Calculate the difference between c_p and c_V for a dense gas, where the ideal gas law is no longer valid.
   For a dense gas, we must use the van der Waals equation,

$$\left(p + \frac{a}{v^2}\right)(v - b) = RT$$

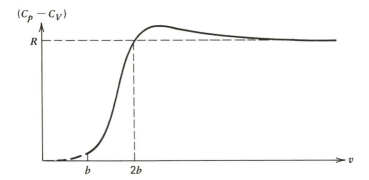

Figure 12.8 Plot of $(c_p - c_V)$ versus molar volume for a van der Waals gas. The ideal gas value, $(c_p - c_V) = R$ is attained at molar volumes large compared to b.

where a and b are constants whose values depend on the particular gas. The differential form of this is

$$\left(dp - \frac{2a}{v^3}\,dv\right)(v - b) + \left(p + \frac{a}{v^2}\right)dv = R\,dT$$

which can be written as

$$p\,dv = \frac{R\,dT - (v - b)\,dp}{\{1 + (a/pv^2)[2(b/v) - 1]\}} \tag{12.19}$$

The difference between c_p and c_V is then given by*

$$c_p - c_V = p\left.\frac{\partial v}{\partial T}\right)_p = \frac{R}{\{1 + (a/pv^2)[2(b/v) - 1]\}} \tag{12.20}$$

There are two things that should be noticed in this result. (See Figure 12.8.) First, when the molecular separations are large, so that

$$p \gg \frac{a}{v^2}$$

and

$$v \gg b$$

we get the ideal gas result, as expected.

$$c_p - c_V = R$$

Second, in the other limit, where the density corresponds to that of the liquid phase, we have

$$p \ll \frac{a}{v^2}$$

---

* If you do not understand how the result (12.20) follows from (12.19), see the footnote on page 207. Also, we are assuming here that we are not at a phase transition, so that any change in the zero-energy reference level $(N\mu)$ is negligibly small. (See page 147.)

and

$$v \approx b$$

so that the denominator in the expression (12.20) becomes very large and

$$c_p - c_V \approx \frac{R}{a/pv^2} \approx 0$$

That is, for the liquid state, the difference between c_p and c_V becomes very small, as we expected.

PROBLEMS

12-10. Steam in a pressure pipe is under very high pressure ($p = 100$ atm) such that its molar volume is small; $v = 0.3$ liters. The van der Waals constants for steam are

$$a = 5.5 \left(\frac{\text{liters}^2}{\text{mole}^2} \text{ atm} \right)$$

$$b = 0.030 \text{ (liters/mole)}$$

What is the difference, $c_p - c_V$, for this gas? (*Hint*. If v is the molar volume, $V = v$. Use Eq. 12.17, where $\partial v / \partial T)_p$ can be gotten from the van der Waals equation.)

12-11. The molar heat capacity at constant volume for some rarified gas is measured to be

$$c_V = 33.3 \ \frac{\text{J}}{\text{mole} - {}^\circ\text{C}}$$

(a)   Does it have monatomic or polyatomic molecules?
(b)   How many degrees of freedom does one molecule have?

12-12. The molar volume of copper is 7.0 cm^3, and its coefficient of thermal volume expansion is

$$\frac{1}{V} \frac{\partial V}{\partial T}\bigg)_p = 4.2 \times 10^{-5}/\text{K}$$

Each copper atom has 6 degrees of freedom.

(a)   At room temperature and atmospheric pressure (10^6 dynes/cm^2), by what percent do you expect c_p and c_V to differ?
(b)   Which is bigger?

F.2   Isothermal Compressibilities

The isothermal compressibility of a system is defined by

$$\kappa \equiv -\frac{1}{V} \frac{\partial V}{\partial p}\bigg)_T \tag{10.14}$$

which can equivalently be expressed in terms of molar volumes

$$-\frac{1}{V}\frac{\partial V}{\partial p}\bigg)_T = -\frac{1}{v}\frac{\partial v}{\partial p}\bigg)_T \tag{12.21}$$

From our everyday experiences, we know that gases are rather compressible, and solids and liquids are not.

EXAMPLE

Find an expression for the isothermal compressibility of an ideal gas.
For an ideal gas

$$pv = RT$$

or

$$p\,dv = R\,dT - v\,dp$$

From this, we see

$$p\frac{\partial v}{\partial p}\bigg)_T = -v$$

or

$$\kappa = -\frac{1}{v}\frac{\partial v}{\partial p}\bigg)_T = \frac{1}{p} \tag{12.22}$$

According to this result, the higher the pressure (or the denser the gas) the less compressible it is. That makes sense.

EXAMPLE

Find an expression for the isothermal compressibility of a gas that is so dense that the ideal gas law is no longer valid.
For a dense gas, we must use the van der Waals equation as we did in deriving Eq. 12.19. From this result, we see*

$$\frac{1}{v}dv = \frac{R\,dT - (v - b)\,dp}{pv\{1 + (a/pv^2)[2(b/v) - 1]\}}$$

or

$$\kappa = -\frac{1}{v}\frac{\partial v}{\partial p}\bigg)_T = \left(\frac{1}{p}\right)\frac{1 - (b/v)}{\{1 + (a/pv^2)[2(b/v) - 1]\}} \rightarrow \begin{cases} 1/p & \text{for } v \gg b \\ 0 & \text{for } v \approx b \end{cases} \tag{12.23}$$

In the limit of low densities ($v \gg b$, $p \gg a/v^2$) this reduces to the ideal gas result, but in the liquid limit ($v \rightarrow b$, $p \ll a/v^2$) this becomes vanishingly small. Although we don't expect the van der Waals equation to be exactly correct quantitatively, it does show the qualitative feature that the compressibility of a liquid is much smaller than that of a gas. (See Figure 12.9.)

---

* If x is a function of two variables $x = x(y, z)$, and if $dx = f\,dy + g\,dz$, then since $dx = dx/dy)_z dy + dx/dz)_y dx$, we see that $dx/dy)_z = f$ and $dx/dz)_y = g$.

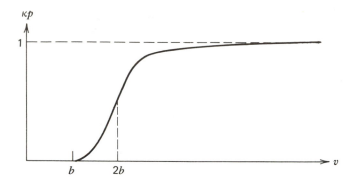

Figure 12.9   Plot of the product of isothermal compressibility times pressure as a function of the molar volume for a van der Waals gas. The ideal gas value, $\kappa p = 1$, is attained at molar volumes large compared to b.

PROBLEMS

12-13.   Suppose the equation of state for some system was

$$p^2 T^{-1/3} e^{aV} = b$$

where a and b are constants.

(a)   Write this in differential form, expressing dV in terms of dT and dp.

(b)   Express the isothermal compressibility of this system in terms of (T, V, p).
$\kappa = -(1/V)\,\partial V/\partial p)_T$

(c)   Express the coefficient of volume expansion for this system in terms of (T, V, p).
$\beta = (1/V)\,\partial V/\partial T)_p$

12-14.   Consider a system that behaves according to the van der Waals equation of state, with appropriate values of the constants a and b.

(a)   Write this in differential form, expressing dv in terms of dT and dp.

(b)   Show that the isothermal compressibility of this system is given correctly by Eq. 12.23.

(c)   Express the coefficient of volume expansion for this system in terms of (T, v, p).
$\alpha = (1/V)\,\partial V/\partial T)_p$

✳ chapter 13

NATURAL
CONSTRAINTS

With such a great deal of interdependency among the various thermodynamic variables, any constraint placed on one of them must affect the interdependencies of others. For example, we saw that the fact that entropy cannot decrease demands that heat flows from hot toward cold, and not vice versa. This is only one of a large number of interrelationships resulting from the second law's constraint on the entropy. The first law and third law also place constraints on system variables, which result in additional interdependencies. Since these laws are inviolate, the resulting interrelationships must apply to all macroscopic systems and all processes.

The tools of statistical mechanics and thermodynamics are so completely general that they can be applied to almost any system imaginable. This is the single most impressive feature of this particular subject.

Unfortunately, it is this same feature that causes the greatest stumbling block to those being introduced to the subject for the first time. There is such a wide range of parameters describing a system, including internal energy, temperature, pressure, entropy, volume, chemical potential, and many more. Individually, each of these parameters has a definite physical meaning, but when combined with other parameters in mathematical expressions, the physical meanings of the various terms frequently seem a bit obscured.

It is readily understood that in principle all these variables are interdependent. For most processes, all parameters are functions of just a few (usually 1, 2, or 3) "independent" variables. However, the interdependence of the parameters varies from one system to the next, so they are usually not given explicitly, leaving the students with expressions for internal energy, entropy, and so on, each involving many parameters whose interrelationships are either vague or unknown.

The use of models, such as those in the previous chapters, help us understand these interrelationships for particular systems, and equations of state give us the needed information for making specific calculations.

In addition, many of the parameters are constrained in ways that are independent of the particular system or process being studied. That is, you can rely on them even if your particular model or equation of state is wrong. These "natural" constraints fall into three basic categories.

1. Those arising from nature's desire to maximize entropy (second law).
2. Those arising from the entropy going to some finite value as $T \to 0$ (third law).
3. Those interrelationships that must exist when there are more parameters than independent variables (Maxwell's relations).

This chapter is devoted to studying the natural constraints arising from these considerations.

A. SECOND LAW CONSTRAINTS

The second law states that when a system is in equilibrium, its entropy must be a maximum. Consequently, as two interacting subsystems approach equilibrium, the entropy of the combined system must increase. In Chapter 11 we saw that this leads to the following constraints on systems interacting appropriately.

1. Heat flows toward the cooler system.
2. Volume is gained by the system having the greater pressure.
3. Particles flow toward the system with lower chemical potential.

Further constraints can be obtained from the observation that when two interacting systems are displaced slightly away from equilibrium, then the resultant change in entropy must be negative.

Consider, for example, a small system A_1 initially in equilibrium with a huge reservoir A_2. (See Figure 13.1.) If these two systems are displaced away from equilibrium by some small transfer of energy, volume, or particles between the two, then the entropy of the combined system must decrease.

$$\Delta S_0 = \Delta S_1 + \Delta S_2 < 0$$

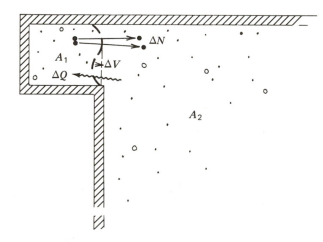

Figure 13.1   Consider a system A_1 initially in equilibrium with a huge reservoir A_2. If they are slightly displaced from equilibrium through the exchange of small amounts of heat, volume, or particles, what will happen to the temperature, pressure, or chemical potential of system A_1?

Since system A_2 is a huge reservoir, these small changes will have no effect on its temperature, pressure, or chemical potential (by definition of a reservoir), but this is not true in general for the smaller system, for which the respective changes are ΔT_1, Δp_1, and $\Delta \mu_1$, respectively. During the displacement, the average temperature, pressure, and chemical potential for the small system are given by

$$\bar{T}_1 = T_1 + \frac{1}{2}\Delta T_1$$

$$\bar{p}_1 = p_1 + \frac{1}{2}\Delta p_1$$

$$\bar{\mu}_1 = \mu_1 + \frac{1}{2}\Delta \mu_1 \tag{13.1}$$

From the first law, we know that the change in entropy of a system is given by

$$\Delta S = \frac{1}{T}(\Delta E + p\,\Delta V - \mu\,\Delta N)$$

so the change of entropy for the combined system, A_0, is given by

$$\Delta S_0 = \Delta S_1 + \Delta S_2$$

$$= \frac{1}{\bar{T}_1}(\Delta E_1 + \bar{p}_1\,\Delta V_1 - \bar{\mu}_1\,\Delta N_1) + \frac{1}{T_2}(\Delta E_2 + p_2\,\Delta V_2 - \mu_2\,\Delta N_2) < 0 \tag{13.2}$$

If we notice that*

$$\frac{1}{\bar{T}_1} = \frac{1}{T_1[1 + \frac{1}{2}(\Delta T/T_1)]} \approx \frac{1}{T_1}\left(1 - \frac{1}{2}\frac{\Delta T_1}{T_1}\right)$$

and that since A_1 and A_2 are interacting only with each other,

$$\Delta E_1 = -\Delta E_2$$

$$\Delta V_1 = -\Delta V_2$$

$$\Delta N_1 = -\Delta N_2$$

the above expression reduce to (See homework Problem 12.1)

$$\Delta S_0 = \left(\frac{1}{T_1} - \frac{1}{T_2}\right)\Delta E_1 + \left(\frac{p_1}{T_1} - \frac{p_2}{T_2}\right)\Delta V_1 - \left(\frac{\mu_1}{T_1} - \frac{\mu_2}{T_2}\right)\Delta N_1$$

$$+ \frac{1}{2}\left[\Delta\left(\frac{p_1}{T_1}\right)\Delta V_1 - \Delta\left(\frac{\mu_1}{T_1}\right)\Delta N_1 - \frac{\Delta T_1 \Delta E_1}{T_1^2}\right]$$

$$< 0 \tag{13.3}$$

where we ignore the third-order terms in incremental changes (e.g., $\Delta T_1\,\Delta p_1\,\Delta V_1$, etc.).

---

* $(1 + \varepsilon)^x = 1 + x\varepsilon +$ terms of order $(x\varepsilon)^2$ and higher. $(1 + \varepsilon)^x \approx 1 + x\varepsilon$, for $x\varepsilon \ll 1$. In particular, $(1 + \varepsilon)^{-1} \approx (1 - \varepsilon)$.

The requirement that ΔS_0 be negative demands that the three first-order terms in the above expression must be zero.* This is equivalent to the requirement that when a function is at its maximum, its first derivatives are zero. This gives us the familiar results (11.5) that when the two are in equilibrium,

$$T_1 = T_2 \text{ ("thermal equilibrium")}$$

$$p_1 = p_2 \text{ ("mechanical equilibrium")} \tag{11.5}$$

$$\mu_1 = \mu_2 \text{ ("diffusive equilibrium")}$$

With these requirements, the expression for the change in entropy becomes

$$\Delta S_0 = \frac{1}{2}\left[\Delta\left(\frac{p_1}{T_1}\right)\Delta V_1 - \Delta\left(\frac{\mu_1}{T_1}\right)\Delta N_1 - \frac{\Delta T_1 \Delta E_1}{T_1^2} \right] < 0 \tag{13.4}$$

Since (E_1, V_1, N_1) are independent variables, each of these terms individually must be less than zero. These are new constraints, which we can summarize as follows.

For a system in equilibrium:

$$\Delta\left(\frac{p}{T}\right)\Delta V \Big|_{E,N} < 0 \tag{13.5}$$

$$\Delta\left(\frac{\mu}{T}\right)\Delta N \Big|_{E,V} > 0 \tag{13.6}$$

$$\Delta T \,\Delta E|_{V,N} > 0 \tag{13.7}$$

In words, these conditions state that for a system initially in equilibrium:

1. Increased volume must result in decreased p/T (E, N constant).
2. Increased number of particles must result in increased μ/T (E, V constant).
3. Increased internal energy must result in increased temperature (V, N constant).

Of particular importance in the treatment of diffusive interactions is the second law requirement that particles tend toward configurations of lower chemical potential. The chemical potential is the zero-energy reference point from which the potential energy of a particle is measured. The zero-energy reference point for a system of N particles is $N\mu$, and that for a system having more than one kind or particle would be the appropriate sum over the various particle types, $\sum_i N_i \mu_i$.

Since the combined system is isolated, any thermal energy released by a downward adjustment in the reference level ($\sum_i N_i \mu_i$) stays within the system. But according to the equipartition theorem, some of this released thermal energy will go into kinetic degrees of freedom, so the net *potential* energy (zero-energy reference level plus the thermal energy stored in potential energy degrees of freedom) will decrease. Therefore, the

---

* To see this look at any one of the terms and notice that the change in the independent variable could be of either sign, one of which would violate the condition $\Delta S_0 < 0$, if it were not multiplied by zero.

Figure 13.2 Particles flow from regions of high chemical potential toward regions of low chemical potential. (Frederick D. Bodin/Stock, Boston)

second law requirement that particles tend toward configurations that lower their chemical potentials is a way of saying that systems tend toward configurations of lower potential energy (Figure 13.2). Hopefully, we learned that already in our freshman physics courses.

The zero-energy reference level, $N\mu$ or $\sum_i N_i\mu_i$, is sometimes called the "thermodynamical potential, φ," and sometimes the "Gibbs free energy, G."* Consequently, when dealing with diffusive interactions, there is a variety of expressions used that mean the same thing. "Minimizing the system's potential energy," "minimizing the zero-energy reference level," "minimizing the thermodynamical potential," and "minimizing the Gibbs free energy," are all equivalent ways of stating this second law requirement.

SUMMARY

The second law requirement that particles tend to flow toward regions of lower chemical potential means that diffusively interacting systems tend toward configurations that minimize the zero-energy reference level ($N\mu$, or $\sum_i N_i\mu_i$). This zero-energy reference level is sometimes called the "thermodynamical potential," and sometimes the "Gibbs free energy." Equivalently, systems tend toward configurations of lowest potential energy.

---

* The Gibbs free energy is more commonly defined as $G \equiv E - TS + pV$, but this can be shown to be equivalent to the above. See Section B, Chapter 22.

PROBLEMS

13-1.   Using Eq. 13.1 show that Eq. 13.2 leads to Eq. 13.3.

13-2.   When water freezes in a closed jar, both its volume and its pressure increase, eventually bursting the jar. Does this violate condition (13.5)? Explain. (*Hint.* Are E and N constant during this process?)

13-3.   When ice melts, temperature remains constant as energy is added. Does this violate the condition (13.7)? (*Hint.* We have used $\mu \, dN$ to represent $\mu_i \, dN_i$ if there's more than one component interacting diffusively. Therefore, constant N really means constant N_i for each component when there are more than one.)

13-4.   Consider particles moving from system A_1 to A_2, due to μ_1 being higher than μ_2. According to condition (13.6) will μ_1 and μ_2 get closer together or farther apart as a result of this transfer of particles? (Assume T constant.)

13-5.   Briefly explain how you could experimentally verify any two of the three constraints (13.5 to 13.7) for any system you wish.

13-6.   Consider a system A_1 having N_1 particles at chemical potential μ_1, and system A_2 having N_2 particles at chemical potential μ_2 with $\mu_1 > \mu_2$. Suppose these two systems are briefly brought into diffusive contact so that a small number of particles, ΔN, goes from one to the other.

   (a)   Which system acquires these particles and why?
   (b)   In terms of (N_1, N_2, μ_1, μ_2) what is the Gibbs free energy of the two systems before the interaction?
   (c)   It is sometimes mistakenly thought that the change in Gibbs free energy during the exchange is given by $(\mu_2 - \mu_1)\Delta N$. Why is this wrong, what has been left out, and what do these two extra terms correspond to physically?

13-7.   Consider the burning of hydrogen to make water,

$$2H_2 + O_2 \rightarrow 2H_2O$$

   with the ratio

$$\Delta N_{H_2} : \Delta N_{O_2} : \Delta N_{H_2O} = -2 : -1 : 2$$

   (a)   The fact that this reaction occurs tells us what about the relative sizes of $(2\mu_{H_2} + \mu_{O_2})$ and $2\mu_{H_2O}$?
   (b)   Physically, by what is this difference in chemical potential caused (e.g., attractions between neighboring molecules? the energies of valence electrons? etc.?)?

B.   MAXWELL'S RELATIONS

So far, we have been introduced to a large number of parameters that characterize a system. Now, we introduce two more, called "Helmholtz free energy" and "enthalpy" (symbols F and H, respectively), which are also occasionally found to be useful concepts in certain types of processes. They are defined as follows.

$$\text{"Helmholtz free energy"} = F \equiv E - TS \tag{13.8}$$

$$\text{"Enthalpy"} = H \equiv E + pV \tag{13.9}$$

which makes their differential forms (using $dE = T dS - p dV + \mu dN$)

$$dF = -S dT - p dV + \mu dN \tag{13.10}$$

$$dH = T dS + V dp + \mu dN \tag{13.11}$$

For our present purposes, the physical meaning of these two additional parameters is of less concern to us than their mathematical properties. In the next chapter we show that they are conserved in certain types of processes.

The Gibbs free energy, introduced in the previous section, has the definition*

$$G \equiv E - TS + pV$$

which makes its differential form,,

$$dG = -S dT + V dp + \mu dN \tag{13.12}$$

This brings the number of parameters commonly used to describe physical systems to 10 (E, T, S, p, V, μ, N, G, H, and F)! We know that the first seven of these are well-defined functions of the present state of the system. That is, their values can be determined by measurements made on the present system, independent of how it arrived there. The last three are defined in terms of the first seven, so they too must be well-defined functions of the present state of the system. This means that the differentials (dE, dT, dS, dp, dV, $d\mu$, dN, dH, dF, dG) are all *exact* differentials.

It is this large number of parameters characterizing a system that sometimes makes the study of thermodynamics seem hopelessly confusing. But we know that there are, at most, three *independent* variables, and there could be fewer, depending on the number of different types of interactions the system engages in. It follows that there must be a great deal of interdependence among the 10 parameters!

This multitude of interdependencies is most easily explored with some rather straightforward mathematical maneuvering. In particular, we exploit the property of exact differentials that if $df(x, y)$ is exact, then

$$\frac{\partial^2 f}{\partial x \, \partial y} = \frac{\partial^2 f}{\partial y \, \partial x} \tag{6.17}$$

We begin by writing down the expressions for the differential forms of the internal energy, Helmholtz free energy, enthalpy, and Gibbs free energy, respectively.

$$
\begin{aligned}
dE &= T dS - p dV + \mu dN \\
dF &= -S dT - p dV + \mu dN \\
dH &= T dS + V dp + \mu dN \\
dG &= -S dT + V dp + \mu dN
\end{aligned} \tag{13.13}
$$

Each of these expressions involves four exact differentials, so each one may be arranged in four different ways to express the dependence of any one exact differential on the other three. For example, the first of the above expressions may be expressed in the following four ways.

---

* This is the same as the definition $G = \mu N$, as is shown in Section B, Chapter 22.

$$dE = T\,dS - p\,dV + \mu\,dN$$

$$dS = \frac{1}{T}\,dE + \frac{p}{T}\,dV - \frac{\mu}{T}\,dN$$

$$dV = -\frac{1}{p}\,dE + \frac{T}{p}\,dS + \frac{\mu}{p}\,dN \qquad (13.14)$$

$$dN = \frac{1}{\mu}\,dE - \frac{T}{\mu}\,dS + \frac{p}{\mu}\,dV$$

Similarly, we could rearrange the expressions for dF, dH, and dG each in 4 different ways, making 16 different expressions for the exact differentials altogether. With each each of these, we can exploit the property (6.17) of exact differentials to find the inter-relationships among the various parameters.

This can be done for each of the 16 differential expressions, yielding 48 inter-relationships (called "Maxwell's relations") altogether. In Table 13.1 we list the 12 simplest and most commonly used of these interrelationships, and we also list separately the 4 of these that are pertinent to the special (but common) case where the system is not interacting diffusively.

We now illustrate how this is done by starting with one of the 16 differential expressions and seeing how we can use it to find interrelationships among the various parameters. A more detailed treatment can be found in Appendix 13A. Consider the very first of the expressions (13.14)

$$dE = T\,dS - p\,dV + \mu\,dN$$

From this we see that

$$\left.\frac{\partial E}{\partial S}\right)_{V,N} = T, \qquad \left.\frac{\partial E}{\partial V}\right)_{S,N} = -p, \qquad \left.\frac{\partial E}{\partial N}\right)_{S,V} = \mu \qquad (13.15)$$

Combining these expressions with the property (6.17) of exact differentials,

$$\frac{\partial^2 E}{\partial S\,\partial V} = \frac{\partial^2 E}{\partial V\,\partial S}$$

$$\frac{\partial^2 E}{\partial S\,\partial N} = \frac{\partial^2 E}{\partial N\,\partial S}$$

$$\frac{\partial^2 E}{\partial V\,\partial N} = \frac{\partial^2 E}{\partial N\,\partial V}$$

we have

$$-\left.\frac{\partial p}{\partial S}\right)_{V,N} = \left.\frac{\partial T}{\partial V}\right)_{S,N} \qquad (13.16)$$

$$\left.\frac{\partial \mu}{\partial S}\right)_{V,N} = \left.\frac{\partial T}{\partial N}\right)_{S,V} \qquad (13.17)$$

$$\left.\frac{\partial \mu}{\partial V}\right)_{S,N} = -\left.\frac{\partial p}{\partial N}\right)_{S,V} \qquad (13.18)$$

Table 13.1   Maxwell's Relations—
             Those Most Commonly
             Used

$$\left.\frac{\partial T}{\partial V}\right)_{S,N} = -\left.\frac{\partial p}{\partial S}\right)_{V,N} \qquad \text{(M1)}$$

$$-\left.\frac{\partial p}{\partial N}\right)_{S,V} = \left.\frac{\partial \mu}{\partial V}\right)_{S,N} \qquad \text{(M2)}$$

$$\left.\frac{\partial T}{\partial N}\right)_{S,V} = \left.\frac{\partial \mu}{\partial S}\right)_{V,N} \qquad \text{(M3)}$$

$$\left.\frac{\partial S}{\partial V}\right)_{T,N} = \left.\frac{\partial p}{\partial T}\right)_{V,N} \qquad \text{(M4)}$$

$$-\left.\frac{\partial p}{\partial N}\right)_{T,V} = \left.\frac{\partial \mu}{\partial V}\right)_{T,N} \qquad \text{(M5)}$$

$$-\left.\frac{\partial S}{\partial N}\right)_{T,V} = \left.\frac{\partial \mu}{\partial T}\right)_{V,N} \qquad \text{(M6)}$$

$$\left.\frac{\partial T}{\partial p}\right)_{S,N} = \left.\frac{\partial V}{\partial S}\right)_{p,N} \qquad \text{(M7)}$$

$$\left.\frac{\partial V}{\partial N}\right)_{S,p} = \left.\frac{\partial \mu}{\partial p}\right)_{S,N} \qquad \text{(M8)}$$

$$\left.\frac{\partial T}{\partial N}\right)_{S,p} = \left.\frac{\partial \mu}{\partial S}\right)_{p,N} \qquad \text{(M9)}$$

$$-\left.\frac{\partial S}{\partial p}\right)_{T,N} = \left.\frac{\partial V}{\partial T}\right)_{p,N} \qquad \text{(M10)}$$

$$\left.\frac{\partial V}{\partial N}\right)_{T,p} = \left.\frac{\partial \mu}{\partial p}\right)_{T,N} \qquad \text{(M11)}$$

$$-\left.\frac{\partial S}{\partial N}\right)_{T,p} = \left.\frac{\partial \mu}{\partial T}\right)_{p,N} \qquad \text{(M12)}$$

For nondiffusive interactions:

$$\left.\frac{\partial T}{\partial V}\right)_{S} = -\left.\frac{\partial p}{\partial S}\right)_{V} \qquad \text{(M1)}$$

$$\left.\frac{\partial S}{\partial V}\right)_{T} = \left.\frac{\partial p}{\partial T}\right)_{V} \qquad \text{(M4)}$$

$$\left.\frac{\partial T}{\partial p}\right)_{S} = \left.\frac{\partial V}{\partial S}\right)_{p} \qquad \text{(M7)}$$

$$-\left.\frac{\partial S}{\partial p}\right)_{T} = \left.\frac{\partial V}{\partial T}\right)_{p} \qquad \text{(M10)}$$

These were derived from abstract mathematical considerations, but they each give us interrelationships which have definite physical meaning. For example, the three above tell us the following.

1. The change in pressure, when heat is added, is related to the change in temperature when the volume is increased.
2. The change in chemical potential, when heat is added, is related to the change in temperature when particles are added.
3. The change in chemical potential, when the volume is increased, is related to the change in pressure when particles are added.

You may wish to consider how you would experimentally verify any of the above interrelationships (because you may be asked to do it in homework problems).Changes in pressure, temperature, and volume (ΔT, Δp, ΔV) are easily measured with standard gauges and meter sticks. Change in number of particles (ΔN) can be determined from the mass of added material, and change in entropy (ΔS) would have to be reckoned from the amount of heat added or released ($\Delta S = \Delta Q/T$).

Determining changes in chemical potential, however, is a little more subtle, because we don't usually have commercial gauges available for this, like we do for temperature and pressure. Recall that the chemical potential (μ) is the zero-energy reference level relative to which the potential energy of a particle is measured (Figure 13.3). The average energy of a particle is given by

$$\bar{\varepsilon} = \mu + \frac{v}{2}kT$$

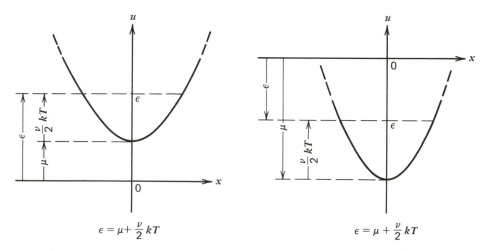

$$\epsilon = \mu + \frac{v}{2}kT \qquad\qquad \epsilon = \mu + \frac{v}{2}kT$$

Figure 13.3   Plot of potential energy versus position for a particle in a parabolic potential well for positive chemical potential (left) and negative chemical potential (right). The average total energy of the particle is the sum of its chemical potential plus its thermal energy. $\varepsilon = \mu + (v/2)kT$, where v is the number of degrees of freedom for the particle. The thermal energy is measured relative to the bottom of the well.

where v is the number of degrees of freedom per particle. The term $(v/2)kT$ is called the particle's "thermal energy," and it includes both the kinetic energy and the potential energy measured relative to μ.

If we add energy ΔE to a system of N particles, then the increase of average energy per particle is given by

$$\Delta \bar{\varepsilon} = \Delta \mu + \frac{v}{2} k \Delta T$$

Since we can measure both the energy added and the change in temperature, we can deduce that any discrepancy between the amount of added energy and the increase in thermal energy must be due to a shift in the zero-energy reference level. That is, since

$$\frac{\Delta E}{N} = \Delta \mu + \frac{v}{2} \Delta T$$

we have

$$\Delta \mu = \frac{\Delta E}{N} - \frac{v}{2} k \Delta T \tag{13.19}$$

An alternative way of looking at this is to consider what would happen if the chemical potential did *not* change. Then all the added energy would go into increasing the particle's thermal energy. In this case, we would expect that

$$\frac{\Delta E}{N} = \frac{v}{2} k \Delta T_{\text{expected}}$$

Any discrepancy between this expected change in temperature and that observed, would tell us that there has been a shift in the chemical potential, given by (see Eq. 13.19)

$$\Delta \mu = \frac{v}{2} k \Delta T_{\text{expected}} - \frac{v}{2} k \Delta T_{\text{observed}}$$

$$= \frac{v}{2} k (\Delta T_{\text{expected}} - \Delta T_{\text{observed}}) \tag{13.20}$$

For the particular (but important) case of chemical interactions, the various reagents tend to combine in ways that minimize the energies of their valence electrons. Hydrogen burns, for example, because the valence electrons in the H_2O molecules have lower energies than they did in the H_2 and O_2 molecules. For chemical reactions, then, the chemical potentials of the reagents reflects how strongly they hold onto the valence electrons. The electron affinities of the reagents can be measured (in volts), which tells us their chemical potentials. For chemical reactions, then, changes in chemical potentials can frequently be gotten by measuring changes in electron affinities, and the more laborious thermal techniques described in the preceding paragraph are not needed.

Because Maxwell's relations are most often and most easily derived from formal mathematical manipulations, and because they are most often and most concisely expressed as interrelationships among partial differentials, they seem to carry an air of abstract formality that makes students wonder how they could possible be of any use.

Before we demonstrate some practical applications, we should point out that their formal appearance is deceiving. Each differential expresses how one easily measured property changes as you change another (Figure 13.4). For example they express how pressure changes as temperature is increased, or how volume changes as heat is added. For any one of these differential relationships, you could manufacture some gauge to put in your system and tell you the value of the differential ratio. Some of these differential ratios are not commonly used, and so appropriate gauges are not available commercially. But they could be, if there was a demand. Remember that common things like temperature and pressure are each really a formal differential ratio among physically measurable parameters.

$$T \equiv \frac{\partial E}{\partial S}\bigg)_{V,N}$$

$$p \equiv T \frac{\partial S}{\partial V}\bigg)_{E,N}$$

The other ratios appearing in Maxwell's relations are, in principle, no different than these, with which everyone is familiar.

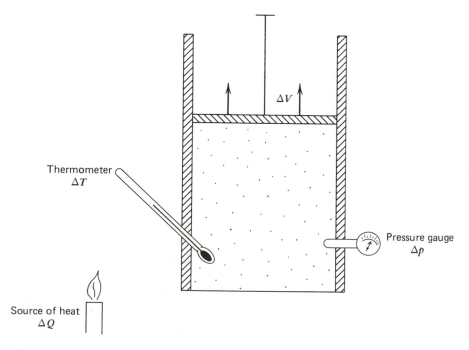

Figure 13.4 Apparatus needed to test Maxwell's relations for nondiffusive interactions.

PROBLEMS

13-8. Show that the differential expressions for the Helmholtz free energy and enthalpy, (13.10) and (13.11), follow from their definitions, (13.8) and (13.9).

13-9. Suppose you wish to experimentally confirm Maxwell's relations for a system of fixed number of particles ($dN = 0$), using an apparatus similar to that of Figure 13.4.

    (a) How would you measure $\partial T/\partial V)_S \approx \Delta T/\Delta V)_S$?

    (b) How would you measure $\partial p/\partial S)_V$?

    (c) How would you measure $\partial S/\partial p)_T$?

13-10. Suppose you wished to experimentally confirm Maxwell's relations for a system that can receive heat, work, or particles from its environment, using an apparatus similar to that of Figure 13.5.

    (a) How would you measure $\partial V/\partial N)_{T,p}$?

    (b) How would you measure $\partial \mu/\partial p)_{T,N}$? (*Hint.* If you allow the system to expand by ΔV, doing work $p\Delta V$, you have to add heat energy ΔQ to keep the temperature constant. If the two (ΔQ and $p\Delta V$) are not equal, then there has been a shift in the internal energy of the system, $\Delta E = \Delta Q - p\Delta V$. If the temperature is the same, then the thermal energies are the same, so the change in internal energy must be due to a shift in potential energy by $N\Delta\mu$.)

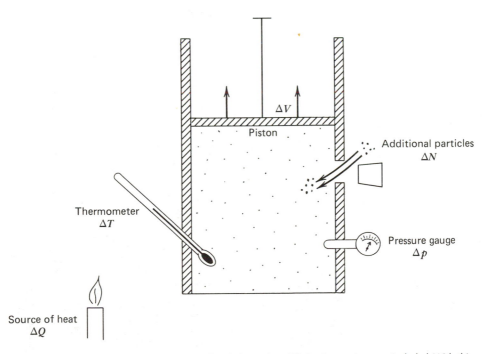

Figure 13.5 Apparatus needed to test Maxwell's relations when diffusive interactions are included. With this apparatus you could measure directly ΔQ, ΔT, Δp, ΔV, and ΔN, but not $\Delta\mu$. How could you measure $\Delta\mu$?

C.   SAMPLE APPLICATIONS—NONDIFFUSIVE INTERACTIONS

The kinds of natural constraints developed in this chapter apply to all thermo-dynamical processes, so it would be impossible in one book to cover the entire spectrum of possible applications. Instead, we will try to illustrate some of the kinds of ways these constraints can be utilized in practical applications through some specific examples.

For simplicity, we will consider applications to systems that are not interacting diffusively ($dN = 0$), which means the number of independent variables in each expression is two, rather than three. The treatment of diffusively interacting systems is no different in principle, but each equation would have to have an additional term (e.g., $\mu \, dN$) to accommodate the additional independent variable. The application of natural constraints to systems interacting diffusively is presented in Chapters 22 and 23, with special attention given to systems undergoing chemical interaction, and change in phase.

In this section, we will use the four Maxwell's relations in Table 13.1, which apply to systems that are not interacting diffusively. In addition we will use the following measurable properties of a system.

1.   Heat capacity at constant volume.

$$C_V = \frac{\partial Q}{\partial T}\bigg)_V = T\frac{\partial S}{\partial T}\bigg)_V$$

2.   Heat capacity at constant pressure.

$$C_p = \frac{\partial Q}{\partial T}\bigg)_p = T\frac{\partial S}{\partial T}\bigg)_p$$

3.   Coefficient of volume expansion.

$$\beta = \frac{1}{V}\frac{\partial V}{\partial T}\bigg)_p$$

4.   Isothermal compressibility.

$$\kappa = -\frac{1}{V}\frac{\partial V}{\partial p}\bigg)_T$$

EXAMPLE: $\Delta E(T, p)$

As an example of how natural constraints may be useful, suppose you were studying some system of fixed size undergoing some process, for which you wished to determine the change in its internal energy.

$$\Delta E(S, V) = T\Delta S - p\Delta V \tag{13.21}$$

Suppose for this particular process it was more convenient for you to measure small changes in temperature and pressure ($\Delta T, \Delta p$) than to measure small changes in entropy and volume ($\Delta S, \Delta V$).

In order to use these two measurements to find the change in internal energy, you would first have to transform the expression (13.21) for ΔE in terms of $(\Delta S, \Delta V)$ to one in terms of $(\Delta T, \Delta p)$. This is done by rewriting ΔS and ΔV each in terms of their dependence on ΔT and Δp.

$$\Delta S(T, p) = \left(\frac{\partial S}{\partial T}\right)_p \Delta T + \left(\frac{\partial S}{\partial p}\right)_T \Delta p$$

$$\Delta V(T, p) = \left(\frac{\partial V}{\partial T}\right)_p \Delta T + \left(\frac{\partial V}{\partial p}\right)_T \Delta p$$

With this, Eq. 13.21 becomes

$$\Delta E(T, p) = T\left[\left(\frac{\partial S}{\partial T}\right)_p \Delta T + \left(\frac{\partial S}{\partial p}\right)_T \Delta p\right] - p\left[\left(\frac{\partial V}{\partial T}\right)_p \Delta T + \left(\frac{\partial V}{\partial p}\right)_T \Delta p\right]$$

$$= \left[T\left(\frac{\partial S}{\partial T}\right)_p - p\left(\frac{\partial V}{\partial T}\right)_p\right]\Delta T + \left[T\left(\frac{\partial S}{\partial p}\right)_T - p\left(\frac{\partial V}{\partial p}\right)_T\right]\Delta p \qquad (13.22)$$

As it stands, this expression may not seem very helpful to you, but all the differential expressions in (3.22) are easily measurable properties of the system:

$$T\left(\frac{\partial S}{\partial T}\right)_p = C_p$$

$$\left(\frac{\partial V}{\partial T}\right)_p = V\beta$$

$$\left(\frac{\partial V}{\partial p}\right)_T - V\kappa$$

and using Maxwell's relation (M10),

$$\left(\frac{\partial S}{\partial p}\right)_T = -\left(\frac{\partial V}{\partial T}\right)_p = -V\beta$$

With these substitutions, the expression for $\Delta E(T, p)$ becomes

$$\Delta E(T, p) = (C_p - pV\beta)\Delta T - (T\beta - p\kappa)V\Delta p \qquad (13.23)$$

and you can determine the change in internal energy in terms of $(\Delta T, \Delta p)$ and readily measurable properties of the system.

EXAMPLE: TESTING THE EQUIPARTITION THEOREM

According to the equipartition theorem, the internal energy of a system having $\mathfrak{N}$ degrees of freedom is given by

$$E = \frac{\mathfrak{N}}{2} kT$$

or

$$\Delta E = \frac{\mathfrak{N}}{2} k\Delta T \qquad (13.24)$$

Strictly speaking, it only applies to the thermal energies of systems, and it is true only for systems in which the energy stored in each degree of freedom can be expressed in the form

$$\varepsilon = bq^2 \tag{13.25}$$

where b is a constant and q is a position or momentum coordinate. Potential energies are measured relative to the bottom of the potential well (μ). This means that the result (13.24) does *not* take into account the possibility that some of the change in internal energy, ΔE, could go into changing the zero-energy reference point, $N \Delta\mu$.

If for some process we were to find the relationship (13.24) wrong, then we know one of the following to be true.

1. For some degrees of freedom, the energy stored cannot be expressed in the form bq^2
2. The location of the bottom of the potential well (μ) shifted during the process.

According to Eq. 13.23, the internal energy of a system varies, in general, both with temperature and pressure.

$$\Delta E = (C_p - pV\beta)\Delta T - (T\beta - p\kappa)V \Delta p \tag{13.23}$$

Comparing this to Eq. 13.24) we see the following result. If Eq. 13.24 is valid,

$$(C_p - pV\beta) = \frac{\mathfrak{N}}{2} k \tag{13.26}$$

$$(T\beta - p\kappa) = 0. \tag{13.27}$$

For many real systems, these last two relationships will be approximately true, but not exactly. For most processes, there will be a small shift in the chemical potential (i.e., the bottom of a particle's potential well), and Eq. 13.24 will be violated. (See implication 2 above.)

We can get a quantitative measure of the degree to which Eq. 13.24 is violated by comparing the two terms in the expression (13.23).

$$\text{degree of violation} \approx \frac{(T\beta - p\kappa)Vp}{(C_p - pV\beta)T} \tag{13.28}$$

For most systems, we find this ratio to be very small, and so Eq. 13.24 is usually reliable. As an example, consider an ideal gas, for which

$$pV = NkT$$

We calculate

$$\kappa = -\frac{1}{V}\frac{\partial V}{\partial p}\bigg)_T = \frac{NkT}{p^2 V}$$

$$\beta = \frac{1}{V}\frac{\partial V}{\partial T}\bigg)_p = \frac{Nk}{pV}$$

From this we see

$$(T\beta - p\kappa) = 0$$

in accord with condition (13.27), and from condition (13.26) we have

$$C_p - Nk = \frac{3N}{2} k$$

or

$$C_p = \frac{5}{2} Nk$$

in agreement with previous calculations.

Of course, this is no surprise, since our definition of an ideal gas is one for which the energy per degree of freedom is $\frac{1}{2}mv_i^2$, and for which there is no potential energy due to interactions at all. Thus, our definition of an ideal gas explicitly satisfies our criterion for the equipartition theorem. In homework Problems 13-13, 14, and 15, you will be asked to repeat the above calculations, but for a van der Waals gas, to see how well it satisfies Eq. 13.24 under various conditions.

EXAMPLE: VARIATIONS IN HEAT CAPACITY

Suppose we have measured the heat capacity, C_V, for a system at one volume, V_1, and we wish to know what it would be at another volume, V_2. We need not make another measurement, providing we know the equation of state and one of Maxwell's relations.

At volume V_2, the heat capacity is given by

$$C_{V_2} = C_{V_1} + \int_{V_1}^{V_2} dC_V = C_{V_1} + \int_{V_1}^{V_2} \left(\frac{\partial C_V}{\partial V} \right)_T dV$$

From the definition of heat capacity, we have

$$\left(\frac{\partial C_V}{\partial V} \right)_T = \frac{\partial}{\partial V} \left(\frac{\partial Q}{\partial T} \right)_V = T \frac{\partial^2 S}{\partial V \, \partial T} = T \frac{\partial}{\partial T} \left(\frac{\partial S}{\partial V} \right)_T$$

Using the Maxwell's relation (M4) this becomes

$$\left(\frac{\partial C_V}{\partial V} \right)_T = T \frac{\partial^2 p}{\partial T^2} \bigg)_V$$

and

$$C_{V_2} = C_{V_1} + \int_{V_1}^{V_2} \left(T \frac{\partial^2 p}{\partial T^2} \right)_V dV \tag{13.29}$$

where the integrand can be calculated from the equation of state.

Following the same procedure, but using a different Maxwell's relation, we can derive a similar result for calculating C_p at any pressure once it is known for any one pressure, p_1. This is (see homework Problem 13-16)

$$C_{p_2} = C_{p_1} + \int_{p_1}^{p_2} \left(T \frac{\partial^2 V}{\partial T^2} \right)_p dp \tag{13.30}$$

where the integrand is calculated from the equation of state.

EXAMPLE: $\Delta S(p, V)$

As another example, suppose you wanted to know how the entropy of a system changed during a process in which the pressure and volume changed by a measured amount (Δp, ΔV). The change in entropy would be given by

$$\Delta S(p, V) = \left.\frac{\partial S}{\partial p}\right)_V \Delta p + \left.\frac{\partial S}{\partial V}\right)_p \Delta V$$

The coefficients $\partial S/\partial p)_V$ and $\partial S/\partial V)_p$ are measurable paramters of a system. For example, you could add a little heat (ΔQ) to the system held at constant volume and observe how the pressure changes,

$$\left.\frac{\partial S}{\partial p}\right)_V \approx \left.\frac{(1/T)\Delta Q}{\Delta p}\right)_V$$

and add a little heat (ΔQ) to the system held at constant pressure to see how the volume changes.

$$\left.\frac{\partial S}{\partial V}\right)_p \approx \left.\frac{1}{T}\frac{\Delta Q}{\Delta V}\right)_p$$

Suppose, however, that for your particular system, you could not add measured amounts of heat for some reason. This means you would not be able to determine the coefficients $\partial S/\partial p)_V$ and $\partial S/\partial V)_p$ directly, so you would have to find alternate means of doing this.

Using Maxwell's relations (M10) and (M7) we see that

$$\left.\frac{\partial S}{\partial p}\right)_V = -\left.\frac{\partial V}{\partial T}\right)_S$$

$$\left.\frac{\partial S}{\partial V}\right)_p = \left.\frac{\partial p}{\partial T}\right)_S$$

These coefficients then could alternately be determined by insulating your system ($\Delta S = 0$), compressing it, and then seeing how the changes in pressure, volume, and temperature are related.

$$\left.\frac{\partial S}{\partial p}\right)_V \approx -\left.\frac{\Delta V}{\Delta T}\right)_S$$

$$\left.\frac{\partial S}{\partial V}\right)_p \approx \left.\frac{\Delta p}{\Delta T}\right)_S$$

The above examples illustrate how Maxwell's relations may be used to express any thermodynamic property of a system in terms of any set of parameters that might be useful to your particular experimental setup or line of interest. They do not reduce the number of independent variables; they just allow you to trade one set for another.

PROBLEMS

13-11. Starting with the first law, $\Delta E = T\Delta S - p\Delta V$, find an expression for the dependence of internal energy on:

(a) T and V.
(b) p and V.
(c) T and S.
(d)   Of the partial derivatives appearing in the above expressions, which could be exchanged for other partial derivatives if desired, using Maxwell's relations (M1, M4, M7, M10)?

13-12. Starting from the first law, find an expression for:

(a) $V(E, S)$.
(b) $V(p, S)$.
(c) $V(p, T)$.
(d) $S(p, T)$.
(e)   Of the partial derivatives appearing in the above expression, which could be exchanged for other partial derivatives, if desired, using Maxwell's relations (M1, M4, M7, and M10)?

13-13. Starting with van der Waals equation of state for a real gas, calculate the following as a function of v, T, and the van der Waals parameters, a and b.

(a)   The coefficient of volume expansion, β.
(b)   The isothermal compressibility, κ.
(c) $(T\beta - p\kappa)$.

$$\left[Ans.\ \beta = R/vA^{-1},\ \kappa = (1 - b/v)A^{-1},\ \text{where}\ A = (RT/(v - b) - 2a(v - b)/v^3).\right]$$

13-14. For water vapor, the van der Waals constants are

$$a_{H_2O} = 5.46\ \text{liters}^2\text{-atm/mole}^2$$

$$b_{H_2O} = 0.0305\ \text{liters/mole}$$

Under a pressure of 3 atm, and at a temperature of 300 K, the molar volume is about 7 liters, $C_p \approx \frac{7}{2}R$, and $v\Delta p/\Delta T)_v \approx R$. For this particular system, estimate the degree of violation of Eq. 13.24 using Eq. 13.28, along with the values of β and κ obtained in Problem 13-13.

13-15. Repeat the above calculation for water vapor under extremely heavy pressure (approximately 50 atm) where its molar volume is only 0.04 liters.

13-16. Derive Eq. 13.30 using the definition of heat capacity, C_p, and one of Maxwell's relations.

13-17. Use Eq. 13.29 and the ideal gas law to show that C_V does not depend on the volume for an ideal gas.

13-18. Give Maxwell's relations corresponding to M1, M4, M7, and M10 for a system involved in magnetic interaction rather than volume expansion. The magnetic moment of the system is μ and the external field is B_0.

13-19. Find an expression for $\Delta E(T, \mu)$ for a system involved in thermal and magnetic interaction only. The system's magnetic moment is μ and the external field is B_0. (Your answer should be similar to Eq. 13.22.)

D. THIRD LAW CONSTRAINTS

The third law states that the entropy of a system goes to some finite constant value as the temperature approaches absolute zero. This value is independent of all other parameters.

Since the entropy is independent of all other parameters at $T = 0$, then in particular it is independent of the volume and the pressure.

$$\left. \frac{\partial S}{\partial V} \right)_{T=0} = 0$$

$$\left. \frac{\partial S}{\partial p} \right)_{T=0} = 0$$

Through Maxwell's relations (M10) and (M4) these become

$$\left. \frac{\partial p}{\partial T} \right)_V = 0 \qquad \text{at} \qquad T = 0$$

$$\left. \frac{\partial V}{\partial T} \right)_p = 0 \qquad \text{at} \qquad T = 0$$

That is, neither the pressure nor the volume of any system varies with the temperature at absolute zero.

Furthermore, since the entropy approaches a finite value as T approaches zero, and since we know entropy is finite for temperatures above zero ($k \log \Omega$), then the change in entropy as T approaches zero must be finite.

$$\left. \frac{\partial S}{\partial T} \right)_{T \to 0} = \text{(something finite)}$$

This means that heat capacities (at constant anything) must vanish as the temperature goes to zero.

$$C = \frac{\partial Q}{\partial T} = T \frac{\partial S}{\partial T} = T \cdot \text{(something finite)}$$

$$C \xrightarrow[T \to 0]{} 0$$

The smallness of the heat capacity has very aggravating consequences for physicists doing experiments at low temperatures. Since

$$C_V = \left. \frac{\Delta Q}{\Delta T} \right)_V = \left. \frac{\Delta E}{\Delta T} \right)_V \xrightarrow[T \to 0]{} 0$$

any small addition of energy, ΔE, to the system will result in large changes in temperature, ΔT. The vibrations due to trucks passing outside the building, or a fly landing on the apparatus, or energy added by changing electromagnetic fields due to some nearby electronic apparatus, or many other unsuspected sources can all add small amounts of energy to the system, suddenly raising the temperature considerably.

SUMMARY

At absolute zero neither the pressure nor the volume of a system varies with temperature, and all heat capacities vanish. As $T \to 0$,

$$\left. \frac{\partial p}{\partial T} \right)_V \to 0$$

$$\left. \frac{\partial V}{\partial T} \right)_p \to 0$$

$$C_p \to 0$$

$$C_V \to 0$$

APPENDIX 13A   MAXWELL'S RELATIONS

Entropy, temperature, pressure, volume, chemical potential, number of particles, and internal energy are all properties of a system which can be determined by measurements made on its present state (i.e., independent of its history). Consequently, the differential forms of these are each exact differentials. Since Helmholtz free energy, enthalpy, and Gibbs free energy are defined in terms of these well-defined functions,

$$F = E - TS$$
$$H = E + pV$$
$$G = E - TS + pV$$

their differential forms are exact, as well.

Using $dE = T\, ds - p\, dV + \mu\, dN$, we find the differential changes in E, F, H, G are as follows:

$$dE = T\, dS - p\, dV + \mu\, dN$$

$$dF = -S\, dT - p\, dV + \mu\, dN$$

$$dH = T\, dS + V\, dp + \mu\, dN$$

$$dG = -S\, dT + V\, dp + \mu\, dN$$

Each of these has the form

$$dz = y\, dx + w\, dv + u\, dt, \tag{13A.1}$$

from which it is seen

$$\left. \frac{\partial z}{\partial x} \right)_{v,t} = y$$

$$\left. \frac{\partial z}{\partial v} \right)_{x,t} = w$$

$$\left. \frac{\partial z}{\partial t} \right)_{x,v} = u \tag{13A.2}$$

Since dz is an exact differential, we have

$$\frac{\partial^2 z}{\partial v \, \partial x} = \frac{\partial^2 z}{\partial x \, \partial v} \qquad \text{or} \qquad \left.\frac{\partial y}{\partial v}\right)_{x,t} = \left.\frac{\partial w}{\partial x}\right)_{v,t}$$

$$\frac{\partial^2 z}{\partial t \, \partial v} = \frac{\partial^2 z}{\partial v \, \partial t} \qquad \text{or} \qquad \left.\frac{\partial w}{\partial t}\right)_{x,v} = \left.\frac{\partial u}{\partial v}\right)_{x,t}$$

$$\frac{\partial^2 z}{\partial t \, \partial x} = \frac{\partial^2 z}{\partial x \, \partial t} \qquad \text{or} \qquad \left.\frac{\partial y}{\partial t}\right)_{x,v} = \left.\frac{\partial u}{\partial x}\right)_{v,t} \qquad (13A.3)$$

Notice that Eq. 13A.1 can be rearranged to express dx as a function of (z, v, t), or dv as a function of (z, x, t), etc.

$$dx = \left(\frac{1}{y}\right) dz - \left(\frac{w}{y}\right) dv - \left(\frac{u}{y}\right) dt$$

$$dv = \left(\frac{1}{w}\right) dz - \left(\frac{y}{w}\right) dx - \left(\frac{u}{w}\right) dt$$

$$dt = \left(\frac{1}{u}\right) dz - \left(\frac{y}{u}\right) dx - \left(\frac{w}{u}\right) dv$$

Each of these is an exact differential and of the form (13A.1), so the same procedures can be applied to get the corresponding relationships (13A.3).

EXAMPLE

From the differential form for the internal energy, find an expression for dS in terms of (E, V, N), and get the Maxwell's relations corresponding to (13A.3).

According to the first law, the differential of the internal energy is

$$dE = T \, dS - p \, dV + \mu \, dN$$

Therefore,

$$dS = \left(\frac{1}{T}\right) dE + \left(\frac{p}{T}\right) dV - \left(\frac{\mu}{T}\right) dN$$

from which we see

$$\left.\frac{\partial S}{\partial E}\right)_{V,N} = \frac{1}{T}$$

$$\left.\frac{\partial S}{\partial V}\right)_{E,N} = \frac{p}{T}$$

$$\left.\frac{\partial S}{\partial N}\right)_{E,V} = -\frac{\mu}{T}$$

Using the property of exact differentials that

$$\frac{\partial^2 f}{\partial x \, \partial y} = \frac{\partial^2 f}{\partial y \, \partial x}$$

we have

$$\frac{\partial^2 S}{\partial V \partial E} = \frac{\partial^2 S}{\partial E \partial V} \quad \text{or} \quad \frac{\partial}{\partial V}\left(\frac{1}{T}\right) = \frac{\partial}{\partial E}\left(\frac{p}{T}\right)$$

$$\frac{\partial^2 S}{\partial N \partial V} = \frac{\partial^2 S}{\partial V \partial N} \quad \text{or} \quad \frac{\partial}{\partial N}\left(\frac{p}{T}\right) = \frac{\partial}{\partial V}\left(-\frac{\mu}{T}\right)$$

$$\frac{\partial^2 S}{\partial N \partial E} = \frac{\partial^2 S}{\partial E \partial N} \quad \text{or} \quad \frac{\partial}{\partial N}\left(\frac{1}{T}\right) = \frac{\partial}{\partial E}\left(-\frac{\mu}{T}\right)$$

These are equivalently expressed as

$$\left.\frac{\partial T}{\partial V}\right)_{E,N} = -T\left.\frac{\partial p}{\partial E}\right)_{V,N} + p\left.\frac{\partial T}{\partial E}\right)_{V,N}$$

$$T\left.\frac{\partial p}{\partial N}\right)_{E,V} - p\left.\frac{\partial T}{\partial N}\right)_{E,V} = -T\left.\frac{\partial \mu}{\partial V}\right)_{E,N} + \mu\left.\frac{\partial T}{\partial V}\right)_{E,N}$$

$$\left.\frac{\partial T}{\partial N}\right)_{E,V} = T\left.\frac{\partial \mu}{\partial E}\right)_{V,N} - \mu\left.\frac{\partial T}{\partial E}\right)_{V,N}$$

respectively.

From the foregoing example we see that three interrelationships were derived from the expression for just one of the differentials (dS) appearing in dE. The same can be done for each of the four differentials appearing in this expression, giving us 12 interrelationships. Twelve more come from the expression for dF, 12 more from dH, and 12 more from dG, giving us 48 interrelationships altogether.

✳ chapter 14

IMPOSED CONSTRAINTS

In addition to the "natural" constraints that must apply to all systems, individual systems may incur additional constraints imposed by such things as their natural environment, or our particular experimental setup. For example, a system we study might have a fixed number of particles or might be held at a constant temperature or pressure. Each such constraint makes one variable a constant, thereby reducing the number of independent variables. Each constraint, then, should make the interdependences among the remaining variables considerably simpler. But how do we transform a constraint on any one variable into interrelationships among the other variables that might be of interest to us?

Frequently, we carry out our studies on systems for which one or more parameters are held constant. For example, many laboratory experiments are carried out at atomspheric pressure (p = constant), and many times we study systems that are neither gaining nor losing particles (N = constant).

Each such constraint reduces the number of independent variables by one. If one parameter is held constant, all thermodynamical properties depend on only two parameters, rather than three.* If two are held constant, then there is only one independent variable left, and if any three parameters are fixed, no thermodynamical property of the system may change at all.

There are a variety of adjectives listed in Table 14.1 used to describe the various common imposed constraints. "Nondiffusive" processes are those for which the number of elements of the system doesn't change. "Isovolumic," or "isochoric" processes are those for which the volume is constant. "Adiabatic" processes are those for which no heat is added or removed, which is equivalent to no change in entropy ($dQ = T\,dS$) if the system remains in equilibrium. "Isobaric" processes are those carried out at constant pressure, and "isothermal" processes are those during which the temperature remains contant.

A. THE REDUCTION OF INDEPENDENT VARIABLES

More often than not, the conditions under which we wish to study a system involve some constraints. (See Figure 14.1.) Quite frequently, the system of interest contains a

---

* For convenience we continue to represent all appropriate forms of work, $\mathscr{F}_i\,dx_i$ as $p\,dV$, and the exchange of all the different kinds of particles in the system $\mu_i\,dN_i$ as $\mu\,dN$.

Table 14.1   Constraints

Adjective	Constraint
Nondiffusive	$dN = 0$
Isovolumic or isochoric	$dV = 0$
Adiabatic[a]	$dQ = 0$
Isobaric	$dp = 0$
Isothermal	$dT = 0$

[a] If the system is in equilibrium throughout the process, then $dQ = T\,dS$, and so the constraint $dQ = 0$ is equivalent to $dS = 0$ for equilibrium processes.

Figure 14.1   During the boiling of a pot of water, the system of liquid water is subject to which of the following constraints: $dN = 0$, $dV = 0$, $dQ = 0$. $dp = 0$, $dT = 0$? (Kathy Bendo and James Brady)

fixed number of particles ($dN = 0$). For example, in studying the properties of a piece of metal, or a bucket of water, we are dealing with such a system. Many of the examples used in this chapter have this constraint. Often, our experiments are carried out under atmospheric pressure, in which case the pressure is constant ($dp = 0$). The system may also be in thermal contact with its environment, which holds its temperature constant ($dT = 0$).

Constraints such as these, imposed either by the system's natural environment or by our laboratory setup, exist for most systems we study. In this section, we explore how to most effectively utilize these constraints in reducing the complexity of a problem.

In either the experimental or theoretical study of any process, we should first ask ourselves the following questions.

1. What property do we wish to study?
2. What are the constraints?
3. What properties can we most easily measure?

Whatever our answer to question 1 is, that property can be expressed as a function of any three others. In principle, these three could be any three of a large number, including (E, S, T, p, V, μ, N, F, H, G) and perhaps others. But to make our study as simple as possible, we should make sure that the three we choose include the following.

1. All those involved in the constraints (question 2).
2. The remaining properties should be among those most easily measured.

For example, suppose we were interested in how parameter w varies with parameters x and y, under the constraint that parameter z be held constant. Then we should write

$$\Delta w = \frac{\partial w}{\partial x}\bigg)_{y,z} \Delta x + \frac{\partial w}{\partial y}\bigg)_{x,z} \Delta y + \frac{\partial w}{\partial z}\bigg)_{x,y} \cancel{\Delta z}^{\,0}$$

This expresses the dependence of Δw on Δx and Δy.

For some processes, the coefficients $\partial w/\partial x)_{y,z}$ and $\partial w/\partial y)_{x,z}$ may be inconvenient for us to use. In these cases we may be able to find alternate expressions either through Maxwell's relations (Table 13.1) or by finding alternate, equivalent interrelationships from the first law.

EXAMPLE

Suppose we have a beaker of water at atmospheric pressure ($dN = 0, dp = 0$), and we wish to know how its internal energy varies as measured amounts of heat are added (Figure 14.2).

For this problem we wish to express dE in terms of (N, p, S) as the first two of these are constrained, and the latter is measured. We have

$$dE(N, p, S) = \frac{\partial E}{\partial N}\bigg)_{p,S} \cancel{dN}^{\,0} + \frac{\partial E}{\partial p}\bigg)_{N,S} \cancel{dp}^{\,0} + \frac{\partial E}{\partial S}\bigg)_{N,p} dS$$

Figure 14.2 If we have a system of fixed size $(dN = 0)$ at atmospheric pressure $(dp = 0)$, how does its internal energy change as measured amounts of heat are added?

or

$$dE = \left.\frac{\partial E}{\partial S}\right)_{N,p} \frac{dQ}{T}$$

EXAMPLE

Suppose that in the above problem it was inconvenient for us to determine the coefficient $\partial E/\partial S)_{N,p}$. Is there any other equivalent coefficient we could use?

None of Maxwell's relations involve this particular differential, but the first law expresses interrelationships among parameters that could be used to find an alternate expression. The first law is customarily written in terms of the parameters (E, S, V, N)

$$dE(S, V, N) = T\,dS - p\,dV + \mu\,dN^{\nearrow^0} \tag{14.1}$$

We need to transform this into an expression involving (E, S, p, N) to get $\partial E/\partial S)_{N,p}$. We can do this by rewriting dV in terms of these parameters, via any of several different ways.

$$dV = \left.\frac{\partial V}{\partial E}\right)_{S,p} dE + \left.\frac{\partial V}{\partial S}\right)_{E,p} dS + \left.\frac{\partial V}{\partial p}\right)_{E,S} dp^{\nearrow^0}$$

$$dV = \left.\frac{\partial V}{\partial E}\right)_{p,N} dE + \left.\frac{\partial V}{\partial p}\right)_{E,N} dp^{\nearrow^0} + \left.\frac{\partial V}{\partial N}\right)_{E,p} dN^{\nearrow^0}$$

$$dV = \left.\frac{\partial V}{\partial S}\right)_{p,N} dS + \left.\frac{\partial V}{\partial p}\right)_{S,N} dp^{\nearrow^0} + \left.\frac{\partial V}{\partial N}\right)_{S,p} dN^{\nearrow^0} \tag{14.2}$$

Inserting any of the above into the first law will give an alternate expression for $\partial E/\partial S)_{N,p}$. For example, using the last expression for dV in the first law (14.1) we

have

$$dE(S,\,p,\,N) = \left[T - p \frac{\partial V}{\partial S}\right)_{p,N} \right] dS - p \frac{\partial V}{\partial p}\right)_{S,N} dp^{0} + \left[\mu - p \frac{\partial V}{\partial N}\right)_{S,p} \right] dN^{0}$$

For our particular case ($dN = 0$, $dp = 0$) this gives

$$dE = \left[T - p \frac{\partial V}{\partial S}\right)_{p,N} \right] \frac{dQ}{T} \tag{14.3}$$

If this doesn't please us, we can use Maxwell's relation (M7) to change this partial differential for another, and if that still isn't helpful; we can find other alternate expressions relating dE to dQ by using one of the other expressions (14.2) for dV.

EXAMPLE

Suppose we were interested in the variation in temperature of an air mass as it rises to go over a mountain range (Figure 14.3). Suppose we could assume that the air mass stays pretty well intact ($dN = 0$) and that there is little or no heat energy put into or removed from the air during this process ($dS = 0$). The pressure is a known function of altitude, so what we wish is an expression for dT in terms of dp.

We wish an expression for dT in terms of (p, S, N) since the first parameter is easily determined and the last two involve the constraints

$$dT(p,\,S,\,N) = \frac{\partial T}{\partial p}\right)_{S,N} dp + \frac{\partial T}{\partial S}\right)_{p,N} dS^{0} + \frac{\partial T}{\partial N}\right)_{S,p} dN^{0}$$

or

$$dT = \frac{\partial T}{\partial p}\right)_{S,N} dp$$

If this differential isn't convenient for us, we can transform it using Maxwell's relation (M7).

Figure 14.3.   As an air mass rises to go over a mountain range, it expands and cools, because the pressure on it diminishes as it rises. Exactly how are changes in temperature, ΔT, related to changes in pressure, Δp, assuming that the air mass stays intact ($\Delta N = 0$) and that no heat is added or removed ($\Delta S = 0$)?

$$dT = \frac{\partial V}{\partial S}\bigg)_{p,N} dp \tag{14.4}$$

PROBLEMS

14-1. Suppose you have a system of a fixed number of particles confined to a fixed volume, and you wish to know how the pressure varies with temperature.

   (a) What are the constraints?
   (b) Write down an expression for dp in terms of dT.
   (c) Write this in an alternate way through the use of one of Maxwell's relations.

14-2. Suppose you have a system of a fixed number of particles held at fixed volume, and you wish to know how the entropy varies with the temperature.

   (a) What are the constraints?
   (b) Write down an expression for dS in terms of dT.
   (c) In the first law, $dE = T\,dS - p\,dV + \mu\,dN$, one of the differentials is neither among the two involved in the constraints nor among the two whose interrelationship we wish to know. Rewrite this differential in terms of any three of the four involved in the constraints and interrelationship (e.g., T, V, N), to get from the first law an alternate expression for dS in terms of dT.

14-3. Suppose you want to know how E varies with T and p during a nondiffusive process. Show that

$$\frac{\partial E}{\partial p}\bigg)_T = V(p\kappa - T\beta)$$

$$\frac{\partial E}{\partial T}\bigg)_p = (C_p - pV\beta)$$

(*Hint*. Start with $dE = T\,dS - p\,dV$, and rewrite dS and dV each in terms of dp and dT. Use Maxwell's relation M10.)

B. ISOBARIC PROCESSES

Most common thermodynamical processes we study are carried out under atmospheric pressure. Some isobaric processes of interest may occur at larger or smaller pressures, such as chemical processes in outer space (Figure 14.4), biological processes on the ocean bottom, or physical processes deep within the earth. But in any case, isobaric processes are extremely common.

Quite often we deal with systems of fixed size ($dN = 0$) undergoing isobaric processes. In these cases the expression for the change in enthalpy (see Eq. 13.11)

$$dH = T\,dS + V\,\cancel{dp}^{\,0} + \mu\,\cancel{dN}^{\,0}$$

reduces to

$$dH = T\,dS \tag{14.5}$$

Figure 14.4 In any local region of this nebula the pressure is small but constant. Therefore, the processes are isobaric. (© 1965 California Institute of Technology and Carnegie Institution)

which is just the amount of heat energy added to the system under equilibrium conditions.

> For the common case of isobaric, nondiffusive processes, the change in enthalpy is just the amount of heat energy added to or removed from the system.

For this reason, enthalpy is frequently referred to as the "heat function."

Equations of state become simpler for isobaric processes, since pressure is usually one of the variables in these equations. For example the ideal gas law reads

$$pV = NkT \tag{12.7}$$

so for isobaric processes in ideal gases, the volume and temperature are directly related.

$$\frac{V}{T} = \text{constant} \tag{14.6}$$

For real gases, the relationship is slightly more complicated. From van der Waals equation of state,

$$\left(p + \frac{a}{v^2}\right)(v - b) = RT$$

we have

$$\frac{T}{(v - b)} - \frac{a}{Rv^2} = \text{constant} \tag{14.7}$$

For solids and liquids, the relationship between temperature and volume for isobaric processes must normally be determined empirically, since there is no universal equation of state from which this interdependence can be derived. The two are related through the coefficient of volume expansion,

$$\beta = \frac{1}{V}\frac{\partial V}{\partial T}\bigg)_p$$

which must be measured and tabulated for the materials of interest. Even for one material, the coefficient of volume expansion depends on the temperature and pressure,

$$\beta = \beta(T, p)$$

so we must frequently interpolate between measured values to find its value at the particular temperature and pressure of interest to us.

C. ISOTHERMAL PROCESSES

We also deal fairly frequently with processes carried out at constant temperature. Many common systems are in good thermal contact with their environment, which keeps their temperature fairly constant (Figure 14.5). Air is a poor temperature moderator since its thermal conductivity is so low. But many systems do make good thermal contact with a reservoir, such as being in a water bath, or on the surfaces of metals, etc.

For isothermal processes, the Helmholtz free energy of a system is sometimes a convenient property to work with. Changes in this property are given by

$$dF = -S\,dT - p\,dV + \mu\,dN$$

Consequently, we see that for isothermal ($dT = 0$), nondiffusive ($dN = 0$) processes, the Helmholtz free energy is a measure of the work done on a system, or of the potential energy stored.

$$dF = -p\,dV$$

For this reason, it is sometimes referred to as the "work function."

Temperature is also one of the parameters frequently involved in equations of state, so these equations are simplified for isothermal processes.

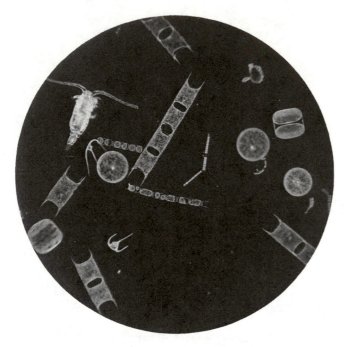

Figure 14.5 Since they are bathed in their ocean environment, all the various physical and chemical processes carried out in these microorganisms are done isothermally. (Douglas P. Wilson, F.R.P.S. Marine Biological Laboratory).

In the case of a gas undergoing isothermal expansion or compression, the relationship between pressure and volume can be derived from the ideal gas law. For isothermal processes,

$$pv = RT = \text{constant} \qquad (v \text{ is the molar volume}) \tag{14.8}$$

If a great deal of accuracy is desired, or if the gas is compressed to a point where the corrections to the ideal gas law are important, then one can find the pressure-volume relationship from the van der Waals equation.

$$\left(p + \frac{a}{v^2}\right)(v - b) = RT = \text{constant} \,(v \text{ is the molar volume}) \tag{14.9}$$

Isothermal processes in solids or liquids are approached differently. For neither of these do we have a single universal equation of state to work with, as we do for gases. The relationship between pressure and volume is determined experimentally in the measurement of the isothermal compressibility.

$$\kappa = -\frac{1}{V}\frac{\partial V}{\partial p}\Big)_T$$

This varies greatly from one material to the next. To make things worse, even for the same material, the isothermal compressibility varies with temperature and pressure.

$$\kappa = \kappa(T, p)$$

So for isothermal processes involving solids and liquids, the relationships between volume and pressure are determined in the laboratory and tabulated. Even then, we must often interpolate between the measured values to get κ at the particular temperature and pressure of interest to us.

D.   ADIABATIC PROCESSES

Another type of process that warrants special consideration includes those processes for which there is no net flow of heat into or out of the system ($dQ = 0$). This is the case for those processes occurring sufficiently rapidly that no significant amounts of heat can be transferred to or from the system during the process (Figure 14.6).

An example of such a process would be the compression of the gases in the cylinder of an automobile. This takes only about 0.02 s, which is too short a time for any appreciable amount of heat energy to be transferred between the gases and cylinder

Figure 14.6   A thunderhead is the result of the rapid adiabatic cooling of a rising air mass. (Grant Heilman)

walls. At the other extreme, some deep ocean currents require thousands of years for circulation, yet the water masses are so huge and conductivities so low that no appreciable quantities of heat are transferred to or from these masses in this long period of time. The gravitational collapse of an interstellar cloud of dust and gas, eventually becoming a new star, is an adiabatic process that may take millions or even billions of years.

Cooling things off to a very low temperatures must be done adiabatically, because they would be heated by their warmer environment if they were not insulated from it. There is a great deal of interest in attaining low temperatures because of the variety of interesting physical phenomena displayed by some systems. Adiabatic expansion is frequently used to cool systems down to around 1 K, but for temperatures lower than this, adiabatic demagnetization of a paramagnetic substance is usually used. These processes are described in detail in Chapter 28.

Equations of state usually do not involve the entropy explicitly, so to use the adiabatic constraint to find simplified relationships among the parameters in the equation of state usually involves some manipulation. Nonetheless, it should be clear that processes that are both adiabatic and nondiffusive, operate under *two* constraints, leaving only *one* independent variable. With some cleverness, we should be able to express the dependence of any one parameter on any other.

D.1   Adiabatic Processes In Gases

For gases that fit the ideal gas model fairly well, we have

$$E = N \frac{v}{2} kT$$

$$pV = NkT \tag{12.8}$$

where N is the number of molecules, and v is the number of degrees of freedom per molecule. The differential forms of these are

$$dE = N \frac{v}{2} k \, dT$$

$$p \, dV + V \, dp = Nk \, dT. \tag{14.10}$$

In this section we will use these two equations, plus the first law

$$dE = dQ - p \, dV$$

which for adiabatic processes ($dQ = 0$) reads

$$dE = -p \, dV \tag{14.11}$$

The relationships (14.10) and (14.11) are three relationships among the four differentials (dE, dT, dp, dV). We can use them to express the relationship between any two of the variables.

The relationships

$$dE = N\frac{v}{2}dT \qquad \text{(ideal gas)} \tag{14.10}$$

$$p\,dV + V\,dp = Nk\,dT \qquad \text{(ideal gas)} \tag{14.10}$$

$$dE = -p\,dV \qquad \text{(adiabatic processes)} \tag{14.11}$$

can be combined to express the interdependence of any two of the variables.

For example, combining the first and last to eliminate dE, we have

$$\frac{v}{2}Nk\,dT = -p\,dV$$

which can be combined with the second relationship to eliminate dT.

$$\frac{v}{2}(p\,dV + V\,dp) = -p\,dV$$

We can rearrange terms to get

$$\left(\frac{v+2}{v}\right)\frac{dV}{V} + \frac{dp}{p} = 0$$

which can be integrated to give

$$\left(\frac{v+2}{v}\right)\ln V + \ln p = \text{constant}$$

or

$$pV^{(v+2)/v} = \text{constant} \tag{14.12}$$

In homework Problem 14-5 you will be asked to perform a similar manipulation of the three relationships in the above box to find the interrelationship between V and T.

$$V^{\gamma-1}T = \text{constant} \qquad \gamma = (v+2)/v \tag{14.13}$$

The exponent $[(v+2)/v]$ appearing in the exponent in Eq. 14.12 is commonly given the symbol γ, and can be related to the heat capacities as follows. We can write the first law in the form

$$dQ = dE + p\,dV$$

from which we see the heat capacities are given by

$$C_V = \left(\frac{\partial Q}{\partial T}\right)_V = \left(\frac{\partial E}{\partial T}\right)_V$$

$$C_p = \left(\frac{\partial Q}{\partial T}\right)_p = \left(\frac{\partial E}{\partial T}\right)_p + p\left(\frac{\partial V}{\partial T}\right)_p$$

From the two relationships (14.10) for an ideal gas, we have

$$\left(\frac{\partial E}{\partial T}\right)_V = \left(\frac{\partial E}{\partial T}\right)_p = N\frac{v}{2}k$$

and

$$p\left(\frac{\partial V}{\partial T}\right)_p = Nk$$

With these, the above heat capacities can be written in the form

$$C_V = \frac{v}{2}Nk$$

$$C_p = \frac{v}{2}Nk + Nk = \left(\frac{v+2}{2}\right)Nk$$

and their ratio gives the factor γ.

$$\frac{C_p}{C_V} = \frac{v+2}{v} = \gamma \tag{14.14}$$

SUMMARY

For adiabatic processes in gases which satisfy the ideal gas law reasonably well, the following interrelationships hold.

$$pV^\gamma = \text{constant} \tag{14.12}$$

$$V^{\gamma-1}T = \text{constant} \tag{14.13}$$

with

$$\gamma = \frac{C_p}{C_V} = \frac{v+2}{v} \tag{14.14}$$

where v is the number of degrees of freedom per molecule.

The ratio of heat capacities, $\gamma = C_p/C_v$, is even more easily measured than are the heat capacities themselves. We could do this by putting the gas in an insulated cylinder with a movable piston. If we move the piston slightly, and record initial and final values of p and V or of p and T, we could solve Eqs. 14.12 or 14.13 for γ, which gives us the number of degrees of freedom per molecule through Eq. 14.14. This is another example of how an easily measured macroscopic property reflects the microscopic structure of a system.

PROBLEMS

14.4   Starting with the result (14.12) and the ideal gas law, show that $V^{\gamma-1}T = \text{constant}$ for adiabatic processes in gases.

14-5.   Prove that $V^{\gamma-1}T = \text{constant}$ for adiabatic processes in gases, starting with expressions (14.10) and (14.11).

14.6. A certain gas is compressed adiabatically. The initial and final values of the pressure and volume are:

$$p_i = 1 \text{ atm} \qquad p_f = 1.1 \text{ atm}$$

$$V_i = 2 \text{ liters} \qquad V_f = 1.85 \text{ liters}$$

What is the number of degrees of freedom per molecule for this gas?
(*Hint.* $(1 + x)^y \approx 1 + yx$ for $|yx| \ll 1$.)

14-7. Consider 0.446 mole of an ideal gas, having 3 degrees of freedom per molecule, initially at a temperature of $0°C$, pressure of 1 atm and volume of 10.0 1. Suppose this gas was expanded to a volume of 10.4 liters. What would be the new values of its temperature and pressure if this expansion were:

(a) Isobaric?
(b) Isothermal?
(c) Adiabatic?

14-8. For each of the above three processes, what would be the:

(a) Heat added to the gas?
(b) Change in internal energy?
(c) Change in enthalpy?
(d) Change in Helmholtz free energy?

14-9. Consider a mole of steam under very high pressure, so that van der Waals equation is the proper equation of state. The van der Waals constants for steam are

$$a = 5.5 \text{ liters}^2\text{-atm/mole}^2$$

$$b = 0.030 \text{ liter/mole}$$

Suppose this steam were initially at a pressure of 100 atm and had a volume of 0.3 liter, and was then expanded to twice that volume.

(a) What is the initial temperature of the steam?
(b) What is the final temperature if the expansion is isobaric?
(c) What is the final pressure if the expansion is isothermal?

D.2 Adiabatic Processes in Liquids and Solids

For solids and liquids we frequently lack adequate models from which equations of state can be derived. We compensate for this theoretical ignorance through experimental determinations of interrelationships such as isothermal compressibilities, heat capacities, coefficients of thermal expansion, etc. As we have seen, adiabatic, nondiffusive processes have two constraints ($dS = 0$, $dN = 0$) and only one independent variable, so we should be able to express the dependence of any one property on any other. For solids and liquids we should be able to do it in terms of these experimentally determined parameters.

For processes performed under equilibrium conditions, there will be no change in entropy if no heat is transferred. Therefore, to find the interrelationship between any two variables, we simply write an expression involving the entropy and the two variables of interest. For example, if we are interested in the dependence of temperature on volume, we could write

$$dS = \frac{\partial S}{\partial T}\Big)_V dT + \frac{\partial S}{\partial V}\Big)_T dV = 0 \qquad (14.15)$$

or if we were interested in the dependence of temperature on pressure we could write

$$dS = \frac{\partial S}{\partial T}\Big)_p dT + \frac{\partial S}{\partial p}\Big)_T dp = 0 \qquad (14.16)$$

The four differential coefficients can be transformed into commonly measured parameters. For example, from the definition of heat capacities

$$\frac{\partial S}{\partial T}\Big)_V = \frac{C_V}{T} \qquad (14.17)$$

and

$$\frac{\partial S}{\partial T}\Big)_p = \frac{C_p}{T} \qquad (14.18)$$

From Maxwell's relation (M10) we have

$$\frac{\partial S}{\partial p}\Big)_T = -\frac{\partial V}{\partial T}\Big)_p = -V\beta \qquad (14.19)$$

and from Maxwell's relation (M4) we have

$$\frac{\partial S}{\partial V}\Big)_T = \frac{\partial p}{\partial T}\Big)_V \qquad (14.20)$$

This last expression can be further simplified by writing the interrelationship between V, p, and T

$$dV = \frac{\partial V}{\partial p}\Big)_T dp + \frac{\partial V}{\partial T}\Big)_p dT = -(V\kappa)\,dp + (V\beta)\,dT$$

or

$$dp = \left(\frac{\beta}{\kappa}\right)dT - \left(\frac{1}{V\kappa}\right)dV$$

With this, our expression (14.20) becomes

$$\frac{\partial p}{\partial T}\Big)_V = \frac{\beta}{\kappa} \qquad (14.21)$$

Using the expressions for the differential coefficients (14.17), (14.18), (14.19), (14.20), and (14.21), our interrelationships between T and V (14.15), and T and p (14.16) become for adiabatic processes ($dS = 0$),

$$\frac{dT}{T} = -\left(\frac{\beta}{\kappa C_V}\right)dV \qquad (14.22)$$

$$\frac{dT}{T} = \left(\frac{V\beta}{C_p}\right)dp \qquad (14.23)$$

Combining these two we find an interrelationship between dV and dp.

$$\frac{dV}{V} = -\left(\kappa \frac{C_V}{C_p}\right)dp \tag{14.24}$$

SUMMARY

For liquids and solids involved in nondiffusive, adiabatic processes,

$$\frac{dT}{T} = -\left(\frac{\beta}{\kappa C_V}\right)dV \tag{14.22}$$

$$\frac{dT}{T} = \left(\frac{V\beta}{C_p}\right)dp \tag{14.23}$$

$$\frac{dV}{V} = -\left(\kappa \frac{C_V}{C_p}\right)dp \tag{14.24}$$

where C_p and C_V are heat capacities at constant pressure and volume, respectively, β is the coefficient of thermal expansion, and κ is the isothermal compressibility.

The parameters β, κ, C_V, and C_p are all positive, so we see that for adiabatic processes the following is true.

1. Increased volume causes decreased temperature.
2. Increased pressure causes increased temperature.
3. Increased pressure causes decreased volume.

Good examples of important adiabatic processes are found in the earth sciences. An important component of our weather is caused by adiabatic processes in air masses. As these masses move across the earth, or change elevations, they encounter different pressures, resulting in changes in volume and temperature as well. Adiabatic ocean convection has a large influence on our climate, and research seems to indicate that convection cells in the earth's mantle (Figure 14.7) may be responsible for the very

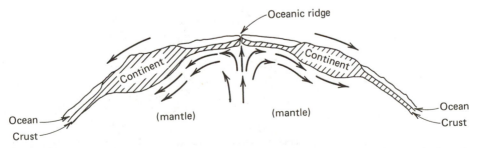

Figure 14.7 Convection cells in the earth's mantle may drive the continents as they drift across the face of the globe. The determination of whether these convection cells could be stable features involves consideration of adiabatic temperature changes in the flowing materials.

existence of the oceans, atmosphere, and many of the other components of our environment.

One problem of current interest is whether convection cells within the earth's mantle would tend to perpetuate themselves, once they are started. That is, for example, once a mass of material deep within the earth begins to rise, will it continue to rise until it reaches the surface? Similarly, once a mass of material begins to sink, will it continue to sink through the mantle?

On the one hand, we know that deeper portions of the mantle tend to be hotter, since they are closer to the heat sources deep within the earth. So as a mass of material rises, its environment becomes cooler, and it may continue to rise, because it finds itself hotter and therefore less dense than the materials around it (assuming homogeneous composition). On the other hand, we know that the rising material experiences lower pressures, therefore expanding and cooling as it rises. If this rate of cooling is greater than that of its immediate environment, then it will soon find itself cooler and denser than its environment, so it will sink again, instead of rising all the way to the surface.

Therefore, to see if the convection will continue to the surface, we must compare the rate of adiabatic cooling as the mass rises to the temperature gradient existing in the mantle (Figure 14.8). If the adiabatic drop in temperature is smaller than the decrease in temperature of the surroundings, then the mass will always be warmer and less dense than its environment, and it will continue rising to the surface. Similar analysis can be applied to the sinking of cooler, denser masses. This is just one example of the many problems involving adiabatic processes in gravity-driven convection of current interest. Similar considerations are employed in studying the stability of convective patterns in the oceans, planetary atmospheres, stellar interiors, and many other areas.

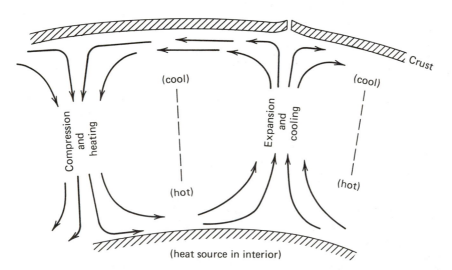

Figure 14.8 Convection cells in the earth's mantle. Does the rising material cool faster or slower than the ambient temperature of the neighboring environment? Equivalently, does the sinking material heat up faster or slower than the ambient temperature of the neighboring environment? The answers to these questions determine whether the convection cells are self-perpetuating.

PROBLEMS

14-10. For most solids and liquids, C_p and C_V are nearly the same. Furthermore C_V, β, κ, and the volume V remain reasonable constant over fairly wide ranges in T and p. Assuming they are fairly constant, integrate Eqs. 14.22, 14.23, and 14.24 to find the relationships between initial and final values of (T, p, V) for adiabatic processes in solids and liquids.

14-11. Near the bottom of some ocean basin (about 5 km down) the temperature of the water is 3°C. Even in the high pressures at that depth, the following parameters are roughly the same as at the ocean's surface.

coefficient of volume expansion	$\beta = 1.2 \times 10^{-4}/\text{K}$
isothermal compressibility	$\kappa = 4.6 \times 10^{-10} \text{ m}^2/\text{N}$
molar heat capacity	$c_v = 75 \text{ J/K}$
molar volume	$v = 1.8 \times 10^{-5} \text{ m}^3$
density	$\rho = 1.03 \times 10^3 \text{ kg/m}^3$

(a) What is the change in pressure with each additional meter's depth?
(b) Measurements taken show that the temperature increases with depth down there at a rate of 0.4×10^{-4} K/m. Is the water stable against vertical convection or not?

14-12. Calculate the parameters β and κ in terms of (p, V, T) for a gas using the ideal gas law.

14-13. Use your results from the preceding problem to do the following for adiabatic processes in gases.

(a) Starting with Eq. 14.22, show that $TV^{\gamma-1} = \text{constant}$.
(b) Starting with Eq. 14.23, show that $Tp^{(1/\gamma)-1} = \text{constant}$.
(c) Starting with Eq. 14.24, show that $pV^{\gamma} = \text{constant}$.

E. REVERSIBILITY

In any thermodynamical process, various systems interact with other systems through the removal or imposition of various constraints. The question we wish to deal with in this section is whether it is possible to reverse the process and bring the entire body of interacting systems back to their original state.

We know that during any thermodynamical process, the entropy of the combined system must either increase or remain constant.

$$\Delta S_0 \geq 0$$

The fact that the total entropy cannot *decrease* means that if the total entropy of all interacting systems increases, the original state cannot be regained, for to do so, the entropy would have to decrease, in violation of the second law. Therefore, only those processes for which there is no change in total entropy are reversible ($\Delta S_0 = 0$).

An incorrect, but unfortunately common, extrapolation from this statement is that since the change in entropy of a system is related to heat added ($dQ = TdS$), then adiabatic processes must be reversible. This is wrong for two reasons. First of all, many nonadiabatic processes are reversible. In a reversible process, heat energy may flow between two systems, as long as the amount of entropy lost by one equals that gained by the other. It is the total entropy of the combined system that is governed by the second law. (In the next chapter we will examine two important reversible processes involving heat exchange between systems.)

Second, there are many processes during which the entropy increases even though no heat is added. This statement may seem to violate the relationship $dQ = TdS$, because it does! This relationship was derived for the case when all the components of a system are in equilibrium.

EXAMPLE

Consider two systems having different pressures separated from each other by a movable piston as in Figure 14.9. Originally, the piston is anchored in place. If the constraint on the piston is removed and the two systems interact mechanically, is the process reversible?

We have seen that the second law demands that the piston move away from the higher pressure system and toward the one under lower pressure, because this causes the total entropy to increase. Since the total entropy increases, the process is irreversible; we could never see the piston go back to its original position, restoring the original conditions.

EXAMPLE

Two systems at different temperatures are thermally insulated from each other, as in Figure 14.10. If the insulation between them is removed and they are allowed to interact thermally, is the process reversible?

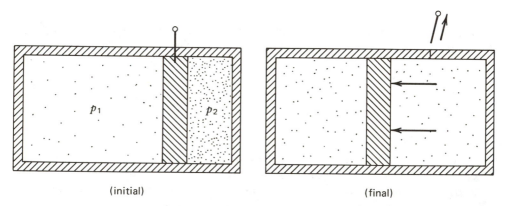

(initial)   (final)

Figure 14.9 An adiabatic process. Both systems are thermally insulated. Initially the two systems are under different pressures and are separated by an anchored piston. When the piston is released, it moves toward the system under lower pressure. Is the process reversible? (That is, even though there is no heat gained or lost by the combined system, is there no change in total entropy?)

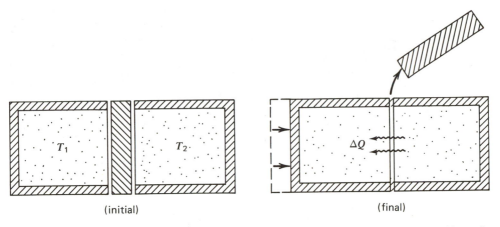

Figure 14.10 An adiabatic process. The combined system is thermally insulated from the outside world. Initially the temperatures of the two systems are unequal. When the insulating barrier is removed, heat flows from the hotter side to the cooler side. Is the process reversible? (That is, even though there is no heat gained or lost by the combined system, is there a change in total entropy?)

The second law demands that heat flow from the hotter system toward the cooler one, as this causes the total entropy to increase. Since the total entropy increases, the process is irreversible; we could never see the heat flow from the cooler system back into the hotter one, restoring the original conditions.

EXAMPLE

Two balls of putty flying in opposite directions through space collide head on and stick together (Figure 14.11). Is the process reversible?

Before they collide the molecules of any one ball of putty are pretty much moving in the same direction and at the same speed. During the collision, this kinetic energy gets transformed into heat, or random motion of the molecules. There are far more states available for the molecules moving in random directions than there are for them to all move the same direction at the same time. That is, the entropy of the states of random motion is far greater than that of the coherent collective motion.

Since the entropy has increased, the process is irreversible. We could never expect to see the balls suddenly fly back apart along the path they came, for this would require all the molecules in each ball to happen to be going the same direction at the same time—a state of very low entropy (equivalently, low probability). The entropy just will not decrease again once it has increased.

This last example can be made into a more general statement regarding friction. Friction takes the coherent collective motion of molecules (states of rather low entropy or probability), and transforms it into more random molecular motions, for which there are more available states, or equivalently higher entropy or probability. For example, as a book slides across a table, molecules originally moving together at the

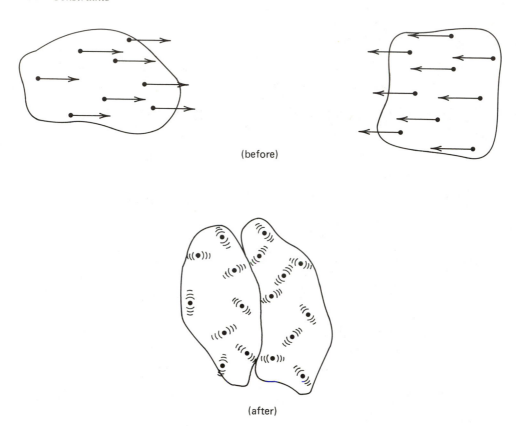

(before)

(after)

Figure 14.11   Two balls of putty before and after collision, with motion of some representative molecules indicated. During the collision, their coherent motion in one direction is transformed into random-direction motion of the molecules. Since there are far more states available for molecules to oscillate in random directions than there are for them to all move the same direction, the entropy increases during the collision, and the process is irreversible. We would never see them fly back apart again once they have stuck together.

same speed and in the same direction, get slowed down, and this energy gets transferred into random molecular vibrations, having much higher entropy. Consequently, friction always increases entropy, and whenever friction is present, the process is irreversible.

SUMMARY

A process is reversible only if the total entropy of all interacting systems remains constant. This may or may not involve the transfer of heat energy between systems. If the interacting systems are not in equilibrium at all times, the process cannot be reversible. Also, if there is any friction, the process cannot be reversible.

The term "quasi-static" is used to describe processes carried out sufficiently slowly that all portions of all interacting systems can be assumed to be in equilibrium at all times. Just how slow this is depends on the process. In most materials, mechanical equilibrium is attained rather quickly and thermal equilibrium rather slowly— especially in gases, since they are poor conductors of heat. So in a gas, for example, quasi-static compression can be accomplished rather quickly. As long as sound waves or other pressure instabilities are not generated, the compression is quasistatic. However, quasi-static heat transfer in gases must be done slowly, as thermal gradients are easily created, and they dissipate slowly.

PROBLEMS

14-14.   Consider a large rotating flywheel connected to a piston as in Figure 15.2. As the flywheel rotates the gas in the cylinder is alternately compressed and expanded isostatically by the moving piston. If the container and piston are thermally insulated and there is no friction, is the process reversible? (*Hint.* Is the entire system back to its starting point after one complete rotation of the flywheel?)

14-15.   Consider an insulated container holding fresh water and seawater, separated by a partition. If they are initially at the same temperature and pressure and the partition is removed, is the process reversible? That is, can the original system be regained by reinserting the partition?

F.   NONEQUILIBRIUM PROCESSES

This is basically a text on equilibrium thermodynamics. In general, nonequilibrium processes are much more difficult (if not impossible) to handle, because there are no general powerful statistical tools, such as the second law, which can be applied to nonequilibrium processes.

Nonetheless, insights gained from studies of equilibrium processes do help us understand some interesting nonequilibrium processes. We examine some of these here.

F.1   Joule-Thompson (Throttling) Process

The cooling of gases is important in our technological society, not only in refrigeration, but also in the liquefied gases used as coolants in our laboratories. In this section we wish to investigate one of the more efficient ways to cool a gas.

One method would be to allow the gas to do work. As shown in Figure 14.12, for example, we could bring a compressed gas into thermal contact with coldest thing around. When the compressed gas is as cold as possible, we then remove it from the reservoir and let the gas expand against the movable piston, doing work, and therefore losing internal energy and cooling. This process works well in theory but not so well in practice. The fault is that gases have rather low heat capacities in comparison to their containers. The containers must be built to withstand large pressure changes, and so

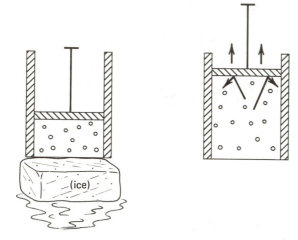

Figure 14.12 One way to cool a gas is to first bring it into thermal equilibrium with the coldest thing around. Then insulate it and expand it against the moving piston. In doing the work, it cools still further.

they are necessarily heavy and have large heat capacities. Much of the energy expended by the expanding gas is returned to it by the walls of the container, and defeats the purpose of expanding the gas in the first place. What is needed is an expansion process where the walls of the container do not release heat to the gas inside. This is accomplished in an adiabatic process called the "Joule-Thompson" process, or "throttling."

The elements of the throttling process are illustrated in Figure 14.13. A gas is forced through a pipe in which there is a constriction of some kind, such as a porous plug, a partition with a small hole in it, etc. Before getting to the constriction the gas is under high pressure. Upon passing through the constriction, it expands under lower pressure to a larger volume, and cools.

Suppose we observe a certain mass of gas going through the constriction. (See Figure 14.13.) Before passing through the constriction it occupies volume, V_i, under pressure, p_i. Upon passing through the constriction, it encounters lower pressure, p_f, and expands to a larger volume, V_f. The work done on it by the gas immediately behind it, in

(porous plug) V_i V_f

Figure 14.13 The elements of the throttling process are an insulated pipe with a constriction (such as a porous plug). The gas is forced through the constriction by high pressure on the left. A part of the gas initially occupying volume V_i, expands to volume V_f on passing through the constriction.

pushing it through the constriction is

$$p_i \int dV_i = p_i V_i$$

Similarly, the work it does on the gas immediately ahead of it by pushing it farther down the part of the pipe beyond the constriction is

$$p_f \int dV_f = p_f V_f$$

Hence, the total work done *by* the mass of interest is

$$\Delta W = p_f V_f - p_i V_i$$

Since the gas near the constriction is thermally insulated from the outside, $\Delta Q = 0$, and we have from the first law,

$$\Delta Q = 0 = \Delta E + \Delta W$$

$$0 = (E_f - E_i) + (p_f V_f - p_i V_i)$$

Rearranging terms gives

$$E_f + p_f V_f = E_i + p_i V_i$$

According to this result, the enthalpy of the gas does not change during the throttling process.

$$H \equiv E + pV = \text{constant}$$

We wish to calculate the change in temperature, ΔT, as a function of the change in pressure, Δp, from one side of the constriction to the other. Clearly, if we write dH in terms of dT and dp,, we can set $dH = 0$, and have the desired relationship between dT and dp. From the first law and the definition of enthalpy above, we get (see Eq. 13.11)

$$dH = 0 = T\, dS + V\, dp^* \tag{14.25}$$

To have the form we wish, we must rewrite dS in terms of dT and dp.

$$dS = \left(\frac{\partial S}{\partial T}\right)_p dT + \left(\frac{\partial S}{\partial p}\right)_T dp$$

Putting this into the expression (14.25) gives

$$T\left(\frac{\partial S}{\partial T}\right)_p dT + \left[T\left(\frac{\partial S}{\partial p}\right)_T + V\right]dp = 0 \tag{14.26}$$

The two partial differentials appearing in this expression may be written in terms of commonly measured properties. From Maxwell's relation (M10) we can write

$$\left(\frac{\partial S}{\partial p}\right)_T = -\left(\frac{\partial V}{\partial T}\right)_p = -V\beta$$

---

* Notice that according to this equation, the entropy changes $[dS = -(V/T)\,dp]$ even though the process is adiabatic (no heat is transferred to the gas by the walls of the pipe). We can conclude that the relationship $dQ = T\,dS$ is wrong for this process. What is it about this process that makes this relationship wrong?

where β is the coefficient of volume expansion, and from the definition of heat capacity, we can have

$$T \frac{\partial S}{\partial T}\bigg)_p = \frac{\partial Q}{\partial T}\bigg)_p = C_p$$

With these substitutions, the above expression relating dT and dp becomes

$$C_p\, dT + V(1 - \beta T)\, dp = 0 \qquad (14.27)$$

SUMMARY

During the throttling process in a gas, the enthalpy remains constant, and the change in the temperature of the gas is related to the change in pressure through

$$dT = \frac{V}{C_p}(1 - \beta T)\, dp \qquad (14.28)$$

where β is the coefficient of volume expansion. It is a nonequilibrium process, so the entropy increases even though no heat energy is added to the system.

PROBLEMS

14-16. A certain gas undergoing the throttling process is initially at a pressure of 100 atm, temperature of 0°C, and the molar volume is 0.25 liters. If the molar heat capacity of this gas is $c_p = 81$ J/mole/K, and the coefficient of volume expansion is $\beta = 2 \times 10^{-3}/°C$, what change in the temperature of the gas do you expect if the pressure is reduced by 1 atm (1 atm = 10^6 dynes/cm².)

14.17. Calculate the coefficient of volume expansion, β, for oxygen gas under a pressure of 100 atm, and a molar volume of 0.25 liters. The van der Waals constants for O_2 are,

$$a = 1.36 \text{ (liters}^2\text{-atm/mole}^2)$$

$$b = 0.0318 \text{ liters/mole}$$

14.18. Calculate β in terms of T for an ideal gas. Put this into Eq. 14.28 to see how the temperature changes with pressure for the throttling process in an ideal gas. Why do you suppose that for real gases, the throttling process works better to cool them if they are under heavy pressures?

F.2   Free Expansion of a Gas

In Section F.1, we studied a gas that was expanding, doing work in the process by pushing the gas ahead of it on down the pipe. It is usually true that gas does work as it expands, and the system on which it does the work can be represented schematically by a moving piston or wall of the container, as in Figure 14.12.

It is possible, however, for a gas to expand without doing any work at all. Consider, for example, a gas in one section of a completely insulated, rigid container, as in Figure

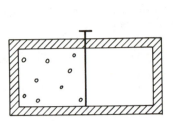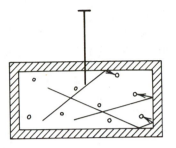

Figure 14.14 In the free expansion of a gas, no work is done (nothing is moved by it). Consequently, its internal energy remains unchanged.

14.14. The other section is completely empty. If the partition is removed, the gas is free to rush into the other section. But because the walls do not move, the gas does no work on them, and because the container is insulated, no heat enters or leaves the gas. Consequently, the internal energy of the gas remains unchanged.

$$\Delta E = \Delta Q - \Delta W = 0$$

Physically, the reason a gas undergoing normal expansion loses energy, and a gas undergoing free expansion does not, involves the movement of the wall. In normal expansion, the gas molecules collide with a receding wall (Figure 14.15), which means that they bounce back with less than their initial energy. In free expansion, however, the wall is stationary and the molecules lose no energy in the collisions.

The temperature of an ideal gas depends only on its internal energy and not at all on its volume (Eq. 12.6). This is just a reflection of the lack of interaction between molecules of an ideal gas, meaning their intermolecular separations do not matter. Consequently, the free expansion of an ideal gas should result in no change in temperature.

For real gases, however, we expect the temperature to change during free expansion. Real gas molecules do interact with each other, so a change in average intermolecular

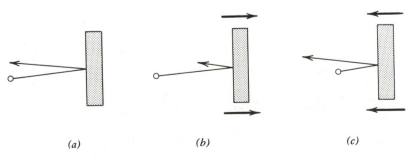

(a)   (b)   (c)

Figure 14.15 Elastic collisions of a gas molecule with a wall. (a) If the wall is stationary, the molecule's kinetic energy remains unchanged. (b) If the wall is receding the molecule loses kinetic energy. (c) If the wall is approaching the molecule gains kinetic energy.

spacings should change their average potential energy (e.g., their chemical potential). If their total internal energy doesn't change, but their potential energy does, then their kinetic energy must also change, resulting in a change in temperature.

To see how temperature changes with volume for free expansion, we must be careful in trying to apply techniques of equilibrium thermodynamics to this nonequilibrium process. Consider a free expansion into an infinitesimal volume, dV, as in Figure 14.16. For this infinitesimal change, we can assume that the pressure and temperature don't change very much. Since the internal energy remains unchanged,

$$dE = 0 = T\,dS - p\,dV$$

we have

$$dS = \frac{p}{T}\,dV \tag{14.29}$$

We want an expression relating (T, V) rather than (S, V) so we must rewrite dS in terms of (T, V).

$$dS = \frac{\partial S}{\partial T}\bigg)_V dT + \frac{\partial S}{\partial V}\bigg)_T dV \tag{14.30}$$

From the definition of heat capacity

$$\frac{\partial S}{\partial T}\bigg)_V = \frac{1}{T}C_V$$

and from Maxwell's relation (M4)

$$\frac{\partial S}{\partial V}\bigg)_T = \frac{\partial p}{\partial T}\bigg)_V$$

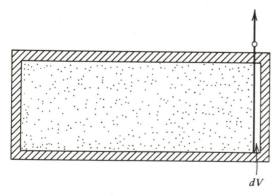

Figure 14.16 When a gas is allowed to expand freely into an infinitesimal volume, dV, the change in entropy is $dS = (p/T)\,dV$.

Putting these into our expression (14.30) gives

$$\frac{1}{T} C_V \, dT + \left(\frac{\partial p}{\partial T}\right)_V dV = \frac{p}{T} dV$$

or

$$dT = \frac{1}{C_V}\left[p - T\left(\frac{\partial p}{\partial T}\right)_V \right] dV \qquad (14.31)$$

The differential $\partial p/\partial T)_V$ can be evaluated from the van der Waals equation of state for the particular gas of interest.

Without going through this calculation (saving it for a homework problem), we should be able to guess how T changes with V, just from physical considerations. There are slightly attractive forces between the molecules of real gases. These forces are the cause of the condensation of gases at sufficiently low temperatures, and were the reason we replaced the pressure in the ideal gas law by $p \rightarrow (p + a/v^2)$. Consequently, a real gas molecule is at any instant in a slight potential well, due to these attractive forces of those molecules around it. At smaller volumes, the molecules are closer together, and the average position of a molecule is farther down the well (Figure 14.17). In free expansion, however, the average separation of molecules increases, and so the average position of a molecule is farther up the potential well. Since potential energy

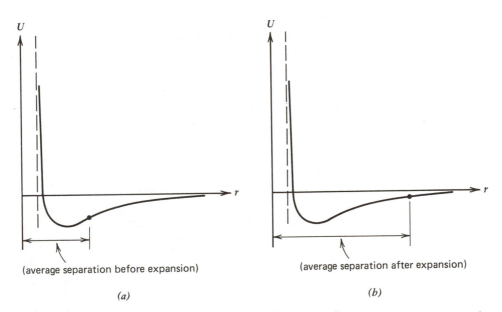

(average separation before expansion)

(a)

(average separation after expansion)

(b)

Figure 14.17 Plots of potential energy versus separation between gas molecules. Upon free expansion, the average molecule will be farther from its neighbors, and consequently not so deep in the potential energy well. An increase in potential energy results in a decrease in kinetic energy, so the temperature should decrease slightly.

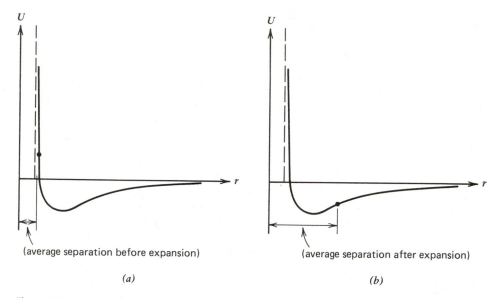

(average separation before expansion)

(a)

(average separation after expansion)

(b)

Figure 14.18 Plots of potential energy versus separation between gas molecules. If the initial state of the gas is sufficiently highly compressed, then free expansion could result in decreased average potential energy, and consequently, increased kinetic energy and higher temperature.

increases, but total energy doesn't change, then the kinetic energy of the molecules will diminish, resulting in lower temperature. That is, for the free expansion of a real gas, increased volume implies:

> increased molecular separation (average)
> increased potential energy (average)
> decreased kinetic energies (average)
> decreased temperature

This scenario would not apply if initially the gas is under such high pressure that the average molecular separations are less than the deepest point in the potential well (Fig. 14.18). Then increased volume (therefore, increased molecular separations) would cause *decreased* potential energy, giving *increased* kinetic energy and *increased* temperature. Hydrogen gas under high pressures is very dangerous for this reason. A ruptured pressure tank could actually result in increased temperature for the escaping hydrogen, which is, of course, explosive.

PROBLEMS

14-19. Use the equation of state for an ideal gas, and Eq. 14.31, to show that in the free expansion of an ideal gas, $dT = 0$.

14-20. (a) Calculate the quantity, $[p - T\,\partial p/\partial T)_V]$, in Eq. 14.31 from the van der Waals equation of state for a real gas, Eq. 12.9.

(b) A mole of steam under high pressure is allowed to expand freely from a volume of 1 liter to a volume of twice that size. The van der Waals constants for steam are

$$a = 5.5 \text{ (liters}^2/\text{mole}^2 \text{ atm)}$$

$$b = 0.030 \text{ (liters/mole)}$$

What is the change in temperature of the steam during the process?

14-21. An ideal gas expands from initial volume, V_i, to final volume, V_f. Calculate the changes ΔE, ΔQ, and ΔW for the gas (in terms of n, T, V_i and V_f):

(a) If the expansion is free expansion.
(b) If the expansion is adiabatic, and starts at temperature T_0. (*Hint. pV^γ = constant.* The value of the constant can be obtained by $p_i V_i = nRT_0$, in terms of T_0 and V_i.)
(c) If the expansion is isothermal (at temperature T_0).

F.3 Removal of Barrier Constraints

The free expansion of a gas is one example of an irreversible process initiated by the removal of barrier constraints. Another example is that illustrated in Figure 14.19,

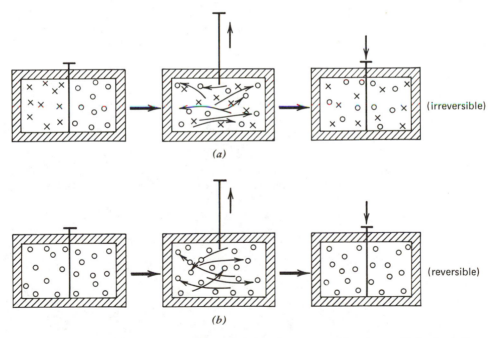

(a)

(b)

Figure 14.19 Two gases of the same temperature and pressure separated by a removable barrier. Is the removal of the barrier a reversible process? That is, can you regain the original system by simply reinserting the barrier? (a) If the molecules of one gas are distinguishable from those of the other, then the process is not reversible. The entropy increases. (b) If the molecules of both gases are the same, then the process is reversible. The entropy does not change. This means that the entropy (or the counting of accessible states) depends on whether or not the particles are distinguishable.

where two gases of equal temperature and pressure are separated by a barrier. If the two gases have different compositions, then the removal of the barrier constraint is an irreversible process.

For example, suppose that the gas on the right was initially oxygen, and that on the left was nitrogen. Once the barrier is removed, the two gases will mix until the mixture is uniform throughout the entire volume. This is because the thoroughly mixed state has the highest entropy. No matter how long you wait, you will never see the two separate again, with all the oxygen moving back to the right side and all the nitrogen to the left. The entropy (or the probability) of this state is far too small. Once the entropy has increased, it cannot again be reduced to its original value (for *macroscopic* systems.)

An important variation of the above experiment is to do it for the case where the two gases not only have the same initial temperature and pressure, but also the same initial composition. For example, suppose both sides of the container have oxygen. When the partition is removed, nothing changes, and the original state can be recovered simply by reinserting the partition. In this case, then, the removal of the barrier caused no change in entropy; the process was reversible.

The above two processes were the same, except for the fact that in the first case the particles were not all identical (some oxygen and some nitrogen), and in the last case they were (all oxygen). In the first case, the entropy increased, and in the second case it didn't. This means that the way we measure entropy, or equivalently the way we count accessible states, depends on whether or not the particles are identical. We'll come back to this thought in later chapters.

PROBLEMS

14-22. (a) Suppose you have two dice, one red and one green. When you shake them up and roll them, how many different distinguishable states are there for the two dice?

     (b) Redo part (a) for the case where the two dice are both white and indistinguishable from each other.

14-23. (a) Suppose you have four pennies that are distinguishable (e.g., different dates on them). When you shake them up, how many distinguishable heads and tails states are there available for them when they land?

     (b) Redo part (a) for the case when the pennies are indistinguishable.

✻ chapter 15

ENGINES AND REFRIGERATORS

The tools we have developed so far apply to systems under equilibrium and subject to well-defined constraints. The operation of real engines can hardly be considered an equilibrium process, because it involves a great deal of turbulence, temperature gradients, and losses. Furthermore, each portion of a cycle in a real engine isn't really following a well defined constraint. Nonetheless, we can study an imaginary idealized engine, called a "Carnot engine," which operates basically the same as real engines, but in equilibrium, without losses, and under a series of well-defined constraints. Studying each portion of the cycle in this idealized engine helps us understand real engines and how they work.

In the last few chapters, we have studied constraints placed on thermodynamical processes, and how we can manipulate these constraints to facilitate analysis and increase our understanding. We divided the various types of constraints between two classes. Natural constraints are those imposed by Nature and are inviolable, no matter what the particular system or process. Imposed constraints are those pertinent to a particular system or process, but do not have to be satisfied by all systems or all processes. These constraints may be fabricated by the experimenter or may be imposed by the peculiar nature of the system or its natural environment.

We have tried to isolate the various common types of constraints and deal with them one at a time, becoming familiar with one before moving on to another. But many systems of interest operate under more than one constraint, and the set of constraints may change with time. So we should also be ready to deal with these cases.

A. THE CARNOT CYCLE

One interesting type of system operating under a sequentially changing set of constraints is an engine. An engine takes in thermal energy at high temperatures, does some work with it, and then exhausts that which remains at lower temperatures (Figure 15.1). Then this process is repeated over and over again.

The operation of real engines usually involves a great deal of turbulence, friction, and other nonequilibrium effects, so it cannot be precisely analyzed with the tools of equilibrium thermodynamics, such as those developed in this book.

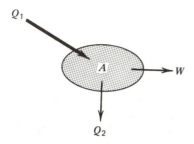

Figure 15.1 An engine is some system, *A*, which receives heat energy Q_1, does some amount of work *W* with it, and exhausts the remainder, Q_2.

Instead, we study a hypothetical, ideal engine, called a "Carnot engine," which is free of any of these nonequilibrium effects, so it can be analyzed with the powerful tools of this course. The Carnot engine serves as a pedagogical instrument to help us understand the operation of real engines. We even have ways of quantifying the degree of variance of any real engine from the ideal Carnot picture.

The entire operating system of a Carnot engine consists of four components (Figure 15.2).

1. A heat reservoir, R_1, at temperature T_1, which provides the thermal energy that runs the engine.
2. A heat reservoir, R_2, at lower temperature T_2 ($T_2 < T_1$), into which thermal energy is exhausted.
3. Some system, A, in a cylinder with a movable piston that converts thermal energy into work.

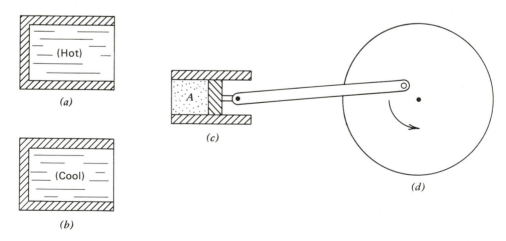

Figure 15.2 The four components of a Carnot engine. (a) A hot reservoir, R_1. (b) A cooler reservoir, R_2. (c) A working system in a cylinder with a movable piston, A. (d) Some system (e.g., a flywheel) on which the work is done.

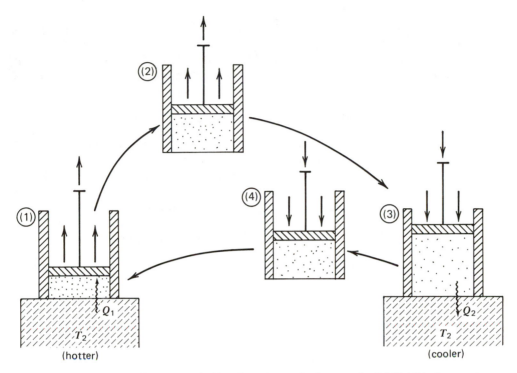

Figure 15.3 The four parts of a Carnot cycle. (1) Isothermal expansion (system gains Q_1). (2) Adiabatic expansion (T falls to T_2). (3) Isothermal compression (system exhausts Q_2). (4) Adiabatic compression (T rises to T_1).

4. Some system on which the work is done. This component is normally left out of sketches of Carnot engines, but it can be thought of as perhaps a large, frictionless flywheel, capable of storing the mechanical energy given it, which could give this mechanical energy back, if we wished.

During any one cycle, these components interact in a particular sequence, which we list here according to the activities of the system in the cylinder, system A. This sequence of interactions is also given schematically in Figure 15.3. Basically, the sequence insures that system A is hotter and therefore under greater pressure on the average during expansion than during compression. Therefore, the work done *by* system A during expansion is greater then that done *on* system A during compression.

1. ISOTHERMAL EXPANSION (THE "IGNITION CYCLE)

Initially at temperature T_1, system A is brought into thermal contact with reservoir R_1. It expands and does work on the piston (and the flywheel) receiving energy Q_1 from the reservoir to replace that spent on the piston, so the temperature remains constant.

2. ADIABATIC EXPANSION (THE "WORK CYCLE")

System A is removed from thermal contact with R_1 and continues to expand ("adiabatically," because there is no transfer of heat to or from system A now). As it expands it does work on the piston, so its internal energy diminishes, as does its temperature.

3. ISOTHERMAL COMPRESSION (THE "EXHAUST CYCLE")

After it has expanded to the point where its temperature has been lowered to T_2, it is placed in thermal contact with reservoir R_2, whereupon the piston (coupled to the flywheel) begins compressing it. This compression would tend to raise its temperature, but since it is in thermal contact with R_2, this thermal energy is exhausted into the reservoir and its temperature remains constant.

4. ADIABATIC COMPRESSION (THE "COMPRESSION CYCLE")

System A is removed from thermal contact with R_2 and the compression continues ("adiabatically," since there is no transfer of heat to or from system A now). As it is compressed, the work done on it by the piston causes its temperature to rise until it reaches T_1, whereupon the cycle is repeated.

The entire cycle is done quasi-statically, so all interacting components are in equilibrium at all times. When heat is transferred between system A and a reservoir, they are at the same temperature (i.e., in thermal equilibrium) at all times. This means there is no net change in entropy for the system as a whole. Whenever heat energy is transferred between system A and either reservoir, we have

$$\Delta Q_A = -\Delta Q_R$$

which means that the total change in entropy is zero.

$$\Delta S_{\text{total}} = \Delta S_A + \Delta S_R = \frac{1}{T}(\Delta Q_A + \Delta Q_R) = 0$$

Similarly, the adiabatic expansion and compression is carried out quasi-statically, and there is no entropy increase due to turbulence or friction.

Since the total entropy of the four-component system does not change during any portion of the cycle, the process is reversible. In the Carnot cycle, thermal energy is removed from the hotter reservoir, some of which is used to do work on an external system, such as the flywheel, and the remainder of which is exhausted into the cooler reservoir. In the *reversed* process, work is done *by* the external system, and thermal energy is taken *from* the cooler reservoir, and exhausted into the hotter reservoir (Figure 15.4). The transfer of heat from cooler to warmer reservoirs is called refrigeration, and so the reversed Carnot cycle is called the "refrigeration cycle."

The common "four-stroke" internal combustion engine found in most automobiles follows the general pattern of the Carnot cycle, with small modifications. In the "ignition cycle," the flammable gases are exploded in the cylinder, and then these hot gases expand pushing the piston outward in the "work cycle." Both of these occur

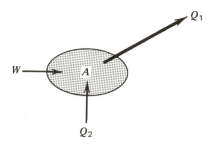

Figure 15.4 A refrigerator is some system A that receives heat energy Q_2 from the cooler reservoir, and with the help of work W performed on it by some external system, it exhausts the total energy received ($Q_1 = W + Q_2$) into the hotter reservoir.

during the first (outward) stroke of the piston. The "exhaust cycle" involves two strokes, as first the hot gases are exhausted, and are then replaced by cool gases coming in through the carburetor. Then these gases are adiabatically compressed on the fourth stroke in preparation for ignition and a repeat of the cycle.

B. THE REFRIGERATION CYCLE

The four components of a refrigerator are the same as those of the Carnot engine, and the external system can also be conveniently thought of as a flywheel. In the Carnot cycle, heat is removed *from* the hotter reservoir and work is done *on* the external system, increasing the flywheel's mechanical energy. In the refrigeration cycle, work is done *by* the external system, decreasing the flywheel's mechanical energy, and heat is transferred *into* the hotter reservoir.

The four parts of the refrigeration cycle are the same as those in the Carnot cycle, except that they are done in reverse, as illustrated in Figure 15.5.

1. ADIABATIC EXPANSION

While isolated from either reservoir, system A is expanded (adiabatically) until its temperature is reduced to T_2.

2. ISOTHERMAL EXPANSION

Then, it is placed in thermal contact with R_2, and the expansion continues. Heat energy is absorbed from R_2 to replace the internal energy lost to work done in the expansion, and its temperature thereby remains constant at T_2.

3. ADIABATIC COMPRESSION

It is then removed from thermal contact with R_2, and work is done *on* it by the external system. This adiabatic compression causes the temperature to rise until it reaches T_1.

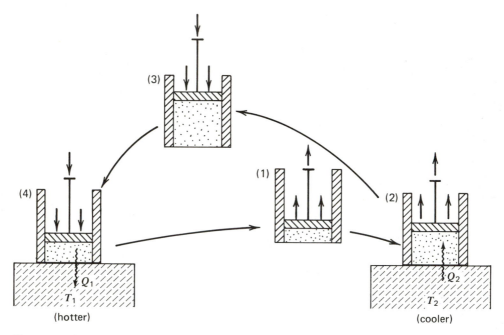

Figure 15.5 The four parts of a refrigeration cycle. (1) Adiabatic expansion (T falls to T_2). (2) Isothermal expansion (system gains Q_2). (3) Adiabatic compression (T rises to T_1). (4) Isothermal compression (system exhausts Q_1).

4. ISOTHERMAL COMPRESSION

At this point it is placed in thermal contact with R_1 and the compression continues. The increased internal energy of system A, due to the continued compression, is exhausted into the reservoir R_1, and its temperature therefore remains constant at T_1. Then it is removed from contact with reservoir R_1, and the cycle is repeated.

In the ideal cycle, each step is performed quasi-statically, and again there is no entropy change for the combined system at any point along the way. The process has the net effect of taking work done *by* an external system and using it to transfer heat from the cooler reservoir to the hotter one.

C. p-V DIAGRAMS

The four parts of a Carnot cycle are conveniently displayed on a *p-V* diagram, such as that of Figure 15.6. Those of a refrigeration cycle are the same, but with the arrows reversed. The temperatures of most materials tend to decrease as they expand, and to increase as they are compressed, and these assumptions are implicit in the description

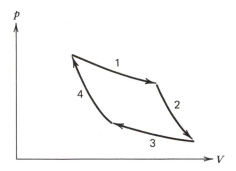

Figure 15.6 The paths followed on a p-V diagram by the four parts of a Carnot cycle. (1) Isothermal expansion. (2) Adiabatic expansion. (3) Isothermal compression. (4) Adiabatic compression.

of the Carnot and refrigeration cycles given previously, and in the p-V diagrams drawn in this section.

There are a few materials for which the reverse is true—that is, they tend to heat up when expanded and cool off when compressed. An example is hydrogen gas under heavy pressure. For such materials, the Carnot cycle would have to be modified from that given here, as would the p-V diagram. The student is encouraged to explore these systems in homework Problems 15-5, 15-6, and 15-7, but here we will concern ourselves only with the more common and familiar case.

For most materials, isothermal expansion results in decreased pressure,* and so on a p-V diagram, the paths of isothermal processes would have negative slopes. The paths of *adiabatic* processes in most systems would have even more negative (i.e., steeper) slopes, as can be seen as follows.

> Consider the expansion of two identical systems, by an identical amount ΔV, one being done isothermally and the other adiabatically. For most systems, adiabatic expansion results in decreased temperature, so the particles of the system expanded adiabatically have lower thermal energies and therefore exert lower pressure than those of the system expanded isothermally.

Consequently, for equal increases in volume, ΔV, the drop in pressure, Δp, is greater for the adiabatically expanded system. That is, the slope, $\Delta p/\Delta V$, is negative for both, but more negative for the adiabatic process.

On the p-V diagram of Figure 15.7, we have qualitatively sketched paths of isothermal processes for a typical system, and paths of adiabatic processes having the

---

* This can be seen from the natural constraint (13.5), $\partial p/\partial V)_E < 0$. From the equipartition theorem, we know internal energy and temperature are related, and so the constraints "constant T" and "constant E" would usually be the same. $\partial p/\partial V)_T \approx \partial p/\partial V)_E < 0$. But you could probably think of phase transitions, or systems not conforming to the model of the equipartition theorem, for which this would not be true.

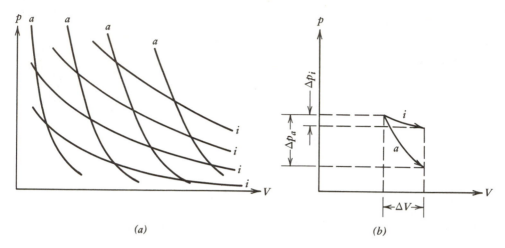

Figure 15.7 (a) Paths followed on a *p-V* diagram for a system undergoing adiabatic expansion or compression (labeled *a*) and for a system undergoing isothermal expansion or compression (labeled *i*). (b) The paths for adiabatic expansion are steeper, because expansion by a certain amount ΔV causes a reduction in temperature for adiabatic expansion, and this drop in temperature causes a greater drop in pressure than for isothermal expansion, where the temperature is held constant.

greater slope. A Carnot cycle could be any path for which the system undergoes: isothermal expansion, followed by adiabatic expansion, followed by isothermal compression, followed by adiabatic compression, and returning to the original starting point.

The *p-V* diagram for a Carnot cycle gives a graphic representation of the amount of work done during one complete cycle. The work done *by* system *A* is the integral

$$\Delta W = \int p \, dV$$

which is positive when the system is expanding (dV positive) and negative when it is being compressed (dV negative). From introductory calculus we know that the integral of $p \, dV$ is the area under the curve in a plot of p versus V, so the net work done *by* system *A* during one complete cycle is equal to the area under the p versus V curve for the system while it is expanding minus the area under the curve while it is being compressed (Figure 15.8). This difference is just the amount of area lying entirely within the closed loop representing the complete cycle. From Figure 15.9 we can see that the amount of work performed during any one cycle can be increased by either increasing the temperature differential between T_1 and T_2, or by increasing the amount of volume change (i.e., the length of the stroke of the piston) during the cycle.

The amount of heat received by system *A* during the ignition cycle, or the amount exhausted during the exhaust cycle, can also be calculated from the *p-V* diagram, provided the equation of state for system *A* is known. During either isothermal process, the amount of heat transferred is given by

$$dQ = T \, dS = T \left[\left(\frac{\partial S}{\partial T} \right)_V \overset{0}{dT} + \left(\frac{\partial S}{\partial V} \right)_T dV \right] \tag{15.1}$$

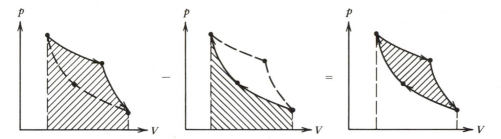

Figure 15.8 Plots of pressure versus volume for a Carnot cycle. The net work done by the system during one complete cycle is the sum of the work done during expansion (area under the curve) plus the work done during compression (negative of the area under the curve), which is the area lying completely within the path of a full cycle on a *p-V* diagram.

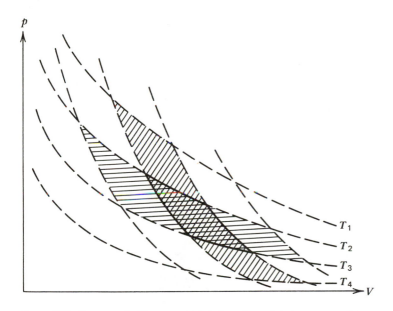

Figure 15.9 Carnot cycle. The amount of work performed during one complete cycle is the amount of area completely within one cycle on a *p-V* diagram. It can be increased by either increasing the change in volume during expansion and compression (horizontal cross-hatching), or by increasing the temperature differential over which the engine operates (diagonal cross-hatching).

Using Maxwell's relation (M4),

$$\left(\frac{\partial S}{\partial V}\right)_T = \left(\frac{\partial p}{\partial T}\right)_V \tag{M4}$$

this become

$$dQ = T\left(\frac{\partial p}{\partial T}\right)_V dV \tag{15.2}$$

The differential coefficient can be calculated from the equation of state, and the total heat transfer can be obtained by integrating.

$$\Delta Q = T \int_{V_1}^{V_2} \left(\frac{\partial p}{\partial T} \right)_V dV \qquad (15.3)$$

SUMMARY

On a p-V diagram for most systems, the paths followed during isothermal processes have negative slopes, and the paths followed during adiabatic processes have slopes even more negative (i.e., steeper).

During one complete Carnot cycle, the work done by the system is just the area inside the curve of the p-V diagram for the cycle. The amount of heat transferred to the system during either isothermal portion of the cycle is given by

$$\Delta Q = T \int_{V_1}^{V_2} \left(\frac{\partial p}{\partial T} \right)_V dV \qquad (15.3)$$

where the integrand is obtained from the appropriate equation of state.

PROBLEMS

15-1.   Consider an ideal gas, whose molecules each have 5 degrees of freedom, which initially has pressure p_0 and volume V_0. If this gas undergoes *isothermal* expansion to volume V_1:

(a)   What is the amount of work done by the gas in terms of p_0, V_0, and V_1? (*Hint.* Integrate $\int p \, dV$, after expressing p in terms of V and the constants p_1 and V_1.)
(b)   What is the pressure after the expansion, p_1, in terms of p_0, V_0, and V_1?

15-2.   Consider the same gas as in the previous problem, which is initially at pressure p_1 and volume V_1. If this gas is expanded *adiabatically* to volume V_2:

(a)   What is the amount of work done by the gas in terms of p_1, V_1, and V_2?
(b)   What is the pressure after the expansion, p_2, in terms of p_1, V_1, and V_2?

15-3.   Combine the results of the above two problems. Consider an ideal gas whose molecules each have 5 degrees of freedom. Suppose this gas starts out at pressure and volume (p_0, V_0), is isothermally expanded to volume V_1, and then adiabatically expanded further to volume V_2. In terms of (p_0, V_0, V_1, V_2):

(a)   What is the total work done by the gas during the expansion?
(b)   What is the final pressure (p_2) of the gas after completing the expansion?

15-4.   Consider an ideal gas having initial pressure and volume (p_0, V_0). For this gas,

(a)   Write an expression for the variation of p with V during isothermal expansion. (*Hint.* The expression should involve V, p_0, and V_0.)
(b)   Write an expression for the variation of p with V during adiabatic expansion. (*Hint.* The expression should involve V, p_0, V_0, and γ.)

(c) Using the fact that $\gamma = c_p/c_V > 1$, show that the slope (dp/dV) of the path on a p-V diagram is steeper for adiabatic processes than isothermal processes.

15-5. Consider a material, such as hydrogen gas under heavy pressure, which gets hotter as it expands adiabatically. Suppose two identical systems of this material have identical initial pressure and volumes (p_0, V_0), and both are expanded identical amounts, ΔV. If one is expanded isothermally and one adiabatically, for which would you expect the decrease in pressure to be greatest? Why?

15-6. Suppose the working system, A, in a Carnot engine is one which heats up when expanded adiabatically. Using the results of the above problem, show that if the system goes through the cycle described in Section A of this chapter (isothermal expansion, followed by adiabatic expansion, followed by isothermal compression, followed by adiabatic compression), then the net work done *by* the system during one complete cycle is negative, and that the exhaust temperature is greater than the ignition temperature. (*Hint.* Sketch the cycle on a p-V diagram.) Would this particular cycle be an engine or a refrigerator?

15-7. Describe a four-step Carnot cycle for the above system that would result in net *positive* work being done during one cycle. Sketch it on a p-V diagram.

15-8. Suppose one mole of an ideal gas at 300 K is expanded isothermally until its volume is doubled. How much heat energy was absorbed by the gas during this process? (*Hint.* Use Eq. 15.3.)

15-9. For a certain material, the equation of state is

$$\frac{pV^{10}}{T^5} = \text{constant}$$

This material starts out at atmospheric pressure, a volume of 0.5 liters, and a temperature of 300 K. It is compressed isothermally until its volume is reduced by 1%. How much heat energy does it release during this process?

D. EFFICIENCIES

The "efficiency" of an engine is defined as the ratio of the net work done, w, to the heat intake, Q_1, during any complete cycle.

$$\text{efficiency} = \eta \equiv \frac{w}{Q_1} \tag{15.4}$$

For the Carnot engine, this can also be expressed in terms of the temperature of the two reservoirs, by using the following information.

After one complete cycle, the working system A is returned to its original state. It is not exchanging particles with any other system, so there are only two independent variables needed to describe system A completely. Since both temperature and volume return to their initial values, and since there are only two independent variables, then all parameters (e.g., $p = p(T, V)$, $E = E(T, V)$, $S = S(T, V)$, etc.) must return to their initial values after one complete cycle.

During one complete cycle, system A absorbs energy Q_1 from reservoir R_1, exhausts energy Q_2 into reservoir R_2, and does work w. The fact that the internal energy of system A is unchanged after one complete cycle, means that these changes all add up to zero.

$$\Delta E_{1\text{ cycle}} = + Q_1 - Q_2 - w = 0 \tag{15.5}$$

The fact that the entropy is unchanged during one complete cycle means that that gained by system A during the ignition cycle is cancelled by that lost during the exhaust cycle.

$$\Delta S_{1\text{ cycle}} = \frac{Q_1}{T_1} - \frac{Q_2}{T_2} = 0 \tag{15.6}$$

(None is gained or lost during the two adiabatic portions of the cycle.) These two equations can be combined to eliminate Q_2 between them, yielding

$$\frac{w}{Q_1} = \frac{Q_1 - Q_2}{Q_1} = \frac{T_1 - T_2}{T_1}$$

Comparing this to the definition of efficiency (15.4), we see that for a Carnot engine, the efficiency is given by

$$\eta_{\text{Carnot}} = \frac{T_1 - T_2}{T_1} \tag{15.7}$$

There are several ways to show that a Carnot engine is the most efficient engine possible operating between reservoirs of temperatures T_1 and T_2. One of the simplest standard ways of proving this is to imagine just any engine, a, coupled to a reversible one, r, in such a way that the work done by engine a drives engine r backward as illustrated in Figure 15.10. (i.e., engine r is in the refrigeration cycle).

Engine a takes heat energy Q_1^a from reservoir R_1, exhausts Q_2^a into R_2 and does work w. With this work, engine r, running in reverse, takes heat Q_2^r from R_2 and transfers Q_1^r into R_1.

According to the second law, there cannot be a net flow of heat to the hotter reservoir from the cooler one because that would cause the total entropy of the combined system to decrease.* Consequently, we require

$$Q_1^a \geq Q_1^r$$

---

* If the net heat transferred from R_2 to R_1 during one cycle is ΔQ, then the net change in entropy for the entrie system during one cycle would be

$$\Delta S_{\text{total}} = \Delta S_1 + \Delta S_2 = \frac{\Delta Q}{T_1} - \frac{\Delta Q}{T_2} = \Delta Q \left(\frac{1}{T_1} - \frac{1}{T_2} \right)$$

Since $T_1 > T_2$, this change in total entropy would be *negative* if ΔQ were positive, and the second law would be violated.

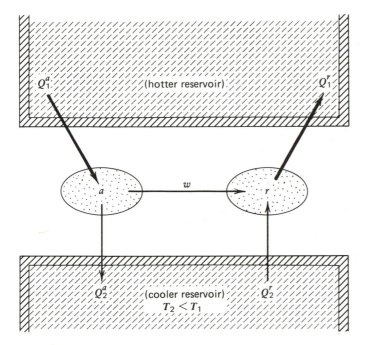

Figure 15.10   Consider any engine, a, coupled to a reversible engine, r, in such a way that the work done by engine a drives engine r in reverse. Because the second law requirement that there can be no net transport of heat from the cooler to the hotter reservoir, engine a cannot be more efficient than the reversible engine r.

which means

$$\frac{w}{Q_1^a} \le \frac{w}{Q_1^r}$$

or

$$\eta_a \le \eta_r$$

This is the proof we sought. No engine a can have an efficiency greater than a reversible one, r. Since a Carnot engine is reversible, it has the maximum possible efficiency operating between the two reservoirs.

In real engines, we expect friction, turbulence, and other nonequilibrium effects to cause their actual efficiencies to fall short of the maximum possible efficiency, or "Carnot efficiency." A measure of how close a real engine comes to achieving its maximum possible efficiency is called its "coefficient of utility." If its efficiency is η, and the efficiency of a Carnot engine running between the same two reservoirs is η_{Carnot}, then

$$\text{(coefficient of utility)} = \frac{\eta}{\eta_{\text{Carnot}}} \tag{15.9}$$

SUMMARY

The efficiency of an engine, η, is defined as the ratio of net work output during one complete cycle, to heat energy intake during the ignition part of the cycle.

$$\eta = \frac{w}{Q_1} \qquad (15.4)$$

The efficiency of a Carnot engine operating between reservoirs of temperatures T_1 and T_2 is

$$\eta_{\text{Carnot}} = \frac{T_1 - T_2}{T_1} \qquad (15.7)$$

No engine operating between reservoirs R_1 and R_2 (or temperatures T_1 and T_2) can be more efficient than a Carnot engine operating between these some two reservoirs. A measure of how close a real engine comes to achieving its maximum possible efficiency is called its "coefficient of utility."

$$(\text{coefficient of utility}) = \frac{\eta}{\eta_{\text{Carnot}}} \qquad (15.9)$$

EXAMPLE

We are going to make a rough estimate of the efficiency of the engine in a typical automobile, given that the oxidation of one gasoline molecule (C_8H_{18}) releases about 57 eV of heat energy, that an engine doing work at a rate of 20 kW (27 hp) burns a liter of gasoline every 6 minutes, and that a liter of gasoline has a mass of about 0.7 kg.

To find the efficiency, we must calculate the total work done by the engine in burning a liter of gasoline, and compare that to the heat energy input.

$$\text{Work} = w = (20 \text{ kW})(6 \text{ min})$$
$$= (2 \times 10^4 \text{ J/s})(360 \text{ s}) = 7.2 \times 10^6 \text{ J}$$

The heat energy input (Q_1) is the product of the heat released by the oxidation of one molecule, times the number of molecules in a liter. The mass of one molecule (8 carbons and 18 hydrogens) in atomic mass units is

$$(8 \times 12 + 18 \times 1)u = 114 \text{ u} = 1.8 \times 10^{-25} \text{ kg}$$

so there are about 3.9×10^{24} molecules in a liter (0.7 kg). The total heat released, then, is

$$Q_1 = (3.9 \times 10^{24})(57 \text{ eV})(1.6 \times 10^{-19} \text{ J/eV}) = 3.6 \times 10^7 \text{ J}$$

and the efficiency is

$$\eta = \frac{w}{Q_1} = .20$$

EXAMPLE

Given that the temperature of the gases during ignition is about 1400 K, and during exhaust is about 800 K, what is the coefficient of utility of the automobile engine in the previous example?

The Carnot efficiency for an engine operating between those temperatures is

$$\eta_{Carnot} = \frac{T_1 - T_2}{T_1} = \frac{1400 - 800}{1400} = .43$$

Therefore, the coefficient of utility is

$$(\text{coefficient of utility}) = \frac{\eta}{\eta_{Carnot}} = .20/.43 = .47$$

PROBLEMS

15-10. A motorcycle engine does work at a rate of about 6 kW. When doing work at this rate, it burns a liter of gasoline every 20 min.

(a) Given that the mass of one gasoline molecule is about 114 atomic mass units ($1\ u = 1.66 \times 10^{-27}$ kg), and the mass of a liter of gasoline is 0.7 kg, how many gasoline molecules are there in a liter?

(b) If each gasoline molecule releases 57 eV of energy upon oxidation, how much heat energy (in joules) is provided by burning one liter of gasoline?

(c) How much work (in joules) is done altogether by the motorcycle engine while it burns 1 liter of gasoline?

(d) What is the efficiency of this motorcycle engine?

15-11. The ignition temperature for the gasoline in the motorcycle is 1300 K and the exhaust temperature is 800 K.

(a) What is the Carnot efficiency for an engine operating between these temperatures?

(b) What is the coefficient of utility for the motorcycle engine? (See Problem 15.10.)

15-12. Most electrical power is produced by burning fossil fuels. The heat produced is used to run engines (steam turbines) that drive generators. The combustion temperature can be controlled to some extent through dilution of combustion gases with extra air passing through the furnaces. To make our limited fossil fuel resources stretch as far as possible, should we make the combustion temperature as high or as low as possible? Why?

15-13. The tropical ocean is a giant collector of solar energy. Surface waters have temperatures around 24°C, and deeper waters have temperatures around 4°C. This temperature differential can be used to run engines that drive electrical generators in a process called "ocean thermal energy conversion," or "OTEC" (Figure 15.11).

(a) What is the Carnot efficiency for engines running between these two temperatures?

(b) The coefficient of utility for a real engine-generator system running between these two temperatures is about .1. What is the efficiency with which this stored solar energy can be converted into electrical energy?

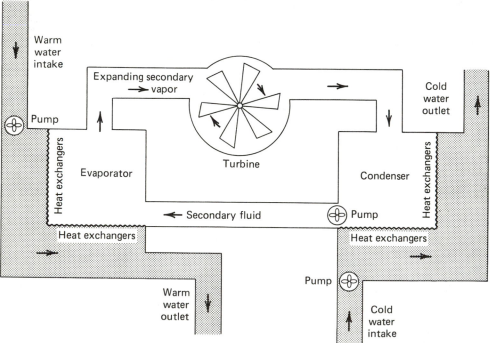

Figure 15.11 Design of an OTEC (Ocean Thermal Energy Conversion) station for the conversion of solar energy, stored in tropical surface waters, into electrical energy. The amount of solar energy stored in this way is very large, but the conversion efficiency is very low, due to the small temperature differential (about 20°C) between surface and deep waters. (Lockheed Missiles and Space Co., Inc.)

PART 7

CLASSICAL STATISTICS

In the part of the book just completed, we focused on gross properties of large systems of particles, such as entropy, pressure, temperature, internal energy, etc. For large systems, these properties are smooth well-behaved functions, which allowed us to apply some very powerful mathematical and statistical tools to their behaviors. With these tools, we can transfer information between whatever set of variables we choose, and manipulate constraints to maximize the amount of useful information we can collect from any experiment.

Now we will study exactly the opposite end of the spectrum of physical phenomena. We will study the individual behaviors of the microscopic components of physical systems, such as individual molecules, atoms, or elementary particles. For these tiny elements, quantum effects are very important. Properties of an individual particle are not smooth, well-behaved functions, but rather they change unpredictably and in discrete jumps.

The fact that these two widely differing areas can be studied with the same underlying principles is a tribute both to the power of the second law, and to the cleverness of some of our predecessors who have been able to envision and develop its ramifications.

chapter 16

PROBABILITIES AND MICROSCOPIC BEHAVIORS

Suppose we were interested in a very small system interacting with a very large one. And suppose we knew the properties of the large system, such as its temperature, pressure, and chemical potential. How could we determine the probabilities for the small system to be found in the various possible states?

Often we would like to know something about a small system that is interacting with a large reservoir. For example, we may wish to know something about a nitrogen molecule in a room full of air, a snowflake in a cold winter's day, a conduction electron in a metal, a hydrogen atom in the sun's photosphere (Figure 16.1), etc.

The smaller the system, the less predictable is its behavior. We cannot predict with certainty which particular state it will be in; we can only deal with the probabilities of its being in each of the various possible states.

A. ENSEMBLES

We can imagine we prepare a very large number of small systems identical to the one of interest to us and have them all interacting with the same reservoir. Then the relative number of these systems in each of the possible states will reflect the probabilities of our system being in these states (Figure 16.2). This large number of identically prepared small systems is called an "ensemble."

For example, suppose we are interested in one small system, A, which could be in any one of three possible states: s_1, s_2, and s_3. We can imagine that we prepare an ensemble of 1000 systems identical to A, and have them all interact with the same reservoir. Suppose that at any given time we find 120 of these in state s_1, 550 in state s_2, and the remaining 330 in state s_3. Then the probability of our system A being in each of these different states at any time must be given by $P_1 = .120$, $P_2 = .550$, and $P_3 = .330$.

Usually, systems of interest will have more than three possible states, and the ensembles will have more than 1000 members. More typical would be our study of a

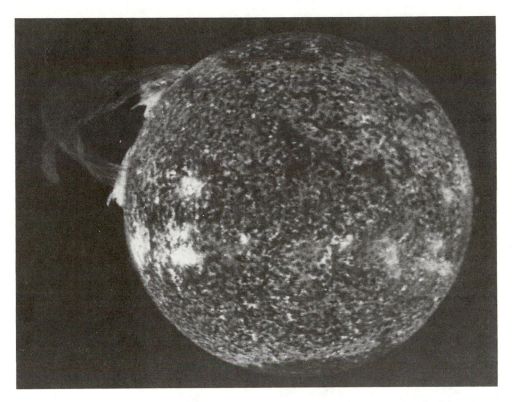

Figure 16.1 Often we would like to know the behavior of a small system which is interacting with a large one. For example, we may wish to know the behavior of a hydrogen atom in the atmosphere of the sun. (Courtesy Naval Research Laboratory on Skylab.)

nitrogen molecule in a room full of air, for instance, where there are a very large number of translational and rotational quantum states available to the molecule. The room would contain some 10^{26} identical molecules, each interacting with the same room full of air, so the ensemble would have some 10^{26} members. Sometimes, the system of interest is unique, in which case the appropriate ensemble exists only in our imaginations.

Various ensembles may be categorized according to how their members interact with the reservoir. As we have seen, the three general types of interactions between systems include the exchange of heat, work, and particles. The members of an ensemble may engage in all of these, none of these, or any combination of these, with the reservoir. Of these various possibilities, three are dealt with rather frequently, and so they are given special names. The members of a "microcanonical ensemble" do not interact at all with a reservoir. Those of a "canonical ensemble" exchange heat and work with the reservoir, and those of a "grand canonical ensemble" exchange heat, work, and particles as well. (See Figure 16.3.)

Figure 16.2 This "ensemble" is a large number of identically prepared pennies. They have all been flipped. The probability that any one penny is in a certain heads-tails state is equal to the relative number of pennies in the ensemble that are in that particular state. (Keith Stowe)

Members of microcanonical ensembles are physically uninteresting because they are never encountered. If they were encountered, they would have to be interacting with something—like ourselves—and therefore they wouldn't be "microcanonical" anymore. The liquid nitrogen in a closed laboratory Dewar flask comes close to being a member of a microcanonical ensemble. It isn't until we use this nitrogen in an experiment—where it is interacting with something—that it becomes interesting, and then it is no longer a member of a microcanonical ensemble.

Most physical systems we encounter here on earth, both macroscopic and microscopic, are interacting thermally and mechanically with their environments, and therefore could be considered members of canonical ensembles. An ice chip in water or

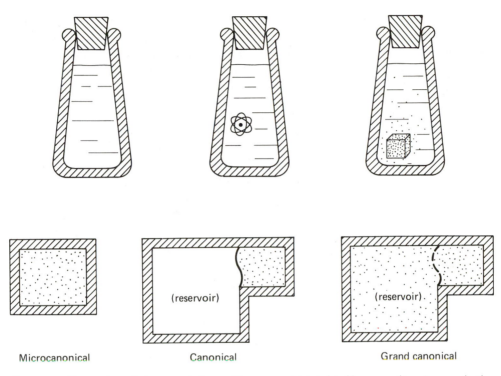

Figure 16.3 The members of microcanonical ensembles are completely isolated from everything. An example of a microcanonecal ensemble would be a large number of identically prepared pots of coffee, stored in perfectly insulating thermos bottles. The members of canonical ensembles interact mechanically and thermally with a reservoir. An example of such an ensemble would be the water molecules in a pot of coffee. The members of a grand canonical ensemble interact thermally, mechanically, and diffusively with a reservoir. An example of such an ensemble would be a large number of identical sugar cubes or grains of sugar in a pot of coffee.

a reagent in a chemical solution could be considered a member of a grand canonical ensemble, as it is exchanging particles as well as heat and work with its environment.

SUMMARY

An ensemble is a very large number of identically prepared systems interacting in the same way with the same reservoir. The ensemble may sometimes be only an imaginary construct. The probability that a given system is in a certain state is the same as the relative number of members of its ensemble that are in that particular state at any one time.

The members of a microcanonical ensemble do not interact at all with a reservoir, those of a canonical ensemble interact thermally and mechanically with a reservoir, and those of a grand canonical ensemble interact diffusively as well as thermally and mechanically with the reservoir.

B.   PROBABILITY OF BEING IN A CERTAIN STATE

We now study the probable behaviors of a microscopic component of a system, using the powerful statistical tools of thermodynamics. At first, this may seem like an impossible task, because these statistical tools apply only to large systems and *not* to the individual microscopic particles.

The task is accomplished through a clever trick. The trick involves the observation that if the whole system is isolated, then whatever energy, volume, or particles the microscopic component gets, it gets at the expense of the rest of the system. The "rest of the system" is very large, so we can apply the powerful statistical tools to this "rest of the system" to infer what the probable behavior of the individual component must be.

In addition to the above trick, the derivation is based on the following three previously encountered concepts.

1. The probability of a system being in a certain configuration is proportional to the number of accessible states having that configuration.

$$\text{probability} \propto \Omega$$

2. The entropy is defined as the following measure of the number of accessible states:

$$S \equiv k \ln \Omega \qquad (\text{or } \Omega \equiv e^{S/k})$$

3. The change in entropy of a system is related to the change in internal energy, volume, and number of particles through the first law.

$$\Delta S = \frac{1}{T}(\Delta E + p\,\Delta V - \mu\,\Delta N)$$

We consider a microscopic system, A_Δ, interacting with a large reservoir, A_R. The combined system, A_0, is isolated from the rest of the universe (Figure 16.4). The number of states accessible to the combined system is the product of the number of states accessible to the two components.*

$$\Omega_0 = \Omega_R \Omega_\Delta$$

If we are interested in the probability that the microscopic system is in one certain state s, then

$$\Omega_\Delta = 1$$

and the probability that the system is in that particular configuration is given by[†]

$$P_s \propto \Omega_0 = \Omega_R \Omega_\Delta = \Omega_R = e^{\ln \Omega_R} = e^{S_R/k} \tag{16.1}$$

---

* See Eq. 7.4 in Section D, Chapter 7.

[†] Notice that we've managed to write the probability of the *microscopic* system being in a certain state in terms of its influence on the reservoir, A_R.

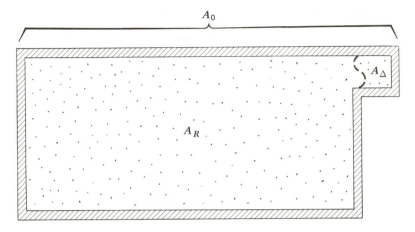

Figure 16.4 Consider a large system A_0, which consists of a microscopic sybsystem, A_Δ, and a large reservoir, A_R. The two subsystems interact thermally, mechanically, and diffusively with each other. If the combined system is completely isolated from the rest of the universe, what is the probability that A_Δ will be in any particular state?

In order to be in state s, the microscopic system takes energy, ΔE, volume, ΔV, and particles, ΔN, from the reservoir, thus reducing the entropy of the reservoir by an amount

$$\Delta S_R = -\frac{1}{T}(\Delta E + p\,\Delta V - \mu\,\Delta N)$$

according to the first law. This means that the actual entropy of the reservoir is given by

$$S_R = S_R^0 - \frac{1}{T}(\Delta E + p\,\Delta V - \mu\,\Delta N)$$

where S_R^0 represents the reservoir's entropy when none of its internal energy, volume, or particles are taken by A_Δ. Putting this expression for the entropy of the reservoir into Eq. 16.1 yields

$$P_s \propto \exp\frac{1}{k}\left[S_R^0 - \frac{1}{T}(\Delta E + p\,\Delta V - \mu\,\Delta N)\right]$$

or

$$P_s = C\,\exp-\frac{1}{kT}(\Delta E + p\,\Delta V - \mu\,\Delta N) \tag{16.2}$$

where C is some constant of proportionality (see Table 16.1).

Table 16.1   Summary of Steps Involved in Deriving Result (16.2)

Consideration	Implication
Entropy and number of accessible states for the reservoir, A_R $S_R = k \ln \Omega_R$	$\Omega_R = e^{S_R/k}$
States accessible to the combined system $\Omega_0 = \Omega_\Delta \Omega_R$	$\Omega_0 = \Omega_\Delta e^{S_R/k}$
One particular state for A_Δ $\Omega_\Delta = 1$	$\Omega_0 = e^{S_R/k}$
First Law $\Delta S = \dfrac{1}{T}(\Delta E + p\,\Delta V - \mu\,\Delta N)$	$\Omega_0 = \exp \dfrac{[S_R^\circ - (1/T)(\Delta E + p\,\Delta V - \mu\,\Delta N)]}{k}$
Probability and number of accessible states $P \propto \Omega_0$	$P = C \exp \dfrac{(\Delta E + p\,\Delta V - \mu\,\Delta N)}{kT}$

SUMMARY

When a microscopic system is interacting with a reservoir, the probability that it will be found in a certain state s is given by

$$P_s = C \exp -\beta(\Delta E + p\,\Delta V - \mu\,\Delta N) \tag{16.2}$$

where

$$\beta = \frac{1}{kT} \tag{16.3}$$

C is a constant of proportionality, and $(\Delta E, \Delta V, \Delta N)$ are the energy, volume, and number of particles removed from the reservoir by the microscopic system when it is in state s.

C.   CLASSICAL AND QUANTUM STATISTICS

Applications of this result can follow either of two general approaches, depending on what you consider to be the nature of the "microscopic system." You can think of it in one of two ways.

1. As a certain particle or group of particles that can occupy any of several different quantum states.
2. As a certain quantum state that could be occupied by various numbers of particles.

In the first case, the number of particles in the microscopic system is fixed; it takes none from the reservoir.

$$\Delta N = 0$$

The term $p\,\Delta V$ is ordinarily negligibly small in comparison to the internal energy, ΔE (See homework Problems 16-2 and 16-3). For example, in excited states atomic volumes increase typically about 10^{-30} m^2, so for a pressure of 1 atm,

$$p\,\Delta V \approx 10^{-6}\,\text{eV} \tag{16.4}$$

This is roughly 10^6 times smaller than typical spacings of energy levels, so we ignore it. For this case we ordinarily write the probability of the microscopic system being in state s of energy E_s as (Figure 16.5)

$$P_s = Ce^{-\beta E_s} \tag{16.5}$$

but the $p\,\Delta V$ term may have to be included for a few rare systems (for example, for systems under 10^6 atm of pressure).

In the second type of approach we consider the "microscopic system" to be a single quantum state, which could be occupied by various numbers of particles. The volume of this quantum state is fixed ($\Delta V = 0$) although the number of particles is not, and we can write the probability of this system being in a certain configuration, s, as

$$P_s = Ce^{-\beta(\Delta E - \mu\,\Delta N)}$$
$$= Ce^{-\beta n(\varepsilon_s - \mu)} \tag{16.6}$$

where n is the number of particles in the state and ε_s is the energy per particle in that state.

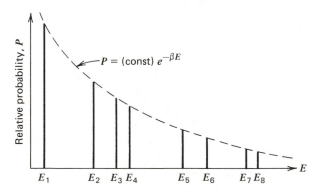

Figure 16.5 Plot of probability of being in a certain state versus the energy of that state for a small system interacting with a reservoir. The relative probabilities for the various states are indicated by the relative heights of the corresponding lines. The lower the energy of a state, the higher the probability that the system is in it.

The first type of approach is commonly referred to as "classical statistics," and the second type of approach is commonly referred to as "quantum statistics." These labels are somewhat misleading, as both approaches involve considerations of quantum effects. In one approach we consider quantum states accessible to a particle (or group of particles), and in the other approach we consider the number of particles occupying a given quantum state. Nonetheless, this nomenclature is customary, so we use it in this book.

SUMMARY

There are two different ways of applying the result (16.2), depending on what we consider to be the nature of the microscopic system.

In the one approach, called "classical statistics," we consider the system to be a certain particle or group of particles that can occupy various quantum states. The probability of the system occupying a certain state s of energy, E_s, is given by

$$P_s = Ce^{-\beta E_s} \qquad \left(\beta = \frac{1}{kT}\right) \tag{16.5}$$

where we have ignored the term $p\,\Delta V$ because it is ordinarily negligibly small in comparison to E_s.

In the second approach, called "quantum statistics," we consider the system to be a certain quantum state s that may be occupied by various numbers of particles. If the energy of a single particle in this state is ε_s, then the probability that there are n particles in this state is given by

$$P_s = Ce^{-n\beta(\varepsilon_s - \mu)} \tag{16.6}$$

The constant of proportionality, C, is determined by the condition that the system must certainly be in one of the accessible states.

$$\sum_s P_s = 1$$

This yields

$$C = \left(\sum_s e^{-\beta E_s}\right)^{-1} \qquad \text{(classical statistics)} \tag{16.7}$$

$$= \left[\sum_n e^{-n\beta(\varepsilon_s - \mu)}\right]^{-1} \qquad \text{(quantum statistics)} \tag{16.8}$$

EXAMPLE

Consider the photons inside an oven at 500 K. If the chemical potential of a photon is zero, and in a certain quantum state a photon has an energy of 0.1 eV, how much more likely is it to find that particular state unoccupied than occupied by one photon?

For $T = 500$ K and $(\varepsilon_s - \mu) = 0.1$ eV, we have

$$\beta(\varepsilon_s - \mu) = 2.32$$

The ratio of the two probabilities is

$$\frac{P_0}{P_1} = \frac{Ce^{-0}}{Ce^{-1\beta(\varepsilon_s - \mu)}} = e^{2.3} = 10$$

It is 10 times more likely that that particular state is unoccupied.

EXAMPLE

In a certain system, particles have chemical potentials of $\mu = -2.1$ eV. If the energy of a particle in a certain quantum state is -2.0 eV, and this state for some reason can have no more than one particle in it at a time, what is the probability that it has one particle in it (P_1) and what is the probability that it has none (P_0), if the temperature is 1000 K?

Using $T = 1000$ K and $(\varepsilon_s - \mu) = 0.1$ eV, we have

$$\beta(\varepsilon_s = \mu) = 1.16$$

which gives us

$$e^{-1\beta(\varepsilon_s - \mu)} = .31$$

Determining the constant C according to Eq. 16.18 gives

$$C = \left[\sum_{n=0}^{1} e^{-n\beta(\varepsilon_s - \mu)} \right]^{-1} = [e^0 + e^{-\beta(\varepsilon_s - \mu)}]^{-1}$$

$$= (1 + .31)^{-1}$$

So the two requested probabilities are

$$P_0 = Ce^0 = \frac{1}{1.31} = .76$$

and

$$P_1 = Ce^{-\beta(\varepsilon_s - \mu)} = \frac{.31}{1.31} = .24$$

PROBLEMS

16-1. Consider a perfectly insulating thermos jug filled with ice water, with the ice water inside engaging in no interactions of any kind with the outside world.

(a) A large number of such identically prepared systems would constitute which type of ensemble?

(b) One of the ice cubes in any one of these jugs would be a member of which type of ensemble?

(c) A single water molecule inside one of these jugs would be a member of which type of ensemble?

16-2.   A certain very delicate organic molecule requires only 0.04 eV of energy to be excited from the ground state. Upon excitation its volume changes by $\Delta V = 2 \times 10^{-30}$ m. (For convenience, we measure its volume relative to its ground-state volume, just as we measure its energy relative to its ground state energy. $E_s = E_1 - E_0 = 0.04$ eV, $V_s = V_1 - V_0 = 2 \times 10^{-30}$ m.) Under what pressure would pV_s be equal to E_s for this excited state?

16-3.   From the Bohr model, the radius of a hydrogen atom in the ground state is 0.53×10^{-10} m, and in the first excited state is four times this amount. The energy of the ground state is -13.6 eV and that of the first excited state is -3.4 eV. If we measure energies and volumes relative to those of the ground state (*i.e.*, how much must be removed from the reservoir, A_R, for system A_A to be in this excited state):

   (a)   What are the values of E_s and V_s for the excited state?
   (b)   At what pressure is the term pV_s comparable to E_s?

16-4.   How do you determine the value of the constant C in Eq. 16.2?

16-5.   Consider the photons inside an oven at 500 K. If the chemical potential of a photon is 0 and the energy of a certain state is 0.2 eV:

   (a)   What is the factor $\beta(\varepsilon_s - \mu)$ for this state?
   (b)   Determine the constant C from Eq. (16.8) accurate to three decimal places.
   (c)   With three-place accuracy, what is the probability of there being three photons in this state at any particular instant?

16-6.   Consider the photons inside an oven. If the chemical potential of a single photon is 0 and the energy of a certain state is 0.1 eV, at what temperature would the ratio of the probability that this state has one photon to that of its being unoccupied be equal to e^{-1}? ($P_1/P_0 = e^{-1}$.)

16-7.   The chemical potential for the conduction electrons in a certain metal is -0.3 eV. A certain electronic state has an energy of -0.26 eV, and can have no more than 1 electron in it ($n = 0$ or 1 only). At room temperature, what are the probabilities that it is

   (a)   Unoccupied?
   (b)   Occupied?

D.   APPLICATION OF CLASSICAL STATISTICS

D.1   Excitation Temperature

The energy of a state may be measured relative to any reference level that you think is convenient. Appropriate adjustments are automatically made in the constant of proportionality C according to Eq. 16.7. Often the reference level is chosen to be the energy of the lowest-lying (i.e., lowest energy) state, called the "ground state."

   Sometimes we are interested in ratios of probabilities for the system to be in various states. In these ratios, the same constant, C, appears in both numerator and denominator, leaving

$$\frac{P_i}{P_j} = \frac{Ce^{-\beta E_i}}{Ce^{-\beta E_j}} = e^{-\beta(E_i - E_j)} \tag{16.9}$$

In other cases, we are interested in absolute probabilities, rather than relative probabilities, and so the constant C must be calculated according to Eq. 16.7.

From Eq. 16.5 we see that states of higher energy will have smaller probability of being occupied. The ratio of the probability of being in the first excited state (E_1) to that of being in the ground state (E_0) is given by

$$\frac{P_1}{P_0} = \frac{Ce^{-\beta E_1}}{Ce^{-\beta E_0}} = e^{-\beta(E_1 - E_0)}$$

From this we see that if $(E_1 - E_0)$ is large compared to kT, then the probability of being in an excited state is negligibly small.

$$\frac{P_1}{P_0} \approx 0 \qquad \text{for } (E_1 - E_0) \gg kT \tag{16.10}$$

It is often convenient to define a temperature, T_e, called the "excitation temperature" of the system.

$$T_e \equiv \frac{1}{k}(E_1 - E_0) \tag{16.11}$$

Then we see that for temperatures small compared to the excitation temperature,

$$T \ll T_e \qquad \text{means} \qquad (E_1 - E_0) \gg kT$$

the system is confined to its ground state. That is, there is negligibly small probability of finding it in an excited state. Only at temperatures comparable to, or larger than the excitation temperature, will there be appreciable probability of finding the system in an excited state.

D.2 Degeneracy

Sometimes, several different quantum states all have the same energy. When this happens, we say that that particular energy level is "degenerate." If n_s states all have energy E_s, we say that the level E_s is "n_s-times degenerate." The probability, P_{E_s}, that the system has energy E_s is the product of the number of such states times the probability that it is in any one of them.

$$P_{E_s} = n_s P_s \tag{16.12}$$

SUMMARY

The energy of a state is usually measured relative to the ground-state energy, but may be measured relative to any other reference level that is convenient.

The ratio of probabilities for the system being in two states is given by

$$\frac{P_i}{P_j} = e^{-(E_i - E_j)/kT} \tag{16.9}$$

If $(E_1 - E_0)$ is the energy spacing between the ground state and the first excited state then

$$(E_1 - E_0) > kT \qquad \text{implies low probability of excitation}$$

$$(E_1 - E_0) < kT \qquad \text{implies high probability of excitation}$$

The "excitation temperature" for a system, T_e, is defined by

$$kT_e \equiv (E_1 - E_0) \tag{16.11}$$

For temperatures much lower than T_e, the probability of being even in the first excited state is negligibly small.

$$\frac{P_1}{P_0} = e^{-(E_1 - E_0)/kT} \approx 0 \qquad \text{for } T \ll T_e \tag{16.10}$$

In this case we say that the system is "confined to the ground state." Only for temperatures near or above the excitation temperature will the system have a reasonable probability of being excited.

If n_s different states all have the same energy E_s, we say that the E_s level is "n_s-times degenerate." The probability that the system has energy E_s equals the product of the number of such states times the probability that it is in one of them.

$$P_{E_s} = n_s P_s \tag{16.12}$$

D.3   Examples

EXAMPLE

The two lowest-lying energy levels of a hydrogen atom are $E_0 = -13.6$ eV and $E_1 = -3.4$ eV. At room temperature, what would be the ratio of hydrogen atoms in the first excited state to the number in the ground state? (Ignore degeneracies.)

The ratio of the two probabilities is

$$\frac{P_1}{P_0} = e^{-\beta(E_1 - E_0)}$$

Normally, we would measure all energies relative to the ground state for convenience (i.e., $E_0 = 0$, $E_1 = +10.2$ eV), but it doesn't matter. With $T = 300$ K we have

$$\beta(E_1 - E_0) = \frac{1}{kT}(10.2 \text{ eV}) = 394$$

so the ratio of the two probabilities is

$$\frac{P_1}{P_0} = e^{-394} = 10^{-171}$$

Clearly, at room temperature all hydrogen atoms will be in the ground state. None will be in even the first excited state, let alone in higher-lying states.

EXAMPLE

At what temperature would we find about half as many hydrogen atoms in the first excited state as in the ground state? (Ignore degeneracies.)
    We solve the following equation,

$$\frac{P_1}{P_0} = \frac{1}{2} = e^{-\beta(E_1 - E_0)}$$

or equivalently, with $(E_1 - E_0) = 10.2$ eV

$$\ln\frac{1}{2} = -\frac{1}{kT}(10.2 \text{ eV})$$

which gives

$$T = 1.7 \times 10^5 \text{ K}$$

Clearly, a significant fraction of hydrogen atoms will be excited only at very high temperatures!

EXAMPLE

In liquid water, a water molecule can dissociate into H^+ and OH^- ions with the help of the highly polarized neighboring water molecules. The energy of this dissociated state is roughly 0.4 eV higher than the nondissociated state. Approximately what fraction of the molecules in a glass of water at room temperature would you expect to be dissociated at any one time (Figure 16.6.)?
    With $(E_1 - E_0) = 0.4$ eV, we have, at room temperature,

$$\beta(E_1 - E_0) = 15.4$$

and so the ratio of molecules in the two states is the ratio of the probabilities,

$$\frac{P_1}{P_0} = e^{-\beta(E_1 - E_0)} = e^{-15.4} = 2 \times 10^{-7}$$

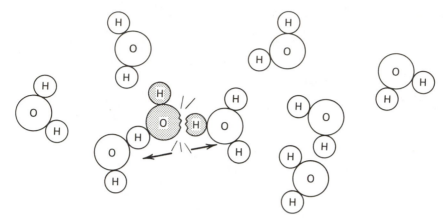

Figure 16.6 A water molecule can be dissociated into H^+ and OH^- ions with the help of neighboring water molecules. If it requires 0.4 eV to dissociate a water molecule into these two ions, what fraction of the water molecules in a glass of water at room temperature would be dissociated?

D.4 Energy Bands

The solutions to the preceding examples involved a bit of a swindle, and that is the reason the terms "rough" or "approximate" were used in giving the answers. We employed the formula

$$\frac{P_1}{P_0} = e^{-\beta(E_1 - E_0)}$$

which is the ratio of probabilities for the system to be in any *single* quantum state of energy, E_1, to that for any other *single* quantum state of energy, E_0. Yet we applied it to systems that had *many* different excited states and *many* different ground states.

For example, the nondissociated water molecule could be in any of a large number of different rotational and translational quantum states. So could the dissociated ions. Also, the outer electron in an atom could be in any of various possible quantum states.

To solve these examples correctly, we should have summed over "ground" states and summed over "excited" states. If we identify excited states by e and ground states by g, then

$$\frac{P_{\text{any excited state}}}{P_{\text{any ground state}}} = \frac{\sum\limits_{e} Ce^{-\beta E_e}}{\sum\limits_{g} Ce^{-\beta E_g}} = \frac{\sum\limits_{e} e^{-\beta E_e}}{\sum\limits_{g} e^{-\beta E_g}}$$

If all ground states are about the same energy and all excited states are about the same energy, then we can consider the two energies (E_e, E_g) to be constants. Bringing the constant factors outside the summation, we get

$$\frac{P_{\text{any excited states}}}{P_{\text{any ground states}}} = \frac{\left(\sum\limits_{e} 1\right)e^{-\beta E_e}}{\left(\sum\limits_{g} 1\right)e^{-\beta E_g}} = \frac{n_e e^{-\beta E_e}}{n_g e^{-\beta E_g}} \tag{16.12}$$

where

$$n_e = \text{number of "excited states"}$$

$$n_g = \text{number of "ground states"}$$

We see that this is just the result that we would get if we just considered the "ground state" as being n_g-times degenerate and the "excited state" as being n_e-times degenerate according to Eq. 16.9. This approximation is justified as long as the spread in "ground-state" energies, and the spread in "excited-state" energies are small compared to the size of the gap between the two regions, $(E_e - E_g)$ (see Figure 16.7). Normally, the exponential factor

$$e^{-\beta(E_e - E_g)}$$

is much more influential than is the linear factor of the ratio of the degeneracies, n_e/n_g.

It is sometimes more convenient to think of atoms or molecules in excited or dissociated states as being different kinds of particles. From this point of view, the excitation or dissociation of a particle would correspond to the particle going from one "system" (eg. those in the ground state) to another (eg. those in the first excited state). This would correspond to diffusive interaction between these systems, and the problem could be approached with the formalism of Chapter 22. However, the approach outlined in this section is often adequate to give results that are reasonably correct.

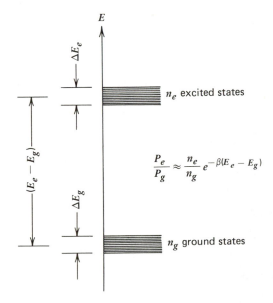

$$\frac{P_e}{P_g} \approx \frac{n_e}{n_g} e^{-\beta(E_e - E_g)}$$

Figure 16.7 If the spread in "ground state" energies, ΔE_g, and the spread in "excited state" energies, ΔE_e, are small compared with the spacing between these two groups of states, $E_e - E_g$, then it is a good approximation to treat all the "ground states" as a single energy level that is n_g-times degenerate, and all the "excited states" as a single energy level that is n_e-times degenerate.

SUMMARY

If the characteristic spread in energies of "ground" states, ΔE_g, and the characteristic spread in energies of "excited" states, ΔE_e, are small compared to the spacing between the two groups of states,

$$\Delta E_g \ll E_e - E_g$$

$$\Delta E_e \ll E_e - E_g \qquad (16.13)$$

then the ratio of probability of being in any of the "excited states" to that of being in any of the "ground states" is

$$\frac{P_{\text{any excited state}}}{P_{\text{any ground state}}} \approx \left(\frac{n_e}{n_g}\right) e^{-\beta(E_e - E_g)} \qquad (16.14)$$

where

$$n_e = \text{number of "excited states"}$$

$$n_g = \text{number of "ground states"}$$

This relationship becomes exact in the limit $\Delta E_g \to 0$, $\Delta E_e \to 0$.

In most ordinary cases, the ratio of the number of states is some number on the order of unity and the ratio of probabilities is dominated by the exponential.

PROBLEMS

16-8. A certain particle is interacting with a reservoir at 500 K, and can be in any of four possible states. The ground state has energy -3.1 eV, and the three excited states all have the same energy, -3.0 eV. Measure energies E_s relative to the ground-state energy.

(a) Is either of the allowed energy levels degenerate? Which one? How many times?
(b) Evaluate the factor $e^{-\beta E_s}$ for each of the allowed states.
(c) What is the probability that it is in the ground state?
(d) What is the probability that it is in a particular excited state?
(e) What is the probability that it is in any state of energy -3.0 eV?

16-9. For a certain molecule, the first excited state lies 0.2 eV above the ground state.

(a) At what temperature would the number of molecules in the ground state be exactly 10 times the number of molecules in the first excited state?
(b) What is the excitation temperature for this molecule?

16-10. For a certain molecule in an ensemble at 500 K, the energies of the various quantum states (measured relative to the ground state, E_0) are given by

$$E = n(0.1 \text{ eV}) \qquad n = 0, 1, 2, \ldots$$

(a) To three-decimal place accuracy, what is the value of C in Eq. 16.7?
(b) What is the probability that this molecule is in the level $n = 1$? $n = 2$?

16-11. For an energy of $E = 1$ eV, at what temperature would the ratio E/kT be equal to 1?

16-12. If the excitation energy of a certain molecule is 0.1 eV, at what temperature would we find 1% of these molecules in an excited state? (You can ignore higher-lying states because if only 1% of them are in the first excited state the number in higher states would probably be negligible.)

16-13. Consider a certain molecule in an ensemble, for which there are five accessible states: the ground state, and four excited states all lying 0.2 eV above the ground state. At what temperature would as many molecules be in excited states as are in the ground state?

16-14. In one example (Figure 16.6), we saw that about 0.4 eV is required to dissociate a single water molecule in liquid water into H^+ and OH^- ions. If the dissociated molecule has twice as many different states accessible to it as does the nondissociated molecule, what fraction of the water molecules in a glass of water at room temperature (300 K) would be dissociated at any instant?

16-15. If 0.03 eV is required to excite the molecules in a certain ensemble into their first excited state, roughly what is the temperature below which there is a rather low probability of excitation and above which there is a relatively high probability of excitation?

16-16. The mass of an average air molecule is about 5×10^{-26} kg. Assuming the atmospheric temperature to be a constant 275 K at all altitudes, and assuming the gravitational acceleration to be a constant 9.8 m/s^2, at what altitude, h, would you expect to find the air density (i.e., the probability of finding an air molecule) to be exactly half that at sea level? (*Hint.* $P_h/P_0 = \frac{1}{2}$; $E_h = mgh$.)

16-17. A certain liquid at room temperature (300 K) has Avogadro's number of molecules, each having spin 1 (i.e., angular momentum quantum number equal to 1), and the magnetic moment of each molecule is one Bohr magneton, μ_0 ($\mu_0 = 0.93 \times 10^{-23}$ J/T). A magnetic field of 1 T is along the z-axis, $B_z = 1$ T. In quantum mechanics it is learned that a spin 1 particle may have only three discrete values for the z-component of its magnetic moment, $\mu_z = (1, 0, \text{ or } -1)\mu_0$.

   (a) What is the probability that a molecule will be in each one of the three possible alignments? (*Hint.* For $x \ll 1$, $e^{\pm x} \approx 1 \pm x$.)
   (b) What will be the average magnetic moment of one molecule?
   (c) What will be the magnetic moment of the entire liquid?

E. MICROSCOPIC WHIM AND MACROSCOPIC FACT

In most cases, the microscopic system of interest to us will be a portion of a large system. For instance, we may be interested in the probable behavior of a single air molecule in a room full of air, or of a single electron state in a semiconductor. In any case, the microscopic system will be sufficiently small that in exchanging thermal energy (classical statistics) or particles (quantum statistics), it causes no noticeable changes in the properties of the rest of the system.

A conceptual problem arises when we try to square our knowledge that we can control the properties of the system as a whole, with our knowledge that the individual elements of the system may take on whatever properties they please. For example, we know that we can control the internal energy of a system by adding or removing thermal energy. How can we have this control if the individual elements take on as much thermal energy as they please?

The answer to this apparent dilemma lies in the observation that the probabilities for the individual particles to be in the various states is dependent on the temperature. When we remove thermal energy from the system, the temperature is reduced, and the reduction in temperatures means reduced probabilities of the particles being in higher energy states. Thus, the connection between our ability to control the internal energy of the system, and the ability of the individual particles to take on whatever energies they please, is through the temperature, which appears as a parameter in the probability distribution.

$$P_s = Ce^{-\varepsilon_s/kT}$$

We can even think of this interrelationship as the way of determining the temperature of the system. The total internal energy of the system is the product of the number of particles times the average energy of each.

$$E = N\bar{\varepsilon} = N \sum_s \varepsilon_s P_s$$

or

$$E = NC \sum_s \varepsilon_s e^{-\varepsilon_s/kT} \tag{16.15}$$

Given the spectrum of states, s, this equation uniquely determines the temperature, T.

A similar concern arises in quantum statistics. We know we can control the number of particles in many systems by adding or removing some. How can we have this control if the individual states can take on as many particles as they please (Figure 16.8)?

The answer to this apparent dilemma is that the probabilities for the individual quantum states to take on various numbers of particles is influenced by the chemical

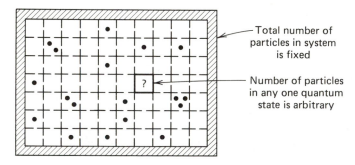

Total number of
particles in system
is fixed

Number of particles
in any one quantum
state is arbitrary

Figure 16.8   How do we square our knowledge that we can control the total number of particles in a system with the result of quantum statistics that each individual quantum state may take on as many particles as it pleases?

potential.* When we remove particles from the system, the chemical potential is lowered and this means reduced probability for the various quantum states to take on particles. Thus, the connection between our ability to control the size of a system, and the apparent ability of the individual quantum states to take on as many particles as they please, is through the chemical potential, μ, which appears as a parameter in the probability for a state to take on various numbers of particles.

$$P_n = Ce^{-n\beta(\varepsilon - \mu)}$$

We can even think of this interrelationship as the way to determine the chemical potential of the system. The average number of particles in a certain quantum state is given by

$$\bar{n} = \sum_n nP_n = C \sum_n ne^{-n\beta(\varepsilon - \mu)}$$

The total number of particles in the system is the sum over all states of the average number in each.

$$N = \sum_s \bar{n}_s = C \sum_s \sum_n ne^{-n(\varepsilon_s - \mu)/kT} \tag{16.16}$$

Given the temperature and the spectrum of quantum states, s, this equation uniquely determines the chemical potential, μ.

SUMMARY

In classical statistics, our ability to control the internal energy of the system is connected to the ability of the individual particles to take on whatever energies they please, through the temperature, which influences the probabilities for the individual particles to be in higher energy states. Changes in internal energy change the temperature, which change these probabilities.

In quantum statistics, our ability to control the size of the system is connected to the ability of the individual quantum states to take on as many particles as they please, through the chemical potential, which influences the probabilities to take on various numbers of particles. Changes in the size of the system changes the chemical potential, which changes these probabilities.

---

* Assume that the temperature is held constant, for example, through contact with a reservoir.

chapter 17

EQUIPARTITION

Because we know the probabilities for any microscopic constituent to be in any state, we can use these probabilities to determine mean values. One important mean value determined in this way is the average energy per degree of freedom.

In the preceding chapter we saw that if a microscopic component of a large system is of a given size, then the probability that that microscopic component is in a certain quantum state, s, having energy, E_s, is given by

$$P_s = Ce^{-\beta E_s} \tag{16.6}$$

where

$$C = \left(\sum_s e^{-\beta E_s}\right)^{-1} \tag{16.7}$$

We can use this result to determine the mean value of the energy stored in any given degree of freedom.

A. AVERAGE ENERGY PER DEGREE OF FREEDOM

Imagine that the microscopic system of interest is a single degree of freedom, and that the energy stored in this degree of freedom is of the usual form

$$E = bq^2$$

where b is a constant and q is a position or momentum coordinate.* For example, our degree of freedom could be the x-portion of the kinetic energy of a particle, $(1/2m)p_x^2$, or the y-portion of the potential energy of an oscillator, $\frac{1}{2}k_s y^2$.

The probability of this microscopic system being in a given state, s, is given above, and the mean value of the energy for this system is

$$\bar{E} = \sum_s P_s E_s = \left(\sum_{s'} e^{-\beta E_{s'}}\right)^{-1} \sum_s e^{-\beta E_s} E_s \tag{17.1}$$

---

* The coordinate q could also be something obtainable from position and momentum coordinates through a suitable transformation of variables. This includes angular momentum and can even be generalized to include electric and magnetic fields.

According to Eq. 2.12 the number of quantum states within coordinate interval dq is directly proportional to the length of this interval;

$$(\text{number of states}) = C\,dq$$

where C is some constant. With this, we can change our sum over states into an integral over the coordinate.

$$\sum_s \rightarrow C \int_{-\infty}^{\infty} dq$$

The sum appears in both the numerator and denominator of the expression (17.1) for $\bar{E}$, so the constant cancels, and we have

$$\bar{E} = \left[\int_{-\infty}^{\infty} dq\, e^{-\beta b q^2} \right]^{-1} \int_{-\infty}^{\infty} dq e^{-\beta b q^2} (bq^2)$$

Integrating by parts,* we have

$$\int_{-\infty}^{\infty} dq\, e^{-\beta b q^2} (bq^2) = 0 + \frac{1}{2\beta} \int_{-\infty}^{\infty} dq\, e^{-\beta b q^2}$$

This new integral is identical to the one in the denominator of the above expression for $\bar{E}$, so they cancel and we have

$$\bar{E} = \frac{1}{2\beta} = \frac{1}{2} kT \qquad (17.2)$$

SUMMARY

With every degree of freedom is associated an average energy of

$$\bar{E} = \frac{1}{2} kT \qquad (17.2)$$

providing the energy at any instant can be written as a function of a position or momentum coordinate, q, in the form

$$E = bq^2$$

where b is a constant.

This is another form of the "equipartition theorem," which we encountered earlier in Chapter 5.

In homework Problem 17-3 you will investigate the average energy stored in a degree of freedom for which the form $E = bq^2$ does *not* apply. You will find that the average energy is still proportional to T, but the constant of proportionality is slightly different.

We now investigate a few specific applications of this result.

---

* Write $(bq^2)dq = (\frac{1}{2}q)d(bq^2)$.

B.   SAMPLE APPLICATIONS

B.1   Heat Capacity of a Gas

In an ideal gas each molecule can move in three dimensions, and so each molecule has 3 degrees of freedom.

$$\bar{\varepsilon}_{1\ molecule} = \overline{\left(\frac{1}{2m}\,p_x^2 + \frac{1}{2m}\,p_y^2 + \frac{1}{2m}\,p_z^2\right)} = \frac{3}{2}\,kT$$

Now let us suppose that the molecules of this gas are diatomic, such as that shown in Figure 17.1. The center of mass of this molecule can move in any of three dimensions so it has 3 translational degrees of freedom as did the ideal gas. But it also has rotational inertia about two axes (the y- and z-axes in the figure), so it is able to store energy in two rotational degrees of freedom as well.

$$\varepsilon_{rot} = \frac{1}{2I_y}\,L_y^2 + \frac{1}{2I_z}\,L_z^2$$

Finally, the distance between the two atoms is fixed by the bottom of the electrostatic potential well caused by their mutual interactions. We have seen that for small displacements from the bottom, a particle in any potential well is a small harmonic oscillator. With appropriate definitions of the relative momentum and separation (p_{rel}, x_{rel}) of the two atoms, we can write

$$\varepsilon_{harm\ osc} = \frac{1}{2m}\,p_{rel}^2 + \frac{1}{2}\,kx_{rel}^2$$

The total number of degrees of freedom for a single diatomic molecule should then be 7:3 kinetic for the center of mass, 2 rotational, and 2 vibrational.

$$\varepsilon_{molecule} = \left[\frac{1}{2m}\,(p_{CM})_x^2 + \frac{1}{2m}\,(p_{CM})_y^2 + \frac{1}{2m}\,(p_{CM})_z^2\right]$$

$$= +\left(\frac{1}{2I_y}\,L_y^2 + \frac{1}{2I_z}\,L_z^2\right) + \left(\frac{1}{2m}\,p_{rel}^2 + \frac{1}{2}\,kx_{rel}^2\right) \tag{17.3}$$

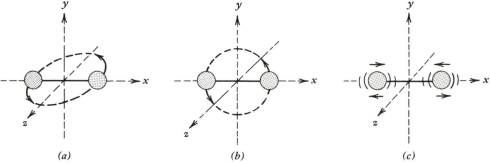

Figure 17.1   In addition to the 3 translational degrees of freedom of the center of mass, a diatomic molecule has 2 rotational degrees of freedom (l_y, $l_z \neq 0$ in the illustrations (a) and (b)), and 2 degrees of freedom for vibrations along the x-axis (c) from potential energy and kinetic energy terms.

A gas of N such molecules would have $7N$ degrees of freedom. Since the average energy per degree of freedom is $\frac{1}{2}kT$, the total internal energy for this gas would be

$$E = \frac{7}{2} NkT$$

This result may be checked in a laboratory by measuring the heat capacity of some diatomic gas. According to the first law

$$\Delta E = \Delta Q - p\,\Delta V$$

and so the heat capacity at constant volume is

$$C_V = \left. \frac{\partial Q}{\partial T} \right)_V = \left. \frac{\partial E}{\partial T} \right)_V = \frac{7}{2} Nk = \frac{7}{2} nR$$

where n is the number of moles of the gas, and R is the gas constant. When the measurements are made, the above result is found to be significantly in error, except at temperatures well above 1000 K.

The reason for this discrepancy is that important quantum mechanical effects have been ignored in our theoretical derivation. For example, the energy of a harmonic oscillator is quantized, and must come in discrete integral multiples of a quantity $\hbar\omega_0$,* which depends on the particular molecule, but is generally around 1 eV. If we calculate the ratio of the number of molecules in their first vibrational level to those in the ground state,

$$\frac{P_{n=1}}{P_{n=0}} = \frac{e^{-\beta\hbar\omega_0}}{e^{-0}} = e^{-\beta\hbar\omega_0}$$

we see that virtually all molecules will be in their ground state if

$$\beta\hbar\omega_0 \gg 1 \quad \text{or} \quad kT \ll \hbar\omega_0$$

For most diatomic gases, this corresponds to temperatures less than a few thousand degrees Kelvin. So at these temperatures the harmonic oscillator states are not available to the molecule, and a molecule has only 5 degrees of freedom rather than 7.

$$C_V = \frac{5}{2} nR \qquad kT \ll \hbar\omega_0 \tag{17.4}$$

We must also consider the accessibility of other quantum states. Angular momentum is quantized too.

$$L = \sqrt{l(l+1)}\,\hbar \qquad l = (0, 1, 2, \ldots)$$

$$\frac{1}{2I} L^2 = \frac{l(l+1)}{2I}\hbar^2$$

---

* As usual, we measure the energy of an excited state relative to the ground state. For linear harmonic oscillators, the ground state energy is $\frac{1}{2}\hbar\omega_0$, and that of the nth excited state is $(n + \frac{1}{2})\hbar\omega_0$. Consequently, measured *relative* to the ground state, the energy of the nth excited state is $n\hbar\omega_0$.

If we calculate the ratio of the number of molecules in their first rotational state to those in the ground state,

$$\frac{P_{l=1}}{P_{l=0}} = \frac{e^{-\beta(\hbar^2/I)}}{e^{-0}}$$

we see that for

$$\beta \frac{\hbar^2}{I} \gg 1 \quad \text{or} \quad kT \ll \frac{\hbar^2}{I}$$

virtually none of the molecules will be rotating. This generally corresponds to a temperature below some tens of degree Kelvin. So below this temperature, the rotational degrees of freedom also are unavailable to a molecule, leaving only the 3 translational degrees of freedom.

$$C_V = \frac{3}{2} nR \quad kT \ll \frac{\hbar^2}{2I_z} \tag{17.5}$$

Finally, at very low temperatures, even the translational part of the heat capacity must go to zero. The third law states that at absolute zero, the entropy goes to some finite constant value, and as we saw in Section J, Chapter 9, this means that the heat capacity must vanish as $T \to 0$. That is, since

$$C_V = T \left(\frac{\partial S}{\partial T} \right)_V = T \times \text{(something finite)}$$

then

$$C_V \xrightarrow[T \to 0]{} 0$$

At sufficiently low temperatures, even the translational degrees of freedom are deprived by quantum effects, but temperatures on the order of 10^{-20} K are required for this to happen. In practice, lots of other things (like condensation)would happen first.

The molar heat capacity of a gas may be written as the sum of the parts due to translational, rotational, and vibrational degrees of freedom, respectively.

$$c_V = c_{\text{trans}} + c_{\text{rot}} + c_{\text{vib}} \tag{17.6}$$

As we have seen, since each degree of freedom carries an average energy of $\frac{1}{2}kT$, each degree of freedom per individual molecule contributes a factor of $\frac{1}{2}N_A k = \frac{1}{2}R$ to the molar heat capacity. Since there could be 3 translational, 2 rotational, and 2 vibrational degrees of freedom per diatomic molecule these contributions would be

$$c_{\text{trans}} = \begin{cases} 0, & T \ll 10^{-20} \text{ K} \\ \frac{3}{2}R, & T \gg 10^{-20} \text{ K} \end{cases}$$

$$c_{\text{rot}} = \begin{cases} 0, & KT \ll \dfrac{\hbar^2}{2I} \\ R, & kT \gg \dfrac{\hbar^2}{2I} \end{cases}$$

$$c_{\text{vib}} = \begin{cases} 0, & kT \ll \hbar\omega_0 \\ R, & kT \gg \hbar\omega_0 \end{cases} \tag{17.7}$$

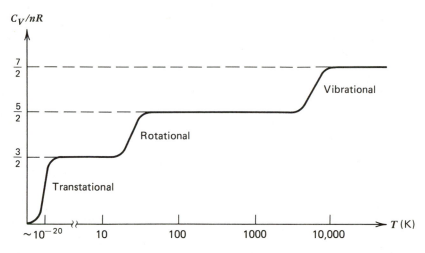

Figure 17.2 Plot of heat capacity versus temperature for a hypothetical noncondensing diatomic gas. At higher temperatures the molecules gain additional degrees of freedom, gaining rotational degrees of freedom typically around 10 K and vibrational degrees of freedom typically at several thousand degrees Kelvin. The disappearance of the translational degrees of freedom at extremely low temperatures has never been observed, of course, because such low temperatures have never been attained.

In Figure 17.2 is plotted the molar heat capacity versus T for a hypothetical diatomic gas in which these changes in heat capacity are seen.

B.2   Heat Capacity of a Solid

As another example of the equipartiton theorem, consider the heat capacity of a solid. Each atom is a tiny harmonic oscillator with energy given by

$$\varepsilon_{\text{atom}} = \frac{1}{2m} p_x^2 + \frac{1}{2m} p_y^2 + \frac{1}{2m} p_z^2 + \frac{1}{2} k_x x^2 + \frac{1}{2} k_y y^2 + \frac{1}{2} k_z z^2 \qquad (17.8)$$

We would therefore expect the entire solid to have $6N$ degrees of freedom, and the heat capacity to be given by

$$C_V = \frac{6}{2} nR = 3nR \qquad (17.9)$$

This result is called the "Dulong-Petit law," being named after the persons who first established its validity experimentally.

Experiments show this to be correct for most solids at or above room temperature (see Table 17.1), although at lower temperatures, quantum mechanical effects again are important. At sufficiently low temperatures ($kT \ll \hbar\omega_0$), even the first excited vibrational level is inaccessible to the atoms of a solid, so they become confined to the ground state. Although at room temperature, the atoms of most solids have 6 degrees of freedom, as the temperature is lowered, they lose these degrees of freedom, causing a

Table 17.1   Molar Heat Capacities of
            Some Common Metals at
            Room Temperature

Solid	$c_V/3R$
Aluminum	0.97
Copper	0.98
Lead	1.04
Silicon	0.85
Silver	1.01
Zinc	1.02

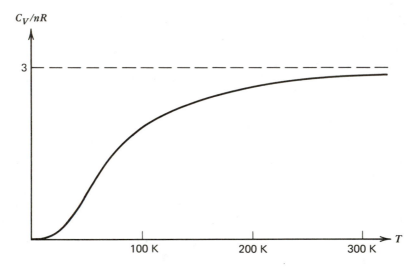

Figure 17.3   Plot of heat capacity temperature for copper.

corresponding reduction in the heat capacity (Figure 17.3). We will study this transition in greater detail in Section A, Chapter 27.

B.3   Brownian Motion

Another direct application of equipartition is to Brownian motion. In 1827 a Scottish botanist, Robert Brown, observed through a microscope that tiny spores jittered about as they floated through air. At first he thought it associated with the fact that it was living matter, but then he found tiny grains of inorganic dust performed the same way.

We find that any small particles immersed in any fluid undergo this "Brownian motion," which we now know to be caused by the bombardment of the particles by the rapidly moving molecules of the fluid. The smaller the particles and the hotter the fluid, the greater will be the motion of the particles.

With the equipartition theorem, we can analyze Brownian motion quantitatively. If the particles can move in three dimensions, then the average translational kinetic energy per particle is given by

$$\bar{\varepsilon}_{trans} = \frac{1}{2}\overline{mv^2} = \frac{3}{2}kT$$

Therefore, the root-mean-square speed is given by

$$v_{rms} = (\overline{v^2})^{1/2} = \left(\frac{3kT}{m}\right)^{1/2} \tag{17.10}$$

Indeed, this increases with temperature and decreases with mass, in agreement with our observations.

PROBLEMS

17-1.   A nitrogen molecule (N_2) has a mass of 4.7×10^{-26} kg. In air at room temperature (300 K):

   (a)   How many translational degrees of freedom does a single nitrogen molecule have?
   (b)   What is the average translational kinetic energy of a single nitrogen molecule?
   (c)   What is the root mean square speed, v_{rms}, of a single nitrogen molecule?
   (d)   If a hydrogen molecule (H_2) has a mass of 3.3×10^{-27} kg, what would be its root-mean-square speed at room temperature?

17-2.   A single nucleon has a mass of 1.7×10^{-27} kg, and that of an electron is 9.1×10^{-31} kg. In violent nuclear reactions, such as those occuring in exploding warheads or in the interiors of stars, temperatures are roughly 10^7 K.

   (a)   What is the root-mean-square speed of a single nucleon in this environment?
   (b)   What is the root-mean-square speed of an electron in this environment?

17-3.   Suppose the energy stored at any instant in a certain degree of freedom is linear in the coordinate, q. That is, suppose $\varepsilon = c|q|$, where c is a constant. Repeat the development between Eqs. 16.15 and 16.16 to show for this case that the average energy for this degree of freedom is $\bar{\varepsilon} = kT$.

17-4.   A small, just barely visible, dust particle has mass of about 10^{-8} g. It falls onto a glass of ice cold water where it is supported by the surface tension, and moves freely in only two dimensions. What is the root-mean-squared velocity of its Brownian motion there?

17-5.   A certain kind of plankton lives in the ocean at temperatures around 8°C. It is unicellular, having a mass of 10^{-10} g, and relies entirely on thermal energy to propel it through the ocean water and bring it into contact with fresh nutrients.

   (a)   What is its root-mean-square mean speed?
   (b)   About what total distance will it travel altogether in one day?

17-6.   A certain grandfather clock was driven by a pendulum of a 0.5-kg weight on the end of a 2-m-long wire of negligible mass. But it hasn't been running for years. The pendulum just hangs there inside the glass case at room temperature. (There is no air convection inside the case.)

(a)   What is the mean thermal energy of the pendulum?

(b)   What is the mean horizontal amplitude of the thermally induced swings of the weight?

17-7.   The atomic mass number of a single nitrogen atom is 14, and the separation between the two atoms in a nitrogen molecule, N_2, is 3.15×10^{-10} m. Rotational kinetic energy is expressible in the form $L^2/2I$, where L is the angular momentum, and I is the rotational inertia.

(a)   What is the rotational inertia, I, of a nitrogen molecule about an axis perpendicular to the line between the centers of the two atoms at the center of mass?

(b)   If the first excited rotational state has an angular momentum of $L = \sqrt{2}\,\hbar$, at what temperature would the ratio of nitrogen molecules in the first excited state to those in the ground state be equal to e^{-1}?

17-8.   Suppose you are determining the molar heat capacity of a gas composed of tetrahedrally shaped molecules. To do this you need to know how many degrees of freedom are available to it. This may depend on the temperature in that if the excited states are inaccessible, then the molecule will not be able to store energy in that particular degree of freedom.

(a)   If the first excited vibrational level of this molecule is at $\hbar\omega_0 = 0.2$ eV, below about what temperature can you ignore vibrational degrees of freedom?

(b)   If the rotational inertia about any axis is 10^{-45} kg-m^2, below about what temperature can you ignore the rotational degrees of freedom? (The angular momentum about any axis of the first excited state is $L = 1\hbar$.)

(c)   At room temperature, what would you expect the molar heat capacity of this gas to be?

chapter 18

MAXWELL DISTRIBUTION FOR GASES

Each particle in a gas interacts with the others, and so it can be considered to be a small system interacting with a large one. Therefore, classical statistics can be used to determine the distribution of particles in a gas as a function of their energy, momentum, or velocity.

We have seen that when a small system is in thermal equilibrium with a large reservoir, the probability of it occupying a certain state, s, having energy E_s, is given by

$$P_s = Ce^{-\beta E_s}, \tag{18.1}$$

where C is a constant determined by setting the sum of probabilities for all states equal to 1.

One important application of this result is to analyze the motion of molecules in a gas. This analysis was first performed by James Clerk Maxwell in 1859, and the resulting description of the motion of molecules in a gas is called the "Maxwell distribution."

A. PROBABILITY DISTRIBUTIONS

Consider the microscopic system to be a single translational degree of freedom for a single gas molecule. If the momentum in the x direction is p_x, then the energy stored in this degree of freedom is $(1/2m)p_x^2$. The probability of this being in a certain quantum state having this amount of energy is according to Eq. 18.1,

$$P_s \propto e^{-(\beta/2m)p_x^2} \qquad \left(\beta = \frac{1}{kT}\right)$$

This probability distribution is Gaussian in the particle's momentum, and is plotted in Figure 18.1.

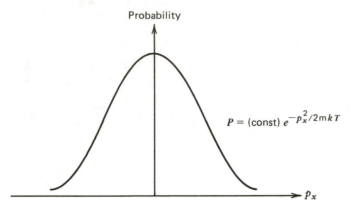

Probability

$$P = (\text{const})\, e^{-p_x^2/2mkT}$$

p_x

Figure 18.1 The probability of occupying a state is Gaussian in the momentum of the gas particle. States of lower momentum (lower energy) have higher probabilities.

If we are interested in the probability of its momentum lying somewhere in the range of dp_x, then the probability is the product of the probability of its being in any one state, times the number of such states.

$$P(p_x)\, dp_x = (\text{probability of being in one state}) \times (\text{number of states}) \qquad (18.2)$$

According to Eq. 2.12, the number of quantum states in the range dp_x is proportional to the size of this interval.

$$(\text{number of quantum states}) \propto dp_x$$

Consequently, we have

$$P(p_x)dp_x = Ce^{-(\beta/2m)p_x^2}\, dp_x$$

The constant C is determined by the requirement that summing the probabilities over all states must equal 1.

$$\int_{-\infty}^{\infty} P(p_x)\, dp_x = C \int_{-\infty}^{\infty} e^{-(\beta/2m)p_x^2}\, dp_x = 1$$

This is a standard integral, given in Appendix 18A, and yields the result

$$C = \left(\frac{\beta}{2\pi m} \right)^{1/2}$$

Our result, then, is that the probability of a given gas molecule having the x-component of its momentum lying in the range between p_x and $p_x + dp_x$ is

$$P(p_x)\, dp_x = \left(\frac{\beta}{2\pi m} \right)^{1/2} e^{-(\beta/2m)p_x^2}\, dp_x \qquad (18.3)$$

If we wish to express this distribution in terms of the molecule's velocity rather than its momentum, we must replace p_x by mv_x everywhere, yielding

$$P(v_x)\,dv_x = \left(\frac{\beta m}{2\pi}\right)^{1/2} e^{-(\beta m/2)v_x^2}\,dv_x \tag{18.4}$$

We can check that the normalization is correct by integrating this probability distribution over all velocities and seeing if it equals 1. (You are asked to do this in homework Problem 18-1.)

As we learned in Chapter 3, and will illustrate again in homework Problems 18-3, 18-4, and 18-5, the probability that a system meets two or more independent criteria is the product of the probabilities that it meets each. As a result, the probability that a given gas molecule has its three momentum components in the ranges dp_x, dp_y, and dp_z, respectively, is the product of the three probabilities. That is,

$$P(p_x, p_y, p_z)dp_x\,dp_y\,dp_z = [P(p_x)\,dp_x][P(p_y)\,dp_y][P(p_z)\,dp_z]$$

$$= \left(\frac{\beta}{2\pi m}\right)^{3/2} e^{-(\beta/2m)(p_x^2 + p_y^2 + p_z^2)}\,dp_x\,dp_y\,dp_z$$

or

$$P(\mathbf{p})\,d^3p = \left(\frac{\beta}{2\pi m}\right)^{3/2} e^{-(\beta/2m)p^2}\,d^3p \tag{18.5}$$

The corresponding probability distribution in terms of the particle's velocity rather than its momentum is

$$P(\mathbf{v})\,d^3v = \left(\frac{\beta m}{2\pi}\right)^{3/2} e^{-(\beta m/2)v^2}\,d^3v \tag{18.6}$$

This is obtained by replacing each p by mv in the distribution (18.6). You will check that this is correct in homework Problem 18-6.

If we are concerned only with the *magnitude* of the molecule's momentum, and not the direction, then it is convenient to express the differential element d^3p in Eq. 18.5 in spherical coordinates

$$d^3p = p^2 dp \sin\theta\,d\theta\,d\phi$$

so we can integrate over all directions (θ and ϕ), giving

$$P(p)\,dp = 4\pi \left(\frac{\beta}{2\pi m}\right)^{3/2} e^{-(\beta/2m)p^2} p^2\,dp \tag{18.7}$$

where p is positive, ranging from 0 to ∞. Similarly, expressing the differential element d^3v in spherical coordinates and integrating over all directions (θ and ϕ) transforms Eq. 18.6 into a probability distribution for the *magnitude* of the molecule's velocity, which ranges from 0 to ∞.

$$P(v)\,dv = 4\pi \left(\frac{\beta m}{2\pi}\right)^{3/2} e^{-(\beta m/2)v^2} v^2\,dv \tag{18.8}$$

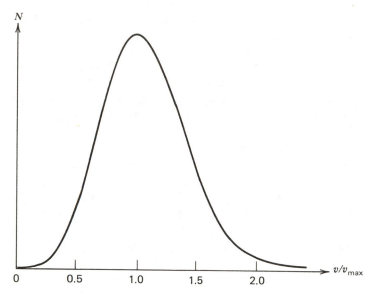

Figure 18.2   Plot of relative number of particles, N, versus speed for particles in a gas.

This probability distribution represents the distribution of molecular speeds you would find at any instant in a gas of a large number of particles, and is sketched in Figure 18.2. Notice that it is *not* Gaussian. The distribution is given by the product of the probability of occupying any given state, which *is* Gaussian in v, times the number of accessible states, which is quadratic in v (Figure 18.3).

$$d^3v = v^2\, dv \sin\theta\, d\theta\, d\phi \rightarrow 4\pi v^2\, dv$$

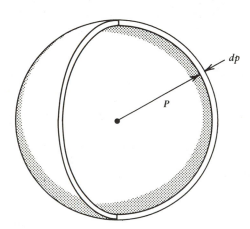

Figure 18.3   The volume of a hollow spherical shell of radius p and thickness dp is given by $(4\pi p^2)\, dp$. Consequently, the number of available quantum states with momenta in the range between p and $p + dp$ increases quadratically with p. (Equivalently, it increases quadratically in the particle velocity, v.)

Consequently, the distribution in speeds is the product of Gaussian and quadratic functions.

SUMMARY

For a molecule of mass m in a gas at temperature T, the probability that the x-component of its momentum is in range dp_x is

$$P(p_x)\,dp_x = \left(\frac{\beta}{2\pi m}\right)^{1/2} e^{-(\beta/2m)p_x^2}\,dp_x \tag{18.3}$$

and the probability that the x-component of its velocity is in the range dv_x is

$$P(v_x)\,dv_x = \left(\frac{\beta m}{2\pi}\right)^{1/2} e^{-(\beta m/2)v_x^2}\,dv_x \tag{18.4}$$

The probability that its total momentum lies in the range d^3p is

$$P(\mathbf{p})\,d^3p = \left(\frac{\beta}{2\pi m}\right)^{3/2} e^{-(\beta/2m)p^2}\,d^3p \tag{18.5}$$

and the probability that its total velocity lies in the range d^3v is

$$P(\mathbf{v})\,d^3v = \left(\frac{\beta m}{2\pi}\right)^{3/2} e^{-(\beta m/2)v^2}\,d^3v \tag{18.6}$$

The probability that the *magnitude* of its momentum lies in the range dp is

$$P(p)\,dp = 4\pi \left(\frac{\beta}{2\pi m}\right)^{3/2} e^{-(\beta/2m)p^2}p^2\,dp \tag{18.7}$$

and the probability that the *magnitude* of its velocity lies in the range dv is

$$P(v)\,dv = 4\pi \left(\frac{\beta m}{2\pi}\right)^{3/2} e^{-(\beta m/2)v^2}v^2\,dv \tag{18.8}$$

where

$$\beta = \frac{1}{kT}$$

PROBLEMS

18-1. The probability that the x-component of a molecule's velocity lies in the range dv_x is given by Eq. 18.4. Check that the normalization is correct by integrating this probability distribution over all v_x, using the standard integrals given in Appendix 18A, to see if it equals one.

18-2. According to Eq. 18.4 the distribution in v_x for particles in a gas is a Gaussian distribution. (See Section B, Chapter 4.) In terms of the particle mass and the temperature of the gas, what is the standard deviation for the x-velocities of these particles?

18-3. Consider a certain gas molecule in your room. At any given instant:

(a) What is the probability that it is in the northern quarter of your room?
(b) What is the probability that it is in the upper half of your room?
(c) What is the probability that it is both in the northern quarter *and* in the upper half of your room?

18-4. Consider a system consisting of a flipped coin and a rolled die.

(a) What is the probability that the coin is "heads?"
(b) What is the probability that the die shows three dots up?
(c) What is the probability that both the coin is "heads" *and* the die shows three dots up?

18-5. Suppose you invest half of your money in each of two businesses. Each business has a 10% chance of going bankrupt. What is the probability that *both* will go bankrupt?

18-6. Check the normalization of the expression (18.6) for $P(v) d^3v$, by expressing d^3v in spherical coordinates, and then integrating over all values of these coordinates. (See Appendix 18A for standard integrals.)

18-7. In terms of the molecular mass, m, and the temperature, T, what is the most probable speed for a molecule in a gas? [*Hint*. "Most probable" means $P(v)$ is a maximum.]

B. MEAN VALUES

We can use the probability distributions (18.3) to (18.8), to calculate the mean value of any function of the molecule's velocity or momentum.

EXAMPLE

What is the average value of v_x for a molecule in a gas at temperature T?
The mean value of the x-component of its velocity is given by

$$\bar{v}_x = \int_{-\infty}^{\infty} v_x P(v_x) \, dv_x = \left(\frac{\beta m}{2\pi}\right)^{1/2} \int_{-\infty}^{\infty} e^{-(\beta m/2)v_x^2} v_x \, dv_x = 0$$

We might have anticipated this result. The molecule is equally likely to be moving in any direction, so the average velocity must be zero.

EXAMPLE

What is the average speed of a molecule in a gas at temperature T?
The speed is the magnitude of the velocity and is independent of direction. Using Eq. 18.8 we have

$$\bar{v} = \int_0^{\infty} v P(v) \, dv = 4\pi \left(\frac{\beta m}{2\pi}\right)^{3/2} \int_0^{\infty} e^{-(\beta m/2)v^2} v^3 \, dv$$

$$= \left(\frac{8}{\pi m \beta}\right)^{1/2} = \left(\frac{8kT}{m\pi}\right)^{1/2} \tag{18.9}$$

where we used Eq. 18A.4 to evaluate the integral. The average speed increases with temperature and decreases with molecular mass, as we would expect.

A particularly interesting mean value is that of the square of the velocity. Since v^2 depends on the magnitude of the velocity and not its direction, we use the probability distribution (18.8).

$$\overline{v^2} = \int_0^\infty v^2 P(v)\, dv = 4\pi \left(\frac{\beta m}{2\pi}\right)^{3/2} \int_0^\infty e^{-(\beta m/2)v^2} v^4 \, dv$$

This integral can be evaluated using Eq. 18A.2, yielding

$$\overline{v^2} = \frac{3kT}{m} \tag{18.10}$$

Multiplying both sides of this result by $m/2$, we have

$$\frac{1}{2}m\overline{v^2} = \overline{E_k} = \frac{3}{2}kT$$

This is exactly the result that the equipartition theorem gives us! With each degree of freedom is an average energy of $\frac{1}{2}kT$, so the three translational degrees of freedom for a molecule give it an average translational kinetic energy of $\frac{3}{2}kT$.

In Figure 18.4 we have indicated the positions of the most probable speed, average speed, and root mean square speed ($v_{rms} = \sqrt{\overline{v^2}}$) on a plot of the Maxwell distribution.

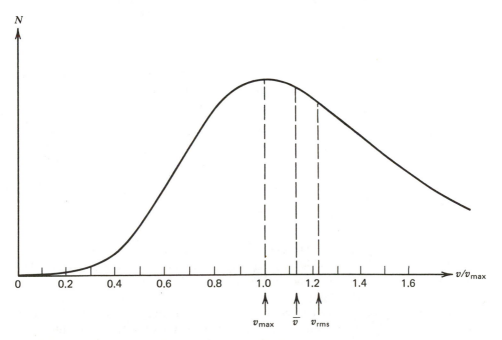

Figure 18.4 The Maxwell distribution for the molecules of a gas, showing the relative values of the most probable speed (v_{max}), average speed ($1.128\, v_{max}$), and root mean square speed ($1.225 v_{max}$).

PROBLEMS

18-8.   For a nitrogen (N_2) molecule at room temperature ($m = 5 \times 10^{-26}$ kg, $T = 300$ K), what is its:

(a)   Average velocity?
(b)   Average speed, $\bar{v}$?
(c)   Root-mean-square speed, $(\overline{v^2})^{1/2}$?
(d)   Most probable speed? (See Problem 18-7.)

18-9.   Suppose you were interested in the average value of the speed cubed ($\overline{v^3}$) for the molecules of a certain gas.

(a)   What is this in terms of the molecular mass, m, and the temperature, T? (See Appendix 18A for standard integrals.)
(b)   What is the cube-root mean speed cubed, $(\overline{v^3})^{1/3}$, for helium ($m = 6.6 \times 10^{-27}$ kg) at room temperature?

18-10.   What would be the ratio of root mean square speeds, $(\overline{v^2})^{1/2}$, for water and carbon dioxide molecules at room temperature?

C.   PARTICLE DISTRIBUTIONS AND FLUX

Consider a gas containing a very large number of identical molecules, N. If this gas occupies a volume, V, then the average density of molecules is given by

$$\rho = \frac{N}{V}$$

The number of molecules in this gas whose x-velocities lie in the range dv_x is the product of the total number of molecules times the probability that any one has an x-velocity in this range.

$$dN = NP(v_x)\,dv_x \tag{18.11}$$

Dividing both sides of this equation by the volume, V, we find that the density of particles whose x-velocities lie in the range dv_x is given by

$$\frac{dN}{V} = d\rho = \rho P(v_x)\,dv_x \tag{18.12}$$

Some interesting applications of this begin by studying the number of particles which cross a unit area per unit time. This is called the particle "flux," and is given the symbol f. For simplicity, we will consider the "unit area" to be in the y-z plane, so we need only consider the x-component of the particle motions.

In particular, we examine the component of the particle flux, df_x, due to only those particles for which v_x is in the range dv_x. This is given by the product of the number of particles per unit volume whose x velocities are in this range, times the number of unit

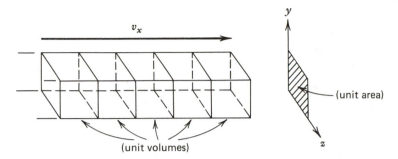

Figure 18.5   The number of unit volumes crossing a unit area per unit time is equal to v_x. Consequently, (number of particles of x-velocity v_x that cross a unit area in the y-z plane per unit time) = (number of such particles per unit volume) v_x.

volumes that cross the unit area per second.* (See Figure 18.5.)

$$df_x(v_x \text{ in range } dv_x) = \text{(number of particles per unit volume with } v_x \text{ in range } dv_x)$$

$$\times \text{ (number of unit volumes that cross the unit area per second)}$$

The number of unit volumes that cross the unit area per second is just v_x (see Figure 18.5), and the number of particles per unit volume with velocities in the range dv_x is given by Eq. 18.12. With this information, the above relationship becomes

$$df_x = (d\rho)v_x = \rho P(v_x)v_x\,dv_x \qquad (18.13)$$

SUMMARY

If $d\rho$ is the density of those particles whose x-velocities are in the range dv_x, then

$$d\rho = \rho P(v_x)\,dv_x \qquad (18.12)$$

and the flux of these particles past any point is given by

$$df_x = \rho P(v_x)v_x\,dv_x \qquad (18.13)$$

As one application of this result, we can integrate it over all $v_x > 0$ to find the total flux of particles moving in the positive x-direction. This total flux, f_x (or equivalently, the total number of particles crossing a unit area per second from one side), would tell us the flux of gas molecules striking the wall of their container.

$$f_x = \int_{v_x=0}^{v_x=\infty} df_x = \rho \int_0^\infty P(v_x)v_x\,dv_x = \rho\overline{v_{+x}} \qquad (18.14)$$

If the wall of a spaceship were ruptured by meteoritic impact, for example, you could use this number to calculate the rate at which the air in the cabin will escape. (See homework Problem 18-12.)

---

* Notice we can ignore the sideways (y and z) components of their velocity because for every molecule moving sideways out of a unit volume, another will move in, replacing it.

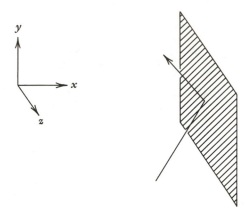

Figure 18.6   Upon elastic collision with a wall in the y-z plane, the x-component of the velocity changes by $\Delta v_x = -2v_x$, while the y- and z-components of the particle's velocity remain unchanged.

As another example of the application of the result (18.12), we can calculate the pressure that a gas exerts on the walls of its container. For convenience, consider the wall to be in the y-z plane, so we need only consider the x-component of a particle's motion. It is collides elastically with the wall, then v_y and v_z remain unchanged, but v_x is reversed (Figure 18.6), so the impulse it delivers to the wall is equal to $2mv_x$.

For the moment, consider only the pressure exerted on the wall by those particles whose x-velocities, v_x, lie in the interval dv_x. Pressure is impulse delivered per unit time per unit area (force per unit area), so the pressure exerted by these particles is the product of the number of particles striking a unit area per unit time, times the impulse delivered by each. That is,

$$dp = df_x(2mv_x)$$

where dp is the pressure due to those particles with v_x in the range dv_x, and df_x is the particle flux given by Eq. 18.13.

To find the pressure exerted by all particles, we must add up the contributions from all those moving in the positive x-direction.

$$p = \int_{v_x=0}^{v_x=\infty} dp = \int_{v_x=0}^{v_x=\infty} df_x(2mv_x)$$

$$= 2m\rho \int_0^\infty P(v_x)v_x^2\, dv_x$$

You will evaluate this integral in homework Problem 18-14, finding

$$\int_0^\infty P(v_x)v_x^2\, dv_x = \frac{1}{2}\frac{kT}{m} \tag{18.15}$$

which means that the pressure exerted by the gas is given by

$$p = \left(2m\frac{N}{V}\right)\left(\frac{1}{2}\frac{kT}{m}\right) = \frac{NkT}{V}$$

or alternatively,

$$pV = NkT$$

We have rederived the ideal gas law using Maxwell distribution of velocities!

PROBLEMS

18-11. Evaluate the integral in Eq. 18.14 using Eq. 18.4 to find the average x-velocity, $\bar{v}_{+x}$, of those particles moving in the $+x$-direction. Express your answer in terms of the molecular mass, m, and the temperature, T.

18-12. The density of air molecules at room temperature and atmospheric pressure is about 2.7×10^{25} molecules/m^3. Use this and your answer to Problem 18-11 to do the following.

   (a) Calculate the flux of particles moving in any one direction past any point in your room.
   (b) If a micrometeorite punctured a hole 0.2 mm in diameter in the wall of a spaceship, at which rate would molecules leave if the air in the spaceship was held at atmospheric pressure and room temperature?

18-13. What is the average value of v_x^2 for gas molecules of mass m at temperature T?

18-14. Evaluate the integral in Eq. 18.15. Why do you suppose the result differs from the answer to Problem 18-13 by a factor of $\frac{1}{2}$?

D. COLLISION FREQUENCY AND MEAN FREE PATH

The Maxwell distribution can also be used to calculate how often the molecules of any particular gas undergo collisions, and how far they travel between collisions, on the average. These are called their "collision frequency" and "mean free path," respectively, and are given the symbols "v_c" and "l."

If the radius of each molecule is R, then anytime that one approaches another to within a center-to-center distance of $2R$, they collide (Figure 18.7). As a molecule moves, we can think of it cutting out a volume each second of length $\bar{v}$ and cross-sectional area $\sigma = \pi(2R)^2$ (Figure 18.8). If the other molecules were sitting still, then the number of collisions undergone by the moving particle per second would be equal to the number of other molecules within this volume.

$$\text{collision frequency} = \rho(\sigma\bar{v})$$

Because the other particles are *not* sitting still, the actual collisional frequency for a molecule will be larger by a factor of $\sqrt{2}$. This is derived in Appendix 18B, and is due to the fact that we must consider the *relative* speeds of colliding molecules, rather than

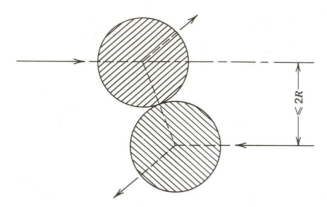

Figure 18.7 For a collision between two particles of radius R to occur, their centers need only come within $2R$ of each other.

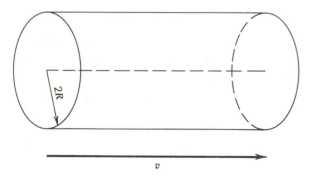

Figure 18.8 As a particle travels, it cuts out a cylindrical volume each second of area $\pi(2R)^2$ and length v. If the other particles were sitting still, then the average number of collisions it would have per second would be the average number of particles in a volume of this size.

the *absolute* speed of one. When all the molecules are moving, their average relative speed is larger than their average absolute speed by a factor of $\sqrt{2}$.*

$$\bar{u}_{rel} = \sqrt{2}\bar{v}_{abs} \tag{18.16}$$

Consequently, the real collision frequency for a gas molecule of mass m, radius R, and at temperature T is given by

$$v_c = \rho(\sigma\sqrt{2}\bar{v})(\sigma = 4\pi R^2) \tag{18.17}$$

where $\bar{v}$ depends on the temperature and particle's mass according to Eq. 18.9.

---

* On a highway, cars going your way don't pass you as frequently as cars going the other way. So the average relative speeds of cars that pass you is quite high.

The mean free path is the product of the average speed times the average length of time between collisions.

$$l = (\bar{v})\left(\frac{1}{v_c}\right) = \frac{1}{\rho(\sigma\sqrt{2})} \tag{18.18}$$

This gives the expected result that the mean free path is shorter if the gas is denser and if the molecules are fatter.

SUMMARY

If the molecules in a gas have mass m and radius R, and if the gas is at temperature T and density ρ, then the collision frequency, v_c, and mean free path, l, are given by

$$v_c = \rho(\sigma\sqrt{2}\bar{v})(\sigma = 4\pi R^2) \tag{18.17}$$

and

$$l = \frac{1}{\rho\sigma\sqrt{2}} \tag{18.18}$$

where $\bar{v}$ is the average speed given by Eq. 18.9.

PROBLEMS

18-15. Suppose you were in a car driving east at 10 m/s. What would be your speed (i.e., *magnitude* of your velocity) relative to a car traveling:

(a) North at 10 m/s?
(b) East at 10 m/s?
(c) South at 10 m/s?
(d) West at 10 m/s?
(e) If cars moving in each of the above directions were equally spaced, which would you pass most frequently?

18-16. If the temperature of a gas is doubled, by what factor do the following things change:

(a) Collision frequency?
(b) Mean free path?

18-17. Assuming they both have the same molecular radius, should the nitrogen (N_2) or water (H_2O) molecules in the air undergo collisions more frequently? By how many times?

18-18. Consider two systems, one of pure helium gas, and one of pure nitrogen gas, at standard temperature and pressure (0°C, 1 atm). Under these conditions, a mole of gas occupies 22.4 liters of volume.

(a) Given the densities of their liquid phases as follows, roughly what is the radius of a molecule of each of these gases?

N_2: density $= 0.808$ g/cm^3 (liquid phase)

He: density $= 0.145$ g/cm^3 (liquid phase)

(b) What is the mean free path of a molecule of each at standard temperature and pressure?
(c) What is the collision frequency for a molecule of each?

APPENDIX 18A   STANDARD INTEGRALS

$$\int_{-\infty}^{\infty} e^{-\alpha x^2}\, dx = 2 \int_0^{\infty} e^{-\alpha x^2}\, dx = \left(\frac{\pi}{\alpha}\right)^{1/2} \tag{18.A.1}$$

$$\int_{-\infty}^{\infty} e^{-\alpha x^2} x^{2n}\, dx = 2 \int_0^{\infty} e^{-\alpha x^2} x^{2n}\, dx = \frac{1 \cdot 3 \cdot 5 \cdots (2n-1)}{2^n \alpha^n} \left(\frac{\pi}{\alpha}\right)^{1/2} \tag{18A.2}$$

$$\int_{-\infty}^{\infty} e^{-\alpha x^2} x^{2n+1}\, dx = 0 \tag{18A.3}$$

$$\int_0^{\infty} e^{-\alpha x^2} x^{2n+1}\, dx = \frac{n!}{2\alpha^{n+1}} \tag{18A.4}$$

$$(n = 0, 1, 2, \ldots)$$

APPENDIX 18B   COLLISION FREQUENCY AND RELATIVE MOTION

Suppose we are interested in the collisions incurred by a certain gas molecule, which we label 1. The probability that it has velocity v_1 at any instant is proportional to $\exp(-\beta m v_1^2/2)$. The probability that it has velocity v_1 *and* some other molecule has velocity v_2 is proportional to the product of the two probabilities.

$$P(v_1, v_2) \propto [e^{-(\beta m/2)v_1^2}][e^{-(\beta m/2)v_2^2}] = e^{-(\beta m/2)(v_1^2 + v_2^2)} \tag{18B.1}$$

Writing v_1 and v_2 in terms of the center of mass velocity, $V = \frac{1}{2}(v_1 + v_2)$, and the relative velocity, $u = v_1 - v_2$, for these two molecules, we have

$$v_1 = V + \frac{u}{2}$$

$$v_2 = V - \frac{u}{2}$$

and on squaring these, Eq. 18B.1 becomes

$$P(V, u) \propto e^{-\beta m V^2} e^{-(\beta m/4)u^2}$$

We are only interested in the relative velocity of the two molecules, not their center of mass velocity, so we sum over all possible center of mass velocities, getting

$$P(u) = C e^{(\beta m/4)u^2} \tag{18B.2}$$

As usual, the constant C is determined by the condition that the sum over all relative velocities must yield 1.

$$\int P(u)\, d^3u = 1$$

An examination of the probability distribution (18B.2) shows that the distribution in relative velocities, u, is the same as the distribution (18.6) in absolute velocities, v, if the mass m is replaced by $m/2$. Therefore, all mean values involving relative velocities will be the same for those involving absolute velocities, but with m replaced by $m/2$.

For example, the average speed is given in Eq. 18.9 as

$$\bar{v} = \left(\frac{8kT}{m\pi}\right)^{1/2}$$

so the average relative velocity would be

$$u = \left[\frac{8kT}{(m/2)\pi}\right]^{1/2} = \sqrt{2}\,\bar{v} \tag{18B.3}$$

chapter 19

TRANSPORT PROCESSES IN GASES

We know the average speed and mean free path for the particles of a gas under equilibrium. Using these, we can make good estimates for the rate of transport of excess particles, excess thermal energy, or excess momentum when the gas is not in equilibrium and there is an uneven distribution of any of these properties.

Suppose that the molecules in one region of a gas have some property that is different from the molecules in neighboring regions. The random thermal motion of the molecules cause them to spread out, until this special property is distributed evenly throughout the entire gas. Examples include the following.

1. Molecules of a different material may be introduced at a certain region in the gas. Their spreading, due to random thermal motion, is referred to as "molecular diffusion."
2. Molecules of one region may carry more thermal energy than those in neighboring regions. As they spread, they carry this extra energy with them, and we refer to this property as the "heat transport," or the "thermal conductivity" of the gas.
3. Molecules of one region may carry a different average momentum than those of neighboring regions, as for example when different layers of the fluid are flowing at different speeds. As the molecules in any one layer diffuse into neighboring layers, they carry this momentum with them, and the flow speeds of neighboring layers tend to even out. This effective "drag" that neighboring layers having differing flow speeds seem to exert on each other is called the fluid's "viscosity."

All these transport processes are caused by the diffusion of gas molecules into surrounding regions. As we would expect, the rate of this diffusion depends on the average molecular speed, $\bar{v}$, and the average distance they go between collisions, or the "mean free path," l. As we saw in the last chapter, if a molecule has mass m and radius R, and if the gas has temperature T and density ρ, then these properties are given by

$$\text{average speed:} \quad \bar{v} = \left(\frac{8kT}{\pi m}\right)^{1/2} \tag{19.1}$$

$$\text{mean free path:} \quad l = \frac{1}{\sqrt{2}\rho\sigma(\sigma = 4\pi R^2)} \tag{19.2}$$

The ratio of these is the collision frequency, given by

$$\text{collision frequency:} \quad v_c = \frac{\bar{v}}{l} = \sqrt{2}\rho\sigma\bar{v} \tag{19.3}$$

In this chapter, we will frequently be interested in the net transport of some property in one particular direction. Consequently, we will also be interested in the answer to the following question: "Of the particles moving in the positive z-direction, what will be the average value of the z-component of their velocity?"

Of all the particles of a gas, only half of them will be moving in the $+z$-direction at any instant. These will have their velocities distributed over all directions in the $+z$-hemisphere (Figure 19.1). Averaging over this hemisphere gives the average value of v_{+z}

$$\bar{v}_{+z} = \overline{v\cos\theta} = \left(\int_{\theta=0}^{\pi/2}\int_{\phi=0}^{2\pi} \bar{v}\cos\theta\sin\theta\,d\theta\,d\phi\right)\left(\int_{\theta=0}^{\pi/2}\int_{\phi=0}^{2\pi}\sin\theta\,d\theta\,d\phi\right)^{-1}$$

$$= \frac{1}{2}\bar{v} \tag{19.4}$$

Similarly, if we wanted to find the average value of the z-component of the mean free path for those particles moving in the positive z-direction, we would do the same

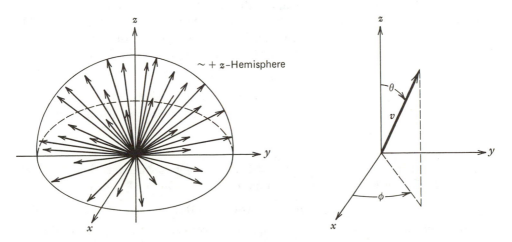

Figure 19.1   At any instant, half of the gas molecules are moving in the positive z-direction, with their velocities distributed uniformly over all directions in the positive z-hemisphere ($\theta < \pi/2$).

integration, and find

$$l_{+z} = \frac{1}{2} l$$

In summary, if we consider the motion of molecules in one particular dimension, at any instant half the molecules will be moving in the positive direction and half in the negative direction. For that half moving in any particular direction, we can write

$$\bar{v}_{+x} = \bar{v}_{+y} = \bar{v}_{+z} = \frac{1}{2} \bar{v} \tag{19.5}$$

$$l_{+x} = l_{+y} = l_{+z} = \frac{1}{2} l \tag{19.6}$$

We should be aware that whenever the properties of one region in a gas differ from those in another region, the gas is *not* in equilibrium. Our computations of mean speed, mean free path, and collision frequency ($\bar{v}$, l, v_c) were done for a gas in equilibrium, and therefore won't be exactly correct for the processes described in this chapter. However, detailed experimental and theoretical investigations have shown the assumption of equilibrium velocity distributions to be satisfactory in obtaining first-order solutions to most transport problems, including those we investigate here.

A. MOLECULAR DIFFUSION

Molecules diffuse from regions of higher concentrations toward regions of lower concentrations at a rate proportional to their density gradient, $\partial \rho / \partial x$. If f_x represents the net flux of particles in the x-direction, then

$$f_x = -D \frac{\partial \rho}{\partial x} \tag{19.7}$$

where D is a constant of proportionality called the "diffusion coefficient."

We can use our intuition to guess that the gas molecules will diffuse more quickly if they move faster and go longer distances between collisions. Consequently, we expect the diffusion coefficient to be proportional to the product of $\bar{v}$ and l.

$$D \propto \bar{v} l$$

It is fairly easy to demonstrate that our intuition is correct. The net flux of particles in the x-direction is the difference between those flowing to the right, and those flowing to the left.

$$f_x^{\text{net}} = f_{+x} - f_{-x} \tag{19.8}$$

At any instant, half the gas molecules will be moving toward the right and half toward the left. For either set, their average x-velocity will be $\frac{1}{2}\bar{v}$ (Eq. 19.4).

$$f_{\pm x} = \rho_{\pm x} \bar{v}_{\pm x} = (\tfrac{1}{2}\rho)(\tfrac{1}{2}\bar{v}) = \tfrac{1}{4}\rho\bar{v} \tag{19.9}$$

Obviously, particles coming from the right must come from the region to the right, and those coming from the left must come from the region to the left. Assuming that each particle gets its present velocity from its last collision, these two regions would be separated by an average distance (Δx) equal to the mean free path. (See Figure 19.3 and Eq. 19.6.)

$$\Delta x = l_x = \tfrac{1}{2}l \tag{19.10}$$

Combining the results (19.8), (19.9), and (19.10) we have

$$f_x^{net} = f_{+x} - f_{-x} = \tfrac{1}{4}\rho(x)\bar{v} - \tfrac{1}{4}\rho(x + \Delta x)\bar{v}$$

$$= -\tfrac{1}{4}\bar{v}[\rho(x + \Delta x) - \rho(x)] = -\tfrac{1}{4}\bar{v}\frac{\partial \rho}{\partial x}\Delta x$$

$$f_x^{net} = -\frac{1}{8}\bar{v}l\frac{\partial \rho}{\partial x} \tag{19.11}$$

Comparing this result to Eq. 19.7, we see that our intuition was correct.

$$D = \frac{1}{8}\bar{v}l \tag{19.12}$$

Implicit in this result are several assumptions, including the assumption that the present velocity of a particle is entirely determined by its last collision (rather than by its last three or four collisions, for example). The main effect of a more complete treatment would simply be a modification of the factor of $\tfrac{1}{8}$ appearing in the above result, giving

$$D = (const)l\bar{v} \tag{19.13}$$

where the constant is dimensionless and on the order of unity.

By referring to Eqs. 19.1 and 19.2 we can see how the rate of diffusion depends on molecular mass and radius, and on the temperature and density of the gas.

$$D = (const)\frac{1}{\rho R^2}\left(\frac{T}{m}\right)^{1/2} \tag{19.14}$$

That means that larger molecular masses, larger collisional cross sections ($\sigma = 4\pi R^2$), and higher gas densities all reduce the rate of diffusion. Higher temperatures, however, increase the rate of diffusion.

B. THERMAL CONDUCTIVITY

Heat flows from hotter regions toward cooler regions at a rate proportional to the temperature gradient. If Q_x represents the heat flux along the x-direction, then

$$Q_x = -K\frac{\partial T}{\partial x} \tag{19.15}$$

where K is a constant of proportionality called the "coefficient of thermal conductivity." It will be our objective in this section to see how we would expect this constant to depend on such things as molecular masses and radii, density, and temperature.

From the equipartition theorem, we know that the average thermal energy carried by one molecule is given by

$$\bar{\varepsilon} = \frac{v}{2}kT \tag{19.16}$$

where v is the number of degrees of freedom per molecule. If we write c_m as the "molecular heat capacity," we have

$$\bar{\varepsilon} = c_m T$$

where

$$c_m = \frac{v}{2} k \tag{19.17}$$

We wish to study the transport of this molecular thermal energy from one region of a gas to another.

Of all the molecules in the gas, only half will be moving in the positive x-direction at any instant, and their velocities will be distributed randomly in the $+x$ hemisphere. Using this plus the result (19.5) we have

$$(\text{particle flux})_{+x} = \left(\frac{1}{2}\rho\right)(\bar{v}_{+x}) = \frac{1}{4}\rho\bar{v}$$

Each of these carries average energy $c_m T$, so we can write the "energy flux" carried by these particles as

$$(\text{energy flux})_{+x} = (\text{particle flux})_{+x}(c_m T) = \left(\frac{1}{4}\rho\bar{v}\right)(c_m T) \tag{19.18}$$

Since at any instant as many particles are flowing in the $-x$-direction as in the $+x$-direction, there will normally be no *net* energy flux.

Suppose now that there is a temperature gradient, such that particles coming from the left carry average energy $c_m(T + \Delta T)$ and those from the right carry energy $c_m T$. Then there will be a net energy flux in the $+x$-direction given by (Figure 19.2)

$$Q_x = (\text{energy flux})_{+x} - (\text{energy flux})_{-x}$$

$$= \frac{1}{4}\rho\bar{v}c_m(T + \Delta T) - \frac{1}{4}\rho\bar{v}c_m T$$

$$= \frac{1}{4}\rho\bar{v}c_m \Delta T \tag{19.19}$$

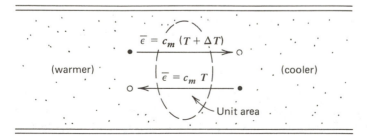

Figure 19.2   Diffusion of particles in a gas with a temperature gradient. Each second, as many particles cross a unit area from the left as from the right. But those coming from the left each carry slightly more thermal energy than those coming from the right, so there is a net transport of energy toward the right.

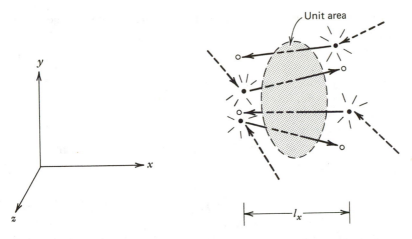

Figure 19.3 Particles of a gas crossing an imaginary unit area in the y-z plane. On the average, the particles crossing from the left last interacted in a region that is a distance l_x farther to the left than did those coming from the right. Therefore, the last interaction of those coming from the left was in a region that was hotter by $\Delta T = -(\partial T/\partial x)\, l_x$ than the region of last interaction for those coming from the right.

The particles coming from the left carry more energy because their last interaction was in a region of higher temperature than the region where those from the right last interacted (Figure 19.3). These two regions of "last interaction" are separated by the distance a particle goes between interactions.

$$\Delta x \approx l_x = \frac{1}{2}l$$

The temperature difference is then

$$\Delta T = -\frac{\partial T}{\partial x}\Delta x \sim -\frac{\partial T}{\partial x}\left(\frac{1}{2}l\right)$$

Putting this into Eq. 19.19 we have

$$Q_x = -\left(\frac{1}{8}\rho\bar{v}c_m l\right)\frac{\partial T}{\partial x}$$

and comparing this with the heat transport equation (19.15), we see

$$K = \frac{1}{8}\rho\bar{v}c_m l$$

Again, we have made some assumptions in this result that aren't strictly correct. For example, we assumed that a molecule acquires its energy characteristics in a single interaction, rather than many. Considerations such as these would require adjustment of the constant factor $(\frac{1}{8})$ in the above result, but the rest would remain the same. Therefore, we can write

$$K = (\text{const})\rho c_m \bar{v} l \qquad (19.20)$$

where the constant is dimensionless, and on the order of unity.

Again, this result agrees with our intuition. The rate of molecular diffusion increases with both $\bar{v}$ and l, and the amount of heat carried increases with both the density of particles, ρ, and the heat capacity of each, c_m.

We can express this in terms of molecular mass, radius, and the density and temperature of the gas through the relationships (19.1) and (19.2)

$$K = (\text{const})\frac{c_m}{R^2}\left(\frac{T}{m}\right)^{1/2} \qquad (19.21)$$

Larger mass and collisional cross-sectional area ($\sigma = 4\pi R^2$) both reduce the thermal conductivity of the gas, because they reduce the average molecular speed and distance of travel between collisions, respectively. Higher temperatures cause faster molecular motion and therefore faster diffusion of thermal energy. Notice that the thermal conductivity does not depend on the density of the gas. On the one hand, higher densities means there are more gas molecules per unit volume that can transport the thermal energy, but on the other hand, higher densities means each molecule undergoes more collisions and has greater difficulty getting anywhere. The two effects just cancel.

C.   VISCOSITY

Consider a gas flowing in the x-direction with velocity varying in the y-direction, as indicated in Figure 19.4.

$$v_x = v_x(y)$$

Because the fluid in one layer has a different flow speed than that in the next layer, the two layers will exert some frictional drag on each other. This drag is quantified in the term "shear stress," which is defined as the net force in the x-direction exerted across a

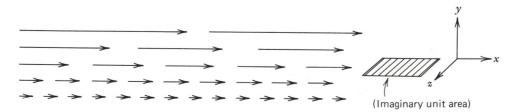

(Imaginary unit area)

Figure 19.4   Consider a fluid flowing in the x direction with a velocity that changes with altitude y. The viscosity of the fluid is defined in terms of the frictional drag per unit area exerted on each by two neighboring layers in the fluid. This frictional drag per unit area will be proportional to both the coefficient of viscosity and to the velocity gradient.

unit area in the x-z-plane (for the case illustrated here). The amount of shear stress will be proportional to the velocity gradient.

$$\frac{F_x}{A} = \eta \left| \frac{\partial v_x}{\partial y} \right| \tag{19.22}$$

The constant of proportionality, η, is called the "coefficient of viscosity." As in the previous sections, we will use our knowledge of the equilibrium properties of gases to gain insight into this nonequilibrium process. It will be our objective to see how we would expect the coefficient of viscosity to vary with molecular mass and radius, and the density and temperature of the gas.

The flux of particles moving in the $+y$-direction across an imaginary plane would be given by the product of the density of particles moving in the $+y$-direction, times their average y-velocity.

$$\text{(particle flux)}_{+y} = \left(\frac{1}{2} \rho \right) \bar{v}_{+y} = \frac{1}{4} \rho \bar{v}$$

Each carries with it an average x-momentum of mv_x. The flux of x-momentum carried by these particles moving in the positive y-direction across an imaginary plane would be (Figure 19.5)

$$\text{(x-momentum flux)}_{+y} = \frac{1}{4} \rho \bar{v}(mv_x)$$

Now suppose that the fluid just above this region is flowing slightly faster, so these particles have average x-momentum of $m(v_x + \Delta v_x)$. Therefore, those particles crossing

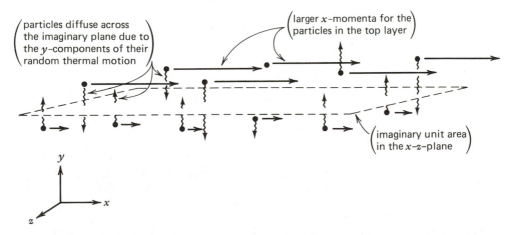

Figure 19.5 Consider two adjacent layers of a fluid that are flowing at different speeds, as illustrated above. Those particles diffusing down from above carry with them more x-momentum than those diffusing upward from below. Consequently, there is a net downward transfer of x-momentum. The flow speed of the lower layer will increase and that of the upper layer will decrease until they are both flowing at the same speed.

the imaginary plane going in the $-y$-direction carry with them slightly more x-momentum than those going in the $+y$-direction. Therefore, there will be a net flow of x-momentum across the imaginary plane, given by

$$(\text{net } x\text{-momentum flux}) = (x\text{-momentum flux})_{+y} - (x\text{-momentum flux})_{-y}$$

$$= \frac{1}{4}\rho\bar{v}mv_x - \frac{1}{4}\rho\bar{v}m\,(v_x + \Delta v_x)$$

$$= -\frac{1}{4}\rho\bar{v}m\,\Delta v_x \tag{19.23}$$

The particles coming from above carry more x-momentum because their last interaction was in a region of higher flow velocity. Their "last interaction" was a distance of

$$\overline{\Delta y} = l_y = \frac{1}{2}\,l$$

above those of the region below to which they transfer this extra momentum. The velocity difference is then

$$\Delta v_x = \frac{\partial v_x}{\partial y}\,\overline{\Delta y} = \frac{1}{2}\,l\,\frac{\partial v_x}{\partial y}$$

If we combine this with Eq. 19.23 we get for the net vertical flux of x-momentum across the imaginary plane

$$(\text{net } x\text{-momentum flux}) = -\left(\frac{1}{8}\,\rho\bar{v}ml\right)\frac{\partial v_x}{\partial y}$$

The "x-momentum flux" is the rate of transfer of x-momentum per unit area, or F_x/A. Therefore, this result is

$$\frac{F_x}{A} = -\left(\frac{1}{8}\,\rho\bar{v}ml\right)\frac{\partial v_x}{\partial y}$$

Comparing this with Eq. 19.22 we see

$$\eta = (\text{const})\rho m\bar{v}l \tag{19.24}$$

where the constant is dimensionless and on the order of unity, but depends on the approximations made.*

Again, the result is roughly what we would have guessed. When a momentum differential exists, the rate of diffusion of this momentum differential (i.e., the viscosity) depends on the rate of diffusion of the particles, $\bar{v}l$, their density, ρ, and the mass carried by each, m.

With the help of Eqs. 19.1 and 19.2 we can write this in terms of the molecular mass, collisional cross section ($\sigma = 4\pi R^2$) and the density and temperature of the gas.

---

* For example, we assumed the momentum transfer was complete after one collision. If more collisions are required, the constant would be different.

$$\eta = (\text{const})\frac{(mT)^{1/2}}{R^2} \tag{19.25}$$

The interesting thing in this result is that the viscosity of a gas *increases* with temperature and *decreases* with collisional cross-sectional area, rather than vice versa, as might at first be expected. The reason is that higher temperature and lower collisional cross section both increase the speed and distance over which the mixing of layers occurs, thereby evening out the differential motion more quickly.

SUMMARY

Consider a gas of density ρ and temperature T, whose molecules each have mass m, radius R, and v degrees of freedom. The average speed, mean free path, and molecular heat capacity for these molecules are given by:

$$\text{average speed:} \quad \bar{v} = \left(\frac{8kT}{\pi m}\right)^{1/2} \tag{19.1}$$

$$\text{mean free path:} \quad l = \frac{1}{\sqrt{2}\rho\sigma(\sigma = 4\pi R^2)} \tag{19.2}$$

$$\text{molecular heat capacity:} \quad c_m = \frac{v}{2}k \tag{19.17}$$

If there exists a density gradient, these molecules diffuse from regions of higher concentrations toward regions of lower concentrations with a particle flux proportional to the density gradient.

$$f_x = -D\frac{\partial\rho}{\partial x} \tag{19.7}$$

If there exists a thermal gradient, heat travels from higher temperatures toward lower temperatures with a heat flux proportional to the temperature gradient.

$$Q_x = -K\frac{\partial T}{\partial x} \tag{19.15}$$

If there exists a velocity gradient (i.e., different layers flowing at different velocities) then momentum is transferred from layers of higher momentum to layers of lower momentum with a momentum flux ("stress") proportional to the velocity gradient.

$$\frac{F_x}{A} = \eta\left|\frac{\partial v_x}{\partial y}\right| \tag{19.22}$$

The constants D, K, and η are called the "diffusion constant," "coefficient of thermal conductivity," and "viscosity," respectively, and are given by

$$D = (\text{const})\bar{v}l \tag{19.13}$$

$$K = (\text{const})\rho c_m \bar{v}l \tag{19.20}$$

$$\eta = (\text{const})\rho m\bar{v}l \tag{19.24}$$

where the "(const)" in each case is dimensionless and on the order of unity. The dependence of these on the temperature and pressure of the gas, and on the molecular mass, radius, and heat capacity is as follows.

$$D \propto \frac{1}{\rho R^2} \left(\frac{T}{m}\right)^{1/2} \qquad (19.14)$$

$$K \propto \frac{c_m}{R^2} \left(\frac{T}{m}\right)^{1/2} \qquad (19.21)$$

$$\eta \propto \frac{(mT)^{1/2}}{R^2} \qquad (19.25)$$

PROBLEMS

19-1.   Calculate the average value of the following functions, averaged over the $+z$-hemisphere ($\theta < \pi/2$).

(a) $\cos \theta$.
(b) $\sin \theta$.
(c) $\sin^2 \theta$.
(d) $\cos \phi$.

19-2.   Suppose the particle distribution, ρ, increases in the $+x$-direction.

(a)   What is the sign of $\partial\rho/\partial x$?
(b)   According to Eq. 19.7, in which direction will the net particle flux, f_x, be? (D is positive.)

19-3.   If the temperature of a gas is doubled, is the rate of diffusion increased or decreased? By what factor?

19-4.   All other things being equal, would the rate of diffusion be larger or smaller in gases of larger molecules? (See Eq. 19.14.) Can you give a physical explanation for this?

19-5.   According to Eq. 19.14, the rate of diffusion is slower in gases of higher densities (all else being equal). Why do you suppose this is?

19-6.   What are the dimensions of the "(const)" in Eq. 19.14? Give an estimate of its value. (*Hint.* Start with Eqs. 19.13, 19.1, and 19.2.)

19-7.   Suppose the temperature of a gas increases in the positive x-direction.

(a)   What is the sign of $\partial T/\partial x$?
(b)   Which way will the net heat flux flow according to Eq. 19.15? (K is positive.)

19-8.   What is the molecular heat capacity, c_m, for a molecule having 5 degrees of freedom?

19-9.   Give a rough estimate of the value of the "(const)" in Eq. 19.21. (*Hint.* Use Eqs. 19.20, 19.1, and 19.2.)

19-10.   In our discussion of viscosity, we were concerned with the flow speed for various layers of the gas in the x-direction. If the concern is the motion of particles in the x-direction, why is the flux of particles in the y-direction relevant?

19-11. Give a rough estimate of the "(const)" in Eq. 19.25. (*Hint*. Use Eqs. 19.24, 19.1, and 19.2.)

19-12. Given that the radius of an air molecule is about 10^{-10} m, its mass is about 5×10^{-26} kg, it has 5 degrees of freedom, and its density is about 2.7×10^{25} molecules/m^3, give an estimate for the values of the following parameters at room temperature (300 K) for air.

(a) Diffusion constant, D.
(b) Coefficient of thermal conductivity, K.
(c) Coefficient of viscosity, η.
(d) From Eqs. 19.7, 19.15, and 19.22 find the dimensions needed for each of the constants. Do they agree with the dimensions in your answers to parts (a), (b), and (c)?

19-13. From your answer to part (b) of Problem 19-12, estimate the net flux of heat past any point in your room if the temperature gradient is 0.5 K/m.

chapter 20

MAGNETIC PROPERTIES OF MATERIALS

Individual atoms have tiny magnetic moments due to the circulation and spin of electrical charges. They tend to line up with an external magnetic field because that is the state of lowest energy. Because we know how the probabilities for various states depend on the energy, we can calculate just how strong this tendency to line up with external fields is, which determines how strongly magnetized the material becomes.

Moving charges generate magnetic fields. Within an atom are positively charged protons moving about in the nucleus, and negatively charged electrons swirling about in their electronic orbits. These generate rather large magnetic fields within an atom (see Figure 2.11).

A. THE NATURE OF THE ATOMIC MAGNETS

As we saw in Chapter 2, the magnetic moment due to the current loop, generated as one of these particles revolves in its orbit, is directly proportional to the angular momentum of that orbit. The z-component of the magnitude moment, for example, is given by (see Eq. 2.19)

$$\mu_z = \frac{q}{2m} L_z \tag{2.19}$$

If l_z is the z-component of the angular momentum of the particle in units of $\hbar$, g is the charge of the particle in units of e, and μ_0 is given by

$$\mu_0 = \frac{e\hbar}{2m} \tag{20.1}$$

then this can be written as

$$\mu_z = \frac{ge}{2m}(l_z\hbar) = gl_z\mu_0 \tag{20.2}$$

As we have seen, angular momentum is quantized, so that l_z takes on integer values only. (See Figures 2.8, 2.9, 2.10.)

$$l_z = 0, \pm 1, \pm 2, \ldots, \pm l \tag{20.3}$$

Charge is also quantized, so the value of the "gyromagnetic ratio" is

$$g = +1 \qquad \text{for proton in orbit}$$

$$g = 0 \qquad \text{for neutron in orbit}$$

$$g = -1 \qquad \text{for electron in orbit} \qquad (20.4)$$

The value of μ_0 is inversely related to the mass of the particle, so it is much larger for an electron than for a proton. For the electron it is called the "Bohr magneton," μ_B, and for a proton it is called the "nuclear magneton," μ_N.

$$\mu_B = 9.27 \times 10^{-24} \text{ J/T}$$

$$\mu_N = 5.05 \times 10^{-27} \text{ J/T} \qquad (20.5)$$

Even if the particles had no orbital angular momentum, an atom may still have a magnetic moment due to the intrinsic spin of the particles. In a spinning charged sphere, for example, the charges move in circular paths generating a magnetic field (Figure 2.12). Measurements confirm that if a particle has the z-component of its spin angular momentum given by s_z (in units of $\hbar$), then its magnetic moment is given by

$$\mu_z = g s_z \mu_0 \qquad (20.6)$$

Because the atom is composed of spin-$\frac{1}{2}$ particles, s_z takes on only the values

$$s_z = \pm \frac{1}{2}$$

for each. For electrons, μ_0 is the Bohr magneton, and for nucleons it is the nuclear magneton. In relating a particle's intrinsic angular momentum and intrinsic spin, the gyromagnetic ratio is *not* the same as the charge of the particle in units of e. This suggests that the distribution of mass and charge are *not* the same within a particle. The gyromagnetic ratio, g, for particle spin is found to have the following values.

$$g = -2 \qquad \text{for electron spin}$$

$$g = -3.82 \qquad \text{for neutron spin}$$

$$g = +5.58 \qquad \text{for proton spin} \qquad (20.7)$$

Notice that the gyromagnetic ratio for a spinning neutron is not zero, even though it is a neutral particle! This suggests that although its net charge is zero, there must be some charge distribution within it, with more positive charge near the center and more negative charge farther out (Figure 20.1). Such a distribution would make the area of the current loop larger for the negative charge, and so the magnetic field it generates would dominate over that due to the inner positive charge.

Because the Bohr magneton is nearly 2000 times larger than the nuclear magneton, the magnetic moment of an atom is almost entirely due to its electrons. We normally ignore any contribution from the nucleus unless doing nuclear physics, high-resolution spectroscopy, or studying the small residual magnetism of atoms whose electrons have zero net angular momentum.

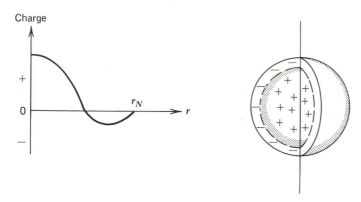

Figure 20.1   The net charge of a neutron is zero. But if it has some distribution of charges, with positive charges nearer the center and negative charges nearer the outside, then the area of the current loops due to the negative charges would be larger than those due to the positive charges. This would give the neutron a net magnetic moment (with negative gyromagnetic ratio) even though it has no net electrical charge.

Combining the results of (20.2), (20.4), (20.6), and (20.7), we see that if we ignore the minor nuclear contributions, the magnetic moment of an atom is given by (Figure 20.2)

$$\mu_z = -(l_z + 2s_z)\mu_B \tag{20.8}$$

where l_z and s_z are the quantum numbers for the z-components of the orbital angular momentum and spin angular momentum, respectively, of all the electrons combined.

$$l_z = 0, \pm 1, \pm 2, \ldots, \pm l$$

$$s_z = \begin{cases} 0, \pm 1, \pm 2, \ldots, \pm s & \text{(even number of electrons)} \\ \pm\frac{1}{2}, \pm\frac{3}{2}, \ldots, \pm s & \text{(odd number of electrons)} \end{cases} \tag{20.9}$$

The allowed range of values (indicated by l and s) tends to be quite limited, because the electrons in an atom tend to take combinations of orbits and configurations that give the atom zero total angular momentum. In fact, completed electronic shells have no net angular momentum, and only electrons in the outer unfilled electronic shells may make any net contribution at all.

The magnetic properties of most compounds are due to the valence electrons shared by the atoms making up each individual molecule. These valence electrons are subject to the same quantum restrictions as are the electrons of individual atoms ($l_z = 0, \pm 1, \pm 2, \ldots \pm l$; $s_z = 0, \pm\frac{1}{2}, \pm 1, \ldots, \pm s$), so the treatment of magnetism in compounds is similar to that for materials made of individual atoms.

B.   DIAMAGNETISM, PARAMAGNETISM, AND FERROMAGNETISM

With this background, consider what happens when we subject a material to an external magnetic field. On an atomic or molecular level, we can consider each electron orbit to be a tiny current loop. When we try to change the magnetic flux through any

$$\mu_z^{\text{Orbit}} = -I_z\mu_{\text{B}}$$

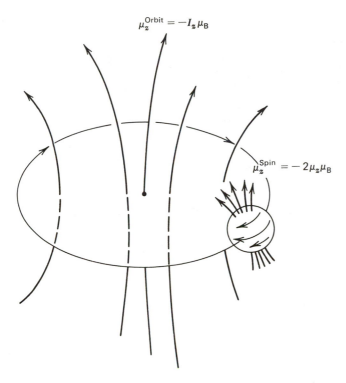

$$\mu_z^{\text{Spin}} = -2\mu_z\mu_{\text{B}}$$

Figure 20.2 An atomic electron makes two contributions to the magnetic moment of the atom, one being due to its spin and the other being due to its orbit. Both create tiny current loops, and the gyromagnetic moment is negative in both cases since the electron carries negative charge.

loop, we know from Lenz's law that we induce an electromotive force in that loop in such a way as to oppose the intruding magnetic field. Consequently, the orbital speeds of the electrons are adjusted slightly in such a way as to oppose the intruding field.* This slight magnetization of the material in opposition to the external field, is called "diamagnetism," and is displayed by all materials.

In most materials, the orientation of the atomic or molecular magnetic moments is not restricted, so when placed in an external magnetic field, they tend to line up with that field, like tiny magnetic compass needles (Figure 20.3). This gives the material as a whole a net magnetization in the direction of the external field. This effect is called "paramagnetism," and it is normally a much stronger effect than diamagnetism. Consequently, diamagnetism is only directly observed in those materials whose atoms

---

* Instead of referring to Lenz's law, this can also be understood in terms of the added force (in addition to their Coulomb attraction toward the nucleus) on the orbiting electrons. This added force requires an appropriate adjustment in their orbital sizes and speeds, which leads to changes in the magnetic moment of the atom, which oppose the external field.

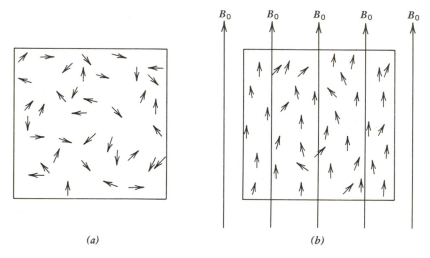

Figure 20.3 (a) Normally, the magnetic moments of the particles in a system have ramdom alignment. (b) When placed in an external magnetic field, the magnetic moments-tend to become aligned, because it is a state of lower energy and therefore higher probability. Thermal agitation prevents perfect alignment of all the particles at any instant.

or molecules either have no magnetic moments, or whose magnetic moments are not free to change their orientations. Most materials exhibit paramagnetism, however, which is the dominant effect. The alignment of large numbers of tiny magnetic moments with an external field is a statistical process, and so the tools of statistical mechanics are appropriate.

In a few materials, the magnetic moments of neighboring atoms are very strongly coupled, so that the orientation of one is controlled by the orientation of its neighbors. Macroscopic regions, called "domains," contain atoms whose magnetic moments are nearly all aligned in the same direction. These materials become very strongly magnetized when placed in external magnetic fields, because the domain as a whole lines up with the external field and the strong coupling between neighboring atomic magnetic moments discourages any one from changing its orientation (Figure 20.4). Thus, the strong coupling between atoms effectively gives each atom a great deal of "magnetic inertia," greatly reducing its tendency to change orientations with thermal agitation. This effect is called "ferromagnetism." Ferromagnetic materials become strongly magnetized at room temperatures, in spite of the significant thermal agitation.

Paramagnetism is the most appropriate challenge for the application of our statistical tools, not only because it is the most common of the magnetic effects, but also because each individual atomic or molecular magnetic moment is free to change its orientation in response to thermal agitation. In studying ferromagnetism, we would have to consider the basic independent elements of our system to be the magnetic moments of macroscopic domains, rather than of the microscopic atoms or molecules. Diamagnetism can be understood from classical electricity and magnetism, without the use of statistical tools.

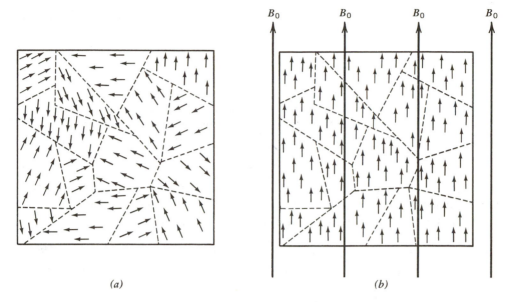

Figure 20.4 In ferromagnetic materials, the magnetic moments of neighboring atoms are strongly coupled, so that all magnetic moments within any one "domain" have the same alignment. When placed in an external magnetic field (b), the domains line up with this field, and the coupling between neighboring magnetic moments is so strong that it greatly reduces the tendency toward random orientation through thermal agitation.

PROBLEMS

20-1. If the spin gyromagnetic ratio of the neutron were positive, what would that imply about the distribution of charge within it?

20-2. Suppose a positively charged spin $\frac{1}{2}$ particle were found that had negative spin gyromagnetic ratio. What would you conclude about the net charge and charge distribution within this particle?

20-3. If the total angular momentum of an atom is given by $[l(l + 1)]^{1/2}\hbar$, and the maximum z-component by $l\hbar$, can the angular momentum ever be lined up exactly along the z-axis?

20-4. If any one electron is a spin $\frac{1}{2}$ particle ($s = \frac{1}{2}$, $s_z = \pm\frac{1}{2}$), how is it possible for the z-component spin angular momentum of some atoms or molecules to be an integer ($s_z = 0$, ± 1, etc.)?

20-5. What causes each of the following:

(a) Diamagnetism?
(b) Paramagnetism?
(c) Ferromagnetism?

20-6. Explain physically why ferromagnetic materials become much more strongly magnetized than paramagnetic materials, under normal conditions.

20-7.   Under what conditions might you expect paramagnetism to be as strong as ferromagnetism?

C.   PARAMAGNETISM

We now consider the magnetization of a paramagnetic material placed in an external magnetic field, $\mathbf{B}_0$. We define the z-direction to be that of the external field, so that the magnetic energy of an individual atom or molecule with magnetic moment μ is given by

$$\varepsilon = -\mu_z B_0 \tag{20.10}$$

We wish to find the average atomic (or molecular) magnetic moment, $\bar{\mu}_z$, because the magnetic moment of the entire material is simply the product of the number of atoms (or molecules) times the average magnetic moment of each.

$$\mu_z^{\text{total}} = N\bar{\mu}_z \tag{20.11}$$

The probability that an atom or molecule is in any given state, s, is given by

$$P_s = Ce^{-\beta\varepsilon_s}$$

where

$$C = \left[\sum_s e^{-\beta\varepsilon_s}\right]^{-1}$$

so the probability that it is in a state s, having magnetic moment μ_z^s is

$$P_s = \left(\sum_{s'} e^{\beta\mu_z^{s'}B_0}\right)^{-1} e^{\beta\mu_z^s B_0}$$

According to Eq. 20.8, the magnetic state of an atom or molecule is determined by the values of its orbital and spin angular momentum quantum numbers, l_z and s_z. So we can say that the probability that it is in a given state is

$$P_{l_z,s_z} = \left[\sum_{l_z',s_z'} e^{-\beta(l_z' + 2s_z')\mu_B B_0}\right]^{-1} e^{-\beta(l_z + 2s_z)\mu_B B_0} \tag{20.12}$$

where we have used Eq. 20.5 to substitute "$-(l_z + 2s_z)\mu_B$" for "μ_z."

The mean value of the atomic or molecular magnetic moment is given by

$$\bar{\mu}_z = \sum_{l_z,s_z} P_{l_z,s_z}\mu_z$$

$$= -\left[\sum_{l_z',s_z'} e^{-(l_z' + 2s_z')\beta\mu_B B_0}\right]^{-1} \left[\sum_{l_z,s_z} e^{-(l_z + 2s_z)\beta\mu_B B_0}(l_z + 2s_z)\mu_B\right] \tag{20.13}$$

The summation is over all accessible states, or equivalently, over all allowed values of the quantum numbers l_z and s_z. This allowed range for l_z and s_z depends on the particular type of atom or molecule, so the average magnetic moment, $\bar{\mu}_z$, differs from one material to the next, even in the same external field, B_0, and at the same temperature, T.

EXAMPLE

A certain material containing 10^{25} atoms is subjected to an external field of 1 T at room temperature. If the electrons of each atom have zero total orbital angular momentum ($l = 0$), and total spin angular momentum quantum number $s = \frac{1}{2}$*, what is the average atomic magnetic moment, and the magnetic moment of the entire material?

In the sums in Eq. 20.13, l_z can take only one value ($l_z = 0$), and s_z only two ($s_z = \pm\frac{1}{2}$). Therefore, the average magnetic moment is given by

$$\bar{\mu}_z = -[e^{-\beta\mu_B B_0} + e^{\beta\mu_B B_0}]^{-1}[e^{-\beta\mu_B B_0}(\mu_B) + e^{\beta\mu_B B_0}(-\mu_B)]$$

where

$$\beta\mu_B B_0 = 2.24 \times 10^{-3}$$

for $T = 300$ K and $B_0 = 1$ T. This gives

$$\bar{\mu}_z = 2.24 \times 10^{-3}\mu_B$$
$$= 2.1 \times 10^{-26} \text{ J/T}$$

and

$$\mu_z^{\text{total}} = N\bar{\mu}_z = 0.21 \text{ J/T}$$

The average magnetic moment, $\bar{\mu}_z$, can always be calculated by straightforward application of Eq. 20.13. However, this calculation can be simplified if the value of $\beta\mu_B B_0$ appearing in the exponent is either much smaller than 1, or much larger than 1.

In Figure 20.5 we have plotted the magnetic moment per mole of a typical substance versus $\beta\mu_B B_0$ using Eq. 20.13. We see that for $\beta\mu_B B_0 \ll 1$, the magnetic moment increases linearly with $\beta\mu_B B_0$, and for $\beta\mu_B B_0 \gg 1$ the magnetic moment approaches some constant maximum value. We now investigate each of these regions in more detail.

CASE 1 $\beta\mu_B B_0 \ll 1$

In the previous example, we saw that for a rather strong magnetic field at room temperature, the factor $\beta\mu_B B_0$ was only 2.2×10^{-3}. In this and most ordinary situations, the factor $\beta\mu_B B_0$ will be much smaller than 1. In fact, only at temperatures lower than a few degrees Kelvin, or in magnetic fields much stronger than are attainable on earth, would this not be the case.

---

* Recall that orbital angular momentum is expressible as $\sqrt{l(l+1)}\hbar$, and spin angular momentum is expressible as $\sqrt{s(s+1)}\hbar$, where l is an integer, and s is integer or half integer. The allowed values of l_z and s_z are $l_z = l, l-1, l-2, \ldots, -l$, $s_z = s, s-1, s-2, \ldots, -s$.

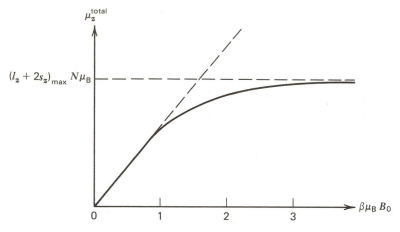

Figure 20.5   Plot of total magnetic moment, μ_z, versus $\mu_B B_0/kT$ for a typical paramagnetic substance. In the range $\beta\mu_B B_0 \ll 1$, the magnetic moment is linear in the applied external field, B_0, in agreement with the Curie law, $\mu_z = (C/T)B_0$. For $\beta\mu_B B_0 \gg 1$, the magnetic moment is given by $\mu_z = (l_z + 2s_z)_{max}N\mu_B$. In the range intermediate between these extremes, it would be necessary to use Eq. 20.13 to calculate the magnetic moment.

We can use the exponential expansion

$$e^{-x} \approx 1 - x \qquad \text{(for } x \ll 1)$$

to transform Eq. 20.13 into the following form

$$\bar{\mu}_z \approx -\left\{\sum_{l'_z,s'_z} [1 - (l'_z + 2s'_z)\beta\mu_B B_0]\right\}^{-1}\left\{\sum_{l_z,s_z} [1 - (l_z + 2s_z)\beta\mu_B B_0](l_z + 2s_z)\mu_B\right\}$$

Because the allowed positive values of l_z and s_z are equal and opposite to their allowed negative values,

$$\sum_{l_z,s_z} (l_z + 2s_z) = 0$$

and we have

$$\bar{\mu}_z \approx \left(\sum_{l_z,s_z} 1\right)^{-1}\left[\sum_{l_z,s_z} (l_z + 2s_z)^2\right]\beta\mu_B^2 B_0 \tag{20.14}$$

In some atoms and molecules there is strong "spin-orbit coupling," which means that the magnetic moment of the electron's spin interacts with that due to its orbit. In such cases, the allowed values of l_z and s_z are not independent of each other. If they were independent of each other, we could separate the two sums

$$\sum_{l_z,s_z} = \sum_{l_z=-l}^{+l} \sum_{s_z=-s}^{+s} \qquad \text{(no spin-orbit coupling)}$$

But we leave our results in the more general form, which will apply to all cases. The specific set of allowed values for (l_z, s_z) depends on the kinds of atoms and molecules involved.

Since the total magnetic moment of N atoms or molecules is $N\bar{\mu}_z$, we can use the result (20.14) to write

$$\mu_z^{\text{total}} = CB_0/T \tag{20.15}$$

where

$$C = \left(\sum_{l_z,s_z} 1\right)^{-1}\left[\sum_{l_z,s_z} (l_z + 2s_z)^2\right]\left(\frac{N\mu_B^2}{k}\right) \tag{20.16}$$

If there are n moles of the material, then $N = nN_A$, and

$$\frac{N\mu_B^2}{k} = n\left(\frac{N_A\mu_B^2}{k}\right) = n(3.75 \text{ J} \cdot \text{K}/\text{T}^2 \cdot \text{mole}) \tag{20.16'}$$

The result (20.15) tells us that for most materials under most ordinary conditions, the magnetization is directly proportional to the strength of the external field, and inversly related to the temperature. It is called the "Curie law" for magnetism, and the constant C is called the "Curie constant." As can be seen from Eq. 20.16 the value of the Curie constant depends on the allowed range of values for l_z and s_z, which in turn depends on the material.

CASE 2 $\beta\mu_B B_0 \gg 1$

The limit $\beta\mu_B B_0 \gg 1$ is attainable only at extremely low temperatures, or in external field strengths much greater than those attainable on earth. Consequently, this limit is primarily of interest to those doing some aspects of low-temperature physics or astrophysics.

In this case, we have

$$e^{\beta\mu_B B_0} \gg 1$$

or

$$e^{n\beta\mu_B B_0} \gg e^{(n-1)\beta\mu_B B_0} \qquad (\text{any } n)$$

This means that in the following sum, we only need consider one value of $(l_z + 2s_z)$

$$\sum_{l_z,s_z} e^{-(l_z + 2s_z)\beta\mu_B B_0} \approx e^{(l_z + 2s_z)_{\text{max}}\beta\mu_B B_0}$$

because all of the other terms in the sum are much smaller in comparison. With this simplification, Eq. 20.13 becomes

$$\bar{\mu}_z = (l_z + 2s_z)_{\text{max}}\mu_B$$

That is, the average value of the magnetic moment of an atom or molecule is its maximum value. All magnetic moments are lined up with the external field, as we would expect in the limit of high magnetic field and low thermal agitation. The total magnetic moment in this limit would be

$$\mu_z^{\text{total}} = (l_z + 2s_z)_{\text{max}}N\mu_B \tag{20.17}$$

SUMMARY

Because the magnetic properties of most materials are due almost entirely to the electrons, and because the gyromagnetic ratio for electron orbit is -1 and for electron spin is -2, we can write the magnetic moment of an atom or molecule as

$$\mu_z = -(l_z + 2s_z)\mu_B \tag{20.8}$$

where μ_B is the Bohr magneton,

$$\mu_B = 9.27 \times 10^{-24} \text{ J/T}$$

and where l_z and s_z are quantum numbers indicating the z-components of the orbital angular momentum and spin angular momentum for the electron cloud. They may take on the values

$$l_z = 0, \pm 1, \pm 2, \ldots, \pm 1$$

$$s_z = \begin{cases} 0, \pm 1, \pm 2, \ldots, \pm s & \text{(even number of electrons)} \\ \pm\frac{1}{2}, \pm\frac{3}{2}, \ldots, \pm s & \text{(odd number of electrons)} \end{cases} \tag{20.9}$$

where the range (determined by l and s) depends on the particular type of atom or molecule. If there is spin-orbit coupling, the allowed values of l_z and s_z may not be independent of each other.

Consider a material at temperature T in an external magnetic field, B_0, oriented along the positive z-axis. Because the probability that an atom or molecule is in a certain state, s, is given by

$$P_s = Ce^{-\beta\varepsilon_s} = Ce^{\beta\mu_z B_0}$$

we find that the average value of μ_z for any atom or molecule is given by

$$\bar{\mu}_z = -\left[\sum_{l_z,s_z} e^{-(l_z + 2s_z)\beta\mu_B B_0}\right]^{-1}\left[\sum_{l_z,s_z} e^{-(l_z + 2s_z)\beta\mu_B B_0}(l_z + 2s_z)\mu_B\right] \tag{20.13}$$

If the material contains N atoms or molecules, then its total magnetic moment is

$$\mu_z^{\text{total}} = N\bar{\mu}_z$$

In the common situation where $\beta\mu_B B_0 \ll 1$, this can be written as the Curie law,

$$\mu_z^{\text{total}} = CB_0/T, \qquad (\beta\mu_B B_0 \ll 1) \tag{20.15}$$

where the Curie constant, C, is given by

$$C = \left[\sum_{l_z,s_z} 1\right]^{-1}\left[\sum_{l_z,s_z}(l_z + 2s_z)^2\right]\left(\frac{N\mu_B^2}{k}\right) \tag{20.16}$$

If there are n moles of the substance. then

$$\frac{N\mu_B^2}{k} = n\left(\frac{N_A\mu_B^2}{k}\right) = n(3.75 \text{ J·K/T}^2\text{·mole}) \tag{20.16'}$$

In the more exceptional case, where $\beta\mu_B B_0 \gg 1$, the result (20.13) becomes

$$\mu_z^{\text{total}} = (l_z + 2s_z)_{\text{max}}N\mu_B, \qquad (\beta\mu_B B_0 \gg 1) \tag{20.17}$$

EXAMPLE

Consider a material made of molecules whose electron cloud has orbital angular momentum quantum number, $l = 1$, and spin angular momentum quantum number, $s = \frac{1}{2}$, so that the allowed values of l_z are $(0, \pm 1)$ and the allowed values of s_z are $(\pm \frac{1}{2})$. If a mole of this material at room temperature were in an external field of 3 T, what would be its magnetic moment?

For this case, the factor $\beta \mu_B B_0$ is given by

$$\beta \mu_B B_0 = 6.7 \times 10^{-3}$$

which is much less than 1. This material, then, obeys the Curie law, with Curie constant given by

$$C = \left[\sum_{l_z = -1}^{+1} \sum_{s_z = -1/2}^{+1/2} 1 \right]^{-1} \left[\sum_{l_z = -1}^{+1} \sum_{s_z = -1/2}^{+1/2} (l_z + 2s_z)^2 \right] \frac{N_A \mu_B^2}{k}$$

$$= \frac{10}{6} \left(\frac{N_A \mu_B^2}{k} \right)$$

$$= 6.25 \ \text{J} \cdot \text{K} / \text{T}^2 \cdot \text{mole}$$

The total magnetic moment is, then,

$$\mu_z^{\text{total}} = C B_0 / T = 0.0625 \ \text{J} / \text{T} \cdot \text{mole}$$

EXAMPLE

If the above material were at a temperature of 0.1 K, what would be its magnetic moment?

In this case

$$\beta \mu_B B_0 = 20 \gg 1$$

so we can use Eq. 20.17 to get

$$\mu_z^{\text{total}} = 2 N_A \mu_B = 11.2 \ \text{J} / \text{T} \cdot \text{mole}$$

PROBLEMS

20-8. Consider sodium metal. Each atom has two completed electronic shells ($l = 0$, $s = 0$ for these) and one electron left over in a state for which $l = 0$, $s = \frac{1}{2}$. If it is sitting in an external field of one tesla ($B_0 = 1 \ T$),

(a) At what temperature would $\beta \mu_B B_0 = 1$?
(b) What is its magnetic moment per mole at room temperature?
(c) What is the Curie constant for a mole of metallic sodium?

20-9. Consider some material for which there is strong spin orbit coupling in the electron cloud. The following are the only allowed sets of values for (l_z, s_z) in a molecule of this material: $(1, -\frac{1}{2})$, $(0, +\frac{1}{2})$, $(0, -\frac{1}{2})$, $(-1 +\frac{1}{2})$.

(a) What is the Curie constant for a mole of this material?
(b) In an external field of 2 T, what would be its magnetic moment per mole at a temperature of 10^3 K? At 10 K?

(c) If sitting in the earth's magnetic field of about 10^{-4} T, beneath what temperature would you expect this material to become fairly completely magnetized ($\mu \sim \mu_{max}$)?

20-10. Consider a material made of molecules for which the electronic clouds each have $l = 1$, $s = 0$. It is placed in an external field $B_0 = 2$ T.

(a) What is the Curie constant for a mole of this material?
(b) What is its magnetic moment per mole at a temperature of 273 K?
(c) If it is at a temperature of 1.4 K, what is the value of $\beta\mu_B B_0$?
(d) At a temperature of 1.4 K, what is its magnetic moment per mole?

20-11. Consider a material made up of molecules for which the electronic clouds each have $l = 0$, $s = 1$. It is placed in an external magnetic field of strength 3 T, and it is at a temperature of 2 K.

(a) What is the value of $\beta\mu_B B_0$
(b) What is the magnetic moment per mole of this material?

20.12. Consider a material whose molecular electron clouds each have $l = 2$, $s = \frac{1}{2}$, and there is no spin orbit coupling. This material is at room temperature and in a field of strength $B_0 = 10^{-4}$ T.

(a) What is the Curie constant for a mole of this material?
(b) What is the magnetic moment per mole of this material?
(c) What is the average magnetic moment per molecule of this material?

20-13. Calculate the value of μ_0 for the ρ-meson, which has mass 1.37×10^{-27} kg.

20-14. A charmed quark is a spin $\frac{1}{2}$ particle whose mass is unknown, but could be around 3.5×10^{-27} kg. What would be the value of μ_0 for this quark?

20-15. Consider an atomic material for which the electron clouds have no net angular momentum ($l = 0$, $s = 0$), the neutrons in the nucleus have no net angular momentum ($l = 0$, $s = 0$), but for which the protons in each nucleus do have net spin angular momentum ($l = 0$, $s = 1$). Therefore, the magnetic properties of this material would be due entirely to the protons in the nucleus.

(a) What would be the value of $N_A\mu_N^2/k$?
(b) What would be the value of the Curie constant for the nuclear magnetization of a mole of this material?
(c) If in an external field of $B_0 = 1$ T and at room temperature (300 K), what would be the magnetic moment per mole of this material?

20-16. What is the Curie constant for a mole of a material whose molecular electron clouds have no spin-orbit coupling (i.e., l_z and s_z are independent of each other) if:

(a) $(l, s) = (0, \frac{1}{2})$?
(b) $(l, s) = (2, \frac{3}{2})$?
(c) $(l, s) = (2, 0)$?

20-17. What is the maximum magnetic moment per mole for each of the three substances in Problem 20-16?

chapter 21

THE PARTITION FUNCTION

The mean value of any appropriate property can be calculated by using the known probabilities for a particle to be in the various accessible states. In practice, however, such calculations can be long and tedious, but they can be done much more expeditiously through the use of the "partition function."

By this point in our studies we are familiar with the idea that the probability of a system being in a state s having energy E_s, is given by

$$P_s = Ce^{-\beta E_s}$$

where C is the appropriate constant of proportionality. Knowing these probabilities, we are able to calculate the mean value of any property of the system. For example, if χ is some property, having the value χ_s when the system is in state s, then the mean value of χ is given by

$$\bar{\chi} = \sum_s P_s \chi_s = \sum_s Ce^{-\beta E_s} \chi_s$$

The trouble with this is that the sum over states can include a very large number of terms, so such calculations can be quite cumbersome, although correct. Furthermore, this calculation must be repeated for every single property (χ) that you wish to study.

To facilitate these otherwise cumbersome calculations, we introduce a function called the partition function. Calculating the partition function for a system does involve a sum over all states, but it is usually a summation that is easily carried out. Furthermore, once the partition function is known, many different properties of the system may be calculated directly from it, and we don't have to repeat the sumation everytime we study a different property.

A. DEFINITION

The partition function for any system is given the symbol Z, and is defined in the following way: if E_s is the energy of the system when it is in state s, then the partition function is defined by

$$Z = \sum_s e^{-\beta E_s} \tag{21.1}$$

We can see that it is closely related to the probability of the system being in a given state. From Eq. 16.8 we have

$$P_s = \left(\sum_{s'} e^{-\beta E_{s'}} \right)^{-1} e^{-\beta E_s} = Z^{-1} e^{-\beta E_s} \tag{21.2}$$

If the system is in any given state, s, we would expect the energy of that state, E_s, to be dependent on the number of particles and their separations, especially if there are any interactions between them. That is, we would expect the energy of any state to be dependent on V and N.

$$E_s = E_s(V, N) \tag{21.3}$$

But we don't have to rely on our "expectations" to show that this is correct. From the first law, we know that the energy of any system is dependent on the *three* variables, (S, V, N), where S is the entropy of the system. The entropy is a measure of the number of states, Ω, and if we specify one certain state s, then $\Omega_s = 1$, and the entropy is determined.

$$S_s = k \ln \Omega_s = k \ln (1) = 0$$

That is, when we specify the state, we reduce the number of variables from three to two, because the entropy is fixed.

With the entropy of state s fixed, the first law becomes

$$dE_s = -p_s \, dV + \mu_s \, dN$$

from which we see that the pressure and chemical of the system when in state s are given by

$$p_s = -\left(\frac{\partial E_s}{\partial V} \right)_N \tag{21.4}$$

$$\mu_s = \frac{\partial E_s}{\partial N} \bigg)_V \tag{21.5}$$

Since the energy of each state, E_s, is a function of V and N, and since the temperature appears explicitly in the factor $\beta = 1/kT$, the partition function is usually most conveniently written as a function of the three variables (T, V, N).

$$Z(T, V, N) = \sum_s e^{-\beta E_s(V, N)}$$

B.   CALCULATION OF MEAN VALUES

It is emphasized that the partition function is introduced through *definition*; it is not derived. It contains no new hidden physical significance. The physics went into our derivation of the probabilities, P_s, in Chapter 16. Anything that can be calculated using the partition function can also be calculated directly from the probabilities, P_s. However, as the following examples will show, a knowledge of the partition of function makes some of these calculations much easier, and that is our justification for dealing with it rather than the probabilities directly.

Suppose we wish to calculate the internal energy of a system whose partition function is known. The mean value of its internal energy is given by (see Eq. 21.2)

$$\bar{E} = \sum_s P_s E_s = Z^{-1} \sum_s e^{-\beta E_s} E_s$$

$$= -Z^{-1} \frac{\partial}{\partial \beta} \left(\sum_s e^{-\beta E_s} \right) = -Z^{-1} \frac{\partial}{\partial \beta} Z$$

$$= -\frac{\partial}{\partial \beta} \ln Z \tag{21.6}$$

Similarly, we can find the mean square of the internal energy by

$$\overline{E^2} = \sum_s P_s E_s^2 = Z^{-1} \sum_s e^{-\beta E_s} E_s^2 = Z^{-1} \frac{\partial^2}{\partial \beta^2} \left(\sum_s e^{-\beta E_s} \right)$$

$$= Z^{-1} \frac{\partial^2}{\partial \beta^2} Z \tag{21.7}$$

Also, the square of the standard deviation for the fluctuations of the internal energy about its mean value* can be gotten through the relationship (see homework Problem 21.4)

$$\sigma^2 \equiv \overline{(\Delta E)^2} = \overline{E^2} - \bar{E}^2 = \frac{\partial^2}{\partial \beta^2} \ln Z \tag{21.8}$$

as you will be asked to prove in homework Problem 21-4.

Similarly, if the pressure of the system in state s is p_s, then using Eq. 21.4 we find the average pressure of the system is given by

$$\bar{p} = \sum_s P_s p_s = -Z^{-1} \sum_s e^{-\beta E_s} \left(\frac{\partial E_s}{\partial V} \right)_N$$

$$= \frac{1}{\beta} Z^{-1} \frac{\partial}{\partial V} \left(\sum_s e^{-\beta E_s} \right)_{T,N} = \frac{1}{\beta} Z^{-1} \left(\frac{\partial Z}{\partial V} \right)_{T,N}$$

$$= \frac{1}{\beta} \frac{\partial}{\partial V} \ln Z \bigg)_{T,N} \tag{21.9}$$

If we replace $p_s = -\partial E_s/\partial V)_N$ by $\mu_s = \partial E_s/\partial N)_V$ in the above development, we would get the analogous result for the chemical potential,

$$\bar{\mu} = -\frac{1}{\beta} \frac{\partial}{\partial N} \ln Z \bigg)_{T,V} \tag{21.10}$$

as you will show in homework Problem 21-5.

---

* Recall that the probabilities, P_s, are for a small system in equilibrium with a large reservoir. Because of this interaction, the internal energy of the small system will fluctuate.

It turns out that our definition of the partition function in Eq. 21.1 was needlessly restrictive. If we multiply the partition function by any arbitrary constant C,

$$Z = C \sum_s e^{-\beta E_s}$$

the results of all our calculations would be the same. We could write the probability of being in state s as

$$P_s = \left(\sum_{s'} e^{-\beta E_{s'}} \right)^{-1} e^{-\beta E_s} = \left(C \sum_{s'} e^{-\beta E_{s'}} \right)^{-1} C e^{-\beta E_s}$$

and because the constant appears in both numerator and denominator, it cancels. If you notice, all our calculated mean values would have this constant appearing in both numerator and denominator.* But since arbitrary multiplicative constants don't matter, the constant C is usually set equal to one, for simplicity.

SUMMARY

Consider some system interacting with a large reservoir. The partition function for this system is defined by

$$Z \equiv \sum_s e^{-\beta E_s} \qquad (21.1)$$

where the sum is over all accessible states, s.

Since entropy is proportional to the logarithm of the number of states, the entropy of any given state is zero ($\log 1 = 0$), and the energy of any given state only depends on two parameters: the volume and the number of particles.

$$E_s = E_s(V, N) \qquad (21.4)$$

The partition function is useful in calculating the mean values of many properties of the system, including the following.

$$\bar{E} = -\frac{\partial}{\partial \beta} \ln Z \bigg|_{V,N} \qquad (21.6)$$

$$\overline{E^2} = Z^{-1} \frac{\partial^2}{\partial \beta^2} Z \bigg|_{V,N} \qquad (21.7)$$

$$\sigma^2 = \overline{(\Delta E)^2} = \overline{E^2} - \bar{E}^2 = \frac{\partial^2}{\partial \beta^2} \ln Z \bigg|_{V,N} \qquad (21.8)$$

$$\bar{p} = \frac{1}{\beta} \frac{\partial}{\partial V} \ln Z \bigg|_{T,N} \qquad (21.9)$$

$$\bar{\mu} = -\frac{1}{\beta} \frac{\partial}{\partial N} \ln Z \bigg|_{T,V} \qquad (21.10)$$

---

* $(\partial/\partial x) \ln Z = (1/Z)(\partial Z/\partial x)$.

PROBLEMS

21-1. A certain system has only two accessible states. Measuring energies relative to the ground state, the energy of the ground state is 0, and that of the excited state is 1 eV. What is the value of the partition function for this system:

(a) At a temperature of 300 K?
(b) At a temperature of 30,000 K?

21-2. If a system is confined to two quantum states, what is its entropy? How many independent variables does its internal energy depend on?
(*Hint.* The entropy is a fixed constant, not a variable in this case.)

21-3. Show that $\overline{E^2} = Z^{-1}(\partial^2/\partial\beta^2)Z$.

21-4. Show that the square of the standard deviation for the energy of a system fluctuating about its mean value, $\sigma^2 = \overline{(\Delta E)^2} = \overline{E^2} - \bar{E}^2$, is given by

$$\sigma^2 = \frac{\partial^2}{\partial\beta^2}\ln Z$$

21-5. Using Eq. 21.5 show that the mean value of the chemical potential of a system is given by

$$\bar{\mu} = -\frac{1}{\beta}\frac{\partial}{\partial N}\ln Z\Big|_{T,V}$$

(*Hint.* Follow a development parallel to that of Eq. 21.9.)

21-6. For a certain system, there is only one accessible state and it has energy

$$E_s = -CV$$

where C is a constant and V is the volume.

(a) Write down the partition function for this system.
(b) Using Eq. 21.9 find the average pressure for this system as a function of the temperature.

21-7. For a certain system, there is only one accessible state and it has energy

$$E_s = -NkT\ln\frac{V}{V_0}$$

where V_0 is a constant.

(a) Write down the partition function for this system.
(b) Using Eq. 21.9 find the average pressure for this system as a function of volume and temperature.

21-8. For a certain system, the partition function is given by $-(V/V_0)^N$, where V is the volume and V_0 is some constant. What is the average pressure of this system as a function of temperature and volume?

21-9. For a certain system, the energy of each state s is given by

$$E_s = f_s(T) - NkT\ln\frac{V}{V_0}$$

where V is the volume, V_0 a constant, and $f_s(T)$ is some function of the temperature only, with the particular function depending on the state.

(a)   Write out the partition function for this system.

(b)   What is the average pressure of this system as a function of temperature and volume?

21-10.   For a certain system, the partition function is given by $-(V/V_0)^N$, where V is the volume and V_0 is some constant. What is the average chemical potential of this system as a function of temperature and volume?

21-11.   For a certain system, the energy of each state s is given by

$$E_s = f_s(T) - NkT \ln \frac{V}{V_0}$$

where V is the volume, V_0 a constant, and $f_s(T)$ is some function of the temperature, with the particular function depending on the state.

(a)   Write out the partition function for this system.

(b)   What is the average chemical potential of this system as a function of the temperature and volume?

21-12.   For a certain system, the energy of each state s is given by

$$E_s = kT\left(C_s + \frac{3}{2}N \ln \frac{\beta}{\beta_0} - N \ln \frac{V}{V_0}\right)$$

where β_0 and V_0 are constants, and C_s is a constant whose value depends on the state.

(a)   Write out the partition function for this system.

(b)   Calculate the average internal energy for this system as a function of (N, V, T).

(c)   Calculate the average pressure for this system as a function of (N, V, T).

(d)   Calculate the average chemical potential for this system as a function of (N, V, T).

(e)   Calculate the standard deviation (σ) for the fluctuations of the internal energy of this system about its mean value.

C.   ENTROPY AND HELMHOLTZ FREE ENERGY

The partition function is simply related to the entropy and the Helmholtz free energy of a system. To demonstrate this, we first look at an incremental change in $\ln Z$.

$$d \ln Z = \frac{\partial}{\partial \beta} \ln Z \bigg|_{V,N} d\beta + \frac{\partial}{\partial V} \ln Z \bigg|_{T,N} dV + \frac{\partial}{\partial N} \ln Z \bigg|_{T,V} dN$$

With the results (21.6), (21.9), and (21.10) this becomes

$$d \ln Z = -E \, d\beta + \beta p \, dV - \beta \mu \, dN$$

If we add $d(\beta E) = E \, d\beta + \beta \, dE$ to both sides, we get

$$d(\ln Z + \beta E) = \beta(dE + p \, dV - \mu \, dN)$$
$$= \beta(T \, dS)$$
$$= \frac{1}{k} dS$$

or equivalently

$$dS = d\left(k \ln Z + \frac{E}{T}\right)$$

Consequently, the entropy is expressed in terms of the partition function as follows:*

$$S = k \ln Z + \frac{E}{T} \qquad (21.11)$$

By rearranging terms, this becomes

$$E - TS = -kT \ln Z$$

but since $E - TS$ is the Helmholtz free energy, we have

$$F = E - TS = -kT \ln Z \qquad (21.12)$$

You may wish to know why we would be interested in anything so seemingly esoteric as the Helmholtz free energy. The reason is that three things that we are often interested in are easily calculated from it: entropy, pressure and chemical potential.

$$S = -\frac{\partial F}{\partial T}\bigg)_{V,N}$$

$$p = -\frac{\partial F}{\partial V}\bigg)_{T,N}$$

$$\mu = \frac{\partial F}{\partial N}\bigg)_{T,V} \qquad (21.13)$$

One way of seeing this is from the definition of Helmholtz free energy ($F \equiv E - TS$), which combined with the first law, gives Eq. 13.16.

$$dF = -S\,dT - p\,dV + \mu\,dN \qquad (13.16)$$

Another way of seeing this is that proposed in homework Problems 21-13 and 21-14, where you will be asked to show that these relationships can be gotten from our results (20.9), (20.10), (20.11), and (20.12) in this chapter.

SUMMARY

The entropy and the Helmholtz free energy of a system can be expressed in terms of the partition function as follows.

$$S = k \ln Z + \frac{E}{T} \qquad (21.11)$$

$$F = -kT \ln Z \qquad (21.12)$$

---

* We can ignore the constant of integration, because Z contains an arbitrary multiplicative constant which is an arbitrary additive constant in the logarithm of Z. Therefore, we can incorporate any arbitrary additive constant in this equation into the partition function, Z.

The Helmholtz free energy is particularly useful to us in this form because it is expressed in terms of (T, V, N), and so the entropy, pressure, and chemical potential may be calculated directly.

$$S = -\frac{\partial F}{\partial T}\bigg)_{V,N} \qquad p = -\frac{\partial F}{\partial V}\bigg)_{T,N} \qquad \mu = \frac{\partial F}{\partial N}\bigg)_{T,V} \qquad (21.13)$$

PROBLEMS

21-13.   Use Eqs. 21.12 and 21.11 to show that $S = -\partial F/\partial T)_{V,N}$.

21-14.   Use Eqs. (21.12), (21.9), and (21.10) to show

(a) $\bar{p} = -\partial F/\partial V)_{T,N}$.
(b) $\bar{\mu} = \partial F/\partial N)_{T,V}$.

21-15.   In deriving Eq. 21.11 from the differential form, we ignored a constant of integration, saying in the footnote that it could be incorporated into the definition of Z. Show that if we define

$$Z \equiv \sum_s e^{-\beta E_s}$$

where all energies are measured relative to the ground state ($E_0 = 0$), and if the ground state is nondegenerate, then the constant of integration is zero. Do this by looking at the low-temperature limit, where there is only one accessible state (the ground state), so that Z contains only one term, and $\bar{E} = E_0$. (*Hint.* Compare it to $S = k \ln \Omega$.)

21-16.   The partition function for a certain system is given by

$$Z = \left(\frac{\beta}{\beta_0}\right)^{-(3/2)N} \left(\frac{V}{V_0}\right)^N \qquad (\beta_0, V_0 \text{ are constants})$$

Find the following things as a function of (N, T, V):

(a)   The average internal energy, using Eq. 21.6.
(b)   The entropy, using Eq. 21.11.
(c)   The Helmholtz free energy, using Eq. 21.12.
(d)   The pressure, using Eq. 21.13.
(e)   The chemical potential, using Eq. 21.13.
(f)   The entropy, using Eq. 21.13.

21-17.   The partition function for a certain system is given by

$$Z = \left(\frac{\beta}{\beta_0}\right)^{-3N} e^{-N(V/V_0)-10} \qquad (\beta_0, V_0 \text{ are constants})$$

Find the following things as a function of (N, T, V).

(a)   The average internal energy, $\bar{E}$.
(b)   The standard deviation for the fluctuations of the internal energy around this mean value, σ.
(c)   The Helmholtz free energy, F.
(d)   The entropy, S.

(e)   The average pressure, $\bar{p}$.

(f)   The average chemical potential, $\bar{\mu}$.

D.   MANY SUBSYSTEMS AND IDENTICAL SUBSYSTEMS

Suppose system S consists of two subsystems, A and B (Figure 21.1). Let s represent the particular state that system S is in, and let a and b represent the particular states that the two subsystems are in. Since the energy of the combined system is the sum of the energies of the two subsystems, and since the sum over all states of the combined system is equivalent to summing over all possible states of the two subsystems, we have

$$Z = \sum_s e^{-\beta E_s} = \sum_{a,b} e^{-\beta(E_a + E_b)} = \sum_a e^{-\beta E_a} \sum_b e^{-\beta E_b} = Z_A Z_B$$

That is, the partition function for the combined system is the product of the partition functions of the subsystems. Extending this result to three or more subsystems gives

$$Z = Z_A Z_B Z_C \dots \tag{21.14}$$

If the subsystems are *not* distinguishable, then this result is *not* correct. The fault is that if the subsystems are not distinguishable, then the sum over all states of the combined system is *not* the same as the sum over all states of the component subsystems. Symbolically, for identical systems,

$$\sum_s \neq \sum_{a,b,c\dots} \tag{21.15}$$

For example, consider the system of three particles depicted in Figure 21.2. Of the possible states of the combined system, the six that are illustrated are all different if

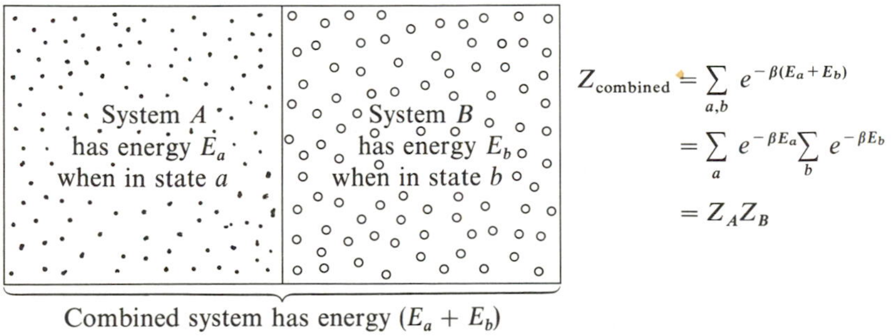

Combined system has energy $(E_a + E_b)$

Figure 21.1   The state of the combined system is determined by specifying states a and b of the subsystems. The sum over states of the combined system is therefore given by summing over all states a of system A and all states b of system B. Because of this, the partition function for the combined system can be written as the product of the partition functions of the small systems (provided they are not identical).

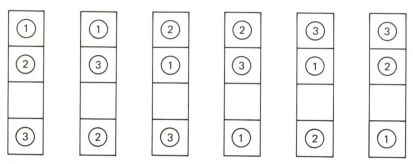

Figure 21.2 Some of the possible arrangements of three particles among four states. These six arrangements are all different if the particles are distinguishable, and are all identical if they are not. Therefore, these six arrangements would correspond to six different terms in the partition function if the particles are distinguishable, and only one term in the partition function if the particles are identical.

the three particles are distinguishable, but they are all identical if the three particles are indistinguishable. Therefore, in calculating the partition function, the various arrangements in this figure would correspond to six different terms in the sum over s if the particles 1, 2, and 3 are distinguishable, but only one term if they are not distinguishable.

As is illustrated in Figure 21.3 there are $N!$ different ways of distributing N particles among N boxes. If the particles are distinguishable then these are $N!$ different arrangements, but if the particles are indistinguishable, then these are all the same.

Therefore, when a system consists of N identical subsystems, there are $N!$ different rearrangements of the subsystems all giving the identical result for the system as a

$1! = 1$ $\boxed{1}$

$2! = 2$ $\boxed{1\,2} + \boxed{2\,1}$

$3! = 6$ $\boxed{1\,2\,3} + \boxed{1\,3\,2} + \boxed{2\,1\,3} + \boxed{2\,3\,1} + \boxed{3\,1\,2} + \boxed{3\,2\,1}$

$4! = 24$ $\boxed{1\,2\,3\,4} + \boxed{1\,2\,4\,3} + \boxed{1\,3\,2\,4} + \boxed{1\,3\,4\,2} + \boxed{1\,4\,2\,3}$

$+ \boxed{1\,4\,3\,2} + \boxed{2\,1\,3\,4} + \boxed{2\,1\,4\,3} + \boxed{2\,3\,1\,4} + \boxed{2\,3\,4\,1}$

$+ \boxed{2\,4\,1\,3} + \boxed{2\,4\,3\,1} + \boxed{3\,1\,2\,4} + \boxed{3\,1\,4\,2} + \boxed{3\,2\,1\,4}$

$+ \boxed{3\,2\,4\,1} + \boxed{3\,4\,1\,2} + \boxed{3\,4\,2\,1} + \boxed{4\,1\,2\,3} + \boxed{4\,1\,3\,2}$

$+ \boxed{4\,2\,1\,3} + \boxed{4\,2\,3\,1} + \boxed{4\,3\,1\,2} + \boxed{4\,3\,2\,1}$

(etc.)

Figure 21.3 There are $N!$ ways of permuting N things among N boxes.

whole. All these different permutations are in fact identical, and there are $N!$ terms in the sum on the right-hand side of Eq. 21.15 for each distinguishable state for the entire system. Symbolically for N identical subsystems,*

$$\sum_s = \frac{1}{N!} \sum_{a,b,c,\ldots} \tag{21.16}$$

Using this to calculate the partition function for a system consisting of N identical subsystems, we have

$$Z = \sum_s e^{-\beta E_s} = \sum_s e^{-\beta(E_a + E_b + \cdots)}$$

$$= \frac{1}{N!} \sum_{a,b,c,\ldots} e^{-\beta(E_a + E_b + \cdots)} = \frac{1}{N!} \sum_a e^{-\beta E_a} \sum_b e^{-\beta E_b} \cdots$$

$$= \frac{1}{N!} Z_A Z_B Z_C \cdots$$

Of course, if the subsystems are all identical, then

$$Z_A = Z_B = Z_C = \cdots$$

and we can write

$$Z = \frac{1}{N!} \zeta^N \tag{21.17}$$

where ζ is the partition function for any one of the identical subsystems.

We most frequently encounter systems made up of identical subsystems. The fact that the subsystems are identical implies that the composition of any one of them cannot change, so we can think of the subsystems as particles. Examples would include identical molecules in a gas, identical electrons in a metal, or identical nuclei in a reactor.

In the result (21.17) we have displayed the explicit dependence of the partition function for the entire system on the number of particles N. We can rewrite this result as

$$Z = \frac{1}{N!} \left(\sum_i e^{-\beta E_i} \right)^N \tag{21.18}$$

where the sum is over the one-particle states, i. In this form we see both variables N and T ($\beta = 1/kT$) appearing explicitly. The volume dependence, however, is hidden in the sum over states and in the energy of each. That is, both the spectrum of states accessible to a system and the energy of each, may in general be dependent on the system's volume.

---

* Strictly speaking, this result is true only if no two identical systems also occupy the same state. If this happens, further modification would be needed. See Chapter 24, or homework Problem 21-22.

Because the dependence on N is explicit, it is easy to calculate the chemical potential, μ. We use

$$\mu = \frac{\partial F}{\partial N}\bigg)_{T,V} \tag{21.13}$$

where

$$F = -kT \ln Z = -kT \ln \left(\frac{1}{N!}\,\zeta^N\right) = kT \ln N! - NkT \ln \zeta$$

Using Stirling's formula (3A.1), for large N,

$$\ln N! \approx N \ln N - N \tag{3.13}$$

this becomes

$$F = NkT \ln N - NkT - NkT \ln \zeta$$

and the chemical potential is

$$\mu = \frac{\partial F}{\partial N}\bigg)_{T,V} = -kT \ln \frac{\zeta}{N} \tag{21.19}$$

SUMMARY

For a system consisting of distinguishable subsystems, $A, B, C, \ldots$, the partition function for the combined system is the product of the partition functions for the individual subsystems.

$$Z = Z_A Z_B Z_C \ldots \tag{21.14}$$

If the system consists of N *in*distinguishable subsystems, then

$$Z = \frac{1}{N!}\,\zeta^N \tag{21.17}$$

where ζ is the partition function for any one individual subsystem. That is,

$$\zeta = \sum_i e^{-\beta E_i} \tag{21.20}$$

where the sum is over states of the individual subsystem. Identical subsystems can properly be referred to as "particles."

The partition function for the combined system is most conveniently written as a function of the variables (T, V, N). Both the spectrum of accessible states and the energy of each are dependent on V, in general.

$$Z(T, V, N) = \frac{1}{N!} [\zeta(T, V)]^N$$

$$= \frac{1}{N!} \left[\sum_i e^{-\beta E_i} \right]^N \qquad (21.18)$$

Using Stirling's formula, we can calculate the chemical potential for a system of a large number of identical particles from

$$\mu = \frac{\partial F}{\partial N} \bigg)_{T,V} \qquad (21.13)$$

with

$$F = -kT \ln Z \qquad (21.12)$$

This gives

$$\mu = -kT \ln \frac{\zeta}{N} \qquad (21.19)$$

PROBLEMS

21-18. How many different ways can you arrange two particles among three boxes (no more than one per box):

(a) If the particles are distinguishable?
(b) If the particles are indistinguishable?

21-19. How many different ways can you arrange two particles among four boxes (no more than one per box):

(a) If the particles are distinguishable?
(b) If the particles are not distinguishable?

21-20. How many different ways can you arrange three particles among three boxes (no more than one per box):

(a) If the particles are distinguishable?
(b) If the particles are indistinguishable?

21-21. How many different ways can you arrange three particles among four boxes (no more than one per box):

(a) If the particles are distinguishable?
(b) If the particles are indistinguishable?

21-22. The results of this section are *not* correct if there is significant probability of two systems occupying the *same* state simultaneously. Show this for the following simple case: Suppose there are two subsystems (e.g., two "particles") and the same two accessible states for each (e.g., two "boxes"). If both subsystems may occupy the same state:

(a) How many different arrangements of the system are there if the two subsystems are distinguishable?

(b) How many different arrangements of the systems are there if the two subsystems are indistinguishable?

(c) Do the answers to parts (a) and (b) differ by a factor of 2!?

21-23. Consider a system at temperature 400 K, consisting of N molecules, each having only two accessible states; $\varepsilon_0 = 0$ and $\varepsilon_1 = 0.1$ eV.

(a) What is the probability for a molecule to be in the excited state?

(b) What is the average energy per molecule?

21-24. Consider a system of N molecules, each having only two accessible states; $\varepsilon_0 = 0$ and $\varepsilon_1 = kT$. Give the following as a function of (T, V, N):

(a) The partition function.

(b) The Helmholtz free energy.

(c) The chemical potential.

21-25. Repeat Problem 21-24 for the case where $\varepsilon_0 = 0$, $\varepsilon_1 = 0.1$ eV.

E. THE PARTITION FUNCTION FOR A GAS

In order to illustrate the results of this chapter, we apply them to the specific example of a gas composed of N diatomic molecules. The partition function for the entire gas, Z, is related to that for a single molecule, ζ, through Eq. 21.17

$$Z = \frac{1}{N!} \zeta^N$$

so our problem is to find the partition function for one single molecule.

The energy of a single molecule, ε, can be stored in translational (ε_t), rotational (ε_r), and vibrational (ε_v) degrees of freedom. It is also conceivable that some energy could be stored in the excitation of electrons to higher energy levels (ε_e), or in excited nuclear states (ε_n), and perhaps other ways as well.

$$\varepsilon = \varepsilon_t + \varepsilon_r + \varepsilon_v + \varepsilon_e + \varepsilon_n + \cdots$$

In calculating the partition function for one molecule, the sum over all states would have to include all translational, rotational, vibrational, electronic, and nuclear states.

$$\zeta = \sum_{t,r,v,e,n,\ldots} e^{-\beta(\varepsilon_t + \varepsilon_r + \varepsilon_v + \varepsilon_e + \varepsilon_n + \ldots)}$$

$$= \sum_t e^{-\beta\varepsilon_t} \sum_r e^{-\beta\varepsilon_r} \sum_v e^{-\beta\varepsilon_v} \sum_e e^{-\beta\varepsilon_e} \sum_n e^{-\beta\varepsilon_n} \ldots$$

$$= \zeta_t \zeta_r \zeta_v \zeta_e \zeta_n \ldots \tag{21.21}$$

That is, the partition function for the molecule can be separated into a product of the partition functions for the individual degrees of freedom.

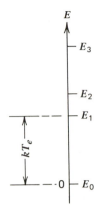

Figure 21.4 The "excitation temperature" for a system is defined by $kT_e \equiv E_1$, where E_1 is the energy of the first excited state (relative to the ground state). For temperatures small compared to T_e, the factor $e^{-E_1/kT}$ is negligibly small, and there is little probability of the system being in an excited state.

We will first simplify our calculation by showing that we can ignore some of the terms appearing in the product of terms in Eq. 21.21. If we measure energies relative to the ground state,* the energy of the ground state is zero, and that of any excited state is positive. The partition function for any degree of freedom, then, becomes

$$\zeta = \sum_i e^{-\beta\varepsilon_i} = e^{-0} + e^{-\beta\varepsilon_{1st\ excited}} + e^{-\beta\varepsilon_{2nd\ excited}} + \cdots \tag{21.22}$$

If the energy of the first excited state is very large compared to kT, then $\beta\varepsilon$ is a very large number, and Eq. 21.22 becomes[†]

$$\zeta = \sum_i e^{-\beta\varepsilon_i} = 1 + 0 + 0 + \cdots$$
$$= 1 \qquad (\varepsilon_{1st\ excited} \gg kT) \tag{21.23}$$

That is, if the energy of the first excited state is large compared to kT, then the excited states are inaccessible (they have negligible probability of being occupied) and the system is confined to the ground state. You needn't consider such degrees of freedom in calculating the partition function, or in doing any statistical or thermodynamic calculations, for that matter. In fact they are normally not referred to as "degrees of freedom," since there is no freedom to reach excited states.

You may recall the definition (16.11) of "excitation temperature," T_e, given by (Figure 21.4)

$$T_e = \frac{1}{k}\varepsilon_{1st\ excited} \tag{21.24}$$

---

* It is customary to measure energies relative to the ground state because it is easiest. However, you may measure them relative to any level you wish.

[†] If the ground state is n-times degenerate, then the sum over ground states gives a factor of n instead of 1. But the partition function can always be adjusted by constant factors without changing any results obtained from it, so we ignore any constant factor, such as n.

We know that for temperatures small compared to this we can ignore this degree of freedom, since

$$\varepsilon_{1st\ excited} \gg kT \qquad \text{for } T \ll T_e$$

and the probability of being in an excited state is negligibly small.

EXAMPLE

The first excited state for the translational motion of an air molecule in a shoebox is about 10^{-38} J. How cold would the air have to be before you could ignore the translational degrees of freedom of the molecules?

The value of T_e according to Eq. 21.24 is

$$T_e = \frac{1}{k}(10^{-38}\ \text{J}) \approx 10^{-15}\ \text{K}$$

so only at temperatures small compared to this could we ignore these degrees of freedom.

In homework Problems 21-26 and 21-27 you will show that for typical gases, the temperature must be thousands of degrees Kelvin for excited electronic states to be accessible and billions of degrees Kelvin for excited nuclear states to become accessible. So under normal conditions we would only expect translational, rotational, and vibrational degrees of freedom to contribute to the partition function.

$$\zeta = \zeta_t \zeta_r \zeta_v$$

The translational portion, ζ_t, can be calculated by converting the sum over translational quantum states to an integral, according to Eq. 2.12.

$$\sum_t \to \iint \frac{d^3r\, d^3p}{h^3}$$

Doing this gives

$$\zeta_t = \sum_t e^{-\beta(p^2/2m)} = \iint \frac{d^3r\, d^3p}{h^3} e^{-\beta(p^2/2m)} = V\left(\frac{2\pi mkT}{h^2}\right)^{3/2} \qquad (21.25)$$

To calculate the rotational portion of the partition function ζ_r, we assume that the rotational inertia of the diatomic molecule is I. The rotational kinetic energy is given by

$$\varepsilon_r = \frac{L^2}{2I}$$

where the angular momentum is quantized according to (see Eq. 2.18)

$$L^2 = l(l+1)\hbar^2, \qquad l = 0, 1, 2, \ldots$$

We also learn in quantum mechanics that any state of given l is $(2l + 1)$-times degenerate, since there are $(2l + 1)$ possible values of L_z

$$L_z = m\hbar \qquad m = 0, \pm 1, \pm 2, \ldots, \pm l$$

Therefore, our sum over rotational states is given by

$$\zeta_r = \sum_r e^{-\beta(L^2/2I)} = \sum_l \sum_m e^{-\beta(\hbar^2/2I)l(l+1)}$$

$$= \sum_l (2l + 1)e^{-\beta(\hbar^2/2I)l(l+1)}$$

If the first excited state is inaccessible,

$$\varepsilon_{1\text{st excited}} = \frac{\hbar^2}{2I} 1(1 + 1) \gg kT \qquad (\text{or } T \ll T_e)$$

then only the $l = 0$ term contributes significantly and the sum is equal to 1. But in the other extreme, this sum can be converted to an integral. Since

$$(2l + 1)\, dl = d(l^2 + l)$$

we have

$$\zeta_r = \int_0^\infty dl(2l + 1)e^{-\beta(\hbar^2/2I)(l^2 + l)}$$

$$= \int_0^\infty dx\, e^{-\beta(\hbar^2/2I)x} = \left(\frac{2IkT}{\hbar^2}\right)$$

We can summarize this result by saying

$$\zeta_r = \begin{cases} \sum_l (2l + 1)e^{-\beta(\hbar^2/2I)l(l+1)} & (\text{always}) \\ 1 & (T \ll T_e) \\ (2IkT/\hbar^2) & (T \gg T_e) \end{cases} \qquad (21.26)$$

where

$$T_e = \frac{1}{k}\left(\frac{\hbar^2}{I}\right)$$

For a typical diatomic gas molecule (e.g., N_2), the rotational inertia is roughly 1.4×10^{-46} kg-m^2/s^2, giving a value of about 6 K for T_e.

Finally, we consider the vibrational portion of the partition function for this diatomic molecule. Each molecule has some characteristic vibrational frequency, ω, that depends on the masses of the atoms and the forces between them. The vibrational energies of these oscillators are quantized according to*

$$\varepsilon_n = \left(n + \frac{1}{2}\right)\hbar\omega \qquad n = 0, 1, 2, \ldots$$

If we measure energies relative to the ground state, as is our custom, then the energy of any vibrational state is given by

$$\varepsilon_v = \varepsilon_n - \varepsilon_0 = n\hbar\omega$$

and the vibrational portion of the partition function is

$$\zeta_v = \sum_v e^{-\beta\varepsilon_v} = \sum_n e^{-n\beta\hbar\omega}$$

---

* This result is derived in introductory quantum mechanics courses.

As is shown in Appendix 21A, this sum can be performed exactly for any positive value of $(\beta\hbar\omega)$, yielding

$$z_v = \sum_n e^{-n\beta\hbar\omega} = \frac{1}{1 - e^{-\beta\hbar\omega}}$$

In the limit that $\hbar\omega \gg kT$ (or $T \ll T_e$) this gives $\zeta_v = 1$, and in the other limit, $\hbar\omega \ll kT$ (or $T \gg T_e$) this gives $\zeta_v = kT/\hbar\omega$. (See homework Problem 21-34.) That is,

$$\zeta_v = \begin{cases} \dfrac{1}{1 - e^{-\beta\hbar\omega}} & \text{always} \\[2ex] 1 & T \ll T_e \\[2ex] \dfrac{kT}{\hbar\omega} & T \gg T_e \end{cases} \tag{21.27}$$

where

$$T_e = \frac{1}{k}\hbar\omega$$

For a typical diatomic gas molecule (e.g., N_2), the characteristic frequency, ω, is about 2×10^{14} s^{-1}, giving a value of about 1500 K to T_e.

With these values of z_t, z_r, and z_v, we can write down the partition function for a gas containing N identical diatomic molecules.

$$Z = \frac{1}{N!}\zeta^N = \frac{1}{N!}(\zeta_t^N \zeta_r^N \zeta_v^N) \tag{21.28}$$

We have summarized our results for the contributions to the partition function from the various degrees of freedom in Table 21.1. There it is seen that so long as $T \gg T_e$, each single degree of freedom contributes a factor proportional to $T^{1/2}$ to the partition function of the molecule. One consequence of this is that the average energy per degree of freedom is given by ($\zeta_1 \propto T^{1/2}$)

$$\bar{\varepsilon} = -\frac{\partial}{\partial\beta}\ln\zeta_1 = \frac{1}{2}kT \tag{21.28}$$

in accordance with the equipartition theorem.

The partition function for one molecule is given by

$$\zeta = \zeta_t\zeta_r\zeta_v = (\text{const})T^{v/2} \tag{21.29}$$

where v is the number of degrees of freedom per molecule, and the partition function for the entire gas should be given by

$$Z = \frac{1}{N!}\zeta^N = (\text{const})\frac{1}{N!}T^{Nv/2} \tag{21.30}$$

where the product Nv is the total number of degrees of freedom for the entire gas ($Nv = \mathfrak{N}$).

Table 21.1   The Partition Function for a Molecule of a Diatomic Gas

Degrees of Freedom		$T_e \left(\dfrac{1}{k} \varepsilon_{\text{1st excited}} \right)$	Contribution to ζ	
Kind	Number		General	For $T \gg T_e$
Translational	3	10^{-15} K	$\sum_t e^{-\beta (p^2/2m)}$	$\left(\dfrac{2\pi mkT}{h^2} \right)^{3/2} = (\text{const}) T^{3/2}$
Rotational	2	10 K	$\sum_r e^{-\beta (L^2/2I)} = \sum_{l=0}^{\infty} (2l + 1) e^{-\beta (\hbar^2/2I) l(l+1)}$	$\left(\dfrac{2IkT}{\hbar^2} \right) = (\text{const}) T^{2/2}$
Vibrational	2	1000 K	$\sum_n e^{-\beta n \hbar \omega} = \dfrac{1}{(1 - e^{-\beta \hbar \omega})}$	$\left(\dfrac{kT}{\hbar \omega} \right) = (\text{const}) T^{2/2}$

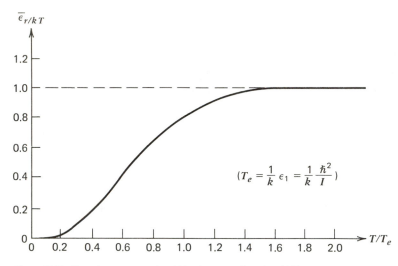

Figure 21.5   Plot of average rotational kinetic energy (in units of kT) versus temperature (in units of T_e) for diatomic molecules of rotational inertia 1.

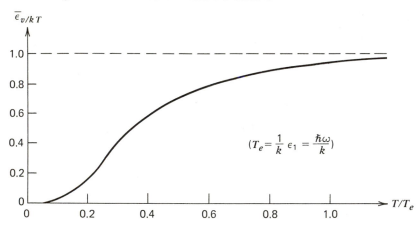

Figure 21.6   Plot of average vibrational energy (in units of kT) versus temperature (in units of T_e) for diatomic molecules.

If the gas were monatomic, it wouldn't have rotational or vibrational degrees of freedom, and we would have

$$Z_{\text{monatomic gas}} = \frac{1}{N!}\,\zeta_t^N = \frac{V^N}{N!}\left(\frac{2\pi mkT}{h^2}\right)^{3/2N} \tag{21.31}$$

The same would hold for a diatomic gas if the temperature were sufficiently cold that rotational and vibrational excited states were inaccessible.

The average translational kinetic energy per molecule is simply

$$\bar{\varepsilon}_t = \frac{3}{2}\,kT \tag{21.32}$$

because temperatures comparable to the excitation temperature can never be obtained. This is not true for the rotational or vibrational degrees of freedom, however, so to calculate the average energy per molecule held in these degrees of freedom, we may have to use Eq. 21.6 and our results for ζ_r and ζ_v from Table 21.1. The results for $\bar{\varepsilon}_r$ and $\bar{\varepsilon}_v$ are displayed in Figures 21.5 and 21.6.

SUMMARY

The partition function for a single diatomic gas molecule can be written as the product of partition functions for the translational, vibrational, rotational, electronic, nuclear, etc. degrees of freedom.

$$\zeta = \zeta_t \zeta_r \zeta_v \zeta_e \zeta_n \cdots \tag{21.21}$$

Whenever the energy of the first excited state is large compared to kT, then we can ignore contributions to the partition function from that degree of freedom.

$$\zeta_m \approx 1 \qquad \text{if } \varepsilon_{\text{1st excited}} \gg kT \tag{21.23}$$

An equivalent condition requires the calculation of a characteristic "excitation temperature," T_e, given by

$$T_e = \frac{1}{k} \varepsilon_{\text{1st excited}} \tag{21.24}$$

Then for $T \ll T_e$ we can ignore this degree of freedom.

Under most normal conditions, the excited electronic and nuclear states are inaccessible ($\varepsilon_{\text{1st excited}} \gg kT$), so we need only consider translational, rotational, and vibrational degrees of freedom. If the molecule has mass m, rotational inertia I, and characteristic frequency ω, these are given as listed in Table 21.1.

The partition function for a system of N identical diatomic molecules is

$$Z = \frac{1}{N!} \zeta^N = \frac{1}{N!} (\zeta_t^N \zeta_r^N \zeta_v^N) \tag{21.28}$$

For a monatomic gas (or a diatomic gas at temperatures sufficiently low that rotational and vibrational excited states are inaccessible) there are no rotational or vibrational contributions and we have

$$Z_{\text{monatomic gas}} = \frac{1}{N!} \zeta_t^N = \frac{V^N}{N!} \left(\frac{2\pi m k T}{h^2} \right)^{3/2N} \tag{21.31}$$

Every single degree of freedom makes a contribution to the partition function proportional to $T^{1/2}$, so if the gas molecules each have v degrees of freedom then the gas of N molecules has Nv degrees of freedom altogether and the partition function is given by

$$Z = \frac{1}{N!} \zeta^N = (\text{const}) \frac{1}{N!} T^{Nv/2} \tag{21.30}$$

As you may recall, the reason for dealing with partition functions is to simplify the calculation of some properties of a system. We illustrate this with a couple of examples for a system of N gas molecules at temperature T.

EXAMPLE

Suppose each gas molecule has mass m, and three translational degrees of freedom only. What is the chemical potential for this gas?
According to Eqs. 21.19 and 21.25, this is given by

$$\mu = -kT\ln\left(\frac{\zeta}{N}\right) = -kT\ln\left[\frac{V}{N}\left(\frac{2\pi mkT}{h^2}\right)^{3/2}\right] \tag{21.34}$$

EXAMPLE

If each molecule has v degrees of freedom, what is the average internal energy of the gas?
According to Eq. 21.6,

$$\bar{E} = -\frac{\partial}{\partial\beta}\ln Z \tag{21.6}$$

and according to the result (21.30)

$$Z = (\text{const})\frac{1}{N!}\,T^{Nv/2} = (\text{const})\frac{1}{N!}\,\beta^{-Nv/2}$$

which gives

$$\ln Z = \ln\left[(\text{const})\frac{1}{N!}\right] - \frac{Nv}{2}\ln\beta$$

Putting this into Eq. 21.6 gives

$$\bar{E} = \frac{Nv}{2}\left(\frac{1}{\beta}\right) = \frac{v}{2}NkT$$

The gas as a whole has Nv degrees of freedom, and our result shows that with each of these is associated an average energy of $\frac{1}{2}kT$, in agreement with the equipartition theorem.

PROBLEMS

21-26.   For typical molecules, the energies of excited electronic states are measured in electron volts. Calculate the excitation temperature, T_e, for a gas molecule whose first excited electronic state lies 1 eV above the ground state.

21-27.   The energies of excited nuclear state are measured in MeV. Calculate the excitation temperature, T_e, for a nucleus whose first excited state lies 2 MeV above the ground state.

21-28. The oxygen (O_2) molecule is composed of two oxygen atoms each having a mass of 2.7×10^{-26} kg, and having center-to-center separation of 1.24×10^{-10} m.

    (a)   What is the rotational inertia, I, of an oxygen molecule about an axis perpendicular to a line joining the two atoms, and midway between them?

    (b)   What is the energy of the first excited ($l = 1$) rotational state for this molecule?

    (c)   What is the rotational excitation temperature for this molecule?

21-29. The vibrations of an oxygen molecule have a characteristic frequency of about $\omega = 2.98 \times 10^{14}$ s^{-1}.

    (a)   What is the energy of the first excited vibrational state (relative to the ground state)?

    (b)   What is the vibrational excitation temperature for this molecule?

21-30. The partition function for the vibrational degrees of freedom at all temperatures for a single molecule is given by Eq. 21.27. Using Eq. 21.6, find the average energy per molecule stored in vibrational degrees of freedom, as a function of the fundamental frequency, ω, and the temperature, T, for all values of T.

21-31. The partition function for a monatomic gas is given in Eq. 21.31. From this, find the average internal energy of a monatomic gas as a function of N and T. (*Hint*. Use Eq. 21.6.)

21-32. Equation 21.26 expresses the partition function for the rotational degrees of freedom of a diatomic gas molecule for all temperatures in terms of a sum.

$$\zeta_r = \sum_l (2l + 1)e^{-\beta(\hbar^2/2I)l(l+1)}$$

Using this, find an expression for the average energy per molecule stored in the rotational degrees of freedom, in terms of its rotational inertia, I, and the temperature, T.

21-33. Using the values of ζ_t, ζ_r, and ζ_v listed in Table 21.1, calculate the values of the average energy per molecule for each of these modes in the high-temperature limit ($\bar{\varepsilon}_t$, $\bar{\varepsilon}_r$, and $\bar{\varepsilon}_v$). How does each of these compare with $\bar{\varepsilon}_m = (v_m/2)kT$, where v_m is the number of degrees of freedom for mode "m"?

21-34. Show that $\zeta_v \approx kT/\hbar\omega$ (Eq. 21.27) for $T \gg T_e$.

APPENDIX 21A   THE SERIES $\sum_n e^{-nx}$

Consider the series

$$\sum_{n=0}^{\infty} a^n \qquad \text{for } |a| < 1$$

If we multiply this series by $(1 - a)$, we have

$$(1 - a)\sum_{n=0}^{\infty} a^n = (1 - a)(1 + a + a^2 + a^3 + a^4 + \cdots)$$

$$= 1(1 + a + a^2 + a^3 + \cdots) - a(1 + a + a^2 + \cdots)$$

$$= (1 + a + a^2 + a^3 + \cdots) - (a + a^2 + a^3 + \cdots)$$

$$= 1$$

Dividing both sides of this equation by $(1 - a)$ we have

$$\sum_{n=0}^{\infty} a^n = \frac{1}{1 - a} \qquad \text{(for } |a| < 1\text{)} \tag{21A.1}$$

In particular, if

$$a = e^{-x}$$

then

$$a^n = e^{-nx}$$

and

$$\sum_{n} a^n = \sum_{n} e^{-nx} = \frac{1}{1 - e^{-x}} \tag{21A.2}$$

In this chapter we use

$$x = \hbar\omega$$

In Chapter 24, we use

$$x = \beta(\varepsilon - \mu)$$

chapter 22

CHEMICAL EQUILIBRIUM

Suppose you mix together various reactants that undergo chemical interaction. After the initial flurry has subsided and reaction has reached equilibrium, how much of each of the initial reactants will remain, and how much of the various products will be produced?

In chemical reactions, various groups of atoms interact through the exchange of electrons to form other groups of atoms. We can think of each distinct type of particle (atom, ion, molecule, radical, etc.) involved in the reaction as a subsystem. For example, in the burning of hydrogen,

$$2H_2 + O_2 \rightleftarrows 2H_2O$$

we can think of the combined system being composed of three distinct subsystems: the H_2 molecules, the O_2 molecules, and the H_2O molecules. Similarly, in the dissociation of water molecules

$$H^+ + OH^- \rightleftarrows H_2O$$

or of nitric acid molecules

$$H^+ + NO_3^- \rightleftarrows HNO_3$$

for example, we can think of the combined system as being composed of the appropriate subsystems. Each subsystem consists of all the particles of one type.

The question we wish to answer in this chapter is this: After any given chemical reaction has reached equilibrium, how will the reactants be distributed among the various subsystems (Figure 22.1)? For example, in the reaction

$$H^+ + OH^- \rightleftarrows H_2O$$

will most of the particles be as H^+ and OH^- ions, or as H_2O molecules? What will be the relative numbers of each?

The way we will answer this question is by investigating the second-law requirement that a system will tend to adopt the configuration of lowest potential energy. This constraint will determine how the reactants are distributed among the various possible subsystems when in equilibrium.

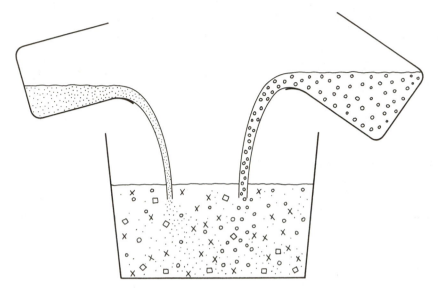

Figure 22.1   After the reactants are mixed and equilibrium is attained, how can we know what the equilibrium concentrations of the various interacting chemicals will be?

A.   THE THERMODYNAMICAL POTENTIAL

We have seen that the chemical potential is the potential energy of a particle sitting at the lowest point in the potential well caused by its interactions with its neighbors. It is the zero-energy reference point from which potential energy stored in thermal degrees of freedom is measured. For example, the average potential energy of an atom engaging in simple harmonic motion in one dimension would be

$$\bar{\varepsilon}_{\text{potential}} = \mu + \frac{1}{2} kT$$

where T is the temperature of the system. Similarly, the average potential energy of an atom undergoing simple harmonic oscillations in three dimensions would be

$$\bar{\varepsilon}_{\text{potential}} = \mu + \frac{3}{2} kT$$

The potential energy of N such three-dimensional harmonic oscillators would be

$$\text{potential energy} = N\mu + \frac{3}{2} NkT$$

For a system at given temperature T, the potential energy will obviously be lowest if it adopts the configuration that minimizes the product $N\mu$. We call this product the "thermodynamical potential" of the system. If the system consists of several sub-systems, the ith one having N_i particles with chemical potential μ_i, the thermodynamical potential of the combined system is just the sum of the subsystems' thermodynamical potentials.

$$\text{thermodynamical potential} = \sum_i N_i \mu_i \tag{22.1}$$

When the system reaches equilibrium, it will arrive at some final temperature and pressure. As long as the temperature is steady, the statement that the system adopts the configuration of lowest potential energy is equivalent to saying that it adopts the configuration of lowest chemical potential. This is what we will use to determine the relative numbers of the various reactants in equilibrium.

SUMMARY

The chemical potential of a particle is caused by interactions with its neighbors, and is the zero-energy reference point relative to which energy stored in potential energy degrees of freedom is measured. At any given temperature, the average potential energy of a particle will be lowest when its chemical potential is lowest.

The sum of the chemical potentials of all particles of a system is called the "thermodynamical potential." If there are several subsystems, each having N_i particles with chemical potential μ_i, then

$$\text{thermodynamical potential} = \sum_i N_i \mu_i \tag{22.1}$$

At any given temperature, the system will be in the configuration of lowest potential energy when its thermodynamical potential is lowest. This is the condition we will use to determine the relative numbers of the various reactants in chemical equilibrium.

PROBLEMS

22-1. Suppose the chemical potential of particle A is -1.3 eV, that of particle B is -0.4 eV, and that of particle C is -1.5eV. Particles A and B could possibly combine to form particle C. Would they? Why or why not?

22-2. Particles of types A, B, and C could interact to form particles of types D and E according to the chemical equation

$$3A + B + 4C \rightleftarrows 2D + 3E$$

If the respective chemical potentials were $\mu_A = -0.4$ eV, $\mu_B = -0.1$ eV, $\mu_C = +0.2$ eV, $\mu_D = +0.1$ eV, $\mu_E = -0.2$ eV, which way would this reaction tend to go (to the right $\rightarrow$, or to the left $\leftarrow$)? Why?

22-3. A chemical reaction among reactants A, B, and C is proceeding according to

$$3A + B \rightarrow 2C$$

(a) From this fact, can you write down a relationship among the chemical potentials μ_A, μ_B, and μ_C? (*Hint.* How does the thermodynamical potential of three A's and one B compare to that of two C's?)

(b)   The chemical potential of a type of a particle increases as the number of particles of that type increases. Explain how this fact could stop the chemical reaction from continuing even before all the particles of type A or B were used up.

B.   CHANGES IN CHEMICAL POTENTIAL

In dealing with diffusive interactions it is convenient to use a quantity called the "Gibbs free energy," defined by

$$G \equiv E - TS + pV \tag{22.2}$$

If we divide the system up into a myriad of tiny subsystems (Figure 22.2), we can write the energy, entropy, volume, and number of particles as the sum of the energies, entropies, volumes and particles in the subsystems.

$$E = \sum_j \Delta E_j \qquad V = \sum_j \Delta V_j$$

$$S = \sum_j \Delta S_j \qquad N = \sum_j \Delta N_j$$

We can build up the entire system by adding on the small pieces one at a time, and so the Gibbs free energy of the system is given by

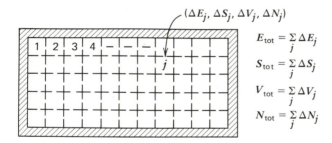

$$E_{tot} = \sum_j \Delta E_j$$

$$S_{tot} = \sum_j \Delta S_j$$

$$V_{tot} = \sum_j \Delta V_j$$

$$N_{tot} = \sum_j \Delta N_j$$

Figure 22.2   Consider a large system that is composed of many interacting subsystems. The total internal energy, entropy, volume, and number of particles is the sum of those in the subsystems. In equilibrium, the temperature and pressure are the same throughout. The Gibbs free energy of the entire system is therefore

$$G = E - TS + pV = \sum_j (\Delta E_j - T\Delta S_j + p\Delta V_j)$$

According to the first law, each contribution ΔE_j to the total internal energy of the system is

$$\Delta E_j = T\Delta S_j - p\Delta V_j + \mu \Delta N_j$$

With this, the above expression for the Gibbs free energy of the system becomes

$$G = \sum_j \mu \Delta N_j = \mu N$$

$$G = E - TS + pV = \sum_j \Delta E_j - T \sum_j \Delta S_j + p \sum_j \Delta V_j$$
$$= \sum_j (\Delta E_j - T \Delta S_j + p \Delta V_j)$$

Using the first law for the changes when the jth piece is added on,

$$\Delta E_j = T \Delta S_j - p \Delta V_j + \mu \Delta N_j$$

the above expressions becomes

$$G = \sum_j \mu \Delta N_j = \mu N \tag{22.3}$$

If this system was one of the reactants in a chemical reaction, then the Gibbs free energy of all the reactants together would be

$$G = \sum_i \mu_i N_i \tag{22.3'}$$

Comparing this to our previous result (22.1) we see that the Gibbs free energy is the same thing as the thermodynamical potential.

Hopefully, it is clear to you that the chemical potential is of central importance in treating diffusive interactions, just as temperature is of central importance in thermal interactions and pressure is of central importance in mechanical interactions. For this reason, we wish to investigate the dependence of μ on the other parameters, and we can use the result (22.3) and the definition (22.2) to do this.

From these two equations we have

$$G = E - TS + pV = \mu N$$

If we consider a system to be under no constraints at all, so that any of these parameters may vary, we have the variation in the Gibbs free energy given by

$$dG = dE - TdS - S\,dT + p\,dV + V\,dp = \mu\,dN + N\,d\mu$$

Invoke the first law

$$dE = TdS - p\,dV + \mu\,dN$$

this becomes

$$dG = -S\,dT + V\,dp + \mu\,dN = \mu\,dN + N\,d\mu \tag{22.4}$$

Subtracting $\mu\,dN$ from both sides of the last equation, we see that the variation in the chemical potential is given by

$$N\,d\mu = -S\,dT + V\,dp$$

or

$$d\mu = -\sigma\,dT + v\,dp$$

$$\sigma = \frac{S}{N} = \text{entropy per particle}$$

$$v = \frac{V}{N} = \text{volume per particle} \tag{22.5}$$

SUMMARY

We define a quantity called the Gibbs free energy by

$$G \equiv E - TS + pV \tag{22.2}$$

which can be shown to be identical to the thermodynamical potential.

$$G \equiv E - TS + pV = \mu N \tag{22.3}$$

By looking at differential changes in the parameters of this last equation, we find that the chemical potential varies with temperature and pressure according to

$$d\mu = -\sigma\, dT + v\, dp$$

where

$$\sigma = \frac{S}{N} = \text{entropy per particle}$$

$$v = \frac{V}{N} = \text{volume per particle} \tag{22.5}$$

PROBLEMS

22-4. The entropy of water at 25°C and one atmosphere of pressure is 45.11 calories/mole·K. Given that the molecular weight of water is 18, and that its specific heat is 1 cal/g·K,

    (a) What is the entropy of water at 27°C? (*Hint.* Use $\Delta Q = T\Delta S$.)
    (b) What is the entropy per molecule for water at 25°C? At 27°C?
    (c) When you change the temperature of water by 2° from 25 to 27°C at atmospheric pressure, by how much does the chemical potential change? (*Hint.* Use Eq. 22.5.)

22-5. At standard temperature and pressure, a mole of water has entropy of 45.11 cal/mole·K, and a volume of 18 cm³. If you raise the temperature by 1°C (or 1 K), by how much would you have to increase the pressure in order to keep the chemical potential of a water molecule unchanged?

C. THE LAW OF MASS ACTION

Imagine a chemical reaction taking place under conditions of constant temperature and pressure. This would happen, for example, if the reactants were immersed in a water bath reservoir at atmospheric pressure. But it always happens if you wait patiently until after the initial flurry of the reaction has subsided and the reactants are in equilibrium.

    Under these conditions, the chemical potential of any of the reactants is fixed, as can be seen from Eq. 22.5.

$$d\mu_i = -\sigma_i\, dT + v_i\, dp = 0 \quad \text{(for } T \text{ and } p \text{ constant)}$$

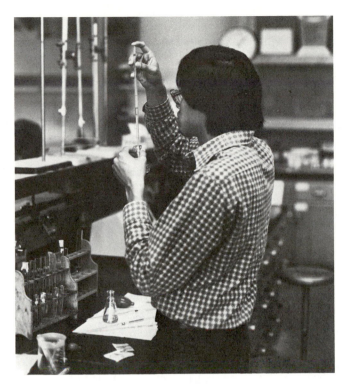

Figure 22.3 After the chemicals have reached equilibrium and the temperature and pressure of the system have stabilized, the chemical potentials of the reagents are constant and the thermodynamical potential (i.e., Gibbs free energy) of the system is a minimum. Since it is at a minimum, differential changes in the concentrations of the reagents lead to no change in the thermodynamical potential. $\sum_i \mu_i \Delta N_i = 0$. (Keith Stowe)

Since the chemical potentials are fixed, the condition that the thermodynamical potential be a minimum becomes* (Figure 22.3)

$$\Delta\left(\sum_i \mu_i N_i\right) = \sum_i \mu_i \Delta N_i = 0 \tag{22.6}$$

We know that the changes in the numbers of the particles of the various reactants is in proportion to their stoichiometric coefficients. For example, in the burning of hydrogen to make water,

$$2H_2 + O_2 \longrightarrow 2H_2O$$

---

* When a function is at a maximum or a minimum, its derivatives are zero and it does not change with differential changes in the variables.

for every two hydrogen (H_2) molecules that disappear, one oxygen (O_2) molecule disappears, and two water (H_2O) molecules appear. So in this example, the ratio of the changes in numbers is given by

$$\Delta N_{H_2} : \Delta N_{O_2} : \Delta N_{H_2O} = -2 : -1 : +2$$

and the numbers $(-2, -1, +2)$ are the stoichiometric coefficients for this reaction.

If b_i represents the stoichiometric coefficient for the ith reactant in a process, then the ratio of the changes in the numbers of particles for the various reactants is given by

$$\Delta N_1 : \Delta N_2 : \cdots : \Delta N_m = b_1 : b_2 : \cdots : b_m$$

and we can write Eq. 22.6 in the more convenient form

$$\sum_i \mu_i b_i = 0 \tag{22.7}$$

EXAMPLE

Write out explicitly the equilibrium condition (22.7) for the following chemical reaction (burning of paraffin):

$$2C_{20}H_{42} + 61O_2 \rightleftarrows 40CO_2 + 42H_2O$$

For this process, the stoichiometric coefficients are given by $(-2, -61, 40, 42)$. (You could equivalently consider them to be $(2, 61, -40, -42)$, since you could convert one into the other by multiplying both sides of Eq. 22.7 by "-1.") Equation 22.7, then, becomes

$$-2\mu_{C_{20}H_{42}} - 61\mu_{O_2} + 40\mu_{CO_2} + 42\mu_{H_2O} = 0$$

If the system ever reaches equilibrium before all the paraffin and oxygen is exhausted, then this condition will be satisfied.

In the previous chapter, we saw how to calculate chemical potentials. For a system of N particless at temperature T, their chemical potential is given by*

$$\mu = -kT \ln \frac{\zeta}{N} \tag{21.19}$$

where ζ is the partition function for a single particle, defined by the following sum over all states s, having energy ε_s, accessible to that particle:

$$\zeta = \sum_s e^{-\beta \varepsilon_s}$$

---

* If you have not yet covered the previous chapter, a brief derivation of this result is given in Appendix A, at the end of this chapter.

Therefore, the chemical potential for particles of the ith type is given by

$$\mu_i = -kT \ln \frac{\zeta_i}{N_i}$$

and Eq. 22.7 becomes

$$\sum_i b_i \left(-kT \ln \frac{\zeta_i}{N_i} \right) = 0$$

If we divide through by $(-kT)$ and rearrange the terms, we get

$$\sum_i b_i \ln \frac{\zeta_i}{N_i} = \sum_i (\ln \zeta_i^{b_i} - \ln N_i^{b_i}) = 0$$

or

$$N_1^{b_1} N_2^{b_2} \cdots N_m^{b_m} = \zeta_1^{b_1} \zeta_2^{b_2} \cdots \zeta_m^{b_m} \tag{22.8}$$

This is the equation we were looking for, which relates the relative numbers of the various reactants when the system is in equilibrium. The stoichiometric coefficients (b_i) are known from the appropriate chemical equation describing the reaction of interest, and the partition functions appearing on the right are calculable functions of the volume and temperature.

Instead of giving the absolute number (N_i) of particles of each type in equilibrium, it is sometimes more convenient to use particle densities, and units of "moles per liter" are frequently used. We can convert N_i into moles per liter through

$$n_i = \frac{N_i}{N_A V}$$

where N_A is Avogadro's number and V is the volume in liters. In this case, the result (22..8) would have the following form:

$$n_1^{b_1} n_2^{b_2} \cdots n_m^{b_m} = A(T, V) \tag{22.9}$$

where

$$A(T, V) = \frac{\zeta_1^{b_1} \zeta_2^{b_2} \cdots \zeta_m^{b_m}}{(N_A V)^{b_1 + b_2 + \cdots + b_m}} \tag{22.10}$$

This is known as the "law of mass action." The function $A(T, V)$ is a calculable function of temperature and volume, but it is usually much easier to determine it experimentally than from theoretical calculations. It is called the "equilibrium constant" for the particular chemical process, and is listed in tables for common chemical processes. It is usually listed as a function of T and p rather than of T and V, and unless otherwise stated, the pressure is understood to be 1 atm.

SUMMARY

For a system at constant temperature and pressure, the chemical potentials of the various reactants are fixed. The condition that the thermodynamical potential be a minimum gives

$$\sum_i \mu_i \Delta N_i = 0 \qquad (22.6)$$

The changes in numbers of particles (ΔN_i) for the various reactants are in proportion to their stoichiometric coefficients (b_i) appearing in the appropriate chemical equation, so the above condition becomes

$$\sum_i \mu_i b_i = 0 \qquad (22.7)$$

Inserting into this equation the expression (21.19) for the chemical potential for particles of the *i*th type,

$$\mu_i = -kT \ln \frac{\zeta_i}{N_i}$$

we get the following equation interrelating the numbers of particles, for the various reactants involved:

$$N_1^{b_1} N_2^{b_2} \cdots N_m^{b_m} = \zeta_1^{b_1} \zeta_2^{b_2} \cdots \zeta_m^{b_m} \qquad (22.8)$$

This is called the "law of mass action," which is frequently written in terms of concentrations (in moles per liter) as

$$n_1^{b_1} n_2^{b_2} \cdots n_m^{b_m} = A(T, V) \qquad (22.9)$$

where $A(T, V)$ is called the "equilibrium constant" and can be calculated or measured for reactions of interest.

PROBLEMS

22-6. (a)   The equilibrium constant for the dissociation of water

$$H_2O \rightarrow H^+ + OH^-$$

is $10^{-16.745}$ l/mole at 24°C. What is the ratio of the H^+ ion concentration to the water molecule (H_2O) concentration (*Hint.* $n_{H^+} = n_{OH^-}$ and $n_{H_2O} = 55.6$ moles/l.)

(b)   The negative exponent of the H^+ ion concentration is called the "pH value." For example, if $n_{H^+}/n_{H_2O} = 10^{-2}$, the pH value is $+2$. What is the pH of pure water at 24°C?

(c)   The equilibrium constant for the dissociation of water at 60°C is $10^{-14.762}$ l/mole. What is the pH of pure water at this temperature?

22-7.   The following chemical reaction was concluded at standard temperature and pressure, for which the equilibrium constant is 10^5 $(\text{mole}/l)^{-3}$.

$$3A + B + 2C \rightarrow D + 2E$$

If in equilibrium the concentrations are $n_A = 0.2$ mole/liter, $n_B = 0.05$ mole/liter, $n_C = 0.1$ mole/liter, and $n_D = 0.2$ mole/liter, what is the concentration of the reactant E?

APPENDIX 22A $\mu = -kT \ln \dfrac{\zeta}{N}$

Very briefly, this derivation involves the use of the relationship $\mu = \partial E/\partial N)_{S,V}$. For a system having possible states s, the average value of μ is (For any given state s, the entropy is fixed, $k \ln 1 = 0$, so we needn't be concerned with holding the entropy constant.)

$$\bar{\mu} = \sum_s \mu_s P_s = \left(\sum_s e^{-\beta E_s} \right)^{-1} \sum_s \mu_s e^{-\beta E_s}$$

$$= \left(\sum_s e^{-\beta E_s} \right)^{-1} \left(-kT \frac{\partial}{\partial N} \right) \left(\sum_s e^{-\beta E_s} \right)_{T,V}$$

$$= -kT \frac{\partial}{\partial N} \ln Z$$

where Z is defined by

$$Z = \sum_s e^{-\beta E_s}$$

The sum over all states of the system is equivalent to the sum over all states accessible to each individual particle in the system. If we label the individual particle states by $(a, b, c, \ldots)$ and define ζ to represent the sum over individual particle states as indicated below, then Z becomes

$$Z = \sum_{a,b,\ldots} e^{-\beta(E_a + E_b + \cdots)} = \sum_a e^{-\beta E_c} \sum_b e^{-\beta E_b} \cdots$$

$$= \zeta_a \zeta_b \zeta_c \cdots$$

If the particles are identical, then there are $N!$ identical rearrangements of the N particles among the individual particle states for each distinguishable state of the entire system. (That is, there are $N!$ ways of rearranging N particles among N states. and all of these rearrangements will be the same if the particles are indistinguishable.) So for identical particles we have $\zeta_a = \zeta_b = \zeta_c = \cdots$, and the above sum has $N!$ too many terms.

$$Z = \frac{1}{N!} \zeta_a \zeta_b \zeta_c \cdots = \frac{1}{N!} \zeta^N$$

Putting this into the above formula for determining $\bar{\mu}$, and using Stirling's formula for determining the logarithm of $N!$, we have

$$\bar{\mu} = -kT \frac{\partial}{\partial N} \ln \left(\frac{1}{N!} \zeta^N \right)$$

$$= -kT \frac{\partial}{\partial N} (-N \ln N + N + N \ln \zeta)$$

$$= -kT \ln \frac{\zeta}{N}$$

chapter 23

EQUILIBRIUM BETWEEN PHASES

As water vapor is cooled, it condenses into liquid water, and at an even lower temperature, the water freezes. What causes phase changes such as these, and how are they dependent on temperature and pressure?

We are familiar with the distinct gaseous, liquid, and solid phases of the materials of our environment (Figure 23.1). Most materials can be made to change phases by appropriate changes in temperature and/or pressure. The change from solid to liquid

Figure 23.1 Water is found abundantly in solid, liquid, and gaseous phases on earth. (George Holton/Photo Researchers)

phase is marked by a discontinuous change in the shear strength of the material, and the change to the gaseous phase is marked by a discontinuous change in density.

There are also other discontinuous changes in the properties of some materials, that are also properly called "phase transitions." The transition from a gas to a plasma is marked by a sudden change in electrical conductivity, as is the transition from a normal solid to a superconductor. Under certain conditions, liquid helium loses all viscosity and becomes a "superfluid."

All phase transitions are caused by attractive forces among the constituents of the systems. For example, it is an attractive force that causes the distinct separation between ground state and excited states for electrons in superconductors and for helium atoms in superfluids. It is an attractive force that causes the charged ions of a plasma to combine to form a neutral gas, that causes gases to condense into solids and liquids, and that causes liquids too condense into solids.

Attractive forces between particles tend to hold them together, trapping them into bound states. This is opposed by the thermal energy of the system that tends to keep the motion of the particles random and free. When sufficient thermal energy is removed, the particles become trapped in the bound states, unable to collect enough thermal energy to break free (Figure 23.2). This is what happens in plasma-gas, gas-liquid, gas-solid, and liquid-solid phase transitions, for example.

Once trapped in bound states, particles usually find many discrete bound states available to them. The remaining thermal energy of the particles may permit rather random transitions among the various bound states, even though not enough thermal energy is available to break completely free. But as still more thermal energy is removed, the particles may eventually become confined to the ground state, and the system as a whole may not have enough thermal energy for it (or some portion of it) to be excited even into the first excited state. This thermal imprisonment of the system in the ground state may also be accompanied by some distinct change in a physical

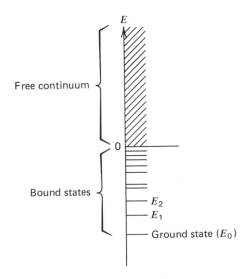

Figure 23.2 Schematic representation of the states available to the particles of some system. For sufficiently high temperatures, the energy of the system is large enough that most of the particles are free. At lower temperatures, the particles are confined to discrete bound states, created by attractive forces. At sufficiently low temperatures, the particles may become trapped in the ground state with no chance of excitation to even the first excited state.

property, revealing another phase transition. This is what happens in super-conductivity and superfluidity, for example.

Transitions such as these, involving imprisonment in ground states, are often gradual. That is, they take place over a range of temperatures (or over a range of some other external parameter, such as pressure or magnetic field), rather than at a fixed temperature, as more and more particles become entrapped. Such phase transitions are called "second-order phase transitions."

A. PHASE DIAGRAMS

At a pressure of 1 atm and a temperature of 100°C, the liquid and vapor phases of water are in equilibrium. The removal of heat would cause water vapor to condense, and the addition of heat would cause liquid water to vaporize under these conditions. We say that the "boiling point" of water is 100°C at atmospheric pressure. Under these conditions the vapor will remain vapor, and the liquid will remain liquid unless heat is added or removed. At a different pressure, the phase change occurs at a different temperature. For example, at a pressure of 0.7 atm, water boils at 90°C. That is why it is difficult to make boiled potatoes, or a good hot cup of coffee when high in the mountains.

For all materials, the temperature at which a phase change occurs depends on the pressure. If we indicate all the various pressures and temperatures at which a phase change occurs on a grid of p versus T, we have a "phase diagram" for that material. A

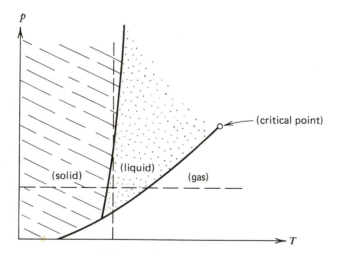

Figure 23.3 The general form of a phase diagram for most substances. The vertical dashed line indicates the changes that would occur if we vary the pressure at constant temperature, and the horizontal dashed line indicates the changes that would occur if we vary the temperature at constant pressure.

phase diagram for the solid, liquid, and gaseous phases of a "typical" material is given in Figure 23.3.

Higher pressures naturally favor the denser phase. For example, higher pressure favors the liquid phase of water over its vapor phase, so greater thermal energy is required to boil water under heavier pressures. Pressure cookers cook foods faster because under pressure the water boils at higher temperatures. For most materials, the solid is denser than the liquid phase, and the liquid phase is more dense than the gas. Since greater pressure favors the denser phase, the melting and boiling points of most materials increase with pressure. This means that on a phase diagram, the lines representing the division between solid-gaseous, solid-liquid, and liquid-gaseous phases all tend to slope upward to the right.

Water is an exceptional material in that the liquid phase is denser than its solid phase, so higher pressures favor the liquid phase for water. Figure 23.4 is a qualitative phase diagram for water. The fact that the line separating the solid and liquid phases slopes upward to the *left* is a reflection of increased pressure favoring the liquid (denser) phase. This is the reason why ice is slippery and other solids are not. Increased pressure melts ice but not most other solids.

The competition between intermolecular attractive forces and thermal motion causes materials to tend to become solids at low temperatures and gases at high temperatures. At sufficiently low pressures, most materials go straight from solid to gaseous phases, skipping the liquid phase entirely. This is called "sublimation." Solid carbon dioxide ("dry ice") sublimes even at atmospheric pressure. Water sublimes at pressures below 0.006 atm, and this is the reason that liquid water could not presently exist on Mars (Figure 23.5).

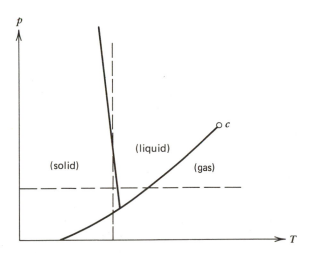

Figure 23.4 The form of the phase diagram for water (not to scale). Following the vertical dashed line, you can see that by increasing the pressure at constant temperature, the sequence of phase changes differs from that of most materials. (Compare with Figure 23.3.)

Figure 23.5 Photo of Mars showing erosion by running water. At present, Mars' atmospheric pressure is too small to permit water in the liquid state on the surface. Water exists only in solid and gaseous phases there. Yet these erosion patterns were caused by the flow of liquid water. Is it possible that Mars' atmosphere has at times been thicker than it is now? (Courtesy NASA)

On both Figures 23.3 and 23.4 it is seen that the line separating the gaseous and liquid phases does not continue indefinitely. Beyond some certain value of the temperature and pressure (T_c, p_c), called the "critical point" for the substance, it is no longer possible to distinguish between liquid and gaseous phases. No discontinuous change in the density of the material occurs in this region, and so there is no phase change.

To illustrate why this happens, suppose we decide to slowly compress a gas until it reaches a density comparable to that of the liquid phase. In order to prevent the molecules from sticking together (condensing) as they are compressed closer and closer together, we must simultaneously increase their thermal motion by increasing the temperature of the gas. Proceeding in this fashion, we could eventually compress the gas to a point where its density is comparable to that of the liquid state, and there would be no discontinuous discrete change in the property of the gas at any time during the process. If the temperature of the material is sufficiently high that the thermal energies of the molecules prevent their sticking together, then there will be no discontinuous change in density as the pressure is changed—therefore, no phase change.

SUMMARY

A discontinuous change in the properties of a material is called a "phase transition." Phase transitions are caused by attractive forces among the particles of a system, which forces the system into bound states when sufficient thermal energy is removed, and may even confine the system to the ground state upon further removal of thermal energy. The imposition or removal of such thermal constraints causes phase transitions.

On a plot of p versus T, the display of all the various pressures and temperatures for which a given substance undergoes phase transitions is called a "phase diagram" for that substance. For sufficiently high temperatures and pressures, there is no clear distinction between liquid and gaseous phases. The "critical point" (p_c, T_c) of a substance marks the temperature and pressure above which the liquid-gas phase transition no longer occurs.

PROBLEMS

23-1. A system of conduction electrons in a metal becomes superconducting at temperatures below 2 K. Roughly what do you estimate to be the amount of energy separating the ground state from the first excited state for this system?

23-2. The latent heat of vaporization for water is 540 cal/g. Roughly how much energy per molecule (in eV) separates the highest bound state from the free continuum for water molecules?

23-3. Water and methane molecules both have about the same mass, yet water has a much higher boiling point at any given pressure. Why do you suppose this is?

23-4. Make a qualitative sketch of the appearance of the phase diagram of a substance for which solid and liquid phases are equally dense.

23-5. The mutual attraction between water molecules is much greater than that between ammonia molecules. For which material do you expect the critical temperature to be higher, and why?

B.   A MODEL

In order to make more specific quantitative calculations involving phase transitions, we must have specific quantitative models. As an example, consider the gas-liquid phase transition for which the van der Waals model can be used.

As you may recall, the ideal gas law is modified in two ways to arrive at the van der Waals equation of state.

1. The volume of the molecules themselves is finite, which means that there is some finite volume, b, beyond which the gas cannot be compressed. This modification requires the replacement of the molar volume, v, by $(v - b)$.

2.   In addition to the external pressure, there is also an intermolecular attraction tending to hold the molecules together. This attractive force is inversely proportional to the sixth power of the average intermolecular separation in the van der Waals model, and requires replacing the external pressure, p, by $[p + (a/v^2)]$, where a is some positive constant.

These two modifications of the ideal gas law give the van der Waals equation of state.

$$\left(p + \frac{a}{v^2}\right)(v - b) = RT \tag{12.9}$$

The reason for the phase transition becomes more apparent if the terms are rearranged in the following way:

$$p = \frac{RT}{v - b} - \frac{a}{v^2} \tag{23.1}$$

or equivalently,

(pressure exerted on container) = (thermal pressure) − (self-attraction)

The thermal pressure tends to keep the molecules dispersed, and the self-attraction tends to pull them together. When the self-attraction term dominates, the gas will condense into the liquid phase.

At large molar volumes, the self attraction term is negligible, and

$$p \approx \frac{RT}{v - b} \qquad \text{for } v \text{ very large}$$

As the volume of the gas is reduced (v smaller) the self-attraction term may become dominant.

$$p \approx -\frac{a}{v^2}, \qquad \text{for } v \text{ small, but} \gg b$$

in which case the gas condenses until its molar volume approaches b, whereupon the thermal pressure term again dominates.

$$p \approx \frac{RT}{v - b} \qquad \text{for } v \approx b$$

These effects are illustrated in Figure 23.6.

We can also see why latent heat must be released upon condensation. Because the force between molecules is attractive, the average potential energy per molecule is reduced when they are closer together. If the temperature remains constant during condensation, the kinetic energies do not change, and the potential energy released is extracted from the system as latent heat. This is illustrated in Figure 23.7.

If the temperature were larger than some "critical temperature," T_c, the thermal pressure term would dominate at all volumes, and no condensation would occur.

$$\frac{RT}{v - b} \gg \frac{a}{v^2} \qquad \text{for all } v > b, \text{ for } T \gg T_c$$

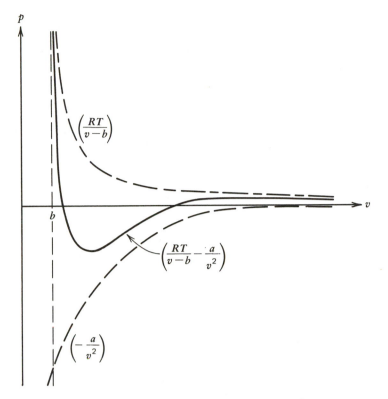

Figure 23.6   Plots of the thermal pressure term $[RT/(v - b)]$, the self-attraction term $[-(a/v^2)]$, and the sum of the two terms for the van der Waals equation of state. In the above case, it is seen that the self-attraction term dominates at intermediate distances (i.e., intermediate molar volumes) and is the reason that gases condense as the volume is reduced isothermally. (Where the pressure is negative, the gas is collapsing inward.)

The critical temperature varies from one material to the next, because the parameter a, which represents the strength of the intermolecular attractive forces, varies from one material to the next. Materials with stronger intermolecular attraction would have correspondingly higher critical temperatures.

These qualitative features of the phase transition, gained by examining the terms in the van der Waals equation of state, are borne out in a more quantitative analysis as follows.

Since our model covers both phases and is written in terms of molar volumes, we consider our system to include both phases with one mole of molecules (N_A) altogether. For diffusive equilibrium, such as the transfer of molecules between phases, we know that the Gibbs free energy (or equivalently, the thermodynamical potential) must be a minimum. (See Section B of Chapter 22.)

$$G \equiv E - TS + pV = \mu N = \text{minimum} \tag{22.3}$$

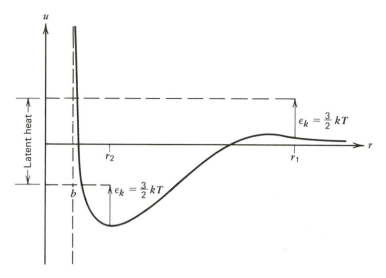

Figure 23.7 Plot of potential energy versus separation for the molecules of a real gas. Suppose that during condensation the average molecular separation changes from r_1 to r_2. If the temperature remains unchanged during this phase change, then the indicated amount of latent heat must be released.

This condition is equivalent to the statement that the system seeks the configuration of lowest potential energy, because $N\mu$ is the zero-energy reference level for the potential energy of the system. Lowering it lowers the potential energy. Since the total number of particles is fixed, this condition requires that the particles seek the configuration of lowest chemical potential.*

$$N\mu = \text{minimum} \rightarrow \mu = \text{minimum } (N \text{ fixed})$$

Using the above definition of the Gibbs free energy, and the first law for differential changes in the internal energy, we find the differential change in the Gibbs free energy is given by

$$dG = -S\,dT + V\,dp^{\dagger} \tag{23.2}$$

Our model does not involve the entropy explicitly, so it is most convenient to use for studying changes in the Gibbs free energy under conditions of constant temperature. For one mole of the substance,

$$dG = v\,dp \qquad (T = \text{constant}, v = \text{molar volume}) \tag{23.3}$$

In Figure 23.8 we have plotted p versus v from the van der Waals equation for several different values of the temperature, T. The scales of the two axes vary from one substance to another because of differences in the parameters a and b, but the general

---

* For chemical interactions the statement is not quite this simple because there are more than one type of particle involved in the interactions, and each type of particle has its own chemical potential.

† The number of particles is fixed by the fact that the model deals with one mole of the substance ($N = N_A$), so the "$\mu\,dN$" term is zero.

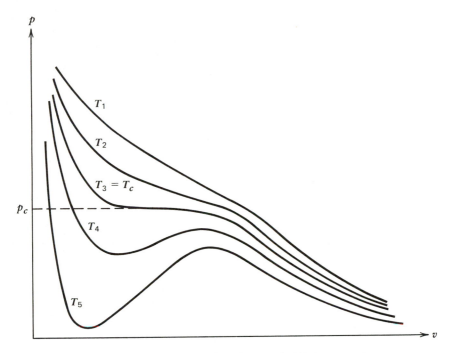

Figure 23.8 Plots of pressure versus molar volume for a van der Waals gas at various temperatures, $T_1 > T_2 > T_3 > T_4 > T_5$. The critical temperature and pressure are indicated.

features are the same for all. Figure 23.9 is a plot of p versus v for just one of the temperatures and the resulting curve is divided into various regions.

Suppose that the Gibbs free energy at position 0 on the curve is given by G_0. Then the Gibbs free energy at any other point i is given by

$$G_i = G_0 + \int_0^i v \, dp \tag{23.4}$$

Consider this integral. From the curve, we see that p is increasing between points 0 and 2, decreasing from point 2 to 3, and increasing again from point 3 on up.

Curve Region	p	$v = \dfrac{dG}{dp} = (slope\ of\ G\ versus\ p)$
0–2	increasing	large
2–3	decreasing	medium
3–5	increasing	small

With this information we can make a plot of G versus p, and we get a curve like that of Figure 23.10. You can see that G is not everywhere a single-valued function of p. If G is to be a minimum, then as the pressure is increased, the system will follow the path 0–(1, 4)–5. The loop 1-2-3-4 will be left out! As a system is compressed or expanded isothermally, it will go directly between points 1 and 4, with the accompanying discontinuous change in volume, which identifies the phase transition. What

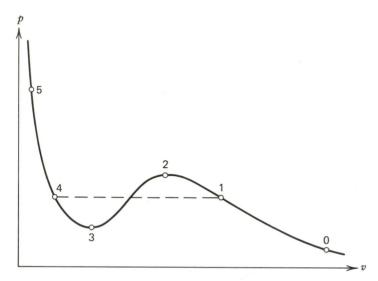

Figure 23.9 Plot of pressure versus volume for a van der Waals gas at fixed temperature. In order that the Gibbs free energy be minimized, the system goes directly from point 1 to point 4 during isothermal compression, and from point 4 to point 1 during isothermal expansion. The (1, 2, 3, 4) portion of the path is skipped.

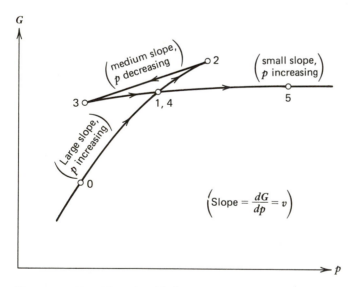

Figure 23.10 Plot of the molar Gibbs free energy as a function of the pressure. It is obtained by integrating $dG = v\,dp$ along the curve of Figure 23.11. The slope of the curve is equal to the molar volume. From Figure 23.11 it is seen that the pressure increases between points 0 and 2, decreases between points 2 and 3, and then increases again from point 3 on up.

is happening physically, of course, is that in the region 1-2-3-4, the self-attraction is dominating over the thermal pressure, resulting in the discontinuous change in volume.

At higher temperatures, the dip in the p-v plot disappears (Figure 23.8). As the volume decreases, the pressure increases for all values of the volume, and so the plot of G versus p does not double back on itself. There will be no discontinuous change in volume, hence no phase transition. The critical temperature, then, is the temperature above which the dip disappears in the p-v diagram. It is indicated along with the critical pressure, p_c, in Figure 23.8.

SUMMARY

The van der Waals model can provide us with a description of the liquid-gas phase transition. It can be written in the form

$$p = \frac{RT}{v - b} - \frac{a}{v^2}$$

$$= \text{(thermal pressure)} - \text{(self-attraction)}$$

At large volumes, the thermal pressure dominates, but at smaller volumes the self-attraction may dominate, causing condensation until the molar volume approaches b, whereupon the thermal pressure term again dominates and prevents further collapse.

Under conditions of constant temperature, the molar Gibbs free energy varies with the pressure according to

$$dG = v\,dp \qquad (T = \text{constant}) \tag{23.3}$$

Integrating this along any isothermal line in the plot of p versus v shows that G is not a single-valued function of p if there is a dip in the p-v plot. The fact that G must be a minimum means that portions of this curve will be skipped, corresponding to a discontinuous change in volume for some value of the pressure. Above some critical value of the temperature, T_c, the dip in the p-v diagram disappears, G is a single-valued function of the pressure, and there is no discontinuous change in the volume at any pressure.

PROBLEMS

23-6.   Figure 23.8 is a p-v plot for various temperatures using the van der Waals model. Notice that for the $T = T_c$ curve, and only for that one curve, there is a point of inflection where both the slope and the curvature of the curve are zero. $\partial p/\partial v = 0$, $\partial^2 p/\partial v^2 = 0$.

    (a)   Show that the volume at which this occurs is given by $v_c = 3b$. (*Hint.* Write the van der Waals equation in the form (23.1), set the first and second derivatives both equal to zero, and then eliminate T and a between the two equations.)

    (b)   What is the critical temperature in terms of the parameters a and b, and the gas constant R?

(c) What is the critical pressure in terms of the parameters a and b?

(d) For water, $a = 5.464$ liters^2-atm/mole^2, $b = 0.03049$ liters/mole. What should be the values of the critical temperature and pressure for water according to the van der Waals model? (The true values $p_c = 218$ atm and $T_c = 374°C$.)

23-7. For methane, $a = 2.25$ liters^2-atm/mole^2, $b = 0.0428$ liters/mole. Use the answers to parts (b) and (c) of Problem 23-6 to estimate the critical temperature and pressure for methane. (The true values for methane are $T_c = -82°C$ and $p_c = 46$ atm.)

C. PRESSURE AND TEMPERATURE DEPENDENCE OF PHASE TRANSITIONS

Suppose we know one temperature and pressure at which some substance undergoes phase transitions. Is there any way we can use this knowledge to help us predict the other temperatures and pressures at which phase transitions occur? For example, given that water boils at 100°C under atmospheric pressure, could we predict the boiling point if the pressure were reduced to 0.6 atm? Or could we predict the pressure under which the boiling point would be 200°C?

To answer these questions, we need to know how changes in temperature and pressure are interrelated when two phases are in equilibrium. If we changed either one (Δp, or ΔT), how much would we have to change the other in order that the two phases remain in equilibrium? The ratio of these changes, $\Delta p/\Delta T$, can be found using the fact that the chemical potentials of the two phases must be equal if they remain in equilibrium. The resulting relationship is called the Clausius-Clapyron equation.

We know that particles diffuse toward regions of lower chemical potential. When two phases are to be in equilibrium, there is no net transfer of particles between the two, so the two chemical potential must be the same.

$$\mu_1 = \mu_2 \qquad \text{(phase equilibrium)}$$

If we change the temperature and pressure (ΔT, Δp), we cause some corresponding change in the chemical potentials as well. But if the two phases are to remain in equilibrium, then the change in one chemical potential must be the same as the change in the other, because the two chemical potentials must remain equal (Figure 23.11.)

$$\Delta\mu_1 = \Delta\mu_2 \qquad \text{(phase equilibrium)}$$

According to Eq. 22.5, these changes in chemical potential are given by

$$\Delta\mu_1 = -\sigma_1 \Delta T + v_1 \Delta p$$

$$\Delta\mu_2 = -\sigma_2 \Delta T + v_2 \Delta p$$

where σ and v are the entropy per particle and volume per particle, respectively.

If the phases remain in equilibrium, these two changes are equal.

$$-\sigma_1 \Delta T + v_1 \Delta p = -\sigma_2 \Delta T + v_2 \Delta p$$

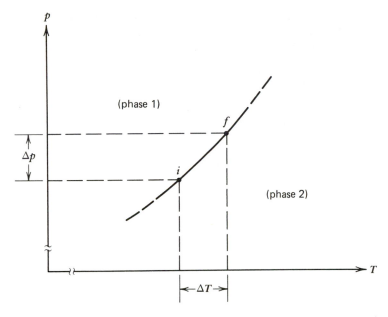

Figure 23.11 Portion of a phase diagram for some substance, showing a portion of the line separating two phases. Along this line the two phases are in equilibrium, so their chemical potentials must be equal at every point on it. However much one chemical potential changes in going from point *i* to point *f*, the $\Delta\mu_1 = \Delta\mu_2$ chemical potential of the other phase must change by the same amount.

This reduces to

$$\frac{\Delta p}{\Delta T} = \frac{\sigma_1 - \sigma_2}{v_1 - v_2} = \frac{\Delta\sigma}{\Delta v} \qquad \text{(phase equilibrium)} \qquad (23.5)$$

where $\Delta\sigma$ is the change in entropy per particle and Δv is the change in volume per particle, when going from one phase to another. This result can be put in a more convenient form by multiplying both numerator and denominator in the ratio $\Delta\sigma/\Delta v$ by Avogadro's number,

$$\frac{\Delta p}{\Delta T} = \frac{\Delta s}{\Delta v} \qquad \text{(phase equilibrium)} \qquad (23.6)$$

where Δs and Δv are the changes in entropy and volume per mole, respectively, in going from one phase to another. If L is the latent heat per mole required to accomplish the phase transition, then

$$T \Delta s = L$$

and we can rewrite the above result in the following form.

$$\frac{\Delta p}{\Delta T} = \frac{L}{T \Delta v} \qquad \text{(phase equilibrium)} \qquad (23.7)$$

In the special case of the vaporization of a liquid, the change in volume is normally quite large. The volume of the liquid phase is negligible in comparison to that of the gas phase, and the latter can be expressed in terms of the pressure and temperature from the ideal gas law.

$$\Delta v \approx v_{\text{gas}} = \frac{RT}{p} \qquad \text{(liquid-gas phase transition)}$$

Putting this into Eq. 23.7 gives

$$\frac{dp}{dT} \approx \frac{L}{T(RT/p)} = \frac{L}{R}\frac{p}{T^2}$$

or

$$\frac{dp}{p} \approx \frac{L}{R}\frac{dT}{T^2} \qquad \text{(liquid-gas equilibrium)}$$

Integrating this result between some initial and final conditions, gives

$$\ln p_f - \ln p_i = -\frac{L}{R}\left(\frac{1}{T_f} - \frac{1}{T_i}\right)$$

or

$$\frac{p_f}{p_i} = \exp\frac{L}{R}\left(\frac{1}{T_i} - \frac{1}{T_f}\right) \qquad \text{(liquid-gas equilibrium)} \qquad (23.8)$$

SUMMARY

When two phases are in equilibrium, their chemical potentials must be equal. For two phases to remain in equilibrium with each other, changes in temperature must be accompanied by corresponding changes in pressure, in order that both chemical potentials change by the same amount. That is,

$$\Delta\mu_1 = \Delta\mu_2 \qquad \text{(phase equilibrium)}$$

which implies

$$-\sigma_1\,\Delta T + v_1\,\Delta p = -\sigma_2\,\Delta T + v_2\,\Delta p$$

or

$$\frac{\Delta p}{\Delta T} = \frac{\sigma_1 - \sigma_2}{v_1 - v_2} \qquad \text{(phase equilibrium)} \qquad (23.5)$$

where σ and v are entropy per particle and volume per particle, respectively. This can be written in the more convenient form

$$\frac{\Delta p}{\Delta T} = \frac{\Delta s}{\Delta v} = \frac{L}{T\,\Delta v} \qquad \text{(phase equilibrium)} \qquad (23.6,7)$$

where Δs and Δv are the changes in entropy per mole and volume per mole, respectively, when going from one phase to the other, and L is the latent heat per mole associated with this phase transition.

For the specific case of liquid-gas phase transitions, we can use the approximations that the molar volume of the liquid phase is negligible compared to that of the gas phase, and that the gas phase obeys the ideal gas law, to write this result in the form

$$\frac{dp}{dT} \approx \frac{L}{R}\frac{p}{T^2} \qquad \text{(liquid-gas equilibrium)}$$

or

$$\frac{p_f}{p_i} = \exp\frac{L}{R}\left(\frac{1}{T_i} - \frac{1}{T_f}\right) \qquad \text{(liquid-gas equilibrium)} \qquad (23.8)$$

where (p_f, T_f) and (p_i, T_i) are any two sets of conditions under which the two phases are in equilibrium.

EXAMPLE

Given that water boils at 100°C under atmospheric pressure, roughly how much pressure is required if a 120°C boiling point is desired?

For this problem we have

$$(p_i, T_i) = (1 \text{ atm}, 373 \text{ K})$$

$$T_f = 393 \text{ K}$$

The latent heat of vaporization for water is 540 cal/g, or 4.07×10^4 J/mole, and the gas constant (R) has the value 8.314 J/mole·K. With these the exponent in Eq. 23.8 has the value

$$\frac{L}{R}\left(\frac{1}{T_i} - \frac{1}{T_f}\right) = 0.668$$

and the pressure of question, p_f, is given by

$$p_f = (1 \text{ atm})e^{0.668} = 1.95 \text{ atm}$$

PROBLEMS

23-8. Given that water boils at 100°C at atmospheric pressure, and that its latent heat of vaporization is 9700 cal/mole, under what pressure (roughly) will water boil at:

(a) 140°C?
(b) 300°C?
(c) 40°C?

23-9. The latent heat of vaporization for water is 4.07×10^4 J/mole, and at atmospheric pressure it boils at 100°C. The value of the gas constant is $R = 8.314$ J/mol·K.

(a) Use Eq. 23.8 to predict the pressure at which its boiling point is 0°C.
(b) The answer to part (a) is found experimentally to be 0.00603 atm. What approximations went into the derivation of Eq. 23.8 that might account for the difference between your answer and this value?

23-10. Why would Eq. 23.8 be especially inaccurate near the critical point? (*Hint.* What approximation did we use regarding the molar volume?)

23-11.   Why does the line separating the solid and liquid phases on the water's phase diagram have negative slope?

23-12.   In Figure 23.10, how do we know that the portion of the curve between points 2 and 3 has "medium" slope? How do we know p is decreasing?

23-13.   Under atmospheric pressure iron melts at 1530°C. The latent heat of fusion is 1.49×10^4 J/mole. At the melting point, the density of the solid phase is 7.80 g/cm^3 and the density of the liquid phase is 7.06 g/cm^3. The atomic weight of iron is 55.8.

   (a)   What are the molar volumes (in m^3) of iron in the solid and liquid phases at the melting point?
   (b)   What is the change in molar volume, Δv, of the iron when solid iron is melted? Is it positive or negative?
   (c)   To force it to melt at a higher temperature, would the pressure on it have to increase or decrease? (If ΔT is positive, is Δp positive or negative?)
   (d)   At what pressure would the melting point be 1600°C? (1 atm = 1.013×10^5 N/m^2)

23-14.   Under atmospheric pressure, water freezes at 0°C. The latent heat of fusion is 6.01×10^3 J/mole. At the melting point, the density of ice is 1.0917 g/cm^3, and that of liquid water is 0.9999 g/cm^3. The molecular weight of water is 18.

   (a)   What are the molar volumes (in m^3) of water in the solid and liquid phases at the melting point?
   (b)   What is the change in molar volume, Δv, of the water when ice melts? Is it positive or negative?
   (c)   To force it to melt at a lower temperature, would the pressure on it have to increase or decrease? (If ΔT is negative, is Δp positive or negative?)
   (d)   At what pressure would the melting point of water be -1°C? (1 atm = 1.013×10^5 N/m^2)
   (e)   At what pressure would the melting point of water be -30°C?

PART 8

QUANTUM STATISTICS

In Part 7 we began our study of a small system in equilibrium with a large reservoir. We saw that since the entropy is a measure of the number of accessible states,

$$S = k \ln \Omega \qquad \text{or} \qquad \Omega = e^{S/k}$$

then the probability that the combined system is in a certain configuration, c is given by

$$P_c \propto \Omega_c = e^{S_c/k}$$

because the probability is proportional to the number of accessible states of that particular configuration.

We saw that if the small system is in a particular state s that removes energy ΔE, volume ΔV, and particles ΔN from the reservoir, then the entropy of the combined system* is reduced to

$$S = S_{\max} - \Delta S$$

where

$$\Delta S = \frac{1}{T} (\Delta E + p \, \Delta V - \mu \, \Delta N)$$

Therefore, the probability that the small system is in that one particular state s is given by

$$P_s = C e^{-(1/kT)(\Delta E + p \, \Delta V - \mu \, \Delta N)}$$

where C is a constant of proportionality. We saw that we could apply this result, using two different approaches, depending on what we consider to be our "small system."

In the first approach, we consider our small system to be a single particle (or group of particles) that could be in any of various possible quantum states. The number of particles is fixed, so none are taken from or returned to the reservoir.

$$\Delta N = 0$$

The change in volume occupied by the particle for the various quantum states was very small, so that for most ordinary circumstances we can ignore the $p \, \Delta V$ term in comparison to the energy held by the particle.

$$p \, \Delta V \approx 0$$

---

* The entropy of the combined system is just that of the reservoir, because the entropy of the small system is zero when its state is specified ($k \ln 1 = 0$).

This gave us

$$P_s = Ce^{-\beta E_s}$$

as the probability that the particle occupies the quantum state s of energy E_s.

In the other approach, we consider our small system to be a single quantum state that could be occupied by various numbers of particles. The volume of a given quantum state is fixed

$$p\,\Delta V = 0$$

but the particles occupying this state must come from the reservoir.

$$\Delta N \neq 0$$

Consequently, the probability that our quantum state is occupied by n particles is given by

$$P = Ce^{-\beta(\Delta E - \mu\,\Delta N)} = Ce^{-\beta n(\varepsilon_s - \mu)}$$

where ε_s is the energy per particle for that quantum state.

These two different ways of looking at problems are referred to as classical statistics and quantum statistics, respectively. This customary nomenclature is unfortunate because quantum considerations are equally deeply engrained in both approaches. In the one case we consider the quantum states occupied by a single particle, and in the other case we consider the particles occupying a single quantum state.

The straightforward application of the classical approach to systems of more than one particle runs into trouble when particle densities (in six-dimensional phase space) become high. The reason is that the states accessible to any one particle may be affected by the presence of others. We will see that for one class of particles, called "fermions," a state becomes inaccessible if occupied by another particle. For the other class of particles, called "bosons," the presence of others does not inhibit access to any state, but rather it changes the number of such states—or equivalently, the way they are counted. We have already encountered this problem of counting states for identical particles in Section D, Chapter 21.

When the number of accessible states is very large compared to the number of particles,* then the probability of any given state being occupied is low, and the problems mentioned in the preceding paragraph do not arise. In this case, the classical approach works just fine. However, when some of the quantum states have a significant probability of being occupied, then the proper application of the classical approach to systems of more than one particle becomes difficult, if not impossible, and the quantum approach should be used instead.

In the last part we developed applications of the classical approach to problems, and in this part we will work from the quantum approach. For many problems, both approaches work equally well and give the same result, but when the quantum states of interest have a significant probability of being occupied, then the quantum approach should be used.

---

* This is equivalent to the condition that the particle density in six-dimensional phase space is low.

chapter 24

THE OCCUPANTS OF QUANTUM STATES

In previous chapters we examined individual particles that could occupy various quantum states. Now we shift our attention to quantum states, which could be occupied by various numbers of particles. How do we determine the average number of particles that would occupy any given quantum state?

In the classical approach, our focus of attention was the individual particle (Figure 24.1). If we could analyze the behavior of a single particle, through the states available to it and the probabilities of its being in each, then we could predict the behavior of an entire system of such particles.

Now our focus will be on the individual quantum state. If we know the characteristics of the various quantum states accessible to the particles, and the average number of particles in each, then we can predict the behavior of the entire system. States containing large numbers of particles have a large influence on the properties of the system, and states containing no particles exert no influence at all.

Problems in quantum statistics, then, can be cleanly divided into two parts: one involves finding the spectrum of states available to the system, and the other involves finding the average number of particles occupying each state (Figure 24.2). That is, we wish to know two things.

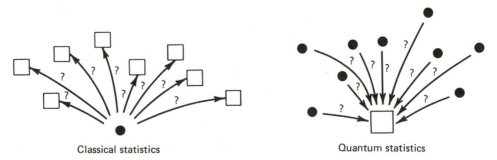

Classical statistics   Quantum statistics

Figure 24.1 In classical statistics, the focus of attention is the individual particle, which could occupy any of several possible quantum states. In quantum statistics, attention is focused on the individual quantum state, which could be occupied by various particles.

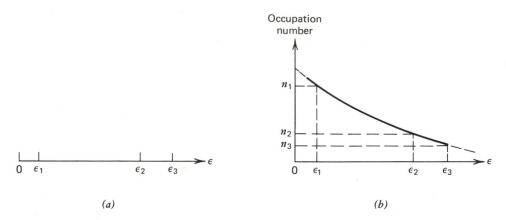

Figure 24.2   Problems in quantum statistics can be divided into two parts: (a) Finding the spectrum of states accessible to the particles of a system and (b) finding the average number of particles occupying a state (the "occupation number") as a function of the energy of the state.

1.   The spectrum of accessible states.
2.   The average number of particles occupying each.

The spectrum of accessible states varies greatly from one system to another. The spectrum of states for electrons in a conductor is quite different than that in a semiconductor or insulator. The spectrum of states accessible to air molecules in your room is quite different from the spectrum of states accessible to the atoms in liquid helium or the atoms in a diamond crystal.

To find the spectrum of accessible states is the subject of intensive research in the various scientific fields. Of course, it would be impossible in one book to adequately describe all this research, and we won't even try. Instead, we will use the results of some of these studies to illustrate the application of quantum statistics to a few representative systems.

The other part of the problem has a simple, uniform answer, applicable to all systems. The average number of particles occupying a given quantum state is called the "occupation number." and is determined simply and uniquely by the energy of a particle in that state. It does not depend on the details of the system. That is, although the spectrum of available states does depend on the detailed nature of the system, the occupation number for any one of those states does not.

A.   THE OCCUPATION NUMBER, $\bar{n}$

According to our result (16.6), the probability that a certain quantum state, s, contains n particles is given by

$$P_n = Ce^{-\beta n(\varepsilon_s - \mu)}$$

(16.6)

where

$$C = \left[\sum_{n'} e^{-\beta n'(\varepsilon_s - \mu)} \right]^{-1}$$

and where ε_s is the energy per particle in state s.

The average number of particles occupying that state is given by

$$\bar{n} = \sum_n P_n n = \left[\sum_{n'} e^{-\beta n'(\varepsilon_s - \mu)} \right]^{-1} \sum_n e^{-\beta n(\varepsilon_s - \mu)} n \tag{24.1}$$

If we let

$$x = \beta(\varepsilon_s - \mu) \tag{24.2}$$

then this can be written as

$$\bar{n} = \frac{\sum\limits_n e^{-nx} n}{\sum\limits_n e^{-nx}} = \frac{-(\partial/\partial x)\sum\limits_n e^{-nx}}{\sum\limits_n e^{-nx}} = -\frac{\partial}{\partial x} \ln \left(\sum_n e^{-nx} \right) \tag{24.3}$$

SUMMARY

If the energy per particle in state s is ε_s, then the average number of particles occupying that state, or the "occupation number" of that state, is given by

$$\bar{n} = \frac{-(\partial/\partial x)\sum\limits_n e^{-nx}}{\sum\limits_n e^{-nx}} = -\frac{\partial}{\partial x} \ln \left(\sum_n e^{-nx} \right) \tag{24.3}$$

where

$$x = \beta(\varepsilon_s - \mu) \tag{24.2}$$

The problem of determining the occupation number is now reduced to evaluating the sum $\sum_n e^{-nx}$ appearing in Eq. 24.3. The upper limit on this sum is the maximum number of particles of that type that may occupy a single quantum state, and Nature provides only two possibilities.

If no more than one particle may occupy a given state, then the particles are called "fermions," and the type of quantum statistics they obey is called "Fermi-Dirac statistics." For this case, the sum is given by

$$\sum_{n=0}^{1} e^{-nx} = 1 + e^{-x}$$

and so from Eq. 24.3 the occupation number of a state for fermions is given by

$$\bar{n}_{\text{fermions}} = -\frac{\partial}{\partial x} \ln (1 + e^{-x}) = \frac{e^{-x}}{1 + e^{-x}} = \frac{1}{e^x + 1} \qquad [x = \beta(\varepsilon_s - \mu)] \tag{24.4}$$

A qualitative plot showing the variation of $\bar{n}$ with ε for fermions is shown in Figure 24.3.

If there is no restriction on the number of particles that may occupy a given state, then the particles are called "bosons," and the type of quantum statistics they obey is

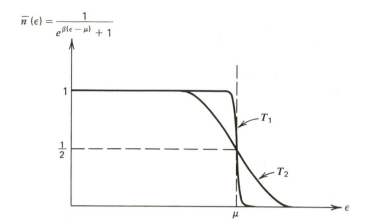

$$\bar{n}(\epsilon) = \frac{1}{e^{\beta(\epsilon - \mu)} + 1}$$

Figure 24.3   Plot of $\bar{n}$ versus ε for fermion systems at two different temperatures $(T_2 > T_1)$. When the energy of the state equals the chemical potential, the occupation number is $\frac{1}{2}$. $\bar{n}(\varepsilon = \mu) = 1/(e^0 + 1) = \frac{1}{2}$.

called "Bose-Einstein statistics." For this case, the sum is a convergent series as long as x is positive, with a value shown in Appendix 21A to be given by

$$\sum_{n=0}^{\infty} e^{-nx} = \frac{1}{1 - e^{-x}} \qquad (x > 0)$$

Putting this into Eq. 24.3 gives the following result for the occupation number of a state for bosons.

$$\bar{n}_{\text{bosons}} = -\frac{\partial}{\partial x} \ln (1 - e^{-x})^{-1} = \frac{\partial}{\partial x} \ln (1 - e^{-x}) = \frac{e^{-x}}{1 - e^{-x}}$$

$$= \frac{1}{e^x - 1} \qquad [x = \beta(\varepsilon_s - \mu) > 0] \tag{24.5}$$

A qualitative plot showing the variation of $\bar{n}$ with ε is shown in Figure 24.4.

One of the first things you should notice about the results (24.4) and (24.5) is their similarity in appearance, differing only by the sign in the denominator. This similarity makes it sometimes convenient to work problems for both types of particles simultaneously, using the occupation number

$$\bar{n} = \frac{1}{e^{\beta(\varepsilon - \mu)} \pm 1}$$

where we just have to remember that the upper sign $(+)$ applies to fermions and the lower one $(-)$ to bosons.

Another thing that should be brought to your attention here is the relationship between the "intrinsic" or "spin" angular momentum of a particle and the type of

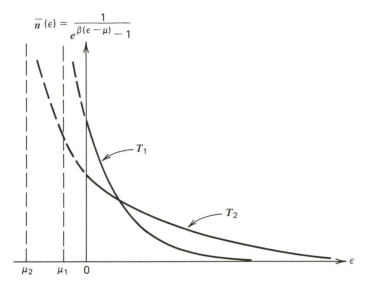

$$\bar{n}(\epsilon) = \frac{1}{e^{\beta(\epsilon - \mu)} - 1}$$

Figure 24.4 Plot of $\bar{n}$ versus ε for boson systems at two different temperatures $(T_2 > T_1)$. In order for the sum in Eq. 24.3 to converge, the chemical potential must be less than the energy of the lowest-lying state. Notice that the occupation number would be infinite at $\varepsilon = \mu$.

quantum statistics it obeys. If the spin quantum number* is half-integer (e.g., $\frac{1}{2}$, $\frac{3}{2}$, etc.) the particles are fermions, and if it is integer (e.g., 0, 1, 2, etc.) they are bosons. Common fermions include electrons, protons and neutrons (all $s = \frac{1}{2}$ or "spin $\frac{1}{2}$"), and common bosons include photons (spin 1) and α-particles (spin 0).

The theoretical proof of this interrelationship is fairly involved and won't be given here. It relies basically on the ideas that whether or not more than one identical particle may occupy a state is related to the behavior of their quantum mechanical wave functions under rotations by 360°, and that the behavior of the wave functions under 360° rotations is determined by the angular momentum of the particle. Suffice it to say that the interrelationship between spin and statistics has been established both theoretically and experimentally.

SUMMARY

Particles having half-integer intrinsic angular momenta are called fermions, and obey Fermi-Dirac statistics. No two identical fermions may occupy the same

---

* The spin angular momentum of a particle is given by

$$S = \sqrt{s(s + 1)}\hbar$$

where the spin quantum number, s, is either integer or half-integer. The z-component of this is given by $S_z = s_z\hbar$ where s_z can take on the following possible values: $s_z = s, s - 1, s - 2, \ldots, -s$.

quantum state. The occupation number for a fermion state is given by

$$\bar{n} = \frac{1}{e^{\beta(\varepsilon - \mu)} + 1} \tag{24.4}$$

where ε is the energy of a particle in that state and μ is its chemical potential.

Particles having integer intrinsic angular momenta are called bosons, and obey Bose-Einstein statistics. Any number of identical bosons may occupy the same quantum state, and the occupation number for a boson state is given by

$$\bar{n} = \frac{1}{e^{\beta(\varepsilon - \mu)} - 1} \qquad (\varepsilon - \mu) > 0 \tag{24.5}$$

PROBLEMS

24-1.   At room temperature, what is the occupation number for a fermion state that is:

   (a)   0.01 eV below the chemical potential?
   (b)   0.01 eV above the chemical potential?
   (c)   0.1 eV below the chemical potential?
   (d)   0.1 eV above the chemical potential?

24-2.   At room temperature, what is the occupation number for a boson state that is:

   (a)   0.01 eV above the chemical potential?
   (b)   0.1 eV above the chemical potential?
   (c)   1.0 eV above the chemical potential?

24-3.   Is it possible for a boson to have the z-component of its angular momentum equal to $\frac{1}{2}\hbar$? Explain. (*Hint.* See footnote on bottom of page 407.)

24-4.   For a temperature of 300 K, compute the ratio of occupation numbers $\bar{n}_{\text{boson}}/\bar{n}_{\text{fermion}}$ for:

   (a) $(\varepsilon - \mu) = 0.002$ eV.
   (b) $(\varepsilon - \mu) = 0.02$ eV.
   (c) $(\varepsilon - \mu) = 0.2$ eV.
   (d) $(\varepsilon - \mu) = 2.0$ eV.

24-5.   A certain quantum state lies 0.1 eV above the chemical potential. At what temperature would the average number of bosons in this state be:

   (a)   5?
   (b)   0.5?
   (c)   0.05?

24-6.   The hydrogen atom is made of two spin $\frac{1}{2}$ particles: a proton and an electron.

   (a)   When the electron is an s-orbital (angular momentum due to the electron's orbit is $L = 0$), what possible values of total angular momentum quantum number (j) could the hydrogen atom have?
   (b)   When the electron is in an orbital of angular momentum $L = \sqrt{2}\hbar$, what possible values could the total angular momentum quantum number (j) of the atom be?
   (c)   The atmosphere of a certain star is atomic hydrogen gas. Would this be a boson gas or a fermion gas?

(d) Deeper down toward the star's interior, due to the high temperatures and pressures,the hydrogen atoms are stripped of their electrons. Would the leftover protons be a gas of bosons or fermions? Would the stripped electrons be a gas of bosons or fermions?

24-7. (a) Draw a plot of $\bar{n}$ versus ε for bosons, showing how it would look in the limits of extremely high temperatures and extremely low temperatures (both on the same plot, and label which is which). Assume that the chemical potential doesn't vary with temperature. (In the next homework problem it will be seen that this is a bad assumption.)

(b) Do the same for fermions.

24-8. Show that for a system of N bosons at very low temperatures (such that they are all in the lowest energy state, $\varepsilon = 0$), the chemical potential varies with the temperature according to

$$\mu \xrightarrow[T \to 0]{} -\frac{1}{N} kT$$

(*Hint.* For small x, $e^x \approx 1 + x$.)

24-9. Suppose that quantum mechanics were wrong, and there is a third type of particle, "goofions," for which up to two particles (but not more) may occupy any one state. Derive the occupation number, $\bar{n}$, as a function of ε, μ, and T for these goofions. (*Hint.* Start with Eq. 24.3.)

B. COMPARISON WITH CLASSICAL STATISTICS

In our study of classical statistics, we found that the probability for a single particle to be in a certain state, s, is given by

$$P_s = Ce^{-\beta \varepsilon_s}$$

where C is an appropriate constant of proportionality. Wouldn't you think that if the system consisted of two particles, then that state would have twice as high a probability of being occupied? Wouldn't you think that the occupation number of a state would be simply the product of the number of particles in the system (N) times the probability (P_s) for any one of them to be in that state?

$$\bar{n}_{\text{classical}} \overset{?}{=} NP_s = N(Ce^{-\beta \varepsilon_s})$$

This line of reasoning is indeed correct, providing particle densities* are sufficiently small that there is little likelihood of two particles attmpting to occupy the same state simultaneously. We see, for example, that the energy dependence of the above classical result for the occupation number of a state is[†]

$$\bar{n}_{\text{cl}} \approx e^{-\beta(\varepsilon - \mu)}$$

---

* That is, densities in six-dimensional phase space.
[†] We measure energies relative to the chemical potential here to facilitate comparison with quantum statistics.

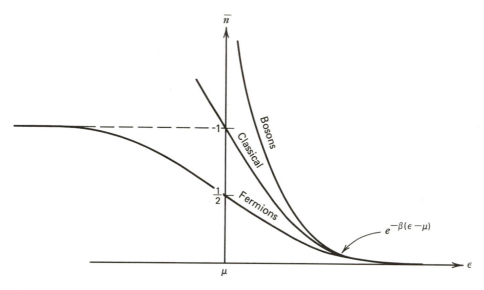

Figure 24.5   Plot of occupation number versus energy of the state for boson, classical, and fermion systems at fixed temperature and chemical potential. For the energy region $(\varepsilon - \mu) \gg kT$, the three are the same, but for $(\varepsilon - \mu) \lesssim kT$, $\bar{n}_{\text{bosons}} > \bar{n}_{\text{classical}} > \bar{n}_{\text{fermions}}$.

By comparison, if we look at the quantum statistics results for bosons and fermions in the region of small occupation numbers, when $(\varepsilon - \mu)$ is large compared to kT, we have

$$e^{\beta(\varepsilon - \mu)} \gg 1 \qquad \text{for} \qquad (\varepsilon - \mu) \gg kT$$

which means*

$$\bar{n}_{\text{bosons}} = \frac{1}{e^{\beta(\varepsilon - \mu)} + 1} \approx e^{-\beta(\varepsilon - \mu)} \qquad \text{for} \qquad (\varepsilon - \mu) \gg kT$$

$$\bar{n}_{\text{fermions}} = \frac{1}{e^{\beta(\varepsilon - \mu)} + 1} \approx e^{-\beta(\varepsilon - \mu)} \qquad \text{for} \qquad (\varepsilon - \mu) \gg kT$$

Consequently, the quantum and classical results are the same for states of sufficiently high energies that their occupation numbers are small. (See Figure 24.5.)

$$\bar{n}_{\text{cl}} \approx \bar{n}_{\text{bosons}} \approx \bar{n}_{\text{fermions}} \approx e^{-\beta(\varepsilon - \mu)} \qquad (\varepsilon - \mu) \gg kT \qquad (24.6)$$

However, for states of lower energy, where we cannot ignore the factor of 1 in the denominator in comparison to $e^{\beta(\varepsilon - \mu)}$, then the classical and quantum results differ. In this region, we have[†]

$$\frac{1}{e^{\beta(\varepsilon - \mu)} - 1} > e^{-\beta(\varepsilon - \mu)} > \frac{1}{e^{\beta(\varepsilon - \mu)} + 1}$$

---

* Since we measure energies relative to the chemical potential in the quantum results, we do the same for the classical result in order to facilitate comparison. We're examining energy dependence and ignoring the overall multiplicative constant.

† To facilitate comparison, we measure energies relative to the same zero-energy reference level (μ) in all three cases.

or

$$\bar{n}_{\text{bosons}} > \bar{n}_{\text{cl}} > n_{\text{fermions}} \qquad (\varepsilon - \mu) \lesssim kT \qquad (24.7)$$

Now we will examine the reasons why the classical and quantum results differ for states of lower energies where occupation numbers are significantly greater than zero. We will find that there are two reasons for these differences.

1. The classical result is based on the assumption that each particle's behavior is independent of all the others. The probability that a particle occupies a certain state,

$$P_s = C e^{-\beta \varepsilon_s}$$

depends only on the energy of the state, ε_s, and the temperature ($\beta = 1/kT$), but not at all on whether the state is already occupied. By contrast, fermions are excluded from states that are already occupied, and this reduces the occupation number for fermions relative to the classical result, especially for those states that have a relatively high probability of being occupied (Figure 24.6).

2. When the particles are identical, then the counting of the number of distinguishable arrangements of the system is altered in such a way that it enhances the number of arrangements that have multiple particles in a single state relative to the number of arrangements that have no more than one particle per state. This applies only to bosons, of course, since fermions do not permit more than one occupant of a state. The enhancement of arrangements with multiparticle states, means the system is more likely to be found in such an arrangement, according to the fundamental postulate. Thus, the occupation number for bosons will be larger than the classical result.

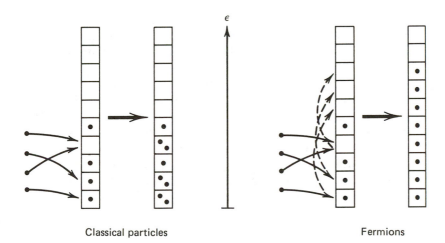

Classical particles   Fermions

Figure 24.6 In classical statistics, the behavior of each particle is completely unaffected by the others. Consequently, many particles may occupy the preferred lower states. Fermions, in contrast, are excluded from states already occupied, so the low-lying states cannot have multiple occupancy. As a result, the low-lying fermion states have an occupation number that is smaller than the classical prediction.

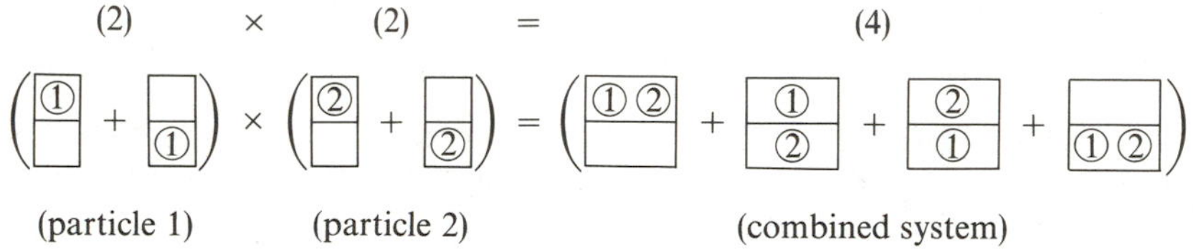

(particle 1)   (particle 2)   (combined system)

Figure 24.7   In classical statistics, the number of states accessible to the combined system is simply the product of the numbers of states accessible to the individual particles (illustrated here for the simple case of two particles each having two accessible states).

We can see that both these effects only occur when the chance for multiple occupancy of a state is rather high, and then they give the effect $\bar{n}_{\text{bosons}} > \bar{n}_{\text{cl}} > \bar{n}_{\text{fermions}}$, as our results (24.7) demand. When particle densities are so low that the chance of multiple occupancy is negligible, then the above two effects play no role, and $\bar{n}_{\text{bosons}} \approx \bar{n}_{\text{cl}} \approx \bar{n}_{\text{fermions}}$, as our results (24.6) demand.

In a quick overview of our classical analysis, we can pinpoint where these flaws first enter. If the particles are distinguishable, the number of states available to a system of N particles is the product of the number of states available to each (Figure 24.7).*

$$\Omega_D = \Omega_1 \Omega_2 \Omega_3 \cdots \Omega_N \tag{24.8}$$

Notice the implicit assumption here that the states available to any one particle do not depend on whether they are already occupied by others. According to Eq. 24.8 the number of states accessible to a system of N distinguishable particles, each having Ω_1 accessible states, is

$$\Omega_D = \Omega_1^N \tag{24.9}$$

For example, if the number of states available to one flipped coin is two, then the number of different states available to a system of three distinguishable flipped coins is $2^3 = 8$.

If the system consists of N *identical* particles occupying N different states, then there are $N!$ different ways of distributing these particles among the states, resulting in the same identical configuration for the system as a whole. This is illustrated in Figures 21.2 and 21.3. Therefore, in applying the result (24.9) to a system of *identical* particles, it must be reduced by a factor of $N!$, corresponding to the number of different distributions of identical particles corresponding to any one distinguishable arrangement of the system as a whole. The classical result for the number of different arrangements for the system is then

$$\Omega_{\text{cl}} = \frac{1}{N!}\Omega_D = \frac{1}{N!}\Omega_1^N \tag{24.10}$$

where Ω_D is the number of states for a system of distinguishable particles, and Ω_1 is the number of states available to any one of them.

* We'll let the subscript D on Ω_D stand for "distinguishable particles."

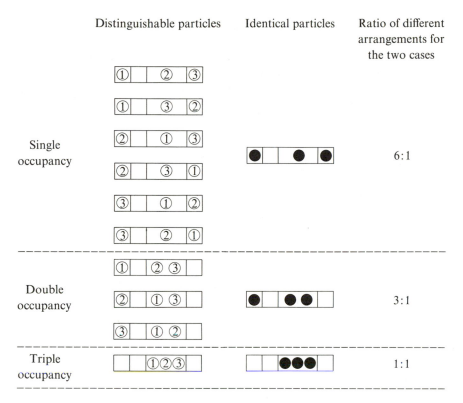

Figure 24.8 Some examples of different arrangements for distinguishable particles that would correspond to the same arrangement if the particles were identical. The ratio of different arrangements for the two cases is $N!:1$ ($N! = 6$ in this case) *only* if there is no more than one particle per state. The ratio is less than this if some states are multiply occupied.

This factor of $1/N!$ is the proper correction provided there is no more than one particle per state, but it is not correct if some states contain more than one particle, as is illustrated in Figure 24.8. Consequently, the correction factor of $1/N!$ restricts too severely the number of distinguishable states of the system when occupation numbers become large.*

---

* We review the counting of states here for illustrative purposes. It is related to the probabilities through the fundamental postulate, and can be seen in the following way. In our derivation of classical statistics, we found the probability of one particle to be in state, a, to be given by $C \exp(-\beta \varepsilon_a)$, and this is assumed to be independent of the behaviors of the other particles. With this assumption, the probability of particle number one to be in state a, particle number two to be in state b, and so on, is simply the product of the individual probabilities,

$$C \exp(-\beta \varepsilon_a) \cdot C \exp(-\beta \varepsilon_b) \cdots .$$

Notice that the multiplicative constant (the product of the C's) is the same whether or not this configuration involves states of multiple occupancy (for example, whether or not states a and b are the same). This is equivalent to multiplying all configurations in the above development by the same factor of $1/N!$, whether or not they involve multiple occupancies, and it leads to the same error.

Thus, we see where the two flaws enter into classical statistics.

1. The assumption that the states accessible to any particle are independent of the other particles, which is not true for fermions.
2. The correction factor of $1/N!$ is too severe when more than one particle occupy the same state. (This is only possible for bosons.)

As a specific example, consider the arrangements of two particles among two states, according to the various types of statistics, as illustrated in Figure 24.9. According to the classical result for distinguishable particles,

$$\Omega_D = 2^2 = 4$$

For identical particles this is corrected to become

$$\Omega_{cl} = \frac{1}{2!} 4 = 2$$

This correction means that there is one distinguishable arrangement with the particles in different states, which is correct, and that there is one distinguishable arrangement with both particles in the same state, which is *wrong*! The $1/N!$ correction is too severe when some states have multiple occupancy.

Notice that when there is no more than one particle per state, all three types of statistics (corrected classical, Bose-Einstein, and Fermi-Dirac) give the same result for the number of distinguishable arrangements of the system. But when there are states of multiple occupancy, the various results differ.

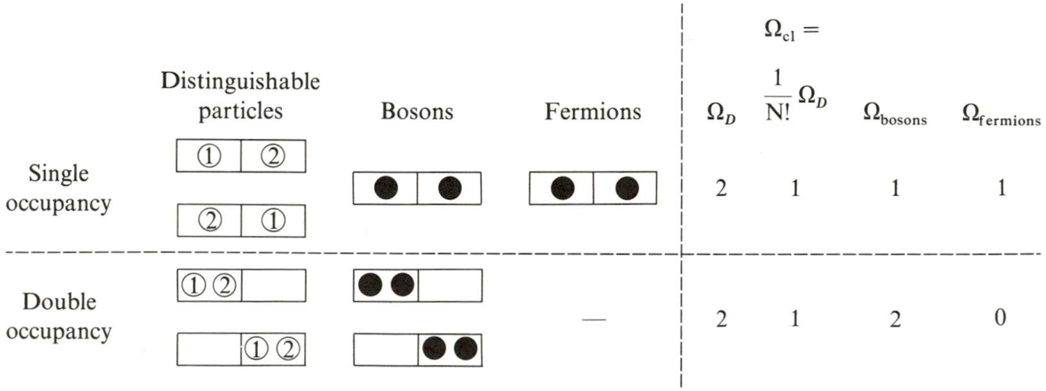

	Distinguishable particles	Bosons	Fermions	Ω_D	$\Omega_{cl} = \frac{1}{N!}\Omega_D$	Ω_{bosons}	$\Omega_{fermions}$
Single occupancy				2	1	1	1
Double occupancy			—	2	1	2	0

Figure 24.9 Illustration of the different possible arrangements of two particles between two accessible states. The number of different possible arrangements are given for the cases where the particles are distinguishable (Ω_D), identical "classical" particles (Ω_{cl}), identical bosons (Ω_{bosons}), and identical fermions ($\Omega_{fermions}$). Notice that for the case of double occupancy, there are more different possible arrangements for identical bosons than the corrected classical result allows, and that the corrected classical result allows more than would be the case for identical fermions (which allow none). This is just one specific illustration of the general case that for single occupancy, $\Omega_{bosons} = \Omega_{cl} = \Omega_{fermions}$ and for multiple occupancy, $\Omega_{bosons} > \Omega_{cl} > \Omega_{fermions} = 0$.

The counting of possible arrangements for a system of three identical particles is illustrated in Figure 24.10. Again it is seen that all three types of statistics give the same result when there is only one particle per state. But again it is also seen that Bose-Einstein statistics give relatively more arrangements having multiparticle states than does the classical approach, and Fermi-Dirac statistics give none.

Since the fundamental postulate states that all different accessible arrangements of the system are equally probable, we would expect if there are relatively large numbers of arrangements having multiparticle states, then the probability of multiple occupancy would be rather high. Similarly, if there are no possible arrangements of the system having multiparticle states, then the probability of multiple occupancy would be zero. This explains why bosons have higher occupation numbers (i.e., average number of particles per state) than the classical predictions, and why fermions have smaller occupation numbers than the classical prediction, and it explains why these differences are most pronounced for low-lying states where the possibility of multiple occupancy would be high.

EXAMPLE

Using the fundamental postulate, what is the probability of the system depicted in Figure 24.9 being in an arrangement having doubly occupied states according to the three kinds of statistics?

The fundamental postulate states that all allowed arrangements are equally probable. Therefore, the probability of being in a configuration having double occupancy is given by

$$P_{\text{double occupancy}} = \frac{\Omega_{\text{double occupancy}}}{\Omega_{\text{all}}}$$

$$= \begin{cases} \frac{2}{3} & \text{for bosons} \\ \frac{1}{2} & \text{for "classical" particles} \\ \frac{0}{1} = 0 & \text{for fermions} \end{cases}$$

SUMMARY

For states of high energy, where occupation numbers are low and where there is little likelihood of two particles attempting to occupy the same state, the classical, Bose-Einstein, and Fermi-Dirac statistics all give the same result for the occupation number of a state.

$$\bar{n}_{\text{bosons}} \approx \bar{n}_{\text{cl}} \approx \bar{n}_{\text{fermions}} \approx e^{-\beta(\varepsilon - \mu)} \qquad (\varepsilon - \mu) \gg kT \tag{24.6}$$

For lower-lying states, occupation numbers are larger. Where there is significant likelihood that more than one particle may attempt to occupy the same state, there is significant disparity between the predictions according to the three kinds of statistics.

$$\bar{n}_{\text{bosons}} > \bar{n}_{\text{cl}} > \bar{n}_{\text{fermions}} \qquad (\varepsilon - \mu) \lesssim kT \tag{24.7}$$

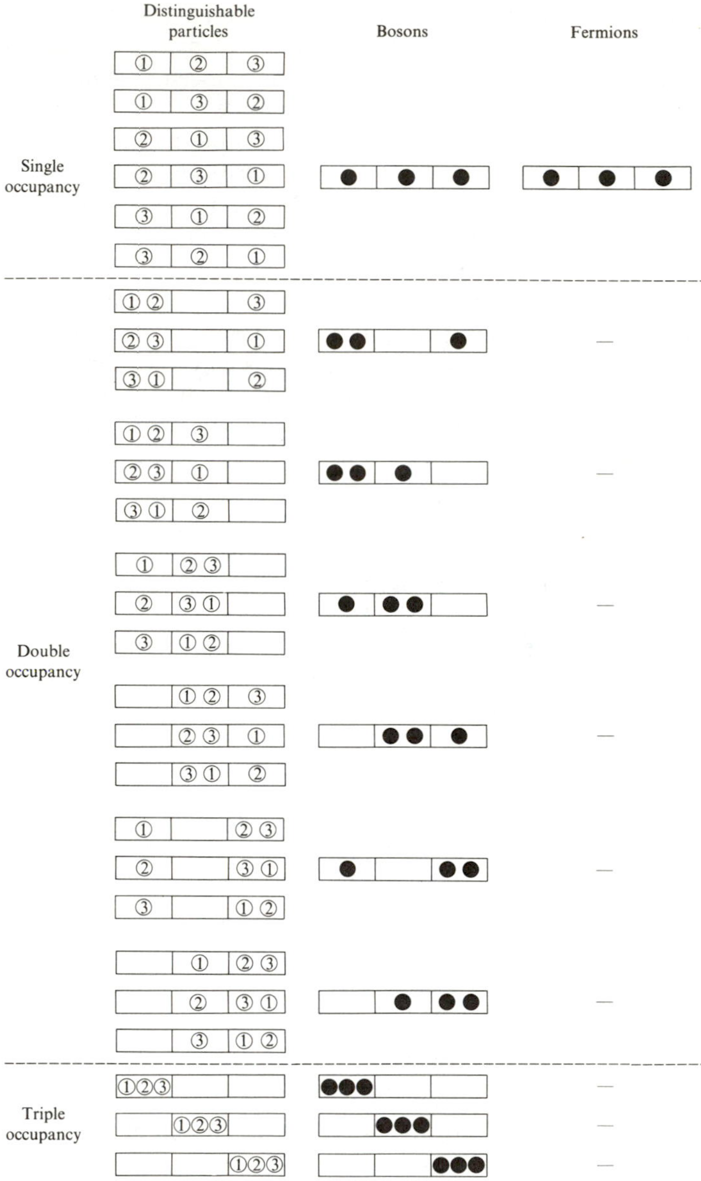

	Distinguishable particles	Bosons	Fermions	Ω_D	$\Omega_{cl} =$ $\frac{1}{N!}\Omega_D$	Ω_{bosons}	$\Omega_{fermions}$
Single occupancy				6	1	1	1
Double occupancy				18	3	6	0
Triple occupancy				3	$\frac{1}{2}$	3	0
			Totals:	27	$4\frac{1}{2}$	10	1

Figure 24.10   Illustration of the different possible arrangements of three particles among three states. The number of different possible arrangements are given for the cases where the particles are distinguishable (Ω_D), identical "classical" particles (Ω_{cl}), identical bosons (Ω_{bos}), and identical fermions (Ω_{ferm}). For the cases of multiple occupancy, Bose-Einstein statistics gives more different arrangements than does classical statistics, and Fermi-Dirac statistics allow none at all. This is just another specific illustration of the general behavior that for single occupancy, $\Omega_{boss} = \Omega_{cl} = \Omega_{ferm}$ and for multiple occupancy, $\Omega_{bos} > \Omega_{cl} > \Omega_{ferm} = 0$.

The reason that $\bar{n}_{cl}$ is larger than $\bar{n}_{fermions}$ in this region is that classical statistics assumes that the behavior of any one particle is independent of the behavior of all the others, whereas a fermion is excluded from states occupied by others. Since "classical" particles are permitted states of multiple occupancy and fermions are not, the average number of particles per state will be higher for "classical" particles than fermions.

The reason that $\bar{n}_{bosons}$ is larger than $\bar{n}_{cl}$ is that in classical statistics we assume that there are $N!$ different arrangements of identical particles for every distinguishable arrangement of the system as a whole. This is true as long as there is no more than one particle per state, but incorrect when multiparticle states are involved. The classical correction factor of $1/N!$ overrestricts the number of different arrangements having multiparticle states. Since the fundamental postulate demands equal probability for all accessible arrangements, the classical prediction gives too low a probability for multiply occupied states, and therefore too low an occupation number.

PROBLEMS

24-10. Given that the classical, Bose-Einstein, and Fermi-Dirac results for the occupation number of a state are all given by $\bar{n} = e^{-\beta(\varepsilon - \mu)}$ when $(\varepsilon - \mu) \gg kT$, what is the occupation number when $\varepsilon = \mu$ according to:

(a) Fermi-Dirac statistics?
(b) classical statistics?
(c) Bose-Einstein statistics?
(d) Is this point represented correctly on the plot of Figure 24.5?

24-11. Using h for "heads" and t for "tails," make tables to show that the number of possible arrangements for a system of N distinguishable flipped coins is given by $\Omega_D = \Omega_1^N$ for systems of $N = 1, 2, 3$, and 4 coins, where Ω_1 is the number of states available to one of them.

24-12. Given that the classical prediction for the occupation number of a state is $\bar{n}_{cl} = e^{-\beta(\varepsilon - \mu)}$, if a system is at 1000 K, what is the occupation number of a state of energy 0.1 eV above the chemical potential according to:

(a) Bose-Einstein statistics?
(b) classical statistics?
(c) Fermi-Dirac statistics?

24-13. Repeat Problem 24-12 for a temperature of 300 K.

24-14. Consider a system of three rolled dice, each having six possible states available to it (six different possible numbers of dots showing upward). According to classical statistics, how many different arrangements are available to this system:

(a) If the dice are distinguishable?
(b) If the dice are identical?
(c) Is the answer to part (b) larger, smaller, or equal to the actual number of distinguishable arrangements available to three identical rolled dice?

24-15. Consider a system of three flipped coins, each having two possible states available to it (heads or tails). According to classical statistics, how many different arrangements are available to this system:

(a) If the coins are distinguishable?
(b) If the coins are identical?
(c) Is the answer to part (b) larger, smaller, or equal to the actual number of distinguishable arrangements available to a system of three identical flipped coins? Demonstrate your answer by listing the actual allowed arrangements.

24-16. Consider the distribution of two particles among three possible states.

(a) List all possible arrangements of the system if the particles are distinguishable.
(b) List all possible arrangements of the system if the particles are identical bosons.
(c) List all possible arrangements of the system if the particles are identical fermions.

24-17. In Problem 24-16, how many possible arrangement of the system are there, having no more than one particle per state according to:

(a) classical statistics for identical particles?
(b) Bose-Einstein statistics?
(c) Fermi-Dirac statistics?

24-18. In Problem 24-16, how many possible arrangements of the system include a doubly occupied state according to:

(a) classical statistics for identical particles?
(b) Bose-Einstein statistics?
(c) Fermi-Dirac statistics?

24-19. In Problem 24-16, what is the probability of finding the system at any instant in an arrangement involving a doubly occupied state according to (*Hint.* See the example on page 416.):

(a) classical statistics for identical particles?
(b) Bose-Einstein statistics?
(c) Fermi-Dirac statistics?

24-20. Consider the distribution of three particles among two possible states.

(a) List all possible arrangements of the system if the particles are distinguishable.
(b) List all possible arrangements of the system if the particles are identical bosons.
(c) List all possible arrangements of the system if the particles are identical fermions.

24-21. In Problem 24-20, how many possible arrangements of the system are there, having one doubly occupied state, according to:

(a) classical statistics for identical particles?
(b) Bose-Einstein statistics?
(c) Fermi-Dirac statistics?

24-22. In Problem 24-20, how many possible arrangements of the system are there, having one triply occupied state, according to:

(a) classical statistics for identical particles?
(b) Bose-Einstein statistics?
(c) Fermi-Dirac statistics?

24-23. In Problem 24-20, what is the probability of finding the system at any instant in an arrangement involving a triply occupied state according to (*Hint.* See the example on page 416.):

    (a)   classical statistics for identical particles?
    (b)   Bose-Einstein statistics?

C. THE LIMITS OF CLASSICAL STATISTICS

To illustrate the differences between the three kinds of statistics, we used particle densities that are rather high. In fact, in the examples of Figures 24.9 and 24.10, we had as many particles as there were quantum states! Most systems in Nature have far more quantum states than particles, so any one quantum state has very low probability of being occupied by even one particle, let alone two. As long as the probability of multiple occupancy is negligibly small, and we can use classical statistics to solve our problems.

The criterion for the validity of the classical approach can then be written as (Figure 24.11)

$$\text{(number of particles)} \ll \text{(number of quantum states)}$$

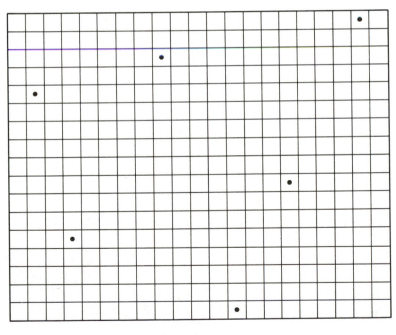

● = Particle   ☐ = Quantum state

Figure 24.11 The classical approach to statistics is quite adequate as long as the probability of having multiply occupied states is small. This would be the case if the characteristic separation of particles is large compared to the dimensions of the individual quantum states.

or

$$\text{(number of quantum states per particle)} \gg 1 \tag{24.11}$$

If the characteristic volume per particle is R^3, and the characteristic allowed range in its momentum is $\langle p \rangle$, then the number of quantum states per particle is

$$\text{(number of quantum states per particle)} = \frac{R^3 \langle p \rangle^3}{h^3}$$

and the condition (24.11) can be written (after taking the cube root of both sides)

$$R \langle p \rangle \gg h$$

or

$$R \gg \frac{h}{\langle p \rangle} \tag{24.12}$$

If we describe the characteristic allowed range in momentum as roughly the root-mean square momentum of the particle, then for nonrelativistic particles

$$\frac{1}{2m} \langle p \rangle^2 = \frac{3}{2} kT$$

and the criterion (24.12) become

$$R \gg \frac{h}{\sqrt{3mkT}} \tag{24.13}$$

If the system consists of N particles and occupies volume V, then the volume per particle is

$$R^3 = \frac{V}{N}$$

and the criterion (24.12) for the validity of the classical approach becomes

$$\left(\frac{V}{N} \right)^{1/3} \gg \frac{h}{\sqrt{3mkT}} \tag{24.14}$$

SUMMARY

Classical results differ from those of quantum statistics only when particle densities are high enough that there is an appreciable likelihood that more than one particle may attempt to occupy the same state. Classical statistics are quite adequate as long as particle densities are low.

$$\text{(number of quantum states per particle)} \gg 1. \tag{24.11}$$

If the characteristic separation between particles is R and their characteristic momentum is $\langle p \rangle$, then this condition becomes

$$R \gg \frac{h}{\langle p \rangle} \tag{24.12}$$

or, for nonrelativistic particles,

$$\left(\frac{V}{N}\right)^{1/3} \gg \frac{h}{\sqrt{3mkT}} \tag{24.14}$$

PROBLEMS

24-24. The kinetic energy of a relativistic particle is given by $E_k = pc$, where p is its momentum and c is the speed of light. If the average energy per degree of freedom for relativistic particles is kT, and each particle can move in three dimensions:

(a) What is the characteristic momentum of a relativistic particle in terms of the temperature?

(b) What would be the condition corresponding to Eq. 24.13, for the validity of applying classical statistics to relativistic particles?

24-25. Consider nitrogen (N_2) gas at room temperature and atmospheric pressure (10^5 N/m^2).

(a) Show whether or not classical statistics would be appropriate for its study. (*Hint.* From the ideal gas law, $N/V = p/kT$.)

(b) At what temperature would the number of accessible quantum states and the number of N_2 molecules become about equal if the particle density remained unchanged?

24-26. Consider the conduction electrons in copper. Suppose each copper atom contributes one electron to the conduction electrons. Copper has atomic weight of 64, and a density of 8.9 g/cm^3.

(a) What is the volume per conduction electron?

(b) What is the characteristic separation of conduction electrons, R?

(c) Can these conduction electrons be treated using classical statistics at room temperature?

(d) Can these conduction electrons be treated using classical statistics at 10,000 K?

24-27. The characteristic separation of molecules of liquid water is about 2 Å.

(a) What is the value of $h/\sqrt{3mkT}$ for water molecules at room temperature?

(b) Could liquid water at room temperature be handled using classical statistics?

chapter 25

SURVEY OF APPLICATIONS

The properties of a macroscopic system are determined by two things: (1) the spectrum of states available to the particles and (2) the average number of particles in each of these states. The second of these is the same for all fermion or boson systems and was analyzed in the preceding chapter. The first, however, varies from one system to the next and is responsible for differences in properties between different systems of similar particles. There are some common and general techniques used to handle the spectra of states accessible to various systems, and we discuss some of these in this chapter.

The properties of a system are determined by the behaviors of its constituent particles. These behaviors of the particles, in turn, are determined by what states are available, and how many particles are in each state (Figure 25.1).

Through the developments in the preceding chapter, we are able to answer the second of these two questions, but not the first. That is, given any state of any energy ε (measured relative to the chemical potential), we are now able to say how many particles occupy it on the average. But we are not able to say what states are available in the first place. The spectrum of available states varies from one system to the next, and so there is no general analysis that covers all possibilities.

Fortunately, there is a tendency for the spectra of many common systems to all display similar characteristics, and so we can group quantum systems into several broad categories according to similarities in the spectra of accessible states. We look at several such categories of systems in this chapter, to illustrate the general methods through which quantum statistics are applied to real systems. Details will vary from one system to the next, but the general idea won't. In succeeding chapters we apply quantum statistics to a few important specific systems with appropriate increase in detail.

A. THE SPECTRA OF ACCESSIBLE STATES

All quantum states are discrete. That is, aside from degeneracies,* there is a discrete separation in energy between states. For many systems, the separation between neighboring states is extremely small compared to kT.

---

* A "degeneracy" is the condition when more than one state has the same energy.

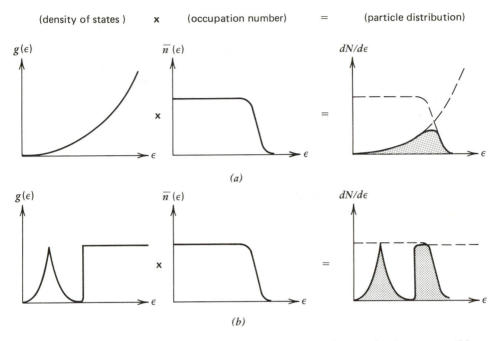

(density of states)   **x**   (occupation number)   **=**   (particle distribution)

Figure 25.1 Distribution of particles in two different fermion systems. These two fermion systems will have very different physical properties because their spectra of accessible states are very different, even though the occupation number (i.e., the average number of particles per state) is the same function of energy for both systems. The particle distribution is the product of the density of states times the occupation number.

$$\Delta\varepsilon_{\text{neighboring states}} \ll kT \qquad (\text{or } \beta\,\Delta\varepsilon \ll 1)$$

In these cases the occupation number changes only slightly from one state to the next, and any summation over discrete states can properly be turned into a continuous integral with little or no loss in accuracy. When the separation of neighboring states is significant in comparison to kT

$$\Delta\varepsilon_{\text{neighboring states}} \gtrsim kT \qquad (\text{or } \beta\,\Delta\varepsilon \gtrsim 1)$$

then there is significant change in occupation number from one state to the next, and sums over states cannot be replaced by continuous integration (Figure 25.2).

For cases where the separation of neighboring states is small, we need to know the density of states,* $g(\varepsilon)$, if we wish to turn the discrete sum over states into continuous integration. The number of states in energy range $d\varepsilon$, is the product of the density of states times the size of the energy increment, $g(\varepsilon)\,d\varepsilon$ (Figure 25.3). The number of particles is the product of the number of states times the occupation number of each.

$$dN = [g(\varepsilon)\,d\varepsilon][\bar{n}(\varepsilon)] = \bar{n}(\varepsilon)g(\varepsilon)\,d\varepsilon \qquad (25.1)$$

---

* The density of states is the number of states per unit energy.

The integration of this over all states must equal the total number of particles in the system.

$$N = \int_0^\infty \bar{n}(\varepsilon)g(\varepsilon)\,d\varepsilon$$

This condition determines the value of the chemical potential appearing in the expression for $\bar{n}(\varepsilon)$.

If the interaction between particles is extremely weak, then the separation of neighboring states is usually very tiny, and the system may be thought of as a gas. For example, the separation between neighboring states for air molecules in your room is about 10^{-23} eV, which is very small compared to kT (2.6×10^{-2} eV at room temperature). Other common systems that can be treated as a gas include photons in an oven (Chapter 26), vibrations in a solid (Chapter 27), electrons in a conductor (Chapter 28), and helium atoms in the liquid state (Chapter 29). For such gases, the sum over discrete states can be replaced by continuous integration according to

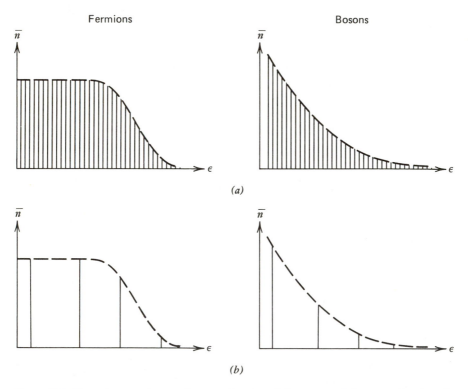

(a)

(b)

Figure 25.2 Plots of occupation number as a function of the energy of the state. (a) When the separation of states is small compared to kT, occupation numbers vary only slightly from one state to the next, and the summation over states can be approximated by continuous integration. (b) When the separation of states is comparable to, or larger than kT, the summation over discrete individual states must be retained.

$\bar{n}\,(\epsilon)$

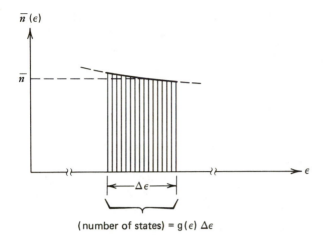

(number of states) = $g(\epsilon)\,\Delta\epsilon$

Figure 25.3 When the states are closely spaced, the number of states within any energy increment, $\Delta\varepsilon$, is equal to the product of the number of states per unit energy times the size of the increment, $\Delta\varepsilon$. The number of particles is the product of the number of states times the occupation number of each.

(number of states) $= (d\Omega/d\varepsilon)\,\Delta\varepsilon = g(\varepsilon)\,\Delta\varepsilon$

(number of particles) $=$ (number of states) $\times$ (occupation number)

$$= [g(\varepsilon)\,\Delta\varepsilon]\bar{n}(\varepsilon)$$

Eq. 2.12

$$\sum_{s} \to \iint \frac{d^3r\,d^3p}{h^3} \tag{25.2}$$

In some systems the spectra of accessible states come in "bands" (Figure 25.4). Within each band the separation of neighboring states is extremely small, but the bands themselves are separated by rather large gaps, called "bands gaps."

$$\Delta\varepsilon_{\text{neighboring states within a band}} \ll kT$$
$$\Delta\varepsilon_{\text{band gap}} \gtrsim kT$$

When this happens, the summation over states within any one band can properly be replaced by continuous integration, but the region of integration must include the width of the band and no more.

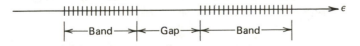

Figure 25.4   In some materials the accessible states come in "bands." Within any band, the states are closely spaced, but between bands are "gaps," having no accessible states at all.

Within any one band, the particles can often be treated as a gas, with the density of states near the band edge in particular being that of a gas. Within any one band, particles having no momentum would have the lowest total energies, so the relationship between energy and momentum would be

$$\varepsilon = \frac{p^2}{2m} + \varepsilon_0 \qquad (25.3)$$

where ε_0 is the energy of the edge of the band.

Systems whose energies come in such bands include the electrons of insulators and semiconductors. In these systems the lowest-energy band is called the "valence band" and the higher band is called the "conduction band." The valence band is normally full (or nearly full) of electrons, so there are no (or few) free states for an electron to go into (Figure 25.5). Consequently, most electrons in the valence band are not free and cannot contribute to the conduction properties of the material. Ordinarily, the conduction band is either empty, or has relatively few electrons in it, so any electrons in this band are free to enter many different states and can contribute to the conduction properties.

For many systems, the separations between all neighboring states are significant in comparison to kT, and so occupation numbers do not vary smoothly from one state to the next. In these cases, the discrete summation must be retained and cannot be replaced by continuous integration. Such systems include many bound states, such as particles in a square well, electrons bound to atoms and molecules, and harmonic oscillators of sufficiently high fundamental frequency that $\hbar\omega \gtrsim kT$.

Some systems are hybrids of the above. For example, electrons bound in atoms have discrete states accessible to them, but once free of the atoms, the electrons have a continuum of accessible states. Thus, in treating materials at sufficiently high temperatures that some fraction of the atoms would be ionized, the spectrum of states accessible to the electrons would have both discrete and continuous components present.

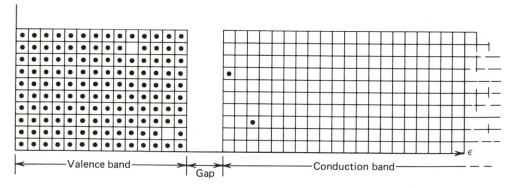

Figure 25.5 In insulators and semiconductors, the valence band is full, or nearly full. Consequently, valence electrons are not free. They cannot change states in response to external stimuli because there are no vacant neighboring states into which they can go. By contrast, the conduction band is empty, or nearly empty. Consequently, any electron in this band would have a great deal of freedom.

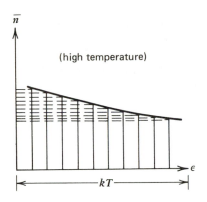
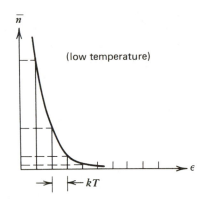

Figure 25.6 Plots of occupation number versus energy for a system of bosons at high and low temperatures. When kT is large compared to the spacing between states, the occupation number varies only slightly from one state to the next. At lower temperatures, such that kT is not large compared to the spacing between states, the occupation number varies discontinuously between neighboring states.

We have seen that whether there is a significant change in occupation number between neighboring states, depends on how the spacing between states compares with kT. At different temperatures, this comparison will be different. Consider, for example, the values of kT at the following two temperatures.

$$kT = 2.6 \times 10^{-2} \text{ eV} \qquad (T = 300 \text{ K})$$

$$kT = 8.6 \times 10^{-6} \text{ eV} \qquad (T = 0.1 \text{ K})$$

Suppose the separation between neighboring states for some system is around 10^{-4} eV. At room temperature, the occupation number would change only very slightly from one state to the next, but at low temperatures, the change would be quite pronounced. At room temperatures excited states would be quite accessible, but at low temperatures they would not! (See Figure 25.6.)

Consequently, there are some materials whose properties vary continuously at normal temperatures, but whose properties show discrete changes, or "quantum effects" when temperatures are lowered. Examples of such low-temperature quantum effects include the superconductivity displayed by some systems of conduction electrons, and superfluidity displayed by liquid helium.

SUMMARY

The spectra of states accessible to the particles of real physical systems are varied, but can be categorized according to general characteristics.

When the interaction between particles is weak, the separation between neighboring states is generally very small compared to kT,

$$\Delta \varepsilon_{\text{neighboring states}} \ll kT$$

the occupation number changes only slightly from one state to the next, and any sum over discrete states could properly be replaced by continuous integration. If the density of states is $g(\varepsilon)$, then the distribution of particles is given by the product of the distribution of states times the occupation number of each.

$$dN = [g(\varepsilon)\, d\varepsilon][\bar{n}(\varepsilon)] = \bar{n}(\varepsilon)g(\varepsilon)\, d\varepsilon \qquad (24.1)$$

If the interaction between particles is very weak, the system may be treated as a gas, and the sum over discrete states can be replaced by continuous integration through

$$\sum_{s} \rightarrow \iint \frac{d^3r\, d^3p}{h^3} \qquad (24.2)$$

Examples of quantum gases include normal gases, photons in an oven, vibrations in a solid, electrons in a conductor, and helium atoms in liquid helium.

The states of some systems come in bands. The spacings of neighboring levels within any one band are small compared to kT, but the spacing between the bands is not. The particles within any one band can often be treated as a quantum gas, providing appropriate limits are placed on the integration and an appropriate transformation of variables is made between energy and momentum. Such systems include the electrons in semiconductors and insulators. In these systems, the lower band is full or nearly full and is called the "valence band." The upper band is empty or nearly empty and is called the "conduction band." Electrons in the valence band are not free and those in the conduction band are.

For many systems, including particles trapped in local potential wells, the separation of states may not be small compared to kT. For these, there may be appreciable change in occupation number between neighboring states, so summation over discrete states may *not* be replaced by continuous integration. The spectra of states available to the particles of some systems are hybrids, including some regions of widely spaced states and other regions where they are closely spaced.

For some systems, the spacing between neighboring states is small compared to kT at room temperatures, but becomes large compared to kT at low temperatures attainable in our laboratories. At these low temperatures the occupation number would vary greatly from one state to the next, and so as we cool such a system down, some physical properties might display discrete changes, called low-temperature "quantum effects." Examples of these include superconductivity and superfluidity.

PROBLEMS

25-1.   Iron atoms have a mass of 9.3×10^{-26} kg, and are bound in place in solid cast iron by electrostatic forces with effective force constants of $k = 2$ N/m.

    (a)   What is the fundamental frequency of vibration for these iron atoms? ($\omega = \sqrt{k/m}$.)

    (b)   What is the excitation temperature?

    (c)   At a temperature of 50 K, is the separation of neighboring states large or small compared to kT?

25-2.   Consider a single conduction electron in a chunk of metal measuring 1 cm on a side.

(a) Use the fact that the area of a quantum state is given by $\Delta p_x \Delta x \approx h$ to determine the amount of momentum, Δp_x, separating neighboring quantum states.

(b) How much energy separates the ground state ($p = 0$) from the first excited state for this electron?

(c) What is the excitation temperature?

(d) At a temperature of 50 K, is the separation of neighboring states large or small compared to kT?

25-3. Consider a system such as electrons in a conduction band, which are confined to volume V and for which the energy and momentum are related through $\varepsilon = \varepsilon_0 + p^2/2m$, where ε_0 is some constant reference level, and m is the mass of a particle. Suppose this system can be treated as a gas.

(a) Start with Eq. 25.2 and integrate over all volume and all angles of the momentum to find an expression for the number of accessible states as a function of the magnitude of the momentum, m, and V. (*Hint.* Write $d^3p = p^2\,dp\sin\theta\,d\theta\,d\phi$, and integrate over all solid angles.)

(b) Convert p into ε to find an expression for the density of states in terms of m, V, ε, and ε_0.

(c) What is the distribution of particles, $dN/d\varepsilon$, for fermions in terms of m, V, ε, ε_0, and μ? (*Hint.* See Eq. 25.1.)

B. QUANTUM GASES

B.1 The Particle Distribution

For a gas, the number of quantum states within the six-dimensional element of phase space, $d^3r\,d^3p$, is given by

$$(\text{number of quantum states}) = \frac{d^3r\,d^3p}{h^3} \qquad (25.4)$$

and the occupation number for each of these states is given by

$$\bar{n} = \frac{1}{e^{\beta(\varepsilon - \mu)} \pm 1} \qquad \begin{cases} + \text{ for fermions} \\ - \text{ for bosons} \end{cases} \qquad (25.5)$$

The number of *particles* within this element of phase space is the product of the number of quantum states times the occupation number of each.

$$dN = \left(\frac{d^3r\,d^3p}{h^3}\right)\bar{n}(\varepsilon) \qquad (25.6)$$

This is the idea that will form the basis of our discussion in this section. First we perform some minor manipulations to put the result (25.6) into a more convenient form.

A system of one type of particle can often be separated into several subsystems, distinguishable by the z-component of their intrinsic angular momentum. A system of otherwise identical spin $\frac{1}{2}$ electrons, for example, can be separated into two distinguishable subsystems, one containing all spin-up ($S_z = +\frac{1}{2}\hbar$) electrons, and the other containing all spin-down ($S_z = -\frac{1}{2}\hbar$) electrons (Figure 25.7).

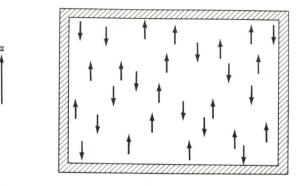

Figure 25.7 A system of otherwise identical spin $\frac{1}{2}$ electrons can be divided into two distinguishable subsystems according to the z-component of their intrinsic angular momentum.

Massive spin j particles* have $2j + 1$ possible orientations of the z-component of their intrinsic angular momenta. If particles are massless, moving at the speed of light, relativistic considerations force them to lose one of these degrees of freedom giving them only $2j$ possible orientations for their intrinsic angular momenta. If we let λ represent the number of different possible values of the z-component of intrinsic angular momentum, then for spin j particles,

$$\lambda = \begin{cases} 2j + 1 & \text{if massive} \\ 2j & \text{if massless} \end{cases}$$

Each system of spin j particles is actually composed of λ distinguishable subsystems.

The result (25.6) applies to the identical particles in any one subsystem, so if we add together the particles in each of the λ distinguishable subsystems, we have

$$dN = \lambda\left(\frac{d^3r\, d^3p}{h^3}\right)\bar{n}(\varepsilon)$$

Since the energy of interaction between gas particles is negligibly small, the energy of the state, ε, is independent of the position, r, and so the integration over d^3r can be done trivially.

$$dN = \lambda\left(\frac{V\, d^3p}{h^3}\right)\bar{n}(\varepsilon)$$

Similarly, the energy does not depend on the *direction* that the particle is moving, so we can write the differential element d^3p in spherical coordinates,

$$d^3p = p^2\, dp\, \sin\theta\, d\theta\, d\phi$$

and integrate over angles, giving

$$dN = \lambda\left(\frac{4\pi V p^2\, dp}{h^3}\right)n(\varepsilon) \tag{25.7}$$

---

* A "spin j particle" is one whose intrinsic angular momentum is $\sqrt{j(j + 1)}\hbar$, and whose z-component may take on the values $j\hbar, (j - 1)\hbar, \ldots, -j\hbar$.

The relationship between energy and momentum for gas particles of mass m is given by*

$$\varepsilon = \frac{p^2}{2m} \qquad \text{(nonrelativistic)}$$

$$\varepsilon = pc \qquad \text{(relativistic)}$$

so we can perform a transformation of variables to write the momentum in terms of energy.

$$p^2\, dp = \frac{1}{2}(2m)^{3/2}\sqrt{\varepsilon}\, d\varepsilon \qquad \text{(nonrelativistic)}$$

$$p^2\, dp = \frac{1}{c^3}\varepsilon^2\, d\varepsilon \qquad \text{(relativistic)}$$

With this substitution, the result (25.7) becomes

$$dN = \left[\frac{\lambda 2\pi V(2m)^{3/2}}{h^3}\right]\left(\frac{\sqrt{\varepsilon}\, d\varepsilon}{e^{\beta(\varepsilon - \mu)} \pm 1}\right) \qquad \text{(nonrelativistic)}$$

$$dN = \left[\frac{\lambda 4\pi V}{h^3 c^3}\right]\left(\frac{\varepsilon^2\, d\varepsilon}{e^{\beta(\varepsilon - \mu)} \pm 1}\right) \qquad \text{(relativistic)} \qquad (25.8)$$

These results express the number of particles as a function of energy for a quantum gas. We obtained it by simply multiplying the number of states by the occupation number of each, and then adding together the particles from the λ subsystems, distinguished by the z-component of the intrinsic angular momentum.

From the results (25.8) we can see how the number of particles per unit energy, $dN/d\varepsilon$, should vary with the particle energy in the low- and high-energy limit. In the low-energy limit, the exponential in the denominator is of little consequence, and

$$\frac{dN}{d\varepsilon} \propto \sqrt{\varepsilon} \qquad \text{(nonrelativistic)}$$

$$\frac{dN}{d\varepsilon} \propto \varepsilon^2 \qquad \text{(relativistic)}$$

In the high-energy limit the exponential in the denominator dominates and we have

$$\frac{dN}{d\varepsilon} \propto e^{-\beta\varepsilon} \qquad \text{(both nonrelativistic and relativistic)}$$

since an exponential function dominates any finite polynomial, for large values of the variable.

The behavior of the number of particles per unit energy, $dN/d\varepsilon$, for all energies can be qualitatively understood by comparing plots of number of states ($\propto \sqrt{\varepsilon}$, or ε^2) and occupation number, $\bar{n}(\varepsilon)$, as is done in Figures 25.8 through 25.11 for fermion and boson

---

* Massless particles are always relativistic.

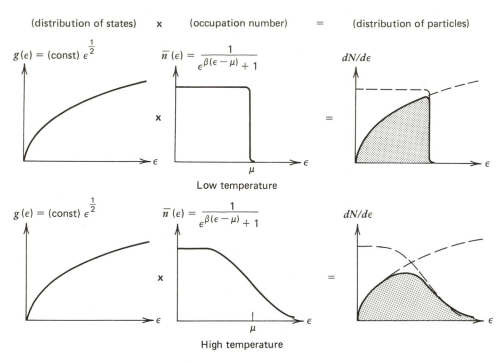

Figure 25.8 Distribution of particles in a nonrelativistic fermion gas at low and high temperatures.

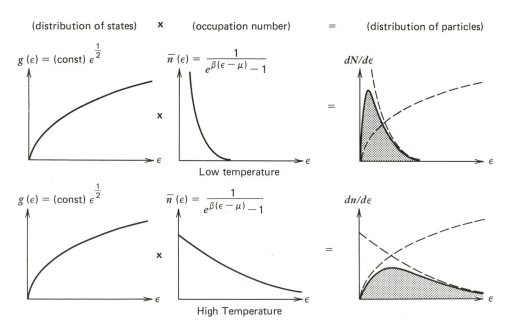

Figure 25.9 Distribution of particles in a nonrelativistic boson gas at low and high temperatures.

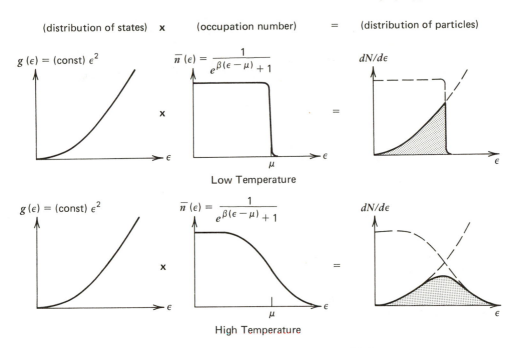

Figure 25.10 Distribution of particles in a relativistic fermion gas at low and high temperatures.

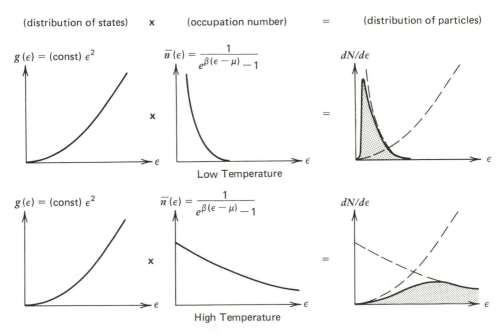

Figure 25.11 Distribution of particles in a relativistic boson gas at low and high temperatures.

gases, in the high- and low-temperature extremes, and in the relativistic and nonrelativistic limits.

The amount of the total internal energy of the system that is held by particles whose energies lie in the range $d\varepsilon$, is the product of the number of such particles times the energy of each.

$$dE = \varepsilon \, dN$$

Using the results (25.8), this becomes

$$dE = \left[\frac{\lambda 2\pi V(2m)^{3/2}}{h^3}\right] \frac{\varepsilon^{3/2} \, d\varepsilon}{e^{\beta(\varepsilon - \mu)} \pm 1} \qquad \text{(nonrelativistic)}$$

$$dE = \left[\frac{\lambda 4\pi V}{h^3 c^3}\right] \frac{\varepsilon^3 \, d\varepsilon}{e^{\beta(\varepsilon - \mu)} \pm 1} \qquad \text{(relativistic)} \qquad (25.9)$$

The total internal energy of the system is obtained by integrating these.

SUMMARY

The distribution of particles is given by the product of the distribution of states times the average occupation number of each. For a gas, the distribution of states is given by

$$(\text{number of quantum states}) = \frac{d^3 r \, d^3 p}{h^3}$$

so the distribution of particles is given by

$$dN = \frac{d^3 r \, d^3 p}{h^3} \, \bar{n}(\varepsilon)$$

If we integrate over all volume and all momentum directions, and sum over all possible orientations of the intrinsic angular momentum of the spin j particles, this becomes

$$dN = \lambda \left(\frac{4\pi V p^2 \, dp}{h^3}\right) \bar{n}(\varepsilon)$$

where

$$\lambda = \begin{cases} 2j + 1 & \text{if massive particles} \\ 2j & \text{if massless particles} \end{cases}$$

The relationship between energy (ε) and momentum (p) for particles in a gas transforms this into

$$dN = \left[\frac{\lambda 2\pi V(2m)^{3/2}}{h^3}\right] \frac{\sqrt{\varepsilon} \, d\varepsilon}{e^{\beta(\varepsilon - \mu)} \pm 1} \qquad \text{(nonrelativistic)}$$

$$dN = \left(\frac{\lambda 4\pi V}{h^3 c^3}\right) \frac{\varepsilon^2 \, d\varepsilon}{e^{\beta(\varepsilon - \mu)} \pm 1} \qquad \text{(relativistic)} \qquad (25.8)$$

The distribution of internal energy is the product of the particle distribution times the energy of each.

$$dE = \varepsilon \, dN$$

PROBLEMS

25-4.   A neutrino is a massless spin $\frac{1}{2}$ fermion. How many different orientations of its intrinsic angular momentum can a neutrino have?

25-5.   The density of states for a nonrelativistic gas can be written as $g(\varepsilon) = C\sqrt{\varepsilon}$. Estimate the value of the constant C for:

   (a)   Air molecules in a room of volume 30 m³.
   (b)   Electrons in a metal of volume 10^{-5} m³.

25-6.   Consider the distribution of *particles*, $dN/d\varepsilon$, among those states in Problem 25-5. For particles of energy $\varepsilon - \mu = kT$, where T is room temperature, estimate $dN/d\varepsilon$ (in units of eV^{-1}) for:

   (a)   Air molecules in a room of volume 30 m³ (assume no intrinsic angular momentum).
   (b)   Electrons in a metal of volume 10^{-5} m³.

B.2   Internal Energy and the Gas Laws

In our derivation of the equipartition theorem, we calculated the average amount of energy stored by a single particle in any one degree of freedom. We found it to be given by*

$$\text{(average energy per degree of freedom)} = \frac{1}{2} kT \qquad (25.10)$$

Since a gas of N particles would have $3N$ degrees of freedom, the internal energy of such a gas would be

$$E = \frac{3}{2} NkT \qquad (25.11)$$

In deriving the result (25.10), we focused on one particle and one degree of freedom only, ignoring all others. We assumed that neither the states accessible to a particle, nor the counting of such states were affected at all by the presence of the others. In the preceding chapter we saw that these two assumptions are wrong for systems of identical fermions and identical bosons, respectively, and these two errors cause differences in the predictions of classical, Fermi-Dirac, and Bose-Einstein statistics, that are particularly pronounced when particle densities are high.

In the case of fermions, the exclusion of particles from states already occupied means that many fermions are forced into higher levels than they would otherwise prefer, because the more desired lower levels are already occupied by others. In the case of bosons, there are actually more configurations of multiple occupancy than classical statistics predicts, so there is a larger probability of finding bosons in low-level, multiply occupied states. (See Figure 25.12.)

Since relatively more fermions are in higher-lying states than the classical prediction, the total internal energy of a system of identical fermions is greater than the classical result.

$$E_{\text{fermions}} > \frac{3}{2} NkT$$

---

* We consider nonrelativistic particles. For the relativistic case, see homework Problem 25-7.

dN/dε

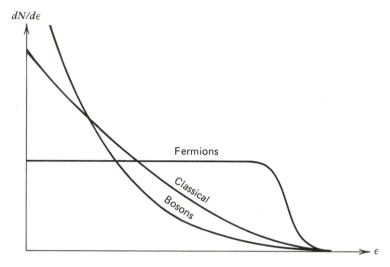

Figure 25.12 Plots of particle distributions for systems of bosons, "classical" particles, and fermions. For illustrative purposes, the density of states is taken to be constant in each case. Each system is at the same temperature and has the same number of particles in it. Notice that compared to the classical prediction, the distribution of bosons is skewed toward lower-lying states, and the distribution of fermions is skewed toward higher-lying states.

Similarly, since relatively more bosons are in lower-lying states than the classical prediction, we expect the total internal energy of a system of identical bosons to be correspondingly lower.

$$E_{\text{bosons}} < \frac{3}{2} NkT$$

Quantitative expressions for the internal energy and number of particles for a quantum gas may be obtained by integrating the expressions (25.8) and (25.9). Because of the similarity between these sets of expressions, we can integrate one by parts to express it in terms of the other. This is done in Appendix 25A with the result that for a quantum gas, the internal energy is given by

$$E = \frac{3}{2} NkT \pm \delta \quad \begin{cases} + \text{ for fermions} \\ - \text{ for bosons} \end{cases} \tag{25.12}$$

where δ is a complicated but positive function of (μ, T), becoming larger when particle densities are higher. Thus, the actual calculations give us the expected result for the differences in the internal energies for systems of "classical particles," fermions, and bosons, and these differences become more pronounced when particle densities are higher.

Similar considerations apply to the pressure exerted by a gas on its container. Particles of higher energies carry more momentum and exert greater pressure. It is easy to show (see Appendix 25B) that the pressure increases with translational kinetic energy according to

$$pV = \frac{2}{3} E_k \qquad (25.13)$$

Using $E_k = \frac{3}{2} NkT$, this gives

$$pV = NkT$$

Since fermions tend to be in higher-lying states than the classical prediction, and since bosons tend to be in lower-lying states than the classical prediction, we expect corresponding differences in the pressures exerted by these gases.

$$p_{\text{fermions}} > p_{\text{cl}}$$

$$p_{\text{bosons}} < p_{\text{cl}}$$

Indeed, putting our result (25.12) for the internal energies into the general relationship (25.13), we have

$$pV = NkT \pm \frac{2}{3} \delta \qquad \begin{cases} + \text{ for fermions} \\ - \text{ for bosons} \end{cases}$$

where δ is a positive function of (μ, T), becoming larger when particle densities are higher. (See Appendix 25A for the derivation of δ.)

SUMMARY

A fermion system tends to have more energy than that predicted classically, because some of the fermions are forced into higher levels when the lower preferred levels are already occupied. A boson system tends to have less energy than the classical prediction because there are actually more arrangements of the system having low-level, multiply occupied states than given by classical theory. If each allowed arrangement has equal probability, then boson systems have higher probability of being found in these low-energy configurations than predicted by classical statistics. Indeed, we find the internal energy of a quantum gas given by

$$E = \frac{3}{2} NkT \pm \delta \qquad \begin{cases} + \text{ for fermions} \\ - \text{ for bosons} \end{cases}$$

where δ is a positive function of (μ, T), becoming larger for higher particle densities, where differences between quantum and classical statistics are more pronounced.

Since the pressure exerted by the gas is proportional to the energy of its particles, we also find the pressure exerted by fermions to be larger than the classical prediction, and by bosons to be smaller than the classical prediction.

PROBLEMS

25-7.   For a relativistic quantum gas, the energy per degree of freedom is $1kT$. How would this change the result (25.12)? (The answer is in Appendix 25A.)

C.   OTHER QUANTUM SYSTEMS

There are many quantum systems for which the spacings between neighboring states are small, but for which interactions among the particles cause the distribution of these states to be significantly different than that for a gas. For making quantitative calculations the density of states is of central importance, because it determines the particle distribution through Eq. 25.1. For these systems, intensive experimental and theoretical research goes into determining the densities of states. The two usually combine in such a way that the experimental data points determine the type of theoretical model that is most appropriate, and these theoretical models, in turn, help us extrapolate our knowledge into the areas between or beyond the data points.

When the spacings between neighboring states are *not* small compared to kT, then we cannot replace sums over states by continuous integration. When the temperature is sufficiently low that the system is confined to the configuration of lowest possible energy, the system is said to be "degenerate." A degenerate system of bosons would have all N particles in the single lowest state, and a degenerate fermion system would have the lowest N states filled, and all higher states empty, since there can be only one fermion per state.

For some systems, the spacings between neighboring states are small compared to kT at room temperature, but large compared to kT for low temperatures attainable in our low-temperature laboratories. At the lower temperatures these systems become degenerate, because there is negligible probability of excitation to even the first excited state. These systems display some strange behaviors because the particles are not able to change states in response to external stimuli as are most system of our experience.

The total number of particles in the system is given by the sum over all states of the occupation number of each.

$$N = \sum_s \left[\frac{1}{e^{\beta(\varepsilon_s - \mu)} \pm 1} \right] \tag{25.14}$$

As we have seen, this determines the chemical potential, μ, as a function of the temperature and number of particles for that system (Figure 25.13).

$$\mu = \mu(T, N)$$

EXAMPLE

What is the chemical potential for a degenerate system of N bosons at temperature T, if the energy of the lowest state is 0 eV?

If it is degenerate, then all particles are in the lowest state, and we need only one term in the sum (25.21)

$$N = \frac{1}{e^{\beta(0 - \mu)} - 1}$$

This gives

$$\mu = -kT \ln\left(1 + \frac{1}{N}\right) \approx -\frac{kT}{N} \quad \text{(for large } N\text{)}$$

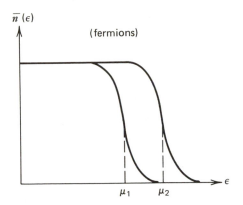
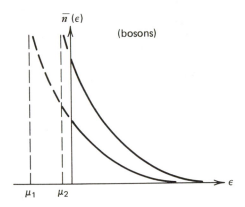

Figure 25.13 Plots of occupation number versus energy for two different values of the chemical potential for both fermions and bosons. In both cases, higher chemical potentials mean larger occupation numbers for the accessible states, and therefore more particles altogether. Therefore, for any given system at any given temperature, the chemical potential is determined by the total number of particles in the system.

Notice that this system would only be degenerate for temperatures sufficiently low that kT is small compared to the energy of the first excited state.

SUMMARY

For many systems having closely spaced states, determination of the density of states, $g(\varepsilon)$, is the subject of great experimental and theoretical interest. It determines the distribution of particles, which determine the properties of the system.

A "degenerate" system is one confined to the configuration of lowest possible energy. Some common systems become degenerate at low temperatures and display strange physical properties, since the particles are no longer free to change states in response to external stimuli.

The chemical potential is determined as a function of (T, N) by the condition that the sum of the occupation numbers of all states must equal the total number of particles in the system.

$$N = \sum_s \left[\frac{1}{e^{\beta(\varepsilon_s - \mu)} \pm 1} \right] \tag{25.21}$$

APPENDIX 25A   THE GAS LAW FOR QUANTUM GASES

Nonrelativistic Gases

According to the results (25.8) and (25.9) the number of particles of a nonrelativistic quantum gas whose energies lie in the range $d\varepsilon$ is given by

$$dN = C \frac{\sqrt{\varepsilon}\, d\varepsilon}{e^{\beta(\varepsilon - \mu)} \pm 1} \tag{25A.1}$$

and their contribution to the internal energy of the system is

$$dE = C \frac{\varepsilon^{3/2} \, d\varepsilon}{e^{\beta(\varepsilon - \mu)} \pm 1} \tag{25A.2}$$

where the constant C is given by

$$C = \left[\frac{\lambda 2\pi V (2m)^{3/2}}{h^3} \right] \tag{25A.3}$$

Integrating Eq. 25A.1 by parts, we have

$$N = C \int_0^\infty \frac{\varepsilon^{1/2} \, d\varepsilon}{e^{\beta(\varepsilon - \mu)} \pm 1} = \frac{2}{3} C \left. \frac{\varepsilon^{3/2}}{e^{\beta(\varepsilon - \mu)} \pm 1} \right|_0^\infty + \frac{2}{3} C \int_0^\infty \frac{\varepsilon^{3/2} e^{\beta(\varepsilon - \mu)} \beta \, d\varepsilon}{(e^{\beta(\varepsilon - \mu)} \pm 1)^2}$$

Evaluating the first term at both limits gives zero, and the second term can be split into two separate integrals by the substitution

$$\frac{e^{\beta(\varepsilon - \mu)}}{(e^{\beta(\varepsilon - \mu)} \pm 1)^2} = \frac{1}{e^{\beta(\varepsilon - \mu)} \pm 1} \left[1 \mp \frac{1}{e^{\beta(\varepsilon - \mu)} \pm 1} \right] \tag{25A.4}$$

This gives

$$N = \frac{2}{3} C\beta \int_0^\infty \frac{\varepsilon^{3/2} \, d\varepsilon}{e^{\beta(\varepsilon - \mu)} \pm 1} \mp \frac{2}{3} C\beta \int_0^\infty \frac{\varepsilon^{3/2} \, d\varepsilon}{(e^{\beta(\varepsilon - \mu)} \pm 1)^2}$$

or

$$\frac{3}{2} NkT = C \int \frac{\varepsilon^{3/2} \, d\varepsilon}{e^{\beta(\varepsilon - \mu)} \pm 1} \mp C \int_0^\infty \frac{\varepsilon^{3/2} \, d\varepsilon}{(e^{\beta(\varepsilon - \mu)} \pm 1)^2}$$

By comparing with Eq. (25A.2), we see that the first of the two terms on the right is just the total internal energy of the system, so this result can be written

$$\frac{3}{2} NkT = E \mp \delta$$

or

$$E = \frac{3}{2} NkT \pm \delta \qquad \begin{cases} + \text{ for fermions} \\ - \text{ for bosons} \end{cases} \tag{25A.5}$$

where δ is a positive function of μ and T, given by

$$\delta = \left[\frac{\lambda 2\pi V (2m)^{3/2}}{h^3} \right] \int_0^\infty \frac{\varepsilon^{3/2} \, d\varepsilon}{(e^{\beta(\varepsilon - \mu)} \pm 1)^2} = C \int_0^\infty \bar{n}^2(\varepsilon) \varepsilon^{3/2} \, d\varepsilon \tag{25A.6}$$

Notice that the correction term, δ, is small if the occupation number $\bar{n}(\varepsilon)$ is small, but becomes increasingly important as the range of energies where $\bar{n}(\varepsilon)$ is significant widens (i.e., as particle densities increase).

Relativistic Gases

According to the results (25.8) and (25.9) the number of particles of a relativistic quantum gas whose energies lie in the range $d\varepsilon$ is given by

$$dN = C' \frac{\varepsilon^2 \, d\varepsilon}{e^{\beta(\varepsilon - \mu)} \pm 1} \tag{25A.7}$$

and their contribution to the total internal energy of the system is

$$dE = C' \frac{\varepsilon^3 \, d\varepsilon}{e^{\beta(\varepsilon - \mu)} \pm 1} \tag{25A.8}$$

where the constant C' is given by

$$C' = \left(\frac{\lambda 4\pi V}{h^3 c^3} \right) \tag{25A.9}$$

As we did for the nonrelativistic case, we can integrate the number of particles (dN) by parts, and split the resulting integral into two, using the substitution (25A.4). The result is

$$3NkT = E \mp \delta'$$

or

$$E = 3NkT \pm \delta' \quad \begin{cases} + \text{ for fermions} \\ - \text{ for bosons} \end{cases} \tag{25A.10}$$

where δ' is a positive function of μ and T given by

$$\delta' = \left(\frac{\lambda 4\pi V}{h^3 c^3} \right) \int_0^\infty \frac{\varepsilon^3 \, d\varepsilon}{[e^{\beta(\varepsilon - \mu)} \pm 1]^2} = C' \int_0^\infty \bar{n}^2(\varepsilon)\varepsilon^3 \, d\varepsilon \tag{25A.11}$$

Again, notice the correction term δ' is positive, and becomes more pronounced when occupation numbers (i.e., particle densities) are larger.

APPENDIX 25B   PRESSURE EXERTED BY A GAS

Suppose dN represents the number of particles having an x-component of their velocities lying in the range between v_x and $v_x + dv_x$. The number of particles crossing a unit area per unit time is called the particle "flux" and is equal to the product of particle density times their velocity, as illustrated in Figure 18.5.

The density of particles whose velocities lie in the range between v_x and $v_x + dv_x$ is given by dN/V, and so the flux of these particles is

$$df_x = v_x \frac{dN}{V}$$

When each of these particles collides with a wall in the y-z plane, it reverses its direction of motion, so the momentum transfer (or impulse delivered) is $2mv_x$, where m is the mass of the particle. The total momentum transfer per unit time per unit area is the force per unit area, or the pressure, and is given by the product of the particle flux times the momentum transferred by each. If dp is the pressure exerted by those particles having x-velocities in the range between v_x and $v_x + dv_x$, then

$$dp = 2mv_x \, df_x = 2mv_x^2 \frac{dN}{V}$$

or

$$V \, dp = 2mv_x^2 \, dN$$

To find the total pressure, we must integrate over all $v_x > 0$. (Those particles with $v_x < 0$ are going the wrong way, and won't strike the wall.)

$$pV = 2m \int_{v_x=0}^{\infty} v_x^2 \, dN$$

Since integrand is symmetric across $v_x = 0$, we can double the region of integration, and divide the result by 2.

$$pV = m \int_{-\infty}^{\infty} v_x^2 \, dN$$

$$= Nm \int_{-\infty}^{\infty} v_x^2 \, \frac{dN}{N}$$

$$= Nm\overline{v_x^2}$$

Since

$$\overline{v_x^2} = \overline{v_y^2} = \overline{v_z^2}$$

and

$$\overline{v_x^2} + \overline{v_y^2} + \overline{v_z^2} = \overline{v^2}$$

we have

$$\overline{v_x^2} = \frac{1}{3} \overline{v^2}$$

and

$$pV = N \frac{1}{3} m\overline{v^2} = N \frac{2}{3} \left(\frac{1}{2} m\overline{v^2} \right) = \frac{2}{3} N\bar{\varepsilon}_k$$

where $\bar{\varepsilon}_k$ is the average kinetic energy per particle. The total energy of the gas is the product of the number of particles times the average kinetic energy of each ($N\bar{\varepsilon}_k$), so this becomes

$$pV = \frac{2}{3} E_k \tag{25B.1}$$

chapter 26

BLACKBODY RADIATION

From our common experiences with ovens we know that the intensity of the electromagnetic radiation within an oven depends on the temperature of the oven walls. We can use quantum statistics to see just how the energy intensity and distribution depends on the oven's temperature. Do you suppose it should also be dependent on any other characteristic of the oven, such as material, texture, or color of the walls?

One important and rather simple application of quantum statistics is the study of the electromagnetic radiation emitted and absorbed by objects. This is, of course, one of the major mechanisms for the transport of energy, and includes the transfer of life-giving solar energy to our planet.

We begin with a study of the electromagnetic radiation contained within an oven held at some temperature T. We will find that the properties of the electromagnetic radiation within the oven depend only on its temperature and not at all on the nature of the oven walls. This gives us a great deal of insight into the absorption and emission of electromagnetic radiation by material objects.

A. PHOTONS IN AN OVEN

We consider the electromagnetic radiation within the oven to be a photon gas* (Figure 26.1). The distribution of states in a gas is given by

$$(\text{number of states}) = \frac{d^3r \, d^3p}{h^3}$$

The distribution of particles is the product of the distribution of states times the occupation number of each.

$$dN = \left(\frac{d^3r \, d^3p}{h^3}\right) \bar{n}(\varepsilon)$$

---

* For a more complete discussion of quantum gases, see Section B, Chapter 25. In this paragraph we review some of the developments of that section.

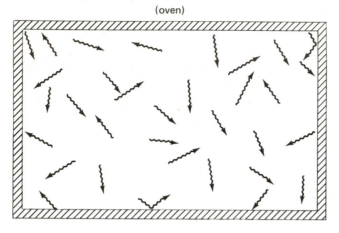

(oven)

Figure 26.1   We consider the electromagnetic radiation inside an oven
to be a photon gas, which is in equilibrium with the oven walls.

If we integrate over volume and momentum directions, and sum over the λ different
possible orientations of the intrinsic angular momentum, this becomes

$$dN = \lambda \left(\frac{4\pi V p^2 \, dp}{h^3} \right) \bar{n}(\varepsilon)$$

A photon is a relativistic spin 1 particle, which means there are only two possible
orientations of its intrinsic angular momentum ($\lambda = 2$) and that its energy and momentum are related through

$$\varepsilon = pc$$

With these substitutions, the distribution of photons in the gas is given by (Figure 26.2)

$$dN = \left(\frac{8\pi V}{h^3 c^3} \varepsilon^2 \, d\varepsilon \right) \bar{n}(\varepsilon) \tag{26.1}$$

The distribution of photon *energies* within the gas is the product of the distribution of
photons times the energy of each.

$$dE = \varepsilon \, dN = \left(\frac{8\pi V}{h^3 c^3} \varepsilon^3 \, d\varepsilon \right) \bar{n}(\varepsilon) \tag{26.2}$$

We usually prefer to speak of "energy density" rather than total energy, because it is
clear to us that the total energy in a gas will be directly proportional to its volume. That
is, the energy in the photon gas depends on the size of the oven, but the energy density
doesn't. If we let u represent energy density,

$$u = \frac{E}{V} = \text{energy density}$$

(distribution of states)   **x**   (occupation number)   **=**   (particle distribution)

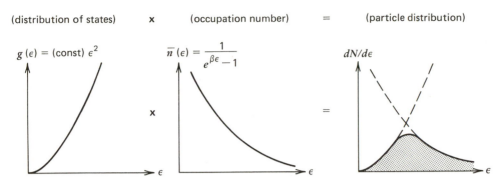

Figure 26.2 The distribution of photons in energy is given by the product of the density of states times the occupation number.

then the distribution of energies (26.2) is given by

$$du = \left(\frac{8\pi}{h^3 c^3} \, \varepsilon^3 \, d\varepsilon \right) \bar{n}(\varepsilon) \tag{26.3}$$

Photons have the convenient property that they do not interact at all with each other. Therefore, their chemical potential is zero.

$$\mu_\gamma = 0$$

Since it is a spin 1 boson with no chemical potential, the occupation number for a photon quantum state is

$$\bar{n}(\varepsilon) = \frac{1}{e^{\beta \varepsilon} - 1}$$

Putting this expression for the occupation number into Eq. 26.3 gives the following distribution of photon energies within a photon gas.

$$du = \left(\frac{8\pi}{h^3 c^3}\right) \frac{\varepsilon^3 \, d\varepsilon}{e^{\beta \varepsilon} - 1} \tag{26.4}$$

The energy carried by most kinds of waves with which we are familiar depends on many things, such as the wave amplitude, frequency, and the length and breadth of the wave train. For some reason that we still don't understand, the energy of a photon depends only on its frequency, ω, and nothing else, being given by

$$\varepsilon = \hbar \omega$$

where $\hbar$ is Planck's constant. As you can guess, this discovery came as quite a shock to early twentieth-century physicists who were well versed in the behavior of ordinary waves.

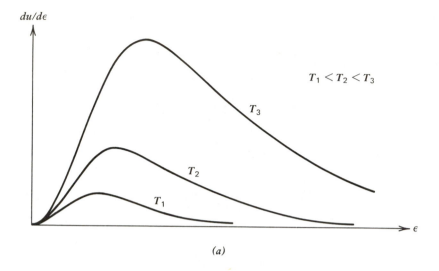

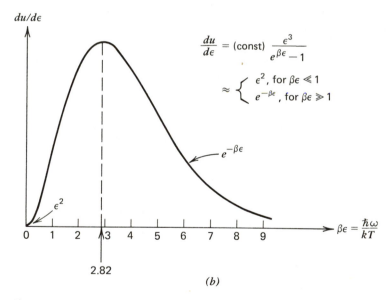

$$\frac{du}{d\epsilon} = (\text{const}) \frac{\epsilon^3}{e^{\beta\epsilon} - 1}$$

$$\approx \begin{cases} \epsilon^2, \text{ for } \beta\epsilon \ll 1 \\ e^{-\beta\epsilon}, \text{ for } \beta\epsilon \gg 1 \end{cases}$$

Figure 26.3 (a) Energy distribution inside a photon gas. At higher temperatures the distribution peaks at higher energies, and more energy is present at all photon wavelengths. (b) Plot of $du/d\epsilon$ (arbitrary units) versus $\beta\epsilon = (\hbar\omega/kT)$. The distribution peaks at $\epsilon = 2.82kT$. At low energies it increases as ϵ^2, and at high energies it decreases as $e^{-\beta\epsilon}$.

Through this relationship, the expression (26.4) above can be written as a distribution in photon frequencies instead of photon energies.

$$du = \left(\frac{h}{\pi^2 c^3}\right) \frac{\omega^3 \, d\omega}{e^{\beta\hbar\omega} - 1} \tag{26.5}$$

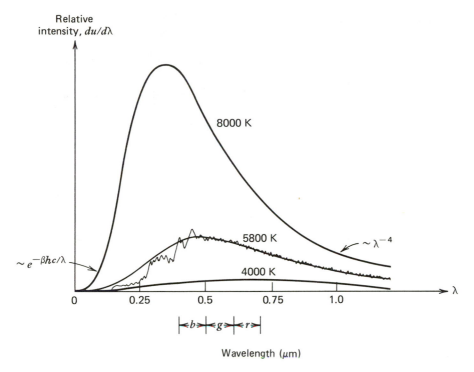

Figure 26.4 Plot of the relative intensities of radiation emitted by blackbodies of various temperatures as a function of photon wavelengths. The red (*r*), green (*g*), and blue (*b*) portions of the visible spectrum are indicated. The wiggly line is the sun's spectrum, which approximates that of a blackbody at 5800 K.

Since frequency and wavelength are related through

$$\omega = \frac{2\pi c}{\lambda}$$

this distribution can also be expressed in terms of photon wavelength.*

$$du = (8\pi hc)\,\frac{\lambda^{-5}\,d\lambda}{e^{\beta hc/\lambda} - 1} \tag{26.6}$$

These results [(26.5), (26.6)] express the distribution of photon energies, or the "energy spectrum," of a photon gas in terms of the angular frequency, and wavelength, respectively. The quantities $du/d\omega$ and $du/d\lambda$ represent the energy per unit of angular frequency, and energy per unit wavelength, respectively, and are plotted in Figures 26.3 and 26.4.

---

* We ignore the minus sign, which simply means increasing frequency means decreasing wavelength, and vice versa.

The distribution (26.4) for $du/d\varepsilon$ peaks at

$$\varepsilon_{max} = 2.82kT$$

as you will show in homework Problem 26-3. This means that at higher temperatures, the average photon energy is greater.

The total energy density for photons of all energies is obtained by integrating Eq. 26.4 over all energy.

$$u = \int du = \left(\frac{8\pi}{h^3 c^3}\right) \int_0^\infty \frac{\varepsilon^3 \, d\varepsilon}{e^{\beta\varepsilon} - 1}$$

By making the substitution $x = \beta\varepsilon$, this integral becomes

$$u = \frac{8\pi(kT)^4}{h^3 c^3} \int_0^\infty \frac{x^3 \, dx}{e^x - 1}$$

This is a standard integral, having the value $\pi^4/15$, so the total energy density is

$$u = aT^4 \qquad (26.7)$$

where the value of the constant a is

$$a = \left(\frac{8\pi^5 k^4}{15 h^3 c^3}\right) = 7.56 \times 10^{-16} \text{ J/m}^3 \cdot \text{K}^4 \qquad (26.8)$$

B. PRINCIPLE OF DETAILED BALANCE*

The interesting thing about the energy spectrum of this photon gas is that it does not depend at all on the nature of the oven. Given the temperature of the oven, T, the energy spectrum is uniquely determined. It doesn't depend, for instance, on whether the walls of the oven are smooth or rough, shiny or black, made of marble or wood, or whether they are red, green, or purple.

You would think, for example, that if you painted the walls of the oven green, then the photon energy spectrum would tend to be enhanced in the greens and depleted in the other wavelengths, since the green walls would reflect the greens and absorb the other colors. But this does not happen. Our results show that the photon energy spectrum does not depend on any property of the oven walls, not even on their color!

The conclusion drawn from this observation is that when the walls of the oven are in equilibrium with the photon gas (i.e., both have the same temperature), then those wavelengths it absorbs more strongly must also be emitted more strongly. At each wavelength, the intensity of radiation absorbed by any material in thermal equilibrium with the photon gas in the oven must be equal to the intensity emitted, since it can have no net effect on the photon spectrum. This is called the "principle of detailed balance."

For example, if the oven walls are green, then they reflect more of the greens and absorb more of the other colors. Therefore they must also emit less of the greens and more of the other colors in order to have no net effect on the photon energy distribution (Figure 26.5).

---

* There are several different "principles of detailed balance" in different scientific fields. Here we investigate the one dealing with photon radiation and absorption.

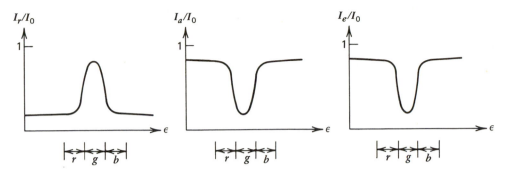

Figure 26.5 Plots of fraction of incident photons reflected (I_r/I_0), absorbed (I_a/I_0), and emitted (I_e/I_0) as a function of the photon energies for photons incident on a green object (i.e., one that would appear green when illuminated by white light). Red, green, and blue portions of the spectrum are indicated. All incident radiation is either reflected or absorbed; $I_r + I_a = I_0$. Furthermore, if the object is in thermal equilibrium with the incident photons, then it can have no effect on the photon distribution. This means that the various kinds of photons must be emitted in the same proportion that they are absorbed; $I_e = I_a$.

Since the rate of photon emission at each wavelength is exactly equal to the rate of absorption, if the walls of an oven could absorb every single incident photon, then they would have to reemit photons with the same distribution in energies as those absorbed. The spectrum of energies incident on (and therefore absorbed by) the walls of such an oven is that of the photon gas of Eq. 26.4, and so this must also be the spectrum of energies reemitted (see Figure 26.6).

A material that absorbs all incident photons would appear black when visible light shines on them, so perfect absorbers are sometimes called "blackbodies." According to the arguments of the above paragraph, the spectrum of photons emitted by a

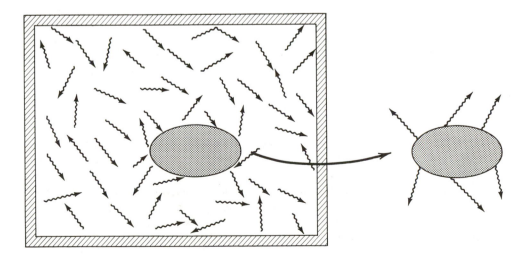

Figure 26.6 A perfect backbody absorbs all incident radiation. When it is in equilibrium with the photons in an oven, the spectrum of photons it emits must be identical to the spectrum of photons it absorbs—the spectrum of photons in the oven. Therefore, the spectrum of photons emitted by a blackbody at temperature T, must be identical to the spectrum of photons inside an oven of that temperature.

Relative
intensity

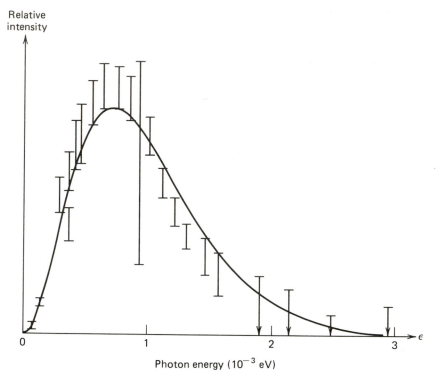

Photon energy (10^{-3} eV)

Figure 26.7 Distribution of photons permeating the universe that appear to be remnants of the primordial "fireball" or "big bang." The solid curve is the best fit to the experimental data points for a blackbody curve, and corresponds to a temperature of 2.96 K. (From *Physics Today*, June 1979.)

blackbody at temperature T must be identical to the spectrum of photons inside an oven at that temperature. For this reason, the distribution of energies in a photon gas as given in Eqs. 26.4 or 26.5 is commonly referred to as "blackbody radiation."

C. ENERGY FLUX

The rate at which energy crosses a unit area from one side is called the energy "flux" in that direction. The energy flux of the photon gas within an oven would be equal to rate at which energy is incident on a unit area of the walls of the oven, and is therefore equal to the flux of energy leaving the surface of a perfectly black body at temperature T.

Consider the energy flux in the $+z$-direction, for example. All the photons are moving at the speed of light, but only half of these are moving toward the $+z$-direction, so only half of the total energy density is attributable to photons moving toward this direction.

$$u_{+z} = \frac{1}{2} u$$

For these, the average z-component of their velocities is given by averaging $v_z = c \cos \theta$ (spherical coordinates) over the $+z$-hemisphere (Figure 19.1).

$$\overline{v_z} = \frac{1}{2\pi} \int_0^{2\pi} d\phi \int_0^{\pi/2} \sin \theta \, d\theta \, (c \cos \theta) = \frac{1}{2} c$$

The energy flux is the product of the energy per unit volume times the number of unit volumes crossing a unit area per time (Figure 18.5), or equivalently, the product of the energy density times the average z-velocity for these photons.

$$\text{flux} = \left(\frac{1}{2} u\right)\left(\frac{1}{2} c\right) = \frac{1}{4} uc$$

Combining this with our expression (26.7) for the energy density, u, gives

$$\text{flux} = \sigma T^4 \tag{26.9}$$

where σ is called the "Stephan-Boltzmann constant," and is given by

$$\sigma = \left(\frac{2\pi^5 k^4}{15h^3 c^2}\right) = 5.67 \times 10^{-8} \text{ J/m}^2 \cdot \text{s} \cdot \text{K}^4 \tag{26.10}$$

EXAMPLE

The surface of the sun acts like a blackbody of temperature 5800 K. What is the rate at which energy leaves each square meter of the sun's surface?

Putting $T = 5800$ K into Eq. 26.9 the energy flux from the sun's surface amounts to

$$\text{flux} = \sigma T^4 = 6.4 \times 10^7 \text{ J/s} \cdot \text{m}^2$$
$$= 6.4 \times 10^4 \text{ kW/m}^2$$

In homework Problem 26-14 you will show that by the time this energy reaches the position of the earth in space it has spread out to the point that the flux is only 1.4 kW/m².

EXAMPLE

If you consider the inside of the sun to be a photon gas of constant temperature 3×10^6 K, how much energy does this gas contain if the volume of the sun is 1.4×10^{27} m³?

The total energy is the product of the energy density times the volume. With a temperature of 3×10^6 K, Eq. 26.7 gives the energy density as

$$u = (7.56 \times 10^{-16} \text{ J/m}^3 \cdot \text{K}^4)(3 \times 10^6 \text{ K})^4$$
$$= 6.1 \times 10^{10} \text{ J/m}^3$$

Multiplying this by the sun's volume gives the total energy in the photon gas inside the sun as

$$E = 8.6 \times 10^{37} \text{ J}$$

SUMMARY

The electromagnetic radiation within an oven can be treated as a photon gas. Photons are spin 1 massless particles, having two possible orientations of their intrinsic angular momentum, no chemical potential, and whose energy and momentum are related by $\varepsilon = pc$. With this information we can write the distribution of particles in this gas (the product of the number of states times the average occupation number of each) as

$$dN = \frac{8\pi V}{h^3 c^3} \frac{\varepsilon^2 \, d\varepsilon}{e^{\beta\varepsilon} - 1} \tag{26.1}$$

and the distribution of energy in the gas as

$$du = \left(\frac{8\pi}{h^3 c^3}\right) \frac{\varepsilon^3 \, d\varepsilon}{e^{\beta\varepsilon} - 1} \tag{26.4}$$

where u is the energy density. The energy, angular frequency, and wavelength of a photon are related through

$$\varepsilon = \hbar\omega = h\frac{c}{\lambda}$$

which enables us to express the above energy distribution in terms of these other variables.

$$du = \left(\frac{\hbar}{\pi^2 c^3}\right) \frac{\omega^3 \, d\omega}{e^{\beta\hbar\omega} - 1} \tag{26.5}$$

$$du = (8\pi hc) \frac{\lambda^{-5} \, d\lambda}{e^{\beta hc/\lambda} - 1} \tag{26.6}$$

Integrating Eq. 26.4 over the entire spectrum we find the total energy density of the gas given by

$$u = aT^4 \tag{26.7}$$

where

$$a = 7.56 \times 10^{-16} \text{J/m}^3 \cdot \text{K}^4 \tag{26.8}$$

Because the distribution of photons in an oven does not depend on the nature of the oven walls (other than their temperature), the spectrum of photons emitted by any material at temperature T must be equal to the spectrum of photons it absorbs when in equilibrium with a photon gas at temperature T. That is, those wavelengths the spectrum absorbs more strongly must also be emitted more strongly. This is the "principle of detailed balance." Perfect absorbers are called "blackbodies," and the spectrum of photons emitted from a blackbody at temperature T must be exactly the spectrum of the photons in a photon gas at that temperature.

The flux of energy emitted from a blackbody at temperature T is given by

$$\text{flux} = \sigma T^4 \tag{26.9}$$

where

$$\sigma = 5.67 \times 10^{-8} \text{ J/m}^2 \cdot \text{s} \cdot \text{K}^4 \tag{26.10}$$

PROBLEMS

26-1. Starting from Eq. 26.5 show that $du/d\omega$ is proportional to:

(a) ω^2, for $\hbar\omega \ll kT$.

(b) $e^{-\beta\hbar\omega}$, for $\hbar\omega \gg kT$.

26-2. Starting from Eq. 26.6 show that $du/d\lambda$ is proportional to:

(a) $e^{-\beta hc/\lambda}$, for $\lambda \ll \beta hc$.

(b) λ^{-4}, for $\lambda \gg \beta hc$.

26-3. Show that the plot of $du/d\varepsilon$ for the energy distribution in a photon gas peaks at $\varepsilon_{max} = 2.82kT$. (*Hint.* Start with the expression (26.4) for $(du/d\varepsilon)$. At the maximum, the derivative of this with respect to ε is zero. $e^{-2.82} = 0.06$.)

26-4. Show that the plot of $du/d\lambda$ for the energy distribution in a photon gas peaks at $\lambda = 0.201 \, \beta hc$.

26-5. From the answer to Problems 26-3 and 26-4, does $\varepsilon_{max} = hc/\lambda_{max}$? If not, why not?

26-6. Given that the distribution $du/d\varepsilon$ peaks at $\varepsilon_{max} = 2.82kT$, we can write an expression in the form $\lambda_{max} = C/T$, where λ_{max} is the wavelength at which the distribution $du/d\varepsilon$ peaks, and C is a constant.

(a) What is the value of the constant C in units of mm · K?

(b) At what wavelength does the energy distribution $du/d\varepsilon$ peak for the photons of the 3-K blackbody radiation which apparently is left over from the primordial "Big Bang" that initiated our present universe?

(c) At what wavelength does the energy distribution $du/d\varepsilon$ from our sun peak, if the visible surface acts like a blackbody of temperature 5800 K?

26-7. A neutrino is a massless particle that travels at the speed of light like a photon, but it has spin $\frac{1}{2}$ instead of spin 1. If neutrinos don't interact with each other ($\mu = 0$), what would be the expression for the distribution of energies, $du/d\varepsilon$, in a neutrino gas in an appropriate oven?

26-8. A helium-neon laser emits orange photons all of very nearly the same energies. Could these photons be in equilibrium with anything?

26-9. A ball painted "perfectly black" has a bright green light shining on it that keeps it heated at temperature T. Describe the distribution of energies absorbed and emitted by the ball. (*Hint.* The green light is *not* in equilibrium with the ball—nor with anything, for that matter—because it doesn't have the equilibrium distribution in frequencies.)

26-10. The sun's surface radiates roughly as a blackbody at a temperature of 5800 K. The sun's radius is 7×10^8 m.

(a) How many joules per second does the sun radiate altogether?

(b) Inside, the sun is much hotter. In fact, averaging T^4 throughout the volume of the sun gives

$$\overline{T^4} = (3 \times 10^6 \text{ K})^4$$

What is the total energy of the photon gas stored inside the sun in joules?

(c) If the thermonuclear fusion in the sun's core stops tomorrow, how many more years could the sun radiate energy at the present rate, before exhausting all the energy stored in the photon gas inside? (1 yr = 3×10^7 s.)

26-11.   If you double the temperature of an oven, by what factor does the total energy of the photon gas increase?

26-12.   (a)   Make a qualitative sketch of a plot of $\bar{n}(\varepsilon)$ versus ε for bosons with $\mu = 0$, for two different temperatures.
     (b)   How does the total number of bosons change as the temperature is increased?
     (c)   How does the average energy per boson change as the temperature is increased?

26-13.   Make qualitative plots of the fractions of reflected, absorbed, and reemitted radiation (I_r/I_0, I_a/I_0, and I_e/I_0, respectively) as a function of photon energy, for an object in equilibrium with a photon gas, if the object looks red when illuminated by sunlight.

26-14.   The radius of the sun is 7×10^8 m, and the radius of the earth's orbit is 1.5×10^{11} m. If the energy flux leaving the sun's surface is 6.4×10^4 kW/m^2, and it spreads out uniformly as it moves out through space, what is the flux of solar energy at the distance of the earth?

26-15.   Suppose you had a half cup of very hot coffee, a half cup of room-temperature water, and only 2 minutes of time in which to get the coffee as cool as possible. Should you add the cool water first and then wait 2 minutes, or first wait 2 minutes and then add the water? Why?

26-16.   If the earth would act like a blackbody of temperature $T = 275$ K:

     (a)   What would be the flux of energy leaving the earth's surface for outer space?
     (b)   What would be the rate of emission of energy by the entire earth, if the earth's radius is 6.4×10^6 m?
     (c)   If the solar constant is 1.4 kW/m^2 at the position of the earth, what fraction of the incident solar radiation is absorbed by the earth if our temperature is to remain constant on the average? (*Hint.* The area absorbing sunlight is not the same as the area emitting energy into outer space.)

26-17.   (a)   Roughly what is the temperature of your skin?
     (b)   What is the energy of flux emitted by your skin?
     (c)   Roughly how much energy is emitted by your skin per second?
     (d)   Roughly how many kilocalories of energy are emitted by your skin per day?
     (e)   If one milkshake provides 200 kcal, how many milkshakes would you have to drink per day if you were naked in outer space (so you only emit, but don't absorb, any radiation), in order to compensate for this energy lost from your skin?
     (f)   Why do you suppose it is biologically advantageous for your blood vessels in your skin to constrict when your skin gets cold?

chapter 27

THE THERMAL PROPERTIES OF SOLIDS

As materials get hotter, atomic motions increase. With the help of quantum statistics and specific models we can study these motions in detail. In this chapter, we are concerned in particular with atomic vibrations in solids and the motions of conduction electrons in metals.

The atoms within solids are anchored in place through electrostatic interaction with neighboring atoms. Although each atom may undergo small amplitude oscillations about its equilibrium point, the interatomic forces are sufficiently strong that they prohibit motion over distances large compared to the interatomic spacings. From a microscopic viewpoint, this is the property that distinguishes solids from liquids and gases.

For small amplitude displacement from the minimum of any potential well, the restoring force is linear in the displacement ($F = -k_s x$), and the potential energy is quadratic in the displacement ($U = \frac{1}{2}k_s x^2$), as is illustrated in Figure 27.1. Consequently each atom behaves as a small three-dimensional harmonic oscillator, and the energy of one atom can be written as six terms, each quadratic in the coordinates.

$$\varepsilon = \frac{1}{2m} p_x^2 + \frac{1}{2m} p_y^2 + \frac{1}{2m} p_z^2 + \frac{1}{2} k_s x^2 + \frac{1}{2} k_s y^2 + \frac{1}{2} k_s z^2$$

Consequently, each oscillator has 6 degrees of freedom, and a solid of N_a atoms has $6N_a$ degrees of freedom.

In addition, some solids have conduction electrons which are free to move in any direction (Figure 27.2). These conduction electrons, then, behave as a gas, each electron being able to move in three dimensions, having 3 kinetic degrees of freedom. If there are N_e conduction electrons, then they should contribute $3N_e$ additional degrees of freedom to the material.

From our study in classical statistics we know that there is associated an average thermal energy of $\frac{1}{2}kT$ with each degree of freedom. Therefore, the classical

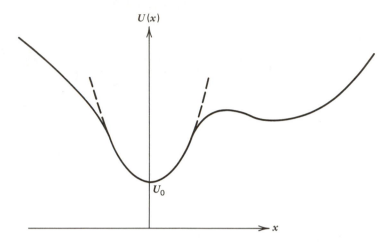

Figure 27.1 Plot of U-versus x for some arbitrary potential well. At the minimum, the first derivative is zero and the second derivative is positive. Therefore, for displacements from the equilibrium position that are small enough that higher-order terms in the Taylor series expansion can be ignored, the potential energy is given as $U = U_0 + \frac{1}{2}kx^2$, which is parabolic. This is true for any local relative minimum in the potential energy.

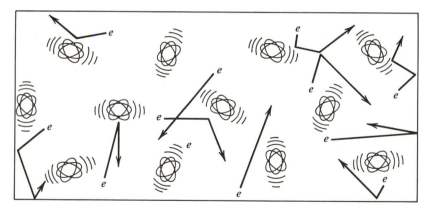

Figure 27.2 A solid metal can be thought of as being composed of a system of atomic oscillators plus a gas of conduction electrons that are confined only by the boundaries of the solid.

prediction for the thermal energy of the solid would be

$$E^{\text{thermal}}_{\text{classical}} = (E - E_0) = 3N_a kT + \frac{3}{2} N_e kT$$

where the first term is due to the vibration of atoms and the second is due to the motion of the conduction electrons. The zero-energy reference level for thermal energies is indicated by E_0.

However, quantum effects cause significant deviation from this classical prediction. We have seen (in Section G, Chapter 2, and Section D, Chapter 12) that the vibrational modes of oscillators are quantized, so if the temperature is not large compared to the excitation temperature

$$T_e = \frac{1}{k} \varepsilon^{(1\text{st excited state})}$$

then many of the oscillators will be confined to the ground state, and they will have no degrees of freedom, contributing nothing to the internal energy of the solid. That is, due to quantum effects, the atomic contribution to the thermal energy will be less than the classical prediction if T is not large compared to the excitation temperature. If μ_a is the zero-energy reference level for any one oscillator then the atomic contribution to the solid's thermal energy is

$$E_a^{\text{thermal}} = (E_a - N_a\mu_a) < 3N_akT \qquad \text{for } T < T_e$$

Furthermore the conduction electrons in most materials are degenerate under normal conditions. This means that the vast majority of these electrons are confined to low-lying states. Since electrons are fermions, no two of them may occupy the same state. All the low-lying states are occupied, so each electron can find no neighboring vacant state to move into. Therefore these electrons have no degrees of freedom and cannot contribute anything to the thermal energy of the system. Only those electrons in the highest energy states (i.e., near the "fermi surface," μ) find any neighboring states are vacant (Figure 27.3), so only these electrons have any freedom. Because most conduction electrons have no freedom, the contribution of the conduction electrons to the thermal energy of most solids will be much less than the classical prediction.

In this chapter we make a more quantitative study of how these quantum effects influence the internal energies of solids. The thermal properties are most easily studied experimentally through the measurement of heat capacities, so the theoretical results will be expressed in a form to facilitate this comparison.

PROBLEMS

27-1. You are going to make a rough estimate of the force constant, k_s, that holds the atoms of a typical solid in place. You will use the following information.

  (a) The average energy per degree of freedom is $\frac{1}{2}kT$, where k is Boltzmann's constant.
  (b) The potential energy stored in the oscillation in one dimension is $\frac{1}{2}k_s x^2$, where $\overline{x^2}$ is the mean square displacement from equilibrium.
  (c) The displacement from equilibrium position must be small compared to the atomic separations. A typical root-mean-square value of the displacement at room temperature is about 0.1 Å.

With this information, estimate the value of k_s for a typical solid.

27-2. Consider the potential energy of a particle in one dimension, which is given by $V(x) = -V_0 e^{-x^2}$.

  (a) At what value of x is this a minimum?
  (b) What are the values of V, $\partial V/\partial x$, and $\partial^2 V/\partial x^2$ evaluated at this position ($V =$ minimum)?

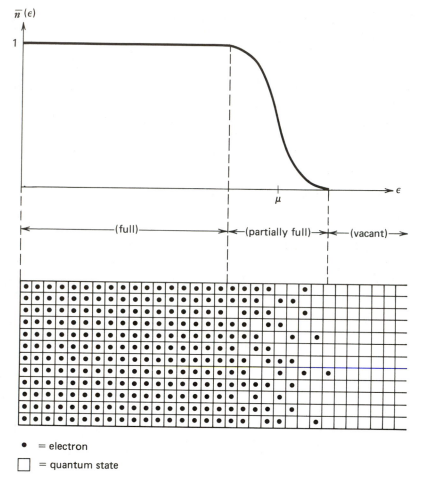

$\bullet$ = electron

$\square$ = quantum state

Figure 27.3 Only electrons near the Fermi surface (μ) have any freedom at all, because only these electrons might be able to find vacant neighboring states into which they can move in response to thermal agitation or some external stimulus.

  (c) Write out the Taylor series expansion in x about the point where $V =$ minimum, keeping only the first three terms in the expansion (zeroth-, first-, and second-order terms in x).
  (d) At $x = 0.1$, by what percentage does the Taylor series expansion differ from the real value of $V(x)$?
  (e) At $x = 0.5$, by what percentage does the Taylor series expansion differ from the real value of $V(x)$?

27-3. If the atoms behaved as classical harmonic oscillators and the conduction electrons as a classical ideal gas, what would be the heat capacity ($C = \partial E/\partial T$) of a metal consisting of 10^{24} atoms and 10^{24} conduction electrons?

27-4. In quantum mechanics, we find the average energy of a simple linear harmonic oscillator is given by

$$\varepsilon = (n + \tfrac{1}{2})\hbar\omega, \qquad n = 0, 1, 2, \ldots.$$

Classically, this harmonic oscillator has 2 degrees of freedom ($\tfrac{1}{2}mv_x^2 + \tfrac{1}{2}kx^2$).

(a) What would be the average energy of this harmonic oscillator as a function of kT, according to the equipartition theorem?

(b) What would be the classical value of the heat capacity ($C = dE/dT$) for this harmonic oscillator?

(c) If the measured values of the heat capacity fell well below this classical prediction for temperatures below 190 K, what would you estimate is the frequency ω for this oscillator? (*Hint.* Consider the relationship between excitation temperature and the difference in energies between the ground and first excited states.)

27-5. You are going to make a rough estimate of the fraction of conduction electrons that are in the "tail" of the fermi distribution at room temperature.

(a) If the characteristic separation of electrons in the degenerate fermion system is 1.2 Å, what is the characteristic momentum of a typical electron? (*Hint.* The product of characteristic separation times characteristic momentum is about the size of a single two-dimensional quantum state, h.)

(b) What is the characteristic kinetic energy of such an electron?

(c) If the width of the tail is roughly kT, what is the spread in energy of quantum states in the tail?

(d) If you assume that the characteristic kinetic energy is roughly half the maximum kinetic energy, and that roughly half the states in the "tail" are filled, then the ratio of electrons in the tail to the total number of electrons would be $(\tfrac{1}{2}kT)/(2\varepsilon_{\text{characteristic}}^k)$, providing the distribution of electron states is constant in energy. What is this ratio?

(e) Considering that the distribution of electron states is *not* constant in energy ($g(\varepsilon) \propto \varepsilon^{1/2}$), would you expect the actual fraction of electrons in the "tail" to be slightly more or slightly less than your answer in part (d)?

A. LATTICE VIBRATIONS

A.1 The Einstein Model

In a paper published in 1907, Albert Einstein proposed a simple model to explain why the measurements of heat capacities in solids differed from the classical prediction. It gave qualitatively the correct behavior, although careful measurements showed deviations from the Einstein model at low temperatures. But the overriding significance of this paper was the demonstration of the importance of quantum statistics. Even with the most simplistic model, the application of quantum statistics yielded nearly the correct answer.

The Einstein model assumes that a solid of N_a atoms can be considered as being $3N_a$ simple harmonic oscillators (each atom in each of three dimensions). Corresponding to these $3N_a$ simple harmonic oscillators are $3N_a$ simple harmonic oscillator quantum

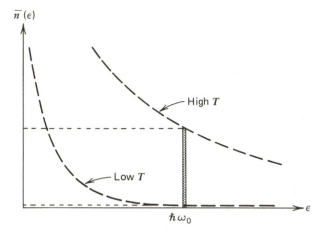

Figure 27.4 According to the Einstein model, there are 3N oscillator states all of the same energy, $\hbar\omega_0$. The occupation number for these states is given by Bose-Einstein statistics, and is small at low temperatures and large at high temperatures.

states, all having the same energy of $\hbar\omega_0$. If the occupation number of a certain state is 0, it means that the oscillator is not oscillating. If the occupation number is 1, it means the oscillator is oscillating with the fundamental frequency ($\varepsilon = 1\hbar\omega_0$). If the occupation number is 2, it means the oscillator is oscillating with twice the fundamental frequency ($\varepsilon = 2\hbar\omega_0$) and so on. Notice the implicit assumption in this model that the fundamental frequency of any oscillator is a constant, independent of the state of motion of its neighbors. This assumption is probably too simplistic, and we should expect the predictions of this model to differ slightly from reality as a consequence.

The number of oscillators in excited states is the product of the number of such states ($3N_a$) times the occupation number of each ($\bar{n}$) (Figure 27.4).

$$(\text{number occupied}) = 3N_a\bar{n}$$

The internal energy is the product of the number of oscillators in excited states, times the energy of each.

$$E = (\text{number occupied})(\hbar\omega_0) = 3N_a\bar{n}(\hbar\omega_0)$$

Since vibrations in one dimension have no intrinsic angular momentum, they are spin 0 bosons. If we assume that their chemical potential is zero (i.e., that the oscillators don't interact with each other), then

$$\bar{n} = \frac{1}{e^{\beta\hbar\omega_0} - 1}$$

and the above expression for the internal energy of the lattice according to the Einstein model is

$$E = \frac{3N_a\hbar\omega_0}{e^{\beta\hbar\omega_0} - 1} \tag{27.1}$$

We can easily check the behavior of this in the high- and low-temperature limits. Because

$$(e^{\hbar\omega_0/kT} - 1) = \begin{cases} e^{\hbar\omega_0/kT} & \text{for } \hbar\omega_0/kT \gg 1 \quad \text{(low } T) \\ \hbar\omega_0/kT & \text{for } \hbar\omega_0/kT \ll 1 \quad \text{(high } T) \end{cases}$$

the Einstein model gives the following prediction for the internal energy in these regions.

$$E = \begin{cases} 3N_a\hbar\omega_0 e^{-\hbar\omega_0/kT} & \text{(low-temperature limit)} \\ 3N_a kT & \text{(high-temperature limit)} \end{cases} \tag{27.2}$$

We can see that it gives the classical result at high temperatures, and goes exponentially to zero at low temperatures, as the oscillators become trapped in their ground states.

With our present understanding of simple harmonic oscillators, we can improve on the Einstein result slightly. We know that the energy in the ground state is $\frac{1}{2}\hbar\omega_0$, and that of the first excited state is $\frac{3}{2}\hbar\omega_0$, from quantum mechanical calculations. Thus, each oscillator has $\frac{1}{2}\hbar\omega_0$ more energy than Einstein assumed, whether it be in the ground state or an excited state.

$$E = \frac{3}{2}N_a\hbar\omega_0 + \frac{3N_a\hbar\omega_0}{e^{\beta\hbar\omega_0} - 1} \tag{27.1'}$$

However, since energies can be measured relative to any reference level that is convenient, this added constant fact, $\frac{3}{2}N_a\hbar\omega_0$, is usually ignored.

A.2 The Debye Model

The success of the Einstein model demonstrated the importance of using quantum statistics, even with the simplest of models. A more realistic model was subsequently developed by P. Debye, in which he abandoned the idea that all quantum states have the same fundamental frequency. We examine the Debye model in this section.

Imagine you have microscopic fingers with which you could reach inside a solid and jiggle one of the atoms. Because of the strong electromagnetic coupling between neighboring atoms in solids, the oscillation of the atom you perturbed would set the neighboring atoms in motion, which would set their neighboring atoms in motion, and so on. The energy you inject at that one point would not stay put, but rather it would travel through the solid. This is how sound travels through solids, and we know that it travels rather fast. The high speed of propagation of such disturbances in solids is simply a reflection of the strong coupling between neighboring atoms.

Since these disturbances can travel from one place to another in the solid, carrying energy with them, we can think of them as some sort of energy-carrying particles, which we call "phonons." Since they are free to travel in any direction across the solid, we can think of these disturbances as a "phonon gas" (Figure 27.5), which is similar in many ways to the photon gas inside an oven. The primary differences between the two are that photons represent electromagnetic waves traveling with the speed of light, whereas phonons represent elastic waves traveling with the speed of sound.

Elastic waves in media have the very convenient property that they obey superposition. That is, two waves can travel right through each other without deflecting or changing each other in any way. In other words, phonons do not interact with each

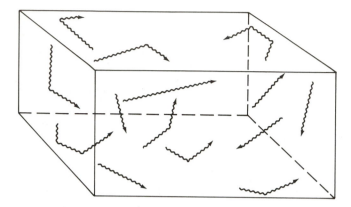

Figure 27.5 We can think of the elastic waves traveling through a solid as individual "phonons" in a "phonon gas."

other at all, which means that the chemical potential of phonons is zero.*

$$\mu_{phonons} = 0$$

Consequently, we can ignore the chemical potential in the occupation number for phonon states.

$$\bar{n}_{phonons} = \frac{1}{e^{\beta\varepsilon} - 1} \tag{27.3}$$

The distribution of particles within a gas is the product of the distribution of quantum states times the average occupation number of each.[†]

$$dN = \left(\frac{d^3r\, d^3p}{h^3}\right)\bar{n}(\varepsilon)$$

Integrating over all volume and all momentum directions, this becomes

$$dN = \left(\frac{4\pi V p^2\, dp}{h^3}\right)\bar{n}(\varepsilon) \tag{27.4}$$

Elastic waves transfer energy from one place to another, but not mass. Therefore, phonons can be treated as massless energy-carrying particles (like photons), for whom the relationship between energy and momentum is

$$\varepsilon = pc_s$$

where c_s is the speed of sound in the material. With this transformation of variables, the expression (27.4) for the distribution of phonons becomes

$$dN = \frac{1}{c_s^3}\left(\frac{4\pi V \varepsilon^2\, d\varepsilon}{h^3}\right)\bar{n}(\varepsilon) \tag{27.5}$$

---

* Similarly we saw that electromagnetic waves obey superposition, which means that photons don't interact with each other. So the chemical potential of photons in a photon gas is also zero.
† For a more thorough discussion of quantum gases, see Section B, Chapter 25.

Elastic waves can be either longitudinal or transverse, and the transverse waves can have two independent possible polarizations. For example, if moving in the z-direction, the transverse oscillations could be in the x-direction or the y-direction (or some combination of the two). This means there are three distinguishable kinds of phonons (three different polarizations), and the result (27.5) is the correct distribution for any one of them. Summing over all three gives a small problem, though, because the speed of sound for longitudinal waves in a material is often different than that for the transverse waves. For convenience we define some sort of average wave speed, c_s, according to

$$\frac{2}{c_{transverse}^3} + \frac{1}{c_{longitudinal}^3} = \frac{3}{c_s^3} \tag{27.6}$$

and then the result (27.5) summed over the three distinguishable types of phonons, becomes (see Figure 25.9 or 26.2)

$$dN = \frac{3}{c_s^3}\left(\frac{4\pi V \varepsilon^2 \, d\varepsilon}{h^3}\right)\bar{n}(\varepsilon) \tag{27.7}$$

This is the distribution of phonons within a phonon gas. It is simply the product of the distribution of states times the occupation number of each. All that we have done so far is to write the distribution of states in a gas in terms of the phonon energy, ε, and sum over the three distinguishable kinds of phonon states.

But there is one important thing left to do, because there is one important way in which phonon gases must differ from other gases. That is that there is an upper limit to the amount of energy that a phonon may have.

Waves of higher energy have shorter wavelength, but the wavelength of elastic waves cannot be less than twice the distance between atoms, as is illustrated in Figure 27.6. Wavelengths shorter than this are indistinguishable from much longer wavelengths. Therefore, the only states available to phonons are those of wavelengths longer than twice the interatomic spacing. To consider wavelengths shorter than this would be double counting. The interatomic spacing is approximately $(V/N_a)^{1/3}$, where N_a is the number of atoms in the solid, so the maximum phonon energy is given by

$$\varepsilon_{max} = \frac{hc_s}{\lambda_{min}} = \frac{hc_s}{2(V/N_a)^{1/3}} = \frac{1}{2}hc_s\left(\frac{N_a}{V}\right)^{1/3} \tag{27.8}$$

A more precise way of determining the maximum allowed energy was suggested by Debye, and goes as follows. From classical physics we know that the motion of one simple harmonic oscillator can be described by one characteristic frequency, that of two coupled simple harmonic oscillators can be described by two characteristic frequencies, etc. To describe the motions of $3N$ coupled simple harmonic oscillators requires no more than $3N$ characteristic frequencies. If we wish to describe these in terms of traveling waves, then the wave numbers are not independent variables, but are related to the frequencies through the speed of sound by

$$k = \frac{\omega}{c_s}$$

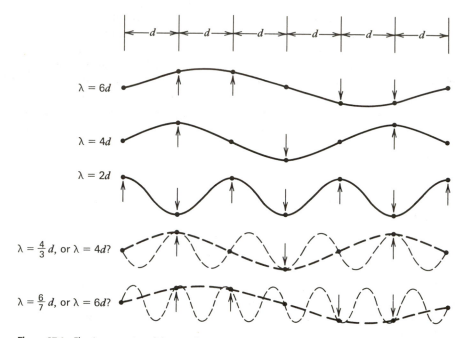

Figure 27.6 Elastic waves in solids must have wavelengths greater than twice the spacing between atoms ($\lambda > 2d$). As is illustrated in the bottom two lines, wavelengths shorter than this are equivalent to longer wavelengths. For illustrative purposes we have used transverse waves. Longitudinal waves have the same problem.

This means that there are only as many independent traveling waves, or phonons as there are fundamental modes of oscillation for the $3N$ harmonic oscillators. Of course the amplitudes may vary (corresponding to different occupation numbers of the phonon states), and they may superimpose on each other in ways that make the motion of any one atom quite complex. But all motion of all atoms in the solid can be described in terms of some combination of $3N$ fundamental modes.

The contribution of Debye, then, was the observation that the number of accessible phonon states in a solid is $3N$, where N is the number of atoms. This condition determines the maximum energy that a phonon state may have. From Eq. 27.7 we see that the distribution of phonon states is given by (see Figure 27.7)

$$(\text{distribution of phonon states}) = \begin{cases} \dfrac{12\pi V \varepsilon^2 \, d\varepsilon}{c_s^3 h^3} & \text{for } \varepsilon \leq \varepsilon_{max} \\[3mm] 0 & \text{for } \varepsilon > \varepsilon_{max} \end{cases}$$

(27.9)

Since the total number of phonon states is $3N_a$, we have

$$3N_a = \int_0^{\varepsilon_{max}} \frac{12\pi V \varepsilon^2 \, d\varepsilon}{c_s^3 h^3} = \frac{4\pi V}{c_s^3 h^3} \varepsilon_{max}^3.$$

(27.10)

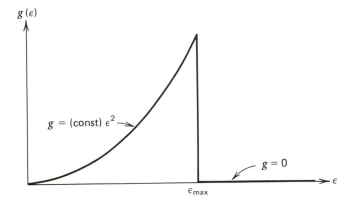

Figure 27.7 Plot of density of states $g(\varepsilon)$ for the Debye model. In the Debye model, there can be no more phonon states altogether than there are fundamental modes of oscillation (3*N*). This puts an upper limit on the energy of accessible quantum states through (total number of quantum states) = $\int_0^{\varepsilon_{max}} g(\varepsilon)\, d\varepsilon = 3N$. Alternately, the upper limit on possible phonon energies is determined by the fact that phonon wavelengths may not be shorter than about twice the interatomic spacings.

Solving this for ε_{max} gives

$$\varepsilon_{max} = \left(\frac{3N_a}{4\pi V}\right)^{1/3} hc_s$$

$$= 0.62 hc_s \left(\frac{N_a}{V}\right)^{1/3}. \tag{27.11}$$

Notice that this is nearly identical with our previous rough estimate (27.8).

This maximum phonon energy depends on the solid, because both the speed of sound (c_s) and the density (N_a/V) vary from one solid to the next. For most common solids it ranges somewhere around 10^{-2} to 5×10^{-2} eV. Table 27.1 lists values of ε_{max} for some common materials.

To summarize, we now know that the distribution of phonon states is that of a phonon gas, up to a maximum energy given by (27.11). There are no phonon states at energies larger than this. The distribution of phonons is the product of the distribution of states times the occupation number of each, which is given by Eqs. 27.9 and 27.3 as (see Figure 27.8)

$$dN = \begin{cases} \left(\dfrac{12\pi V}{c_s^3 h^3}\right) \dfrac{\varepsilon^2\, d\varepsilon}{e^{\beta\varepsilon} - 1} & (\varepsilon \le \varepsilon_{max}) \\[2ex] 0 & (\varepsilon > \varepsilon_{max}) \end{cases}$$

Using the definition of ε_{max} (27.11), this becomes

$$dN = \begin{cases} \left(\dfrac{9N_a}{\varepsilon_{max}^3}\right) \dfrac{\varepsilon^2\, d\varepsilon}{e^{\beta\varepsilon} - 1} & (\varepsilon \le \varepsilon_{max}) \\[2ex] 0 & (\varepsilon > \varepsilon_{max}) \end{cases} \tag{27.12}$$

Table 27.1   Debye Model Values of ε_{max} for Some Common Materials

Material	$\varepsilon_{max}(10^{-2}$ eV)
Sodium	1.36
Magnesium	3.45
Aluminum	3.69
Silicon	5.56
Potassium	0.78
Calcium	1.98
Iron	4.05
Copper	2.96
Arsenic	2.43
Silver	1.94
Gold	1.42
Lead	0.90

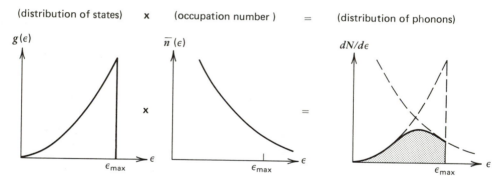

Figure 27.8   The distribution of phonons in a solid is the product of the distribution of phonon states times the occupation number.

The distribution of internal energy among the phonons is the product of the distribution of phonons times the energy of each,

$$dE = \varepsilon \, dN = \begin{cases} \left(\dfrac{9N_a}{\varepsilon_{max}^3}\right) \dfrac{\varepsilon^3 \, d\varepsilon}{e^{\beta\varepsilon} - 1} & (\varepsilon \le \varepsilon_{max}) \\ 0 & (\varepsilon > \varepsilon_{max}) \end{cases} \tag{27.13}$$

and the total internal energy of the system is obtained by integration.

$$E = \left(\frac{9N_a}{\varepsilon_{max}^3}\right) \int_0^{\varepsilon_{max}} \frac{\varepsilon^3 \, d\varepsilon}{e^{\beta\varepsilon} - 1} \tag{27.14}$$

With the substitution of variables $x = \beta\varepsilon$, this becomes

$$E = \left(\frac{9N_a}{\varepsilon_{max}^3}\right)(kT)^4 \int_0^{(\varepsilon_{max}/kT)} \frac{x^3 \, dx}{e^x - 1} \tag{27.15}$$

The value of this integral depends on the temperature through the upper limit, and is best done numerically. In Appendix 27A we solve it analytically in the high- and low-temperature limits with the following results:

$$\int_0^{(\varepsilon_{max}/kT)} \frac{x^3 \, dx}{e^x - 1} = \begin{cases} \dfrac{\pi^2}{15} & \text{for } \left(\dfrac{\varepsilon_{max}}{kT}\right) \gg 1 \quad \text{(low temperature)} \\[4mm] \dfrac{1}{3}\left(\dfrac{\varepsilon_{max}}{kT}\right)^3 & \text{for } \left(\dfrac{\varepsilon_{max}}{kT}\right) \ll 1 \quad \text{(high temperature)} \end{cases} \tag{27.16}$$

Putting these values for the integral into Eq. 27.15 gives

$$E = \begin{cases} \left(\dfrac{3N_a\pi^2 k^4}{5\varepsilon_{max}^3}\right) T^4 & \text{low-temperature limit} \\[4mm] 3N_a kT & \text{high-temperature limit} \end{cases} \tag{27.17}$$

Notice that the result is the same as the classical prediction in the high-temperature limit, but at low temperatures the internal energy decreases rapidly with decreased temperature, reflecting the inability of the oscillators to make the jump to the first accessible quantum state.

You may wonder how the vanishing of the internal energy squares with the results of quantum mechanical computations, which tell us that each oscillator undergoes zero-point oscillation even at absolute zero. The answer is that our phonon gas is superimposed on this background of lattice points undergoing zero-point oscillation. In other words, we choose this zero-point oscillation of the lattice to be our zero-energy reference point for the phonon gas.

Different textbooks use different notations in discussing the Debye model, so in Appendix 27B are listed these results in terms of other popular sets of variables.

SUMMARY

The Einstein model for lattice vibrations in a solid of N_a atoms assumes $3N_a$ oscillator quantum states, each having the same energy, $\hbar\omega_0$, associated with it. The number of oscillators is the product of the number of oscillator quantum states ($3N_a$) times the occupation number of each ($\bar{n}$). The internal energy of the system, in turn, is the product of the number of oscillators ($3N_a\bar{n}$) times the energy of each ($\hbar\omega_0$).

$$E = 3N_a\bar{n}\hbar\omega_0 = \frac{3N_a\hbar\omega_0}{e^{\beta\hbar\omega_0} - 1}$$

The Debye model represents the lattice vibrations as a "phonon gas," where the phonons represent elastic waves traveling with the speed of sound, c_s, through the solid. The distribution of phonons is the product of the distribution of phonon states times the occupation number of each.

$$dN = \left(\frac{d^3r \, d^3p}{h^3}\right)\bar{n}(\varepsilon)$$

If we integrate over all volume, all momentum directions, sum over all three phonon polarizations, and convert momentum to energy through $\varepsilon = pc_s$, for the massless phonons, this distribution becomes

$$dN = \left(\frac{12\pi V \varepsilon^2 \, d\varepsilon}{c_s^3 h^3}\right) \bar{n}(\varepsilon) \tag{27.7}$$

Since phonons are bosons with no chemical potential, the occupation number is given by

$$\bar{n}(\varepsilon) = \frac{1}{e^{\beta\varepsilon} - 1} \tag{27.3}$$

Elastic waves may not have a wavelength shorter than twice the interatomic spacings in the solid. This corresponds to a maximum possible phonon energy, ε_{max}, determined by the fact that the total number of possible phonon states must be equal to $3N_a$, where N_a is the number of atoms in the solid.

$$\int_0^{\varepsilon_{max}} dN = 3N_a$$

This determines the value of ε_{max},

$$\varepsilon_{max} = \left(\frac{3N_a}{4\pi V}\right)^{1/3} hc_s \tag{27.11}$$

which varies from one solid to another through the density (N_a/V) and the speed of sound (c_s). With this, the distribution of phonons within the solid can be written as

$$dN = \begin{cases} \left(\dfrac{9N_a}{\varepsilon_{max}^3}\right) \dfrac{\varepsilon^2 \, d\varepsilon}{e^{\beta\varepsilon} - 1} & \varepsilon \leq \varepsilon_{max} \\[2mm] 0 & \varepsilon > \varepsilon_{max} \end{cases}$$

The distribution of energy among the phonons is the product of the distribution of phonons times the energy of each. The total internal energy is then obtained by summing this product over the entire distribution.

$$E = \int_0^{\varepsilon_{max}} \varepsilon \, dN = \left(\frac{9N_a}{\varepsilon_{max}^3}\right) \int_0^{\varepsilon_{max}} \frac{\varepsilon^3 \, d\varepsilon}{e^{\beta\varepsilon} - 1} \tag{27.14}$$

With the substitution of variables, $x = \varepsilon_{max}/kT$, this becomes

$$E = \left(\frac{9N_a}{\varepsilon_{max}^3}\right)(kT)^4 \int_0^{(\varepsilon_{max}/kT)} \frac{x^3 \, dx}{e^x - 1} \tag{27.15}$$

The value of the integral is a complex function of the temperature, best solve numerically, but gives the following values in the high- and low-temperature limits.

$$E = \begin{cases} \left(\dfrac{3N_a \pi^2 k^4}{5\varepsilon_{max}^3}\right) T^4 & kT \ll \varepsilon_{max} \\[2mm] 3N_a kT & kT \gg \varepsilon_{max} \end{cases} \tag{27.17}$$

The model gives the classical result at high temperatures, but shows the strong suppression by quantum effects at low temperatures.

PROBLEMS

27-6. Show that in the low-temperature limit, the Einstein model predicts the internal energy of a solid to be $3N\hbar\omega_0 e^{-\hbar\omega_0/kT}$.

27-7. Show that in the high-temperature limit, the Einstein model predicts the internal energy of a solid of N atoms to be $3NkT$.

27-8. (a) Write an expression for $E/3NkT$ in terms of the parameter $x = \hbar\omega_0/kT$ from the Einstein model's prediction for E.
(b) Make qualitative plot of $E/3NkT$ versus x. Is $x = 0$ the high- or the low-temperature limit?

27-9. For lead, $E \approx 3NkT$ for temperatures above 90 K.

(a) Roughly what is the value of ω_0 in the Einstein model for lead? (*Hint.* The energy of the first excited state is roughly kT_e.)
(b) If the mass of a lead atom is 3.4×10^{-25} kg, what is the force constant, k_s, holding a lead atom in place? (*Hint.* The frequency of an oscillator is related to the force constant and mass through $\omega_0 = (k_s/m)^{1/2}$.)

27-10. For a diamond, $E \approx 3NkT$ for temperatures above 2300 K.

(a) Roughly what is the value of ω_0 in the Einstein model for a diamond?
(b) If the mass of a diamond atom is 2.0×10^{-26} kg, what is the force constant, k_s, holding a carbon atom in place?

27-11. Suppose the oscillations caused by a phonon traveling in the positive x-direction can be approximated by a plane wave

$$y = A \cos(kx - \omega t)$$

where A is the amplitude of oscillation, k the wave number ($k = 2\pi/\lambda$), and ω the angular frequency ($\omega = 2\pi/T$).

(a) What is the speed of this phonon, c_s, in terms of ω and k?
(b) If the momentum and energy are given by $p = \hbar k$ and $\varepsilon = \hbar\omega$, how are the momentum, energy, and wave speed interrelated?

27-12. Comparing Eqs. 27.5 and 27.7 for the phonon distributions, we see that they differ by a factor of 3. What is different about the distribution (27.7) that causes us to include this factor of 3? (*Hint.* It would be there even if longitudinal and transverse phonons all traveled with the same speed.)

27-13. Make a sketch similar to that of Figure 27.6 for the displacement of atoms when the phonon's wavelength is exactly two-fifths the distance between atoms ($\tfrac{2}{5}d$).

(a) Is this distinguishable from a wavelength that is two-thirds the distance between atoms ($\tfrac{2}{3}d$)?
(b) What phonon wavelength that is longer than $2d$ would these both correspond to?

27-14. In the text it said that the interatomic spacing is "approximately" $(V/N_a)^{1/3}$. Under what conditions might the spacings be greater or smaller than this in certain directions?

27-15. In a certain solid, the spacing between atoms is about 1 Å, and the speed of sound is 10^3 m/s. What is the maximum phonon energy in this solid according to the Debye model? (*Hint.* Use Eq. 27.11.)

27-16. The density of lead is 13.6 g/cm³, and the mass of a single atom is 3.4×10^{-22} g.

  (a) What is $(N_a/V)^{1/3}$ for lead?

  (b) Using the value of ε_{max} for lead in the Debye model from Table 27.1, estimate the speed of sound in lead. (*Hint.* Use Eq. 27.11.)

27-17. With the help of Table 27.1, estimate the temperatures above which you expect the internal energy to start approaching the classical prediction ($3N_a kT$) for the following:

  (a) Magnesium.

  (b) Silicon.

  (c) Potassium.

27-18. Two different solids consist of the same number of atoms, and the speed of sound is the same in both. But the spacings between atoms in solid A are twice as great as the spacings between atoms in solid B. According to the Debye model:

  (a) For which solid will ε_{max} be greatest? By how many times?

  (b) At low temperatures, which solid will have the largest internal energy? By how many times?

B. CONDUCTION ELECTRONS

If the solid is a metal, additional contributions to its thermal properties are made by the conduction electrons. It is rather easy to demonstrate that these conduction electrons are a highly degenerate electron gas (see homework Problem 24-27). In Section C, Chapter 24, we saw that for a gas to be nondegenerate, the number of accessible quantum states must be large compared to the number of particles. If R is their characteristic separation, and $\langle p \rangle$ their characteristic momentum, then this condition can be expressed as follows.

$$\frac{R\langle p \rangle}{h} \gg 1 \qquad \text{(if nondegenerate)}$$

Since the characteristic separation of conduction electrons in metals is about 1 Å, this condition means

$$\langle p \rangle \gg 6 \times 10^{24} \text{ kg-m/s} \qquad \text{(if nondegenerate)}$$

which for electrons corresponds to energies of

$$\varepsilon \approx \frac{\langle p \rangle^2}{2m} \gg 250 \text{ eV} \qquad \text{(if nondegenerate)}$$

For the typical energies of free electrons to be this high, the temperature would have to be well over 2 million degrees!! ($\varepsilon \approx \frac{3}{2}kT$.)

$$T \gg 2 \times 10^6 \text{ K} \qquad \text{(if nondegenerate)}$$

Clearly, this is not satisfied, so the conduction electrons in metals are highly degenerate.

  This means that although the number of conduction electrons in a metal may be large, very few of these are free. By far most of them are trapped in low-lying states with nowhere to go, because all the neighboring states are full.

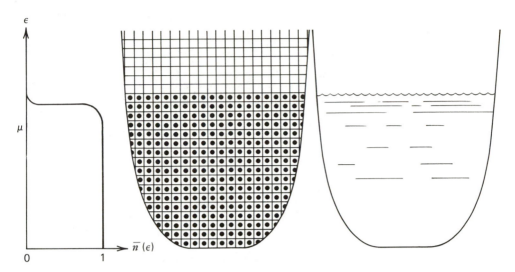

Figure 27.9 The distribution of particles in a degenerate fermion gas is similar to that in a glass of water. All states up to the fermi surface are full, and all those above that are empty.

In the low-temperature limit, the fermion occupation number has the following behavior:

$$n(\varepsilon) = \frac{1}{e^{(\varepsilon - \mu)/kT} + 1} \xrightarrow{T \to 0} \begin{cases} 1 & \text{for } \varepsilon < \mu \\ 0 & \text{for } \varepsilon > \mu \end{cases} \qquad (27.18)$$

Therefore, in the low-temperature limit, all states up to $\varepsilon = \mu$ are full, and all higher lying states are empty. A degenerate fermion system is something like a glass of water, being completely full up to a certain point, and completely empty above that (Figure 27.9). The energy $\varepsilon = \mu$ is sometimes referred to as the "Fermi level" or "Fermi surface," and it divides the states that are full from those that are empty in a degenerate fermion system. The Fermi level is sometimes given the symbol "ε_f" or "μ_f."

The distribution of electrons is the product of the distribution of states times the occupation number of each.

$$dN_e = \left(\frac{d^3 r \, d^3 p}{h^3} \right) \bar{n}(\varepsilon)$$

If we integrate this over all volume and all momentum directions, and add together electrons of both possible orientations of their intrinsic angular momentum (i.e., spin up and spin down), this becomes

$$dN_e = \frac{8\pi V p^2 \, dp}{h^3} \, \bar{n}(\varepsilon)$$

If we write this in terms of electron energies,

$$\varepsilon = \frac{p^2}{2m}$$

it becomes

$$dN_e = \left[\frac{4\pi V(2m)^{3/2}}{h^3}\sqrt{\varepsilon}\,d\varepsilon\right]\bar{n}(\varepsilon) \tag{27.19}$$

Since all states below the Fermi surface are completely full and all those above the Fermi surface are completely empty, this particle distribution is given by (Figure 27.9)

$$dN_e = \begin{cases} \left[\dfrac{4\pi V(2m))^{3/2}}{h^3}\right]\sqrt{\varepsilon}\,d\varepsilon & \text{for } \varepsilon < \mu \\[2ex] 0 & \text{for } \varepsilon > \mu \end{cases} \tag{27.20}$$

The integral of this is the total number of electrons,

$$N_e = \int dN_e = \left[\frac{4\pi V(2m)^{3/2}}{h^3}\right]\int_0^\mu \sqrt{\varepsilon}\,d\varepsilon = \left[\frac{8\pi V(2m)^{3/2}}{3h^3}\right]\mu^{3/2} \tag{27.21}$$

which determines the Fermi level in terms of the particle density

$$\mu = \frac{h^2}{2m}\left(\frac{3N_e}{8\pi V}\right)^{2/3} \tag{27.22}$$

The total internal energy of the system is obtained by integrating the product of the electron distribution times the energy of each.

$$E = \int \varepsilon\,dN_e = \left[\frac{4\pi V(2m)^{3/2}}{h^3}\right]\int_0^\mu \varepsilon^{3/2}\,d\varepsilon$$

$$= \left[\frac{8\pi V(2m)^{3/2}}{5h^3}\right]\mu^{5/2}$$

Comparing this to the expansion (27.21) for N_e above, we see the internal energy is given by

$$E = \frac{3}{5}N_e\mu \tag{27.23}$$

where the Fermi level, μ, is given by Eq. 27.22.

At temperatures somewhat above the $T = 0$ limit, the electrons are not completely degenerate, because the occupation number develops a little "tail" near the Fermi surface, as illustrated in Figure 25.8. Since this "tail" is caused by the exponential factor of (ε/kT) appearing in the expression for the occupation number, the "width" of the tail will be proportional to the temperature. At higher temperatures, the tail will span more quantum states, and so there will be more electrons in the tail.

$$(\text{number of electrons in tail}) \propto T$$

Unlike the lower-lying states, the electrons in the tail do have some freedom since there are some unoccupied neighboring states that they can move into (Figure 27.10). With each of these new degrees of freedom comes an additional thermal energy of $\frac{1}{2}kT$ for the system. Consequently, as the temperature of the system increases, the increase in thermal energy of the system is equal to the product of the number of free electrons

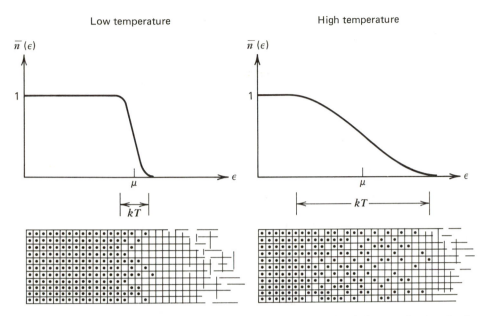

Figure 27.10 The characteristic width of the tail of the Fermi distribution is kT, which means that the tail region widens with increased temperature. Only electrons in the tail region have any freedom, as only they have vacant neighboring states into which they can move.

times the average thermal energy of each. Both of these factors are proportional to the temperature. That is, as the temperature increases,

$$(\text{increase in internal energy}) = (\text{number of free electrons})$$
$$\times (\text{thermal energy of each})$$
$$\propto T \times T = T^2 \tag{27.24}$$

Therefore, from results (27.23) and (27.24) we should be able to write the internal energy of a degenerate fermion gas at temperatures above zero as (Figure 27.11)

$$E = \frac{3}{5} N_e \mu + cT^2 \tag{27.25}$$

where c is some small positive constant. This should be good as long as the gas is reasonably degenerate, which means temperatures smaller than a few million degrees for typical metals.

This result can also be obtained by blindly cranking out the integral

$$E = \int_0^\infty \varepsilon \, dN_e$$

with the appropriate expression for the occupation number at any temperature inserted into the expression (27.19) for dN_e.

$$E = \left[\frac{4\pi V (2m)^{3/2}}{h^3} \right] \int_0^\infty \frac{\varepsilon^{3/2} \, d\varepsilon}{e^{\beta(\varepsilon - \mu)} + 1} \tag{27.26}$$

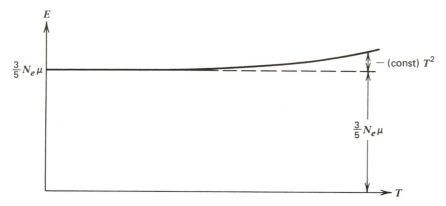

Figure 27.11 Plot of the internal energy of a degenerate fermion gas as a function of temperature. It increases slightly with temperature according to $E = (\frac{3}{5})N\mu + cT^2$, where N is the number of fermions in the gas, μ is their chemical potential, and c is some small positive constant.

This integral is very difficult to do, but gives the following result for degenerate gas. (That is, for $T \ll \mu/k \approx 10^6$ K, which is satisfied by all metals).

$$E = \frac{3}{5} N_e \mu \left[1 + \left(\frac{5\pi^2}{12} \right) \left(\frac{kT}{\mu} \right)^2 \right] \qquad \left(\text{for } T \ll \frac{\mu}{k} \text{ or } T \ll 10^6 \text{ K} \right) \qquad (27.27)$$

where μ is the Fermi level at $T = 0$, given by Eq. 27.22. Comparing this to our result (27.25) we can see the internal energy of the degenerate electron gas increases quadratically with temperature, as expected.

SUMMARY

The conduction electrons in metals are very degenerate at all temperatures lower than a few million degrees Kelvin. The chemical potential is called the "Fermi level" or "Fermi surface." At the low temperature limit all states are full up to the Fermi level, and empty above that. If we consider the electrons to be a degenerate fermion gas, then the Fermi level is related to the electron density through

$$\mu = \frac{h^2}{2m} \left(\frac{3N_e}{8\pi V} \right)^{2/3} \qquad (27.22)$$

and in the low-temperature limit the total internal energy is given by

$$E = \frac{3}{5} N_e \mu \qquad (27.23)$$

where N_e is the number of conduction electrons, and V is the volume they occupy (i.e., the volume of the piece of metal).

At finite temperatures, the "tail" in the Fermi distribution has a width proportional to the temperature. The electrons in this tail are somewhat free to move into vacant neighboring states, so these free electrons each contribute additional energy proportional to kT. The product of the number of relatively free electrons times the thermal energy of each is quadratic in temperature, so as the temperature of the degenerate electron gas increases, the internal energy increases quadratically in T according to

$$E = \frac{3}{5} N_e \mu \left[1 + \left(\frac{5\pi^2}{12} \right) \left(\frac{kT}{\mu} \right)^2 \right] \tag{27.27}$$

where μ is the Fermi level at $T = 0$ given by (27.22).

PROBLEMS

27-19. The spacings between identical fermions (e.g., protons or neutrons) in a nucleus is about 1.3×10^{-15} m.

    (a) What is the characteristic kinetic energy of a nucleon in a nucleus?

    (b) How hot would a system of such fermions have to be in order that it not be degenerate?

27-20. What would have to be the characteristic spacings of conduction electrons in order for them to be nondegenerate at room temperature?

27-21. The characteristic spacing between electrons is given by $R = (V/N_e)^{1/3}$, and the characteristic spacing and momentum are related by $R\langle p \rangle = h$.

    (a) Find an expression for an electron's characteristic momentum in terms of h, and (N_e/V).

    (b) Write an expression for the characteristic kinetic energy ($\langle p \rangle^2/2m$) in terms of the electron mass, h, and (N_e/V).

    (c) How does your result compare with the chemical potential of a system of conduction electrons as given in Eq. 27.22?

27-22. In a certain system of degenerate conduction electrons, only 10^{15} electrons are "free" at $T = 200$ K.

    (a) What is the thermal energy of these electrons?

    (b) At $T = 400$ K, how many electrons are free?

    (c) What is the thermal energy of these electrons at 400 K?

27-23. In a certain solid at a temperature of 40 K, the lattice contribution to the thermal energy is 10^4 times greater than the contribution from the conduction electrons. Assume $kT \ll \varepsilon_{max}$ for the lattice.

    (a) Assuming the Debye model is correct for the energy of the lattice, at what temperature would the contribution from both be the same?

    (b) At what temperature would the electron contribution be 100 times greater than the lattice's contribution?

27-24. According to Eq. 27.27 we can write the internal energy of the degenerate conduction electrons as $E = 3/5 N_e \mu (1 + bT^2)$. If μ/k is on the order of 10^6 K for most metals, roughly what is the value of the constant b?

27-25. The radius of a uranium nucleus is 7×10^{-15} m. It has atomic weight of 238, and atomic number of 92. Protons and neutrons are each spin $\frac{1}{2}$ particles.

(a) What is the density of neutrons in the nucleus (N/V)?

(b) What is the density of protons in the nucleus?

(c) What is the Fermi temperature, μ/k, of each of the two Fermi gases in parts (a) and (b)?

27-26. The density of matter in the center of the sun is about 150 g/cm^3, and the temperature is about 12×10^6 K. About half of this mass is made up of individual protons. Is this proton gas degenerate?

27-27. The ^{3}He isotope of helium is a fermion. (It has two protons, one neutron, two electrons and its net spin is $\frac{1}{2}$.) Hence, a gas of ^{3}He atoms is a fermion gas. At standard temperature and pressure, a mole (6.022×10^{23} atoms) occupies 22.4 liters (1 liter $= 10^3$ cm^3) of volume.

(a) What is its Fermi temperature, μ/k?

(b) Is it a degenerate gas at standard temperature and pressure?

C. HEAT CAPACITIES

A solid contains a fixed number of atoms, and metals contain a fixed number of conduction electrons. With ΔN equal to zero, the first law relates heat added to changes in internal energy and volume through

$$\Delta Q = \Delta E + p \Delta V$$

The heat capacities of solids, then, are given by

$$C_V = \frac{\Delta Q}{\Delta T}\bigg)_V = \frac{\Delta E}{\Delta T}\bigg)_V \tag{27.28}$$

Because the coefficient of thermal expansion for solids is so small, the heat capacity at constant volume and heat capacity at constant pressure are very nearly the same, differing by only about one part in 10^6 for typical solids. (See Section E.1, Chapter 12.) Consequently, for solids we can simply express heat capacity as

$$C = \frac{\Delta E}{\Delta T} \tag{27.28'}$$

and we need not be concerned about whether the measurements are performed at constant volume or at constant pressure, unless extremely high accuracy is desired.

In the previous sections we studied how the internal energies of solids are related to their temperatures. We studied both the Einstein and Debye model predictions for the energy held in lattice vibrations, and the degenerate fermion gas prediction for the energy of the conduction electrons. These results are summarized in Table 27.2.

The heat capacities are obtained simply by taking the derivatives with respect to temperature, according to Eq. 27.28'. We can see that both Einstein and Debye models give the same prediction in the high-temperature limit

$$\frac{dE^{\text{Einstein}}}{dT} = \frac{dE^{\text{Debye}}}{dT} = 3Nk \qquad \text{(high-temperature limit)} \tag{27.29}$$

Table 27.2   Internal Energies of Solids

Component	Prediction	Low-Temperature Limit	High-Temperature Limit
Lattice Vibrations			
Einstein model[a]	$3N_a \dfrac{\hbar\omega_0}{e^{\beta\hbar\omega_0} - 1}$	$3N_a\hbar\omega_0 e^{-\hbar\omega_0/kT}$	$3N_a kT$
Debye model[b]	$9N_a \dfrac{(kT)^4}{\varepsilon_{max}^3} \displaystyle\int_0^{(\varepsilon_{max}/kT)} \dfrac{x^3\,dx}{e^x - 1}$	$3N_a \left(\dfrac{\pi^2 k^4}{5\varepsilon_{max}^3}\right) T^4$	$3N_a kT$
Conduction[c] Electrons	$3N_e(a + bT^2)$	$3N_e(a + bT^2)$	$3N_e(a + bT^2)$

[a] ω_0 is a parameter determined by experiment.
[b] $\varepsilon_{max} = (3N_a/4\pi V)^{1/3} hc_s$.
[c] $a = \mu/5$, $b = \pi^2 k^2/12\mu$, where $\mu = (h^2/2m)(3N_e/8\pi V)^{2/3}$.

in agreement with the classical prediction in this region. However, at low temperatures, the two predictions differ.

$$\frac{dE^{\text{Einstein}}}{dT} \propto \frac{1}{T^2} e^{-\hbar\omega_0/kT} \qquad \text{(low-temperature limit)}$$

$$\frac{dE^{\text{Debye}}}{dT} \propto T^3 \qquad \text{(low-temperature limit)}$$

(27.30)

A comparison with experimental measurement of heat capacities (e.g., Figure 27.12) shows that the Debye model seems to represent Nature most accurately.

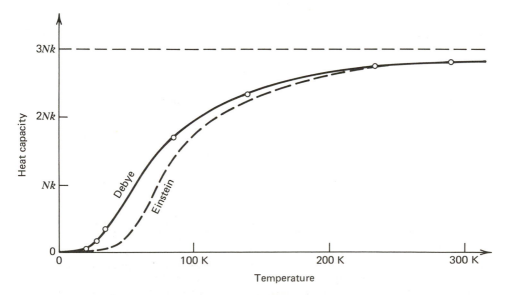

Figure 27.12   Comparison of Einstein and Debye model predictions with data for the heat capacity of copper.

The contribution of the conduction electrons to the internal energy of the system has the form

$$E^{\text{cond. el.}} = 3N_e(a + bT^2)$$

The constant b is very small, because this term represents the contributionss from electrons in the "tail" of the Fermi distribution. The conduction electrons in metals are extremely degenerate, which means that the "tail" is very narrow, and there are few electrons in it.

In computing the heat capacity of this electron gas,

$$C^{\text{cond. el.}} = \frac{dE^{\text{cond. el.}}}{dT} = 6N_e bT \tag{27.31}$$

we see that only those electrons in the "tail" of the Fermi distribution make any contribution. The majority of the electrons are in low-lying states, surrounded by occupied states, and so they have nowhere to go when extra energy is added to the system. Only the electrons in the tail have neighboring vacant states into which they can go, and so only these electrons are able to take on energy added to the system. This is the physical reason why only electrons in the tail make any contribution to the heat capacity.

Since the number of electrons in the tail of the Fermi distribution is so small, they have negligible effect on the heat capacity of metals at higher temperatures where the $6N$ degrees of freedom of the lattice swamp the contributions of the relatively few free electrons. However, at low temperatures, the heat capacity of the lattice goes to zero rapidly as T^3, whereas that of the electron gas goes to zero linearly with T. This means that at sufficiently low temperatures, the electron contribution will dominate. If we add

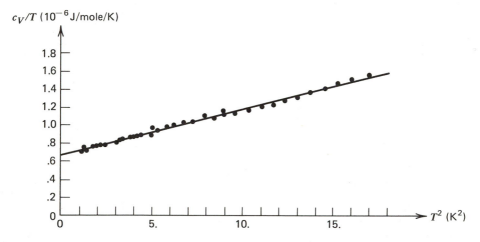

Figure 27.13 Plot of C_V/T versus T^2 for copper at very low temperatures. The intercept at $T^2 = 0$ represents the contribution from the electron gas, and the fact that the plot is linear as T^2 increases demonstrates that the Debye model describes the lattice's contribution to the heat capacity correctly at these low temperatures. (From Corak, Garfunkel, Satterthwaite, and Wexler, *Phys. Rev., 98*, 1699, 1955.)

the contributions of the lattice and the electron gas together, we have for metals at low temperatures

$$C = k_e T + K_l T^3 \qquad \text{(low temperatures)} \qquad (27.32)$$

where K_l is some large constant due to the lattice, and k_e is some very small constant due to the relatively few free electrons. If we plot C/T versus T^2,

$$\frac{C}{T} = k_e + K_l T^2 \qquad (27.33)$$

the plot should be linear in T^2 with the intercept at $T = 0$ representing the free electron contribution. As is seen in Figure 27.13, this is indeed what happens, giving us confidence that our models represent Nature fairly well. Treating lattice vibrations as a phonon gas, and conduction electrons as a degenerate fermion gas seems to work.

SUMMARY

Since the coefficient of thermal expansion for solids is so small, the heat capacities at constant pressure and constant volume are nearly the same.

$$C_p \approx C_V = \frac{dE}{dT}$$

Both the Einstein model and the Debye model give the same prediction for heat capacities of solids at high temperatures, in agreement with the classical prediction.

$$C = 3Nk \qquad \text{(high temperature)} \qquad (27.29)$$

However, the predictions of the two models differ at lower temperatures, where the Debye model seems to do better.

$$C^{\text{Debye}} \propto T^3 \qquad \text{(low temperature)} \qquad (27.30)$$

Only those conduction electrons in the "tail" of the Fermi distribution make any contribution at all to the heat capacities of metals. These "tail" electrons are so few that their contribution to the heat capacity at medium or high temperatures is negligible in comparison to the lattice vibrations. However, because the contribution of the conduction electrons is linear in T, at low temperatures, where the contribution of the lattice vibrations goes as T^3, the electrons may dominate. That is, at low temperatures, the heat capacity of metals should behave as

$$C = k_e T + K_l T^3 \qquad (27.33)$$

where k_e is a small constant associated with the conduction electrons, and K_l is a large constant associated with the lattice vibrations. At sufficiently small T, the electron term will dominate.

PROBLEMS

27-28. From the result (27.17) for the internal energy of a lattice of N_a atoms at low temperatures, find an expression for the heat capacity of the lattice at low temperatures in terms of N_a, $\varepsilon_{\max}$, and T.

27-29. From the result (27.27) for the internal energy of the N_e conduction electrons at low temperatures, find an expression for the heat capacity of these electrons at low temperatures in terms of N_e, μ, and T.

27-30. In a typical piece of metal, $N_a = N_e = 10^{25}$, the chemical potential of the conduction electrons is typically $\mu = 250$ eV, and the value of ε_{max} for the lattice vibrations is typically 0.02 eV.

    (a) With this information and the answers to Problems 27-28 and 27-29, find the temperature at which the heat capacity of the electrons equals that of the lattice for a typical piece of metal.

    (b) At a temperature of one-tenth that calculated in part (a), which contribution to the heat capacity will be larger and by how many times?

27-31. With the information given at the beginning of Problem 27-30, and the answers to Problems 27-28 and 27-29, calculate the values of the constants k_e and K_l appearing in Eq. 27.33 for a typical metal.

27-32. Consider a supermetal that will not melt at any temperature. Carefully, make a qualitative plot of heat capacity versus temperature for this supermetal, which considers both the contributions of the lattice and the conduction electrons for all ranges of temperature. (N = number of atoms = number of conduction electrons.) Label the C_V axis in units of Nk. Be sure your graph depicts the following effects.

    (a) For very low temperatures, $T^3 \ll T$, so only the conduction electrons contribute.
    (b) For medium low temperatures, $T^3 > T$, so the lattice term dominates.
    (c) For medium high T, the lattice term becomes constant, but then the electron term slowly increases linearly in T.
    (d) For very high temperatures, the conduction electrons are no longer degenerate, and the classical value of the metal's heat capacity is reached.

27-33. The velocity of sound (phonon velocity) in copper is 3.5 km/s. Its density is 8.9 g/cm^3, and its atomic mass number is 64.

    (a) What would be its molar heat capacity at high temperatures due to the lattice vibrations?
    (b) Calculate ε_{max} from Eq. 27.11.
    (c) What is the Debye temperature, $\theta = \varepsilon_{max}/k$, for copper?

27-34. According to the equipartition theorem, the number of degrees of freedom associated with $3N$ simple harmonic oscillators is $6N$. (Each one has 2 degrees of freedom. $\varepsilon_i = (1/2m)p_i^2 + (\frac{1}{2}k_i)x_i^2$.) Hence, the energy of this system should be $(6N)\frac{1}{2}kT = 3NkT$, and the heat capacity should be $C_V = 3Nk$. But according to the Debye theory we found this is true only for sufficiently high temperatures. Does this apparent contradiction between the equipartition theorem and the Debye theory mean that one of them must necessarily be wrong at low temperatures? If not, why not?

APPENDIX 27A THE INTEGRAL (27.16)

Consider the integral (27.16).

$$\int_0^{(\varepsilon_{max}/kT)} \frac{x^3\, dx}{e^x - 1}$$

In the limit $T \to 0$, the upper limit goes to ∞, and the integral is the standard integral.

$$\int_0^\infty \frac{x^3 \, dx}{e^x - 1} = \pi^2/15 \qquad (T \to 0)$$

In the other limit of very large T, the upper limit is very small. In this case the variable x is small throughout the range of integration, and

$$e^x - 1 \approx x \qquad (x \ll 1)$$

In this region the integral is

$$\int_0^{(\varepsilon_{max}/kT)} \frac{x^3 \, dx}{x} = \int_0^{(\varepsilon_{max}/kT)} x^2 \, dx = \frac{1}{3} \left(\frac{\varepsilon_{max}}{kT} \right)^3$$

APPENDIX 27B   DEBYE MODEL NOTATION

Scientists often prefer thinking in terms of phonon angular frequencies, ω, rather than energies, ε.

$$\varepsilon = \hbar\omega$$

The maximum possible phonon energy, ε_{max}, corresponds to a maximum angular frequency, called the "Debye frequency," and given the symbol ω_D.

$$\varepsilon_{max} = \hbar\omega_D$$

Also, a "Debye temperature," θ_D, is sometimes defined through

$$\varepsilon_{max} = k\theta_D = \hbar\omega_D$$

With these substitutions, the relationships (27.11), (27.12), (27,15), and (27.17) become

$$\omega_D = \left(\frac{6\pi^2 N}{V} \right)^{1/3} c_s \qquad\qquad (27.11')$$

$$dN_p = \begin{cases} \left(\dfrac{9N}{\omega_D^3} \right) \dfrac{\omega^2 \, d\omega}{e^{\beta\hbar\omega} - 1} & (\omega \le \omega_D) \\[2ex] 0 & (\omega > \omega_D) \end{cases} \qquad (27.12')$$

$$E = \left(\frac{9N\hbar}{\omega_D^3} \right) \int_0^{\omega_D} \frac{\omega^3 \, d\omega}{e^{\beta\hbar\omega} - 1}$$

$$= \left(\frac{9N}{\hbar^3 \omega_D^3} \right) (kT)^4 \int_0^{(\theta_D/T)} \frac{x^3 \, dx}{e^x - 1} \qquad (27.15')$$

$$E = \begin{cases} \left(\dfrac{3N\pi^2 k}{5\theta_D^3} \right) T^4 & T \ll \theta_D \\[2ex] 3NkT & T \gg \theta_D \end{cases} \qquad (27.17')$$

chapter 28

SEMICONDUCTORS AND INSULATORS

Semiconductors and insulators have few conduction electrons because of the rather large energies required to strip the outer valence electrons from the individual atoms. The electronic states, then, come in groups, with the states for conduction electrons higher in energy above the states for valence electrons. With this picture for the groupings of electron states and the known dependence of occupation numbers on electron energies, we can analyze the electrical properties of these materials.

We now turn our attention to the electrical properties of semiconductors and insulators. In these materials, each atom holds onto its electron rather strongly, and only the outer electron has any chance to be stripped from the atom and help contribute to the electrical conductivity of the material. Consequently, the system of particular interest to us in this section will be the system consisting of the outer electron of each atom, and when we refer to the "electrons" in the semiconductor or insulator, it will be this particular set of electrons that we will be referring to. Normally, they are tied to the individual atoms as "valence electrons," but under some conditions they may be stripped from the atoms and flow through the material as "conduction electrons."

A. BAND STRUCTURE

The states accessible to the electrons of semiconductors and insulators come in bands. Within each band, the spacings of states are so small that they can be considered continuous, but the size of the gap between the bands is quite appreciable, often being several electron volts.

If the material contains no impurities and no crystalline imperfections, then these will be the only states available to the electrons. Such perfectly pure materials are called "intrinsic." We will deal with intrinsic materials throughout most of this chapter, saving a discussion of the effect of impurities for the end.

The lower-lying band is called the "valence band," and the upper one is called the "conduction band." Each band contains exactly as many electron states as there are atoms,* so at absolute zero, when the lowest levels are filled, the valence band is completely full, and the conduction band is completely empty.

* Actually, the number of states in a band may sometimes be small integral multiples (e.g., 2 or 3) of the number of atoms, when there are bands that happen to overlap. But such considerations have negligible effect on our discussions here, so we ignore them.

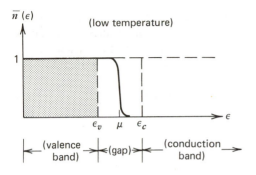

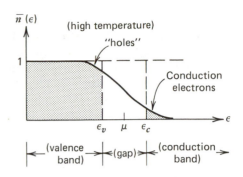

Figure 28.1 Plots of occupation number versus energy of state for semiconductors or insulators. At low temperatures, the valence band is completely full and the conduction band completely empty. At higher temperatures, the "tail" of the Fermi distribution might reach across the band gap, causing some "holes" in the valence band, and some electrons in the conduction band.

As the temperature rises, there will be some finite probability of finding electrons in excited states (Figure 28.1). A quick calculation, however, shows that the excitation temperature,

$$T_e = \frac{1}{k} \Delta \varepsilon_{\text{gap}}$$

is several tens of thousands of degrees Kelvin for a band gap of several electron volts. Clearly, at normal temperatures, the probability of excitation into the conduction band is extremely small. Even though a typical system would have 10^{24} or 10^{25} electrons, we would expect few if any to be in the conduction band at any instant.

If there are no electrons in the conduction band, then these states cannot contribute to the electrical conductivity of the material. In such a case, the valence band is full, so there are no vacant states there for the electrons to move into. Consequently, the electrons are not free, cannot move into new states when electrical fields are applied, and so the valence electrons cannot contribute to the electrical conductivity of the material. It is an insulator.

However, when some electrons are excited into the conduction band, they find many vacant states there that they can move into in response to applied electrical fields. In addition, they leave behind vacancies, or "holes," in the valence band. These holes in the valence band give the remaining valence electrons some limited number of vacant states which they too can move into in response to applied fields. Thus, the excitation of some electrons to the conduction band allows both the conduction band and the valence band to make some contribution to the electrical conductivity of the material.

The electrical conduction by holes in the valence band is illustrated in Figure 12.7. Since the electrons are many and the holes are few, movement of electrons must be sequential; one electron fills a hole but leaves another, then the next electron fills that hole and leaves another, etc. Since the absence of an electron leaves a net positive charge behind, a hole in the valence band can be thought of as some sort of positively charged particle. The sequential movement of electrons in one direction is equivalent to the movement of a positively charged "hole" in the other direction. In instrinsic semiconductors, electrical current is carried both by electrons in the conduction band and by holes in the valence band.

If we treat electrons in the conduction band as a quantum gas, then the energy and momentum of these particles is related through*

$$\varepsilon = \varepsilon_0 + \frac{p^2}{2m} \tag{28.1}$$

where ε_0 is the lowest energy possible, and corresponds to particles having no momentum. That is, the lowest state in the conduction band corresponds to particles with no kinetic energy. Higher states correspond to greater kinetic energy.

The distribution of particles in a gas is given by the product of the number of states times the occupation number of each.

$$dN = \left(\frac{d^3r \, d^3p}{h^3}\right) \bar{n}(\varepsilon) \tag{28.2}$$

After integrating over all volume and all momentum directions, and adding in contributions of both spin-up and spin-down electrons, this becomes (see Eq. 25.7)

$$dN = \left(\frac{8\pi V p^2 \, dp}{h^3}\right) \bar{n}(\varepsilon) \tag{28.3}$$

Converting the momentum to energy according to Eq. 28.1 above, this becomes

$$dN = (C\sqrt{\varepsilon - \varepsilon_0} \, d\varepsilon) \left(\frac{1}{e^{\beta(\varepsilon - \mu)} + 1}\right) \tag{28.4}$$

where

$$C = \left(\frac{4\pi V(2m)^{3/2}}{h^3}\right)$$

The total number of electrons in the conduction band can be obtained by integrating this over all $\varepsilon > \varepsilon_0$. But this is a difficult integral, best being done numerically, and the qualitative features of the result can be seen graphically. If we write this as

$$\frac{dN}{d\varepsilon} = (C\sqrt{\varepsilon - \varepsilon_0})\bar{n}(\varepsilon)$$

we see that the number of particles per unit range in energy is the product of two factors. The first is the number of quantum states, or "phase space" factor, which increases as

---

* This is not necessarily the exact relationship between energy and momentum, but it is a model commonly used, especially for states near the lower edge of the conduction band (and for hole states near the upper edge of the valence band) where most of the conduction electrons (and hole) would be found.

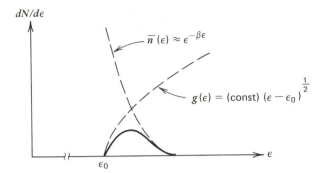

dN/dε

$\bar{n}(\epsilon) \approx \epsilon^{-\beta\epsilon}$

$g(\epsilon) = (\text{const})\,(\epsilon - \epsilon_0)^{\frac{1}{2}}$

ϵ_0

ε

Figure 28.2 Illustration of the distribution of electrons in the conduction band. The electron distribution is the product of the density of states times the occupation number.

the energy (hence the momentum) increases. The second is the occupation number, which falls off exponentially with energy. These two factors, and their product are shown in Figure 28.2, illustrating how the number of particles in the conduction band varies with energy.

Clearly, this cannot be exactly correct, because the conduction band is of finite extent in energy. But the above density of states, $(C\sqrt{\varepsilon - \varepsilon_0})$ extends forever. Fortunately, it is only necessary that the density of states be accurate near the lower edge of the conduction band, because the occupation number, $\bar{n}(\varepsilon)$, falls off exponentially with energy, cutting out any contributions from higher-lying states anyhow. For similar reasons, we are only interested in the density of states for holes near the upper edge of the valence band, and usually a similar expression, $(C\sqrt{\varepsilon_0 - \varepsilon})$ does just fine. The occupation number for holes is the probability that a state is *not* occupied (see Figure 28.3).

$$\bar{n}_{\text{holes}} = (1 - \bar{n}_{\text{electrons}})$$

SUMMARY

The electron states for semiconductors and insulators come in bands, with the spacings between neighboring states within the bands being small, and the gap between the bands relatively large, measured in electron volts. For intrinsic materials, there are no other accessible states.

At low temperatures, the valence band is completely full and the conduction band completely empty. The material is an insulator. At higher temperatures, some of the valence electrons may jump the gap into the conduction band, and electrical conduction may be accomplished by both these conduction electrons and the positively charged "holes" left behind in the valence band. The distribution of both conduction electrons and hole is given by the product of the distribution of accessible states times the occupation number of each.

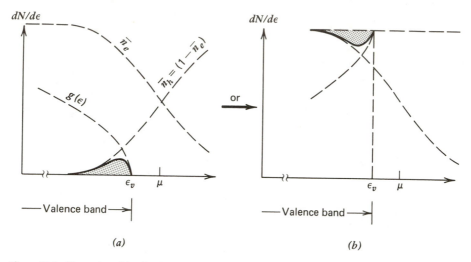

Figure 28.3   Illustration of the distribution of holes in the valence band. (*a*) Since a hole is the absence of an electron, the occupation number for holes ($\bar{n}_h$) is given by $(1 - \bar{n}_e)$, when $\bar{n}_e$ is the electron occupation number. The distribution of holes is the product of the density of electron states times the occupation number for holes. (*b*) A common alternate way of representing the distribution of holes in the valence band.

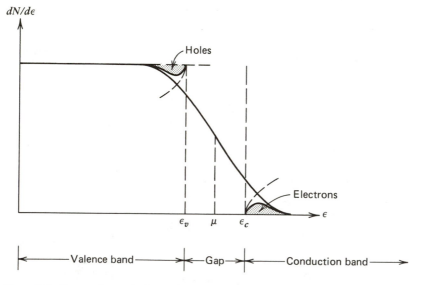

Figure 28.4   If every electron in the conduction band comes from the valence band, then there must be as many holes in the valence band as there are electrons in the conduction band. Consequently, the chemical potential would lie midway between the two.

B. ELECTRICAL PROPERTIES

Since any electron reaching the conduction band must have come from the valence band, there must be one unoccupied state in the valence band for every electron in the conduction band. This means that the chemical potential must lie midway between the two bands for an intrinsic material, as illustrated in Figure 28.4.

If the chemical potential lies between the bands, then the energies $(\varepsilon_c - \mu)$ and $(\varepsilon_v - \mu)$ for states in the conduction band and valence band, respectively, would be measured in electron volts. By contrast, the factor, kT, is very small for normal temperatures,

$$kT \ll 1 \text{ eV} \qquad \text{(normal temperatures)}$$

being about 0.026 eV at room temperatures. This means that the ratios $(\varepsilon - \mu)/kT$ are very large numbers under ordinary circumstances.

$$\frac{\varepsilon_c - \mu}{kT} \gg 1 \qquad \text{or} \qquad e^{\beta(\varepsilon_c - \mu)} \gg 1$$

$$\frac{\varepsilon_v - \mu}{kT} \ll -1 \qquad \text{or} \qquad e^{\beta(\varepsilon_v - \mu)} \ll 1$$

Consequently, the occupation numbers for states in the conduction and valence bands, under normal circumstances, can be written as

$$\bar{n}_c = \frac{1}{e^{\beta(\varepsilon_c - \mu)} + 1} \approx e^{-\beta(\varepsilon_c - \mu)}$$

$$\bar{n}_v = \frac{1}{e^{\beta(\varepsilon_v - \mu)} + 1} \approx 1 - e^{\beta(\varepsilon_v - \mu)} * \tag{28.5}$$

We see that if the chemical potential is midway between the two bands, then

$$e^{-\beta(\varepsilon_c - \mu)} = e^{\beta(\varepsilon_v - \mu)}$$

for states near the inside edge of the two bands, and therefore from Eq. 28.5

$$\bar{n}_v = 1 - \bar{n}_c \tag{28.6}$$

For each particle in the conduction band, there is one missing from the valence band.

We are now prepared to do some quantitative calculations involving the conduction properties of intrinsic semiconductors and insulators. Because the electrons in the conduction band lie very close to the lower edge of this band, ε_c, and because the holes in the valence band lie very close to the upper edge of this band, ε_v, (see Figures 28.2 to 28.4), it is often a good approximation to assume their energies are given by ε_c and ε_v, respectively. We do this in the following examples.

---

* We use the expansion $(1 + \varepsilon)^{-1} \approx 1 - \varepsilon$ for small ε.

EXAMPLE

Assuming a band gap of 2 eV, what is the occupation number of the lowest-lying state in the conduction band at room temperature?

Since the chemical potential is midway between the two bands,

$$\frac{\varepsilon_c - \mu}{kT} = \frac{1 \text{ eV}}{0.026 \text{ eV}} = 38$$

Therefore, according to Eq. 28.5, the occupation number is

$$\bar{n} = e^{-38} = 3 \times 10^{-17} \qquad (\Delta\varepsilon = 2 \text{ eV} \qquad T = 300 \text{ K})$$

EXAMPLE

What is the occupation number in the above example if the temperature is doubled? If the band gap is doubled?

If the temperature is doubled, then the factor in the exponent is halved.

$$\frac{\varepsilon_c - \mu}{kT} = \frac{1}{2} \times 38 = 19 \qquad (\Delta\varepsilon = 2 \text{ eV} \qquad T = 600 \text{ K})$$

If the band gap is doubled, the exponent is doubled.

$$\frac{\varepsilon_c - \mu}{kT} = 2 \times 38 = 76 \qquad (\Delta\varepsilon = 4 \text{ eV} \qquad T = 300 \text{ K})$$

This gives the following occupation numbers.

$$\bar{n}_c = e^{-19} = 6 \times 10^{-9} \qquad (\Delta\varepsilon = 2 \text{ eV} \qquad T = 600 \text{ K})$$

$$\bar{n}_c = e^{-76} = 10^{-33} \qquad (\Delta\varepsilon = 4 \text{ eV} \qquad T = 300 \text{ K})$$

For making rough calculations, we frequently make the approximations that all states in the conduction band have the energy of the lower band edge, and that all states in the valence band have the energy of the upper band edge. We know this is a poor approximation for two reasons. First, typical bandwidths are several electron volts, so a large fraction of the states have energies considerably different than that of the band edge. Second, because of the exponential falloff of the occupation numbers, only those states near the band edge will have appreciable occupation numbers, so only states near the band edge will have appreciable influence on the electrical properties of the material. Both these arguments seem to say that only some fraction of the states in a band should be considered as having the energy of the band edge. More sophisticated treatments of this subject take this into consideration.

But for rough qualitative calculations, this extra work is not necessary. The results are not particularly sensitive to the exact number of states used; they vary only linearly with the number of states, whereas they vary exponentially with other variables, such as energies or temperatures. Furthermore, for making comparisons, such as looking at ratios or relative changes, the number of states used is irrelevant. For these reasons, in addition to simplicity, we will assume in our treatment that all states in a band, rather than some fraction of them, have the band-edge energy.

The "electrons" we deal with in working with semiconductors and insulators are really only the outer electrons of the atoms. Only these have any chance of being stripped from the atoms and contributing to the electrical properties of the material. Equivalently, we could say that these electrons are normally in the valence band but may occasionally jump into the conduction band. There are as many of these electrons as there are atoms of the material, and there are as many states in the conduction band and in the valence band as there are atoms of the material, or as there are of these electrons.*

The number of electrons in the conduction band is the product of the number of states times the occupation number of each.

$$\text{(number of conduction electrons)} = \text{(number of states)} \times \bar{n}$$
$$= \text{(number of electrons)} \times \bar{n}$$

For a typical system of 10^{25} electrons, then, we get the following number of electrons in the conduction band for each of the cases in the previous examples.

$$N_c = 10^{25} \times (3 \times 10^{-18}) = 3 \times 10^7 \qquad (\Delta\varepsilon = 2 \text{ eV} \qquad T = 300 \text{ K})$$
$$10^{25} \times (6 \times 10^{-9}) = 6 \times 10^{16} \qquad (\Delta\varepsilon = 2 \text{ eV} \qquad T = 600 \text{ K})$$
$$10^{25} \times (10^{-33}) = 10^{-8} \approx 0 \qquad (\Delta\varepsilon = 4 \text{ eV} \qquad T = 300 \text{ K})$$

From this example, we see that rather small changes in the temperature or the size of the band gap, will have profound effects in the conductive properties of the material. This is because the occupation numbers, and therefore the electrical conductivity, are *exponential* in these two properties. The difference between a semiconductor and an insulator need be more than just a fraction of an electron volt in the size of the band gap.

The great sensitivity of conductive properties of these materials to the band gap and the temperature, means we can create very large changes in their conductivities through rather minor changes in their physical environment. Consider what happens, for example, when an insulator is placed in a large electric field. The external electric field tends to create a slightly different and somewhat polarized distribution of charge on an atomic level, so some of the electron states are now in regions of higher electrical potential and others in regions of lower electrical potential. Thus, the states within any one band tend to shift slightly, some to higher energies and some to lower energies, causing the band to spread in both directions (Figure 28.5).

As the two bands spread out, the size of the gap between them lessens, and as we have seen, this causes an enormous increase in conductivity. What we observe in our labs is

---

* To show that the number of states in the conduction band or valence band is the same as the number of electrons is not trivial, and is done in some quantum mechanics courses. If you don't wish to accept this on faith, you can use classical statistics. The exchange of electrons between valence and conduction bands is a diffusive interaction, and the tools developed for chemical equilibrium can be applied. If e_v represents a valence electron, e_c a conduction electron, and h a hole in the valence band, then the appropriate "chemical" equation for the diffusive interaction is

$$e_v \rightleftarrows h + e_c$$

The results gained using this approach are the same as those gained using the quantum statistics presented here.

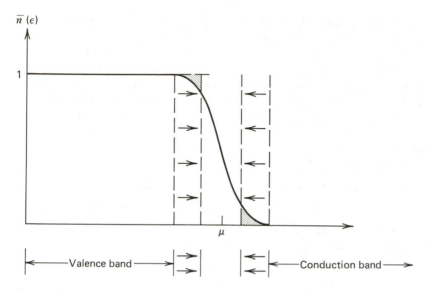

Figure 28.5 Applying pressure or an external electric field may cause a slight shift in the distribution of charge on a microscopic scale within the material. This shift may cause changes in the energies of the various electron states, resulting in increasing band widths. As the bands spread, they may cross over into the "tail" region of the Fermi distribution, and the material may become a conductor.

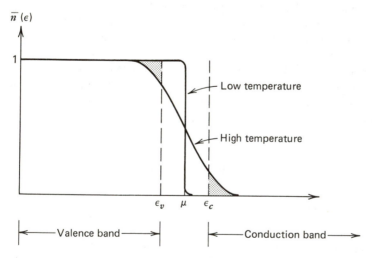

Figure 28.6 At higher temperatures the tail of the Fermi distribution becomes broader, and may cross the edges of the valence and conduction bands. An insulator at low temperatures may become a conductor at high temperatures.

that when the field intensity is increased beyond a certain point, the insulator suddenly breaks down and becomes conducting. Now we know why.

In some materials a similar effect happens when they are subjected to changes in pressure. Increased pressure causes minor shifts in charge distributions on an atomic scale, which causes shifts in the electrical potential energy of electron states—some to higher energies and some to lower energies. Thus, the increased pressure causes the bands to spread, narrowing the gap between them, and causing large increase in conductivity. This is called the "piezoelectric effect," and is used in very sensitive pressure-sensing devices.

Demonstrations of the sensitivity of electrical properties to temperature include insulators that break down and become conducting at high temperatures. Sensitive temperature measuring devices are based on the large variation in a semiconductor's conductivity with changes in temperature (Figure 28.6). This sensitivity of semiconductors to rather modest changes in their environment is the basis of a large and rapidly growing number of semiconductor devices.

SUMMARY

For intrinsic materials, the chemical potential lies midway between the valence and conduction bands, because for each electron in the conduction band there must be a corresponding hole in the valence band.

If we assume the band gap to be large compared to the width of the bands, and the number of states in the conduction band to be the same as the number of states in the valence band (i.e., within small multiplicative factors), then the probability of an electron jumping the gap is roughly the same as the occupation number of a state in the conduction band, being given by

$$\text{probability} \approx \bar{n} \approx e^{-\beta(\varepsilon_c - \mu)} = e^{-\beta(\Delta\varepsilon_g/2)} \tag{28.7}$$

where $\Delta\varepsilon_g$ is the width of the band gap. For a typical band gap of 2 eV and at room temperature, this probability is about 10^{-17}.

Since the probability for an electron to jump the gap varies exponentially with gap width and temperature, small changes in either of these make large changes in the electrical properties of semiconductors. This sensitivity is the basis for a large variety of technical applications.

PROBLEMS

28-1. For an intrinsic material of 10^{25} valence electrons and a band gap of 2 eV, at what temperature will there be one electron in the conduction band?

28-2. For an intrinsic material of 10^{26} valence electrons at room temperature, what is the size of the band gap if there is one conduction electron?

28-3. Consider an intrinsic material having 10^{26} valence electrons and a band gap of 3.0 eV. At what temperature will the number of conduction electrons be:

(a) 1?

(b) 10,000?

(c) What was the percentage change in temperature that caused this 10,000-fold increase in conductivity?

28-4. Consider an intrinsic material having 10^{25} valence electrons and a band gap of 2.0 eV, originally at room temperature. By how many degrees must the temperature increase if a 100% increase in conductivity (i.e., number of conduction electrons) is desired?

28-5. Consider an intrinsic material at room temperature. If under pressure, the size of the band gap is reduced from 2.2 eV to 2.1 eV, by what factor does the electrical conductivity (i.e., number of conduction electrons) increase?

28-6. In a certain insulator at room temperature, only one out of every 6.022×10^{23} of the lowest-lying states in the conduction band is occupied. How many electron volts separate the conduction from the Fermi surface (μ)?

C. IMPURITIES

The band structure in semiconductors is caused by a periodic array of identical atoms or groups of atoms. If impurities are introduced, such as atoms of different types or as

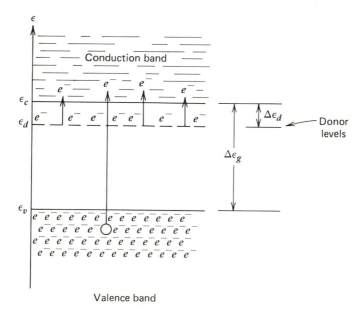

Figure 28.7 Donor impurities may also contribute electrons to the conduction band. The valence band contains far more electrons available for excitation than do the donor levels, but the probability of excitation for any single electron from the valence band is far smaller than that from a donor level, due to the greater energy required.

imperfections in the periodic array, they cause disruptions of the band structure, often leading to electron states lying somewhere between the two bands in the band gap.

Because the probability of excitation falls off exponentially in energy, excitation of electrons to or from these intermediate impurity levels is much more probable than excitation all the way across the band gap. Consequently, a relatively small concentration of impurities (parts per million or parts per billion being typical) may have a dominant influence on the electrical properties of a semiconductor, especially at low temperatures ($kT \ll \Delta\varepsilon_g$) where the band gap is essentially insurmountable.

Some impurities have loosely held electrons that can be released with relatively small thermal agitation. These impurities are called "donors," because they readily donate electrons to the conduction band. On an energy level diagram, these donor states would be found only slightly below the conduction band, reflecting the relatively small amount of thermal energy needed to shake these electrons loose into the conduction band (Figure 28.7).

Other impurities would readily take on an extra electron. They readily "borrow" electrons from neighboring atoms in the material with the help of a little thermal agitation. These are called "acceptor" impurities. On an energy level diagram, these acceptor states would be found only slightly above the valence band, reflecting the relative ease of excitation of a valence electron into one of these acceptor levels (Figure 28.8).

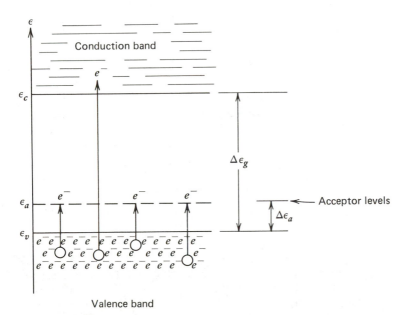

Figure 28.8 Many holes in the valence band may arise from the excitation of valence electrons into acceptor levels. There are far more vacant states in the conduction band than there are in acceptor levels, but the probability of excitation to any one of them is smaller, due to the greater energy required.

If the electrical properties of a semiconductor are dominated by conduction electrons excited from donor levels, it is said to be an "n-type" semiconductor. Similarly, a "p-type" semiconductor is one whose electrical properties are dominated by holes in the valence band caused by excitation of valence electrons to acceptor levels.

We have seen that the probability of an electron jumping the band gap, from valence to conduction bands, is extremely sensitive to the temperature. At low temperatures, the band gap is insurmountable, and all electrical conduction is attributable to impurities. At low temperatures, small excitations may be possible, but large ones are not. All electrons in the conduction band (if any) must have come from donor levels, and all holes in the valence band (if any) must have come from excitation to acceptor levels.

In the other extreme, the band gap becomes easily surmounted at high temperatures. Even though the probability of excitation across the band gap is still smaller than the probability for excitation from an impurity level, there are far more electrons ready to be excited across the band gap than there are electrons in donor levels or vacancies in acceptor levels. Thus, at high temperatures you can ignore impurities and the semiconductor becomes "intrinsic."

We will now make a rough calculation of the number of excitations from level ε_i, originally full, to level ε_j, originally empty, for the case when excitations involving any other levels are negligible. For example, these two levels would be the valence band and the conduction band (ε_v and ε_c) in intrinsic materials, the valence band and acceptor levels (ε_v and ε_a) in p-type materials, and the donor levels and conduction band (ε_d and ε_c) in n-type materials. We will assume that all states in the valence band have the energy of the upper band edge, and that all states in the conduction band have the energy of the lower band edge. As we have seen, this is not always a very good approximation, but it simplifies the calculations a great deal, and the results obtained will still be qualitatively correct.

Suppose that there are N_i lower-lying states of energy ε_i, and N_j higher-lying states of energy ε_j. Also suppose that kT is small compared to the transition energy required, $kT \ll (\varepsilon_j - \varepsilon_i)$, so that most lower-lying states are full and most higher-lying states are vacant. For this case, the Fermi surface (μ) must lie somewhere between the two levels.

The number of electrons in the upper level is the product of the number of states times the occupation number of each, $N_j \bar{n}_j$. Similarly, the number of vacancies in the lower level would be given by the product of the number of states times the probability of one being empty, $N_i(1 - \bar{n}_i)$. Since each electron excited to the higher level leaves a vacancy in the lower level, these two numbers must be the same.

$$N_j \bar{n}_j = N_i(1 - \bar{n}_i) \tag{28.8}$$

It is conservation of *particles* that this relationship guarantees, not conservation of probabilities. This means that the Fermi surface (μ) will lie midway between the two levels only if $N_i = N_j$, as we will see as follows.

If kT is small compared to the energy differences ($\varepsilon_j - \mu$) and ($\mu - \varepsilon_i$), then we can write

$$\bar{n}_j = \frac{1}{e^{\beta(\varepsilon_j - \mu)} + 1} \approx e^{-\beta(\varepsilon_j - \mu)}$$

and

$$1 - \bar{n}_i = 1 - \frac{1}{e^{\beta(\varepsilon_i - \mu)} + 1} \approx e^{\beta(\varepsilon_i - \mu)}$$

With these inserted into Eq. 28.8, we find

$$e^{\beta\mu} = \left(\frac{N_i}{N_j}\right)^{1/2} e^{\beta(\varepsilon_i + \varepsilon_j)/2}$$

Notice that the Fermi surface would lie midway between the two energies ($\mu = (\varepsilon_i + \varepsilon_j)/2$) only if the number of states were the same for both levels ($N_i = N_j$). This was the case for excitation between valence and conduction bands, so we correctly placed the Fermi surface midway between them in the previous section on intrinsic materials. But it is not true when impurity levels are involved (Figure 28.9), as you will investigate further in homework problem 28-7.

We can now use this expression for $e^{\beta\mu}$ to calculate the number of excitations, N_x, which is the product of the number of higher-lying states, N_j, times the occupation number for each, $\bar{n}_j$.

$$N_x = N_j \bar{n}_j \approx N_j e^{-\beta(\varepsilon_j - \mu)} = (N_i N_j)^{1/2} e^{-\beta \Delta \varepsilon/2} \tag{28.9}$$

where

$$\Delta \varepsilon = \varepsilon_j - \varepsilon_i$$

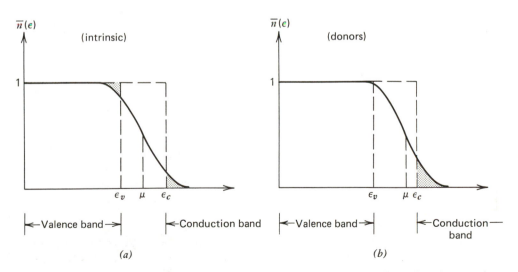

Figure 28.9 (a) In intrinsic semiconductors, the chemical potential (or "Fermi level") lies midway between the valence and conduction bands because each electron excited to the conduction band leaves a hole in the valence band, so there must be equal numbers of each. (b) If some of the conduction electrons come from donor impurities, then there will be more electrons in the conduction band than there are holes in the valence band. Consequently, the chemical potential would lie closer to the conduction band in this case.

SUMMARY

Suppose we assume the valence band can be represented by N_v electron states all having energy ε_v, and that the conduction band can be represented by $N_c \, (= N_v)$ electron states all having energy ε_c. In addition, a semiconductor may have N_a acceptor states of energy ε_a, or N_d donor states of energy ε_d. If the semiconductor is n type, p type, or intrinsic, then transitions between only two of these levels are predominant, and we can write the number of such excitations between levels i and j as

$$N_x \approx (N_i N_j)^{1/2} e^{-\beta \, \Delta\varepsilon/2} \tag{28.9}$$

where

$$\Delta\varepsilon = \varepsilon_j - \varepsilon_i$$

providing that kT is small compared to the energies involved.

For example, the number of holes in the valence band due to excitations of valence electrons to acceptor levels in a p-type material is

$$N_h \approx (N_v N_a)^{1/2} e^{-\beta \, \Delta\varepsilon_a/2} \tag{28.9a}$$

where

$$\Delta\varepsilon_a = \varepsilon_a - \varepsilon_v$$

Similarly, the number of conduction electrons excited from donor levels in a n-type material is

$$N_e \approx (N_d N_c)^{1/2} e^{-\beta \, \Delta\varepsilon_d/2} \tag{28.9b}$$

where

$$\Delta\varepsilon_d = \varepsilon_e - \varepsilon_d$$

and the number of conduction electrons or holes in an intrinsic material is given by

$$N_h = N_e \approx (N_v N_c)^{1/2} e^{-\beta \, \Delta\varepsilon_g/2} \tag{28.9c}$$

where

$$\Delta\varepsilon_g = \varepsilon_c - \varepsilon_v$$

This last result is the same as that obtained for intrinsic materials in the previous section because $N_v = N_c$.

At low temperatures the band gap is insurmountable, and the electrical properties are dominated by impurity contributions. At high temperatures, the material becomes "intrinsic," because there are far more electrons available to be excited across the band gap than there are electrons in donor levels or vacancies in acceptor levels.

Sometimes we would like to have a rough idea of the temperature below which impurities dominate the electrical properties and above which conduction is predominantly intrinsic. We know that at the temperature of transition between these two extremes, the number of charge carriers (electrons or holes) that are intrinsic will about equal those from impurities. Consequently, we can set the formulas for the two types of charge carriers equal to each other and solve for the temperature. Because of the approximations we made in deriving the equations (28.9a,b, and c), we expect our result to be only roughly correct.

For example, suppose our sample has N_d donor impurities, whose energy levels lie $\Delta\varepsilon_d$ below the conduction band. Then at the temperature of transition between "*n*-type" and "intrinsic" electrical properties, the conduction electrons from each of these two sources will be about equal.

$$(N_d N_c)^{1/2} e^{-\beta\, \Delta\varepsilon_d/2} \approx (N_v N_c)^{1/2} e^{-\beta\, \Delta\varepsilon_g/2}$$

Squaring these, taking the logarithms, and using $N_v = N_c$, yields

$$T \approx \frac{\Delta\varepsilon_g - \Delta\varepsilon_d}{k \ln(N_c/N_d)} \qquad (28.10)$$

for the temperature above which conduction is predominantly intrinsic. This formula could also be used to get an idea of the maximum (or minimum) number of donor impurities permissible to keep the material intrinsic (or impurity dominated) at any given temperature. A similar formula can be developed to get a rough idea of the temperature of transition to intrinsic properties in a *p*-type semiconductor. (See homework Problem 28-8.)

PROBLEMS

28-7.   Suppose the donor levels lay 0.2 eV below the conduction band in an *n*-type semiconductor. Suppose there were 10^{17} donor impurities and 10^{24} atoms of the intrinsic material (therefore, $N_v = N_c = 10^{24}$). If the chemical potential lay midway between the donor levels and the conduction band:

   (a)   How many vacancies would there be in the donor levels at room temperature?
   (b)   How many electrons would there be in the conduction band?
   (c)   Would there be one donor vacancy for every electron in the conduction band?
   (d)   In order to make these two numbers equal, would μ have to be closer to ε_d or to ε_c?

28-8.   Derive a formula for the temperature of transition to intrinsic conduction for a *p*-type semiconductor in terms of $\Delta\varepsilon_g$, $\Delta\varepsilon_a$, N_v, and N_a. (*Hint.* Follow a development similar to that for *n*-type materials done above.)

28-9.   A certain semiconductor contains 10^{26} valence electrons and 10^{14} donor impurities. If the band gap is 2.2 eV and the impurity level lies 0.2 eV below the conduction band:

   (a)   How many conduction electrons are there at room temperature?
   (b)   Roughly what is the temperature of transition to intrinsic conduction?

28-10.   A certain semiconductor has 10^{25} valence electrons, 10^{16} acceptor impurities, and a band gap of 1.8 eV. If there are 2×10^{12} holes in the valence band at a temperature of 350 K, how far above the valence band do the impurity levels lie?

28-11.   Consider a semiconductor having 10^{25} valence electrons, a band gap of 2.0 eV, and 10^{15} donor impurities with donor levels lying 0.1 eV below the conduction band. Above roughly what temperature will the conduction become intrinsic?

28-12.   Consider a semiconductor having 10^{24} atoms and a band gap of 1.8 eV. If it is contaminated with donor impurities whose energy levels lie 0.16 eV below the conduction band, roughly what is the maximum number of impurities this sample may contain if it is to remain intrinsic down to room temperature?

---

chapter 29

LOW TEMPERATURES

At low temperatures, the particles of systems tend to become trapped in the configuration of lowest possible energy. The temperature at which this happens depends on the amount of energy required to excite particles out of the lowest energy states. For some systems, this transition occurs at temperatures above room temperature, and for others the transitions occur at temperatures far smaller than are obtainable in laboratories. But for some, the transitions occur at low, but obtainable, temperatures. These systems take on properties that seem quite strange compared to the properties we are familiar with at normal temperatures.

At absolute zero, a system of particles would be in the configuration of lowest energy. For bosons, this would mean all particles in the lowest-lying state. For a system of N identical fermions, this would mean the lowest N states filled and all higher states empty, since no more than one can occupy any one state. When a system is in this configuration of lowest possible energy, we call the system "degenerate."

At temperatures above absolute zero, there will be some probability of finding some of the particles in excited states. If $\Delta\varepsilon$ represents the minimum energy necessary to excite a particle into an excited state, then we can define an "excitation temperature," T_e, by

$$kT_e = \Delta\varepsilon \tag{29.1}$$

For temperatures large compared to this excitation temperature, there is a rather large probability for any particle to be found in an excited state.

However, at temperatures smaller than T_e, the probability of excitation is small, or alternatively, the occupation number of excited states is small. Such a system is degenerate (or nearly degenerate), and the particles are confined to the lowest accessible quantum states.

Quantum effects are important in degenerate systems, as the transition to an excited state is a discrete, discontinuous jump, having a discrete, discontinuous effect on the physical properties of the system. At higher temperatures, by contrast, the very large number of particles in excited states allows very large numbers of rearrangements with a large and seemingly continuous range of effects on the physical properties of the system (Figure 29.1). Therefore, at higher temperatures, the properties of the system vary continuously in response to external conditions, rather than discretely.

We do not need extremely low temperatures to observe quantum effects. Many common ordinary systems are degenerate at room temperature, and exhibit strong quantum behavior. The excitation temperature for vibrations of air molecules is typically several hundred degrees Celsius. That for the excitation of electrons in

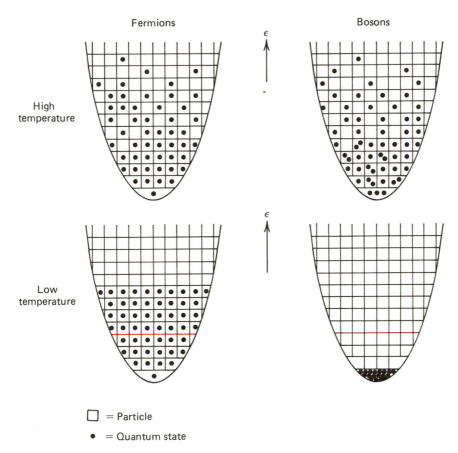

Fermions   Bosons

High
temperature

Low
temperature

☐ = Particle

● = Quantum state

Figure 29.1 When temperature is high, such that *kT* is large compared with the spacings between states, particles have a great deal of freedom, because there are many vacant neighboring states that particles may move into in response to external stimuli. Physical properties of the material seem to vary smoothly in response to external conditions. At low temperatures, particles find no vacant neighboring states. The only possible transitions are upward and require considerable energy (in comparison with *kT*). The physical properties of the system do not vary smoothly, but rather in discrete quantum jumps.

ordinary molecules is typically 10,000 K, and that for nuclear excitation is normally several million degrees Kelvin. The influence on the heat capacity of air or the fact that the electrons and nucleons may move continuously in their orbits without slowing down one bit or radiating any energy at all (Figure 29.2) does not seem strange to us because we are familiar with them.

However, when "normal" familiar systems start exhibiting quantum effects, we consider this behavior to be "strange." It's not that quantum effects in themselves are strange; we are surrounded by many that are quite familiar and seem quite "natural" to us. It is just that when "old, familiar" nondegenerate systems suddenly take on new and different properties, we become alerted and intrigued. It would be like our close friends

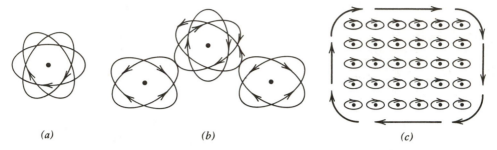

(a)   (b)   (c)

Figure 29.2   In some forms, superconductivity is a common ordinary phenomenon at room temperature. The electrons in atoms (a) and molecules (b) continue in their orbits, without slowing down or stopping, even in the absence of any sustaining electromotive force. Permanently magnetized ferromagnets (c) are one common macroscopic manifestation of this.

suddenly growing fins and scales and taking a fancy toward water. We are quite familiar with fish; we are just not used to our friends taking on that form.

We are not used to liquids that have no viscosity, such that whirlpools in them will spin without stopping, and such that they flow through the tiniest crack without impediment. We are not used to the need for capping the containment vessel so the liquid doesn't flow out by climbing up over the walls. We are not used to conductors having absolutely no electrical resistance at all so that electrical currents could flow forever around a loop without stopping or requiring any sustaining electromotive force. Yet these are the characteristics displayed by liquid helium and many conductors when they are at sufficiently low temperatures.

In this chapter, we examine these two particular kinds of systems, called "superfluids" and "superconductors," respectively, using quantum statistics to investigate these quantum behaviors.

A.   ATTAINING AND MEASURING LOW TEMPERATURES

We find ourselves living in an environment whose temperature is around 300 K. Through rather straightforward mechanical manipulation, we can refrigerate small regions to temperatures that are 10 or 20% lower than this. But because our refrigerators are in an ambient environment of about 300 K, and because the objects we refrigerate start out at these rather high temperatures, it becomes increasingly difficult to attain and maintain systems at lower and lower temperatures. To arrive at temperatures of a few degrees Kelvin and lower requires a great deal of sophistication.

Things can be cooled by transferring heat to other objects, and this is why you put ice cubes in lemonade, of course. But ice cubes won't work if you wished to attain temperatures much lower than 0°C. If the object we are trying to cool is colder than anything we can bring nearby, then we must insulate it from the environment, because any heat transfer would be in the wrong direction.

For this reason, cooling of systems to very low temperatures must be done *adiabatically*, through either diffusive or mechanical interactions. We appreciate the

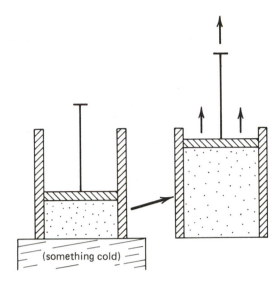

Figure 29.3 If we bring a system under pressure into thermal equilibrium with the coldest thing we can find, and then let the system expand adiabatically, it will cool still further.

(something cold)

effect of diffusive cooling when we step out of a hot shower, and the moisture on our skin begins to evaporate. At very low temperatures, we can put liquid nitrogen, hydrogen, or helium, for example, in a partial vacuum, and the vaporization of some of the liquid cools that which remains.

We are also familiar with the adiabatic cooling of an object through mechanical interactions. As a system under pressure is allowed to expand, it cools. This is how the refrigerant in our refrigerators works, and it is how we are able to get various gases cold enough to condense in the liquid form (Figure 29.3).

For example, we can allow ice-cold air under extreme pressure to expand adiabatically, getting liquid nitrogen to condense. Then we can allow compressed hydrogen gas, originally at liquid nitrogen temperature, to expand adiabatically and cool sufficiently that some liquid hydrogen condenses. Then we can allow compressed helium gas, originally at liquid hydrogen temperature, to expand adiabatically and cool sufficiently that liquid helium forms. This provides a bath of about 4 K. If we pump a vacuum on this liquid helium it begins to vaporize, and through this adiabatic diffusive interaction we can reduce the temperature to about 1 K (see Figure 29.4).

We can immerse systems in cold liquid helium baths, produced in this manner, in order to reduce their temperatures to 1 K. But what do we do to produce temperatures lower than this?

Adiabatic expansion of objects is ineffective at producing temperatures lower than this, because vessels that would hold the system under large pressures and allow it to expand would have to be constructed sturdily. Their high heat capacity and the friction of the moving parts would defeat our efforts to reduce the temperature of the system significantly.

We need a way of allowing the system to cool down by performing work adiabatically, which does not require the use of containers with large heat capacities

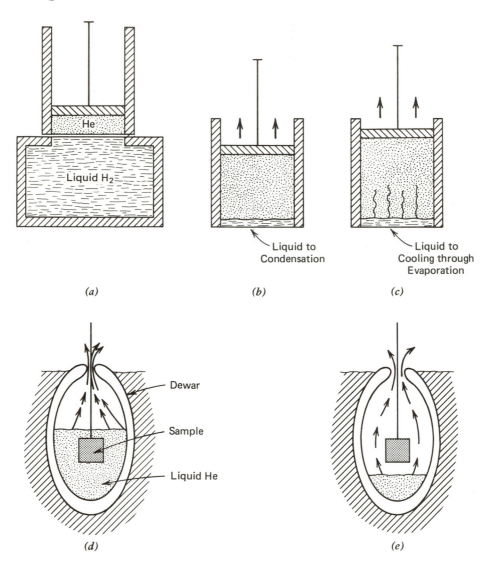

Figure 29.4   (a) If we bring helium under pressure into thermal equilibrium with liquid hydrogen, and then (b) let it expand adiabatically, it will continue to cool, and some liquid helium will condense. (c) If we continue to lower the pressure on the system, the vaporization of some of the liquid will cool the remaining liquid still further. (d) If our sample is initially immersed in this liquid helium, then after the helium has vaporized and has been removed from the Dewar flask (e) the sample will be insulated from the environment and be able to be cooled still further by some adiabatic process.

and moving parts. This is accomplished by allowing the system to do magnetic work, and the process is called "adiabatic demagnetization."

If the system contains some particles with magnetic moments, then these magnetic moments tend to line up in the presence of a strong magnetic field. If the system is insulated from its environment and then the field is turned off, these magnetic moments find themselves in a state of very low entropy (Figure 29.5). A high degree of order (e.g.,

$M = 4\,\mu$

$M = 0\,\mu$

↑↑↑↑

↑↑↓↓; ↑↓↑↓; ↑↓↓↑
↓↓↑↑; ↓↑↓↑; ↓↓↑↑

(a)

(b)

Figure 29.5 A system of four spin $\frac{1}{2}$ particles, each with magnetic moment μ. There is only one arrangement having them all aligned $(M = 4\mu)$, but six different arrangements having two up and two down $(M = 0)$. Therefore, the entropy of the ordered $(M = 4\mu)$ system is smaller.

all magnetic moments pointing in the same direction) corresponds to very low entropy. There is only one arrangement by which they can all point the same direction, but many different arrangements for them pointing many directions.

As the lattice vibrations begin shaking the magnetic moments into random orientations, the additional entropy of the magnetic moments comes at the expense of the lattice vibrations. In other words, thermal energy is transferred from the lattice into the magnetic moment system, so the lattice cools as they come into equilibrium (Figure 29.6).

You can see that a good deal of experimental sophistication is required to attain and maintain these very low temperatures. An equally impressive degree of sophistication is required to measure them. You can't just go into a hardware store and buy a thermometer that works in this region. The properties of a large number of materials, such as conductivities and magnetic susceptibilities, have been measured and recorded at low temperatures. So these materials can now be used as low-temperature thermometers; by measuring the particular physical property, you can determine the temperature. But how was the first such "thermometer" calibrated? When you are investigating a low temperature area that has never been probed before, where no data has previously been taken, and the properties of no materials are known, how do you know what the temperature is?

This fundamental question requires a return to fundamental concepts for an answer. The definition of temperature is given as

$$\frac{1}{T} = \frac{\Delta S}{\Delta E}\Big)_V \tag{29.2}$$

Aligned, low entropy

Random orientation, high entropy

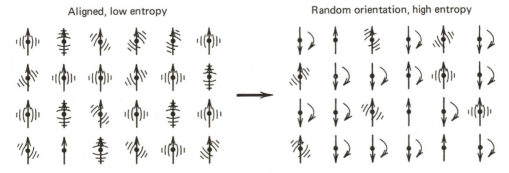

Figure 29.6 If the magnetic moments in a solid are initially all aligned, the energy required to flip some of them and randomize their orientations is supplied by the thermal vibrations of the lattice. This expenditure of thermal energy by the lattice causes it to cool. Alternately, the flow of entropy (or thermal energy) is from the lattice to the system of magnetic moments.

so it is a measure of how the entropy changes as energy is added. At low temperatures, small additions of energy will cause large changes in entropy. Equivalently, at low temperatures, the addition of a small amount of energy will cause the number of accessible states to increase by a large factor.

Therefore, to measure ΔS, we must add a certain small amount of energy (ΔE) to the system and determine by what factor the number of accessible states increased as a result. If the system is in thermal equilibrium then the temperature of all portions of the system is the same, so we need only determine the temperature of any portion of the system to know that of the whole system.

For example, we could determine the temperature from the magnetic degrees of freedom only. If we know the intrinsic magnetic moment of a single particle and the number of such particles in the system, then the measurement of the magnetic moment of the system will determine the number of accessible magnetic states. We could add a small amount of energy, ΔE, measure the change in the magnetic moment of the system, and that would tell us the change in the number of accessible states, which tells us the change in entropy, ΔS.

EXAMPLE

Suppose a system consists of four spin $\frac{1}{2}$ particles, each having magnetic moment μ. What is the entropy of this system if its total magnetic moment in the $+z$-direction is:

(a) 4μ?
(b) 2μ?
(c) 0?

Spin $\frac{1}{2}$ particles can have their magnetic moments either up or down. If $M_z = 4\mu$, all particles have magnetic moment up. If $M_z = 2\mu$, three are up and one down. If $M_z = 0$, two are down and two up.

(a) 1 state: $4\mu = \uparrow\uparrow\uparrow\uparrow$. $S = k \ln 1 = 0$.
(b) 4 states: $2\mu = \uparrow\uparrow\uparrow\downarrow, \uparrow\uparrow\downarrow\uparrow, \uparrow\downarrow\uparrow\uparrow, \downarrow\uparrow\uparrow\uparrow$. $S = k \ln 4$.
(c) 6 states: $0 = \uparrow\uparrow\downarrow\downarrow, \uparrow\downarrow\uparrow\downarrow, \uparrow\downarrow\downarrow\uparrow, \downarrow\downarrow\uparrow\uparrow, \downarrow\uparrow\downarrow\uparrow, \downarrow\uparrow\uparrow\downarrow$. $S = k \ln 6$.

The above example demonstrates that by measuring the magnetic moment of the system, you can determine the number of accessible states, and therefore its entropy. The temperature is determined by how the magnetic moment—and therefore its entropy—changes when small amounts of energy, ΔE, are added. In homework Problem 29-16 you will do this for a more realistic system of 10^{22} particles.

Another result of the definition of temperature (29.2), is that from it we have determined that the probability of any given particle being in a certain state, s, of energy ε_s is given by*

$$P_s = Ce^{-\beta \varepsilon_s}$$

From this, we can determine the temperature from the relative populations in two very closely spaced levels. Examples of such levels might include the orientations of nuclear

---

* Classical statistics is appropriate here, because we are interested in a certain particle, which may occupy various quantum states. Each particle is a separate system, with its own set of accessible states.

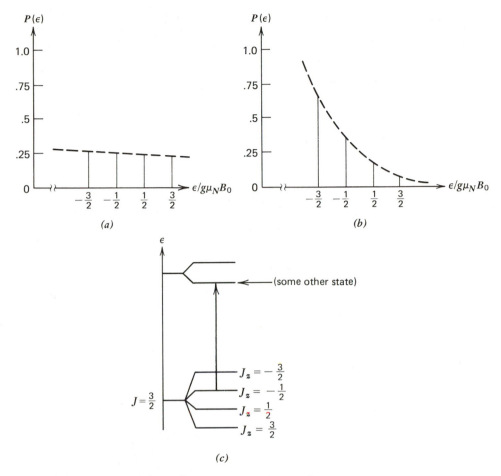

Figure 29.7 One method for measuring low temperatures. A spin $\frac{3}{2}$ nucleus with gyromagnetic ratio g in a magnetic field B_0. The spacing between levels, $\Delta\varepsilon = g\mu_N B_0$ is quite small for ordinary magnetic fields, because of the smallness of the nuclear magneton, μ_N. From the relative occupation of the various levels, we can determine the temperature. (a) At higher temperature, such that $\Delta\varepsilon \ll kT$, the occupation number of all levels will be about the same. (b) At lower temperature, such that $\Delta\varepsilon \lesssim kT$, there is a significant difference in occupation number between the various levels. (c) We can determine the relative occupation of the various states by measuring the intensity of radiation absorbed by the sample when it is illuminated with photons of just the right energy to cause transitions from the level of interest to some other level.

magnetic moments in a weak external magnetic field (Figure 29.7), or two closely spaced electronic levels in some special molecule.

EXAMPLE

Suppose each particle has two very closely spaced levels, separated by 10^{-5} eV, and the number of particles in the upper level is only 10% of the number in the lower level. What is the temperature?

The ratio of probabilities for the upper and lower levels is

$$\frac{P_u}{P_l} = \frac{Ce^{-\beta \varepsilon_u}}{Ce^{-\beta \varepsilon_l}} = e^{-\beta(\varepsilon_u - \varepsilon_l)} = .1$$

Taking the natural logarithm of both sides yields

$$\frac{\varepsilon_u - \varepsilon_l}{kT} = -\ln(.1)$$

Putting in 10^{-5} eV for $(\varepsilon_u - \varepsilon_l)$ yields

$$T = 0.05 \text{ K}$$

SUMMARY

Attaining and maintaining extremely low temperatures requires a great deal of experimental sophistication. When the desired temperatures are colder than any surrounding objects, then the system must be insulated from its environment, as heat transfer would be in the wrong direction. Cooling must be performed adiabatically, either through mechanical or diffusive interactions.

Gases are liquefied by putting them under pressure and bathing them in the coldest available environment. Then they are insulated from the environment and allowed to expand adiabatically, cooling further to the point where some of it liquefies. These cold liquids can be cooled further by further reduction in pressure and allowing some to vaporize.

Bathed in liquid helium, systems can reach temperatures of 1 K or so, but to be cooled beyond this, they must be insulated from this environment, and then allowed to do work adiabatically. Adiabatic demagnetization is the preferred mechanism for this, since volume expansion would require heavy pressure vessels with high heat capacities and significant friction from moving parts.

To measure extremely low temperatures in regions where the properties of materials have not yet been measured, also requires sophistication. This can be done using the definition of temperature

$$\frac{1}{T} \equiv \frac{\Delta S}{\Delta E}\bigg)_V \tag{29.2}$$

If we can determine the change in entropy (ΔS) as a small amount of energy (ΔE) is added, then we know the temperature through the above relationship. If the system is in thermal equilibrium, then we need only determine the temperature of any portion of the system, such as that of the magnetic degrees of freedom.

Also from the definition of temperature, we know the relative population of two closely spaced levels is

$$\frac{P_1}{P_2} = e^{-\beta(\varepsilon_1 - \varepsilon_2)} \tag{29.3}$$

So by measuring the relative populations of two closely spaced levels, we can determine the temperature of the system.

PROBLEMS

29-1. For a certain system, the first excited state lies 1 eV above the ground state. Is the system degenerate at room temperature? Is it degenerate at 10^6 K?

29-2. What is the excitation temperature for a system whose first excited state lies:

(a) 0.01 eV above the ground state?
(b) 0.1 eV above the ground state?
(c) 0.001 eV above the ground state?

29-3. Consider an electron (mass $= 9.1 \times 10^{-31}$ kg) confined to a box of width L.

(a) Use the uncertainty principle ($\Delta x \, \Delta p_x \geq h$), or equivalently, your knowledge that the minimum area of a two-dimensional quantum state is h, to find an approximate expression for the minimum momentum of this electron in terms of L.
(b) What is the kinetic energy of this electron in terms of L?
(c) What is the width of the box if the minimum kinetic energy of this electron equals kT, where T is room temperature?
(d) How does this width compare to the width of a typical atom, 1 Å?

29-4. Repeat Problem 29-3 for a proton of mass 1.67×10^{-27} kg.

29-5. The rate at which an object radiates energy is proportional to T^4, where T is its temperature. If liquid nitrogen, at a temperature of 77 K surrounds the Dewar (i.e., thermos bottle) holding the liquid helium, by what factor is the heat energy radiated into the Dewar reduced over that which would be radiated in if it was surrounded by air at 300 K?

29-6. A low-temperature Dewar is a jar within a jar. The two jars are silvered and there is a vacuum between them.

(a) Why do you suppose they are silvered?
(b) Why do you suppose there is a vacuum between them?

29-7. The rate at which an object radiates energy is proportional to T^4. If two otherwise identical objects are placed side by side, but one is at a temperature of 300 K and the other is at 0.1 K, how much faster does the 300 K object radiate energy toward the 0.1 K object, than the 0.1 K object radiates back toward the 300 K object?

29-8. Repeat Problem 29-7 for objects at 300 K and 100 K, respectively.

29-9. Why can't thermal interactions between systems be used to cool something below the temperature of the coldest available reservoir?

29-10. Give an example of cooling something via mechanical interaction. Diffusive interaction.

29-11. Why is adiabatic demagnetization done adiabatically?

29-12. Why is adiabatic expansion *not* used to attain temperatures much lower than 1 K?

29-13. Consider a system consisting of five spin $\frac{1}{2}$ particles each of magnetic moment μ. What is the entropy of the system if the total magnetic moment is:

(a) -5μ?
(b) 3μ?

29-14.   Consider a system consisting of three spin 1 particles, each of magnetic moment μ. What is the entropy of the system if the total magnetic moment is:

(a) 3μ?
(b) 2μ?
(c) μ?

29-15.   Consider a system that includes three spin $\frac{1}{2}$ particles each having magnetic moment μ. Suppose that in addition to these 3 spin degrees of freedom, the system has 7 other degrees of freedom, making 10 altogether. When 10^{-5} eV of energy is added to the system, the total magnetic moment of the system changes from $+3\mu$ to $+\mu$.

(a)   Assuming equipartition, how much added energy went into magnetic degrees of freedom?
(b)   By how much did the entropy of the magnetic moments change?
(c)   What is the temperature of the system as determined from the magnetic properties of the system?

29-16.   If a system consists of N spin $\frac{1}{2}$ particles, the number of different arrangements of the system for which n of them are spin up, and the remaining $(N - n)$ of them are spin down is given by $N![n!(N - n)!]^{-1}$, where the factorials of large numbers can be approximated by Stirling's approximation, $\ln M! = M \ln M - M$. Consider a system of 10^{22} spin $\frac{1}{2}$ particles each having magnetic moment μ. Suppose that in addition, the system has 2×10^{22} other degrees of freedom, making 3×10^{22} degrees of freedom altogether, one-third of which are magnetic.

(a)   If the magnetic moment of the entire system is $10^{21}\ \mu$, how many particles are spin up?
(b)   If the magnetic moment of the entire system is $1.2 \times 10^{21}\ \mu$, how many particles are spin up?
(c)   What is the change in entropy of the system, if the magnetic moment goes from $1.2 \times 10^{21}\ \mu$ to $10^{21}\ \mu$?
(d)   Suppose this change occurred when $2J$ of energy was added to the system. What is the temperature of the system?

29-17.   In a certain type of molecule, electronic levels j and k are separated by 2×10^{-5} eV.

(a)   If the ratio of molecules in level k to those in level j is $1:20$, what is the temperature of the system?
(b)   If level l is 3×10^{-5} eV above level k, what is the ratio of molecules in level l to those in level j?

B.   SUPERFLUIDITY

B.1   The Phenomenon

One system that takes on surprising properties at very low temperatures is liquid helium. These properties are not displayed by any other liquids because other liquids solidify before the necessary low temperature region is reached.

The reason helium does not solidify is attributable to the very weak forces between helium atoms (Figure 29.8). To illustrate this, consider what would happen if helium

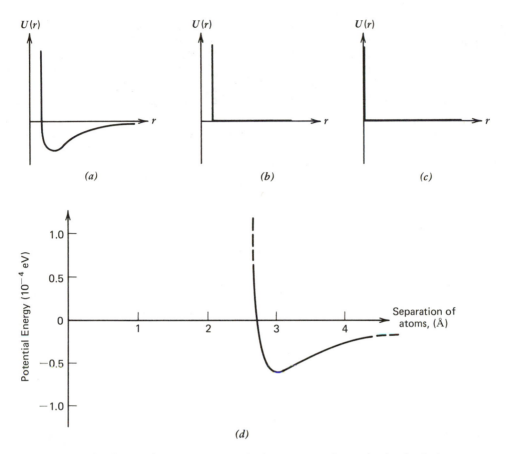

Figure 29.8 (a) Plot of potential energy versus separation between two real gas molecules, showing long-range attraction and short-range repulsion. (b) Plot of potential energy versus separation between billiard balls, showing short-range repulsion, but no long-range attraction. (c) Plot of potential energy versus separation for ideal gas molecules—billiard balls of infinitesimal size. (d) Plot of potential energy versus separation for two helium atoms. Clearly, they resemble hard spheres of radius $2R = 2.7$ Å, but there is a small, slight attraction for separations greater than 3 Å.

were a solid. The zero-point oscillations of an atom in any one direction would have energy

$$\varepsilon = \frac{1}{2} k_s x_0^2$$

where k_s is the "force constant" of the restoring force, and x_0 is the amplitude of the oscillations. In quantum mechanics we learn that these zero-point oscillations have energy

$$\varepsilon = \frac{1}{2} \hbar\omega = \frac{1}{2} \hbar \sqrt{\frac{k_s}{m}}$$

where m is the atomic mass. Equating these two expressions and solving for the amplitude, we have

$$x_0 = \frac{\hbar^{1/2}}{(k_s m)^{1/4}}$$ (29.4)

The interatomic forces are very small, so both the force constant, k_s, and the mass, m, appearing in the above expression would be very tiny for helium, giving rise to large amplitudes of oscillation. In fact the amplitude of the zero-point oscillation for helium would be much larger than the interatomic distances, and therefore it would "melt" from its own zero-point oscillations.

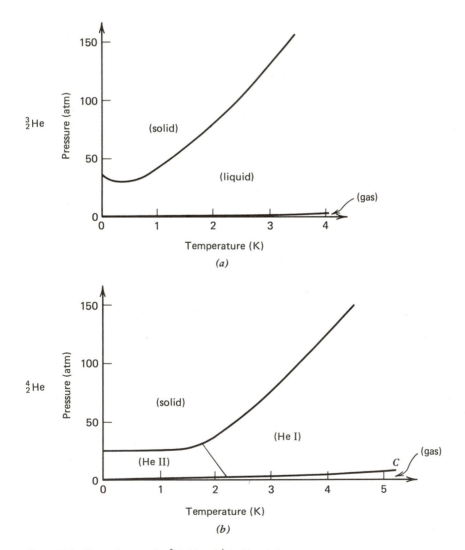

Figure 29.9 Phase diagrams for $^3_2$He (a) and $^4_2$He (b) at low temperatures.

The two isotopes of helium behave quite differently at low temperatures since one is a fermion and one a boson. Their phase diagrams are shown in Figure 29.9. Both have two electrons in a state of zero total angular momentum. He^3 has a nucleus of three spin $\frac{1}{2}$ particles (two protons and one neutron), arranged in such a way that their total spin is also $\frac{1}{2}$. Consequently, it is a fermion. He^4 has four spin $\frac{1}{2}$ particles in its nucleus with a total spin of zero, making it a boson. The He^4 isotope is far more common (by a factor of 10^6), so usually it is understood that the name "liquid helium" refers to the He^4 isotope, unless He^3 is specifically mentioned.

At atmospheric pressure, helium condenses into a normal (but cold) liquid at 4.2 K. But as the temperature is lowered still further, it undergoes another phase transition at 2.17 K—not to a solid, but rather to a liquid of different and surprising properties. We differentiate between these two phases by calling that above 2.17 K "helium I," and that below 2.17 K "helium II." The transition temperature of 2.17 K is often referred to as the "lambda point" (see Figure 29.10).

Helium II consists of two components, a "normal fluid" component and a "super-fluid component," which gives the system the remarkable properties of zero viscosity and infinite thermal conductivity. The infinite thermal conductivity means that heat seems to be disbursed infinitely fast. Regardless of where you heat the liquid, it will evaporate from the top surface, where the pressure is lowest. By contrast, normal fluids, such as water, vaporize mostly on the bottom of the vessel or wherever the heat input is, and then bubble to the surface. The He I → He II phase transition is visually characterized by the disappearance of bubbles and boiling (Figure 29.9). The lack of viscosity means that the superfluid component will flow through the smallest crack without impediment, and if you put it into circular motion around the container, it keeps on going forever, having no friction.

This phase transition at 2.17 K is called a "second-order" phase transition, and it differs from the usual phase transitions in some important ways. Although there are discontinuous changes in some physical properties of the system as this temperature is crossed, no latent heat is released, and the entire system doesn't change at once. That is, at 2.17 K the superfluid component begins to form, but the transformation is not complete until temperatures much lower than this. Although the superfluid component is responsible for the remarkable properties of liquid helium below 2.17 K, the normal fluid component is still present until temperatures much lower than 1 K.

B.2   A Model

The superfluid component of He II must be caused by these bosons being in the ground state. If the ground state is separated from the first excited state by a finite gap, then bosons in this ground state would be trapped, and could not move easily into excited states in response to external stimuli. Indeed, for flow velocities above 7 m/s, viscosity does begin to appear in superfluid He II, whereas below 7 m/s the viscosity is zero. In other words, only for flow velocities above 7 m/s are some of the collisions with the container walls strong enough to excite helium atoms out of the ground state, which implies an energy gap separating the ground state from excited states.

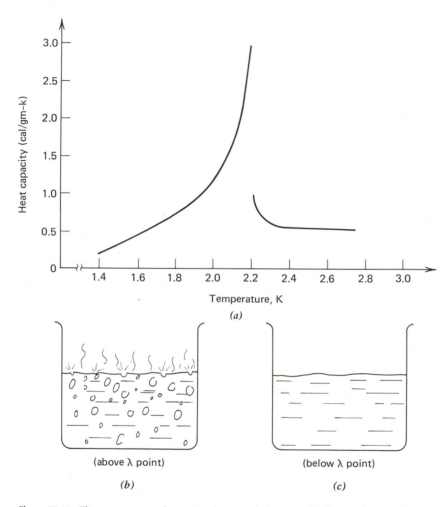

Figure 29.10 The temperature of transition between helium I and helium II phases in liquid helium is called the λ-point because of the shape of the plot of heat capacity versus temperature (*a*). At temperatures above the λ-point, the liquid helium boils (*b*). But below the λ-point (*c*), the bubbles disappear, because the infinite thermal conductivity causes the temperature to be uniform, and it vaporizes from the top surface where the pressure is least, regardless of where the heat energy enters the system.

This energy gap must be caused by the weak attractive forces between helium atoms. Each ground state atom is in a potential well due to the attractive forces of its neighbors. In quantum mechanics we learn that the states accessible to particles in potential wells have discrete energies, separated by finite gaps. Since the attractive forces between helium atoms are so weak, we expect the potential wells to be broad and shallow, which means that the discrete energy levels are close together. Indeed, we know the gaps must be small because the superfluid quantum effect shows up only at very low temperatures.

Albert Einstein proposed a simple model that allows us to apply the tools of quantum statistics to liquid helium, in order to gain insight into its peculiar behavior. According to this model, we assume that the distribution of excited states accessible to the liquid helium atoms is that of a quantum gas. Integrating over volume and momentum directions, this distribution of states would be

$$\int \frac{d^3r \, d^3p}{h^3} = \left(\frac{4\pi V}{h^3}\right) p^2 \, dp = C\sqrt{\varepsilon} \, d\varepsilon \tag{29.5}$$

where the constant C is given by

$$C = \left[\frac{2\pi V (2m)^{3/2}}{h^3}\right] \tag{29.6}$$

We know that it is acceptable to replace a sum over discrete states by an integral, as long as the states are so numerous that their distribution can be considered continuous. But the above distribution shows that the states are fewer and fewer as $\varepsilon \to 0$, so we expect this integration to be least appropriate for the lowest-lying states (Figure 29.11). In fact, the above distribution shows *no* states of zero energy, which is an important error. Since the ground state is presumably responsible for the superfluid properties, it would be a critical mistake if we left the ground state out of our sum over states.

Consequently, in the Einstein model, we treat the ground state separately. The distribution of excited states is assumed to be that of Eq. 29.5, and by elimination we know that those not in excited states must be in the ground state. If there are N_a atoms altogether, of which N_g are in the ground state and N_x in excited states, then

$$N_a = N_g + N_x \tag{29.7}$$

The distribution of atoms in excited states is the product of the distribution of states times the occupation number of each.

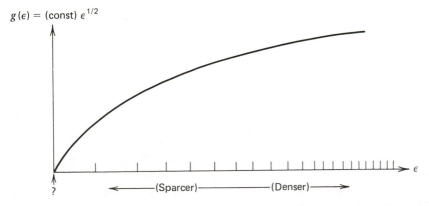

$g(\epsilon) = (\text{const}) \, \epsilon^{1/2}$

────(Sparcer)──────(Denser)────▶

Figure 29.11 In the gas model, the density of states decreases at lower energy as $\sqrt{\varepsilon}$. This means that as ε approaches zero, the states become farther and farther apart, and replacing the discrete sum by an integral is less justified. In fact, the model does not allow for the ground state at all, since $g(\varepsilon) = 0$ at $\varepsilon = 0$.

$$dN_x = C \frac{\sqrt{\varepsilon}\, d\varepsilon}{e^{\beta(\varepsilon - \mu)} - 1} \tag{29.8}$$

For quantitative calculations we need to know the value of $e^{-\beta\mu}$ appearing in the denominator (Figure 29.12). We know that the chemical potential varies with temperature, but for simplicity we assume that it is constant, remaining at its low temperature limit over the small range of temperatures of interest to us. At the low temperature limit, all particles must be in the ground state, so the occupation number of the ground state ($\varepsilon = 0$) must be N_a.

$$\bar{n}_g = \frac{1}{e^{\beta(\varepsilon - \mu)} - 1} = \frac{1}{e^{-\beta\mu} - 1} = N_a \qquad \text{(low-temperature limit)}$$

Inverting this expression, we have

$$e^{-\beta\mu} - 1 = \frac{1}{N_a}$$

or

$$e^{-\beta\mu} = 1 + \frac{1}{N_a} \approx 1 \qquad \text{(low-temperature limit)} \tag{29.9}$$

With this, the distribution of particles in excited states becomes

$$dN_x = C \frac{\sqrt{\varepsilon}\, d\varepsilon}{e^{\beta\varepsilon} - 1}$$

We integrate this to find the total number of particles in excited states. Using the substitution of variables, $y = \beta\varepsilon$, this becomes

$$N_x = C \int_0^\infty \frac{\sqrt{\varepsilon}\, d\varepsilon}{e^{\beta\varepsilon} - 1} = C(kT)^{3/2} \int_0^\infty \frac{y^{1/2}\, dy}{e^y - 1}$$

Numerical evaluation of the integral gives its value of 2.32, so the number of atoms in excited states is

$$N_x = 2.32 C(kT)^{3/2} \tag{29.10}$$

Clearly, there is something wrong with this result, because it implies that the number of atoms in excited states becomes arbitrarily large with arbitrarily large temperature. Yet we know there are a finite number of atoms, and there can be no more in excited states than there are atoms altogether.

$$N_x \leq N_a \tag{29.11}$$

The source of this error is our assumption that the chemical potential remains at its low-temperature limit, not varying at all as the temperature is increased (Figure 29.12).

To get around this obstacle, we assume that the result (29.10) is valid as long as the constraint $N_x \leq N_a$ is not violated. If T_0 is the maximum temperature for which result (29.10) satisfies this constraint, then

$$N_x = 2.32 C(kT)^{3/2} \qquad \text{for } T \leq T_0$$

$$= N_a \qquad \text{for } T > T_0 \tag{29.12}$$

(distribution of states)   x   (occupation number)   =   (distribution of particles)

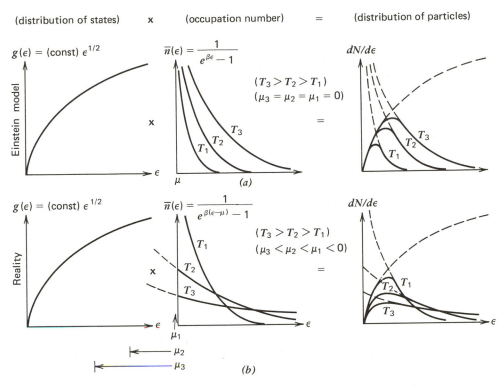

Figure 29.12 The distribution of particles is the product of the density of states times the occupation number. The occupation number depends on the chemical potential, so different chemical potentials give rise to different particle distributions. (a) In the Einstein model we assume that $\mu = 0$, and that it doesn't change with temperature. This means that the total number of particles in excited states (the area under the $dN/d\varepsilon$ versus ε curve) increases indefinitely with temperature. (b) In reality, the chemical potential decreases with increasing temperature, so that the number of particles in excited states does not grow indefinitely.

That is, for low temperatures, the number of atoms in excited states increases as $T^{3/2}$ until all atoms are in excited states at temperature T_0. Above this temperature the number in excited states remains constant as there are none left in the ground state to excite.

When $T = T_0$ all N_a atoms are in excited states, so Eq. 29.10 gives us

$$N_a = 2.32C(kT_0)^{3/2} \tag{29.13}$$

or

$$T_0 = \frac{1}{k}\left(\frac{N_a}{2.32C}\right)^{2/3}$$

Substituting the value of the constant C from Eq. 29.6, this becomes

$$T_0 = \left(\frac{h^2}{2mk}\right)\left(\frac{N_a}{4.64\pi V}\right)^{2/3}$$

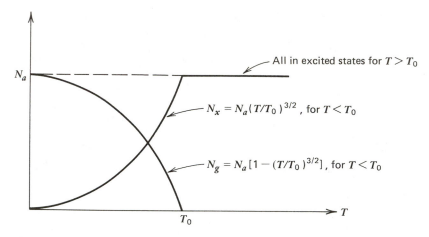

Figure 29.13 The number of helium atoms in the ground state (N_g) and excited states (N_x) according to the Einstein model. N_a is the total number of atoms.

If we put into this expression the values of m and N_a/V for atoms of liquid helium, we get

$$T_0 \approx 3 \text{ K}$$

which is not much different from the observed temperature of 2.17 K for the onset of condensation into the ground state. Consequently, we feel that this model may be giving us a fair representation of what is actually happening.

According to this model, the fraction of atoms in excited states for temperatures below T_0, is given by dividing Eq. 29.12 by Eq. 29.13,

$$\frac{N_x}{N_a} = \left(\frac{T}{T_0}\right)^{3/2} \tag{29.14}$$

and the fraction in the ground state would be

$$\frac{N_g}{N_a} = 1 - \left(\frac{T}{T_0}\right)^{3/2} \tag{29.15}$$

These are plotted as a function of the temperature in Figure 29.13.

SUMMARY

Under atmospheric pressure, helium gas liquefies at 4.2 K, and undergoes another phase transition at 2.17 K, not to a solid, but to another liquid with some remarkably different properties. This is a "second-order" phase transition, marking the onset of condensation of the superfluid component. As the temperature is lowered, further condensation occurs and the superfluid component of He II becomes increasingly prominent.

The superfluid component consists of helium atoms in the ground state, which are separated from excited states by a small but finite energy gap caused by the weak interatomic attractive forces.

Albert Einstein proposed a simple model in which the distribution of excited states is that of a boson gas, and the ground state is treated separately. The total number of particles in excited states is obtained by integrating over the distribution, and is found to increase with temperature as $T^{3/2}$ until that temperature, T_0, at which all N_a particles are in excited states. That is,

$$N_x = \begin{cases} 2.32\left[\dfrac{2\pi V(2m)^{3/2}}{h^3}\right](kT)^{3/2} & \text{for } T \le T_0 \\[2ex] N_a & \text{for } T > T_0 \end{cases} \qquad (29.12)$$

where T_0 is determined by solving Eq. 29.12 for the temperature at which all N_a particles are in excited states. This gives

$$T_0 = \left(\frac{h^2}{2mk}\right)\left(\frac{N_a}{4.64\pi V}\right)^{2/3}$$

which is about 3 K for liquid helium, close to the experimental value of 2.17 K. The fraction of atoms in excited and ground states is given by

$$\frac{N_x}{N_a} = \left(\frac{T}{T_0}\right)^{3/2} \qquad (T < T_0) \qquad (29.14)$$

$$\frac{N_g}{N_a} = 1 - \left(\frac{T}{T_0}\right)^{3/2} \ (T < T_0) \qquad (29.15)$$

PROBLEMS

29-18. Derive Eq. 29.4, relating the amplitude of oscillation x_0, the force constant, k_s, and the atomic mass, m, for an atomic harmonic oscillator undergoing zero-point oscillation. Start with the energy of a harmonic oscillator being related to the amplitude of oscillation through $\varepsilon = \frac{1}{2}k_s x_0^2$, and with the energy of the zero-point oscillation being given by $\varepsilon = \frac{1}{2}\hbar\omega = \frac{1}{2}\hbar(k_s/m)^{1/2}$.

29-19. Suppose you are given that the amplitude for the zero-point oscillation of helium atoms would be about 10 Å, and that for the zero-point oscillation of iron atoms is about 0.1 Å. Use Eq. 29.4 to compare the strengths of the force constants, caused by attraction of neighboring atoms, for iron and helium.

29-20. If helium II has infinite thermal conductivity, why does it vaporize from the top surface, rather than boiling uniformly throughout its volume?

29-21. Consider what happens after you stir helium II to get it to swirl around inside a circular container. If the normal fluid component experiences friction with the walls of the container and the superfluid component doesn't, describe the subsequent motion of the two components.

29-22. Estimate the size of the energy gap separating the ground state and excited states in liquid helium, using your knowledge that the phase transition starts at 2.17 K.

29-23. We are going to examine two of the assumptions in the Einstein model, and see how they would influence the model's prediction for the number of atoms in the ground state.

(a) The model assumes an ideal gas distribution for excited states with no gap separating the first excited state from the ground state. Should the absence of a gap cause the model to predict too many or too few atoms in the ground state? (*Hint*. Is it harder or easier for an atom to get out of the ground state when there is no gap?)

(b) The model assumes the chemical potential does not decrease with increasing temperature, as we know it must. Should this cause the model to predict too many or too few atoms in the ground state? (*Hint*. Draw a plot of $\bar{n}$ versus ε for two values of μ. As μ gets more negative, does the difference in occupation numbers between ground state and neighboring excited states get larger or smaller?)

29-24. Liquid helium contains about 2×10^{25} atoms per liter.

(a) Evaluate the constant C of Eq. 29.6 for one liter of liquid helium.

(b) Use this value of C to solve Eq. 29.13 for T_0.

29-25. At a temperature $T = \frac{1}{2}T_0$, what fraction of the atoms are in the ground state according to the Einstein model?

C. SUPERCONDUCTIVITY

C.1 The Phenomenon

Many ordinary metals become superconductors at temperatures on the order of a few degrees Kelvin. They exhibit no resistance to the flow of small currents. Once started, a current in a superconducting loop keeps going apparently forever. In one experiment, for example, there was no detectable reduction in the current in a superconducting loop over a period of more than one year! We can use the observed properties of superconductors, plus our knowledge of the occupation numbers of states, to gain some insight into the causes of this phenomenon.

The superconducting electrons are fermions. Consequently, large numbers of them cannot fall into the lowest-lying state, as was the case with superfluid helium II, which was composed of bosons. In fact, no more than one electron can occupy any one state, so we must assume that for a system of N identical electrons at temperature $T = 0$, the lowest N states are filled and the others are vacant.

The system remains superconducting from $T = 0$ up to some temperature of transition to the normal conducting state. Since the properties of a system of electrons are dependent on the states they are in, we must conclude that a change of properties is associated with a change in states occupied. In particular, we may conclude that the electron system remains in its ground state up to the transition temperature, and then transition to higher states begins. This is our first clue into what is causing the superconducting phenomenon, because it indicates there is a finite energy gap between the ground state of the system, and the first excited state. In terms of the individual particle states, this can be reworded to say that there is a finite energy gap between the lowest N electron states and the next state above these (Figure 29.14).

For this quantum effect to be observed, the size of the gap, $\Delta\varepsilon$, separating the ground state and first excited state of the system, must be on the order of kT, where T is the transition temperature. A quick calculation (see homework Problem 29-26) shows that

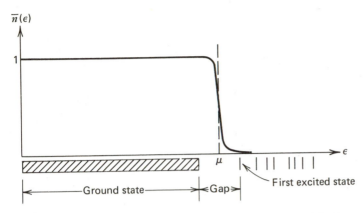

Figure 29.14 If there is a finite gap between the fermi surface and the first excited electron state, then at low temperatures the "tail" does not extend as far as the first excited state. Under these conditions, the system is confined to the ground state, because the probability of finding an electron in an excited state is negligibly small.

if there are no interactions between electrons in a gas, the characteristic separation of states (and consequently, the transition temperature, T) is far too small for quantum effects to be detected. Superconductivity is observed, however, so the electrons cannot be treated as a perfect gas at low temperatures; there must be some mechanism that widens the spacings of the low-lying levels of the system, which we would normally associate with the system sitting in some sort of potential well.

C.2   A Model

The fact that the gap always appears just above the last filled state must be more than coincidence. It tells us that whatever the physical explanation for the gap may be, it must involve the entire system as a whole. Any quantitative evaluation of it is a many-body problem, involving all the conduction electrons. The problem cannot be adequately approached by looking at isolated electrons in some kind of individual potential well, for example. The theory for handling this, developed by Bardeen, Cooper, and Schrieffer, is called the "BCS theory,"[*] and goes beyond the scope of this book. Here we just present a quick intuitive sketch of the reason for the gap.

The fact that there exists a finite energy gap between the ground state and the first excited state of the system indicates that there must be some kind of mutually attractive force among the electrons.[†] How can the interelectron forces be attractive when the superconductor as a whole is neutral? In the BCS theory, such interactions are shown

---

[*] See R. D. Feynman's lectures in *Statistical Mechanics*, Benjamin, 1972, for example, for an explanation of the BCS theory.

[†] Actually, it need be only less repulsive than random interactions would be.

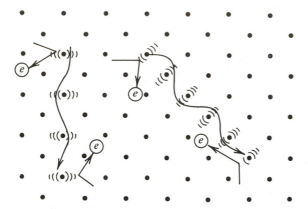

Figure 29.15 In the BCS theory, the electron-electron interactions are mediated by phonons. That is, one electron interacts with one lattice point, and the elastic wave generated in this manner travels through the solid until it transfers this energy to some other electron.

to be possible when mediated by phonons (Figure 29.15). One electron interacts with one atomic lattice point, and the lattice excitation travels through the lattice until giving the impulse back to some other electron at some other point in the lattice. Of course, the interaction could be attractive or repulsive, but the BCS theory can demonstrate that when in the ground state, the forces between such electrons near the Fermi surface must be attractive.

This is not as coincidental as it might sound. Of all the various kinds of phonon-mediated interactions that could occur between electrons, some will be attractive and some repulsive, giving rise to a wide range of accessible states. At low temperatures, the electrons will naturally prefer states of lower energy, and these lower energy states correspond to interelectron forces that are predominantly attractive. As a consequence, the lowest-energy configuration for the system demands an extremely high degree of correlation among the behaviors of the electrons. Among other things, the electrons must be paired so that for each electron, there exists another "partner" with opposite momentum and spin.

C.3   Stability of the Superconducting State

The interaction of an electron with phonons generated by other electrons in this highly correlated arrangement, leads to a net reduction in its energy. By contrast, its interaction with random thermally generated phonons could have many possible effects. As a result, thermally generated phonons tend to disrupt the highly correlated electron behaviors necessary for their lowered energies, and so the width of the gap decreases as temperature increases.

Consequently, there are two reasons why it becomes easier for the electrons to jump into excited states as the temperature of a superconductor is increased. One is that the

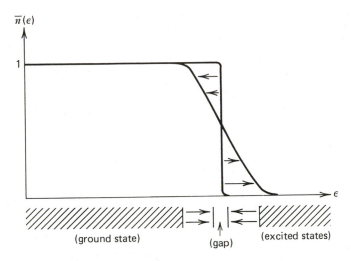

$\bar{n}(\epsilon)$

(ground state)   (gap)   (excited states)

Figure 29.16 There are two reasons why the superconducting state becomes less stable as the temperature rises. One is that the tail of the Fermi distribution broadens, and the other is that the gap narrows, due to the effects of thermally generated phonons. Both these things make the occupation of excited states more likely.

"tail" of the Fermi distribution broadens, making occupation of excited states increasingly probable. The other is that the thermally generated lattice vibrations tend to narrow the width of the gap (Figure 29.16).

According to Faraday's law, when we try to change the magnetic flux through a circuit, an electromotive force is induced, and a current will flow in such a way as to oppose the change in flux. For a superconductor, the resistance is zero. Any finite electromotive force would result in infinite current. Therefore, a superconducting circuit is able to oppose any change in magnetic flux, perfectly.

When we turn on a magnet near a superconductor, a current will be created around the perimeter of the superconductor such that it is shielded perfectly from the magnetic field,* as indicated in Figure 29.17. We write

$$B = B_0 + B_M$$

where B_0 is the external field imposed by currents in wires ($\mu_0 H$), and where B_M is the field caused by the magnetism of the material ($B_M = \mu_0 M$). Inside the superconductor,

$$B = 0 \quad \text{or} \quad B_M = -B_0$$

The magnetic work per unit volume done on the superconductor in the process is

$$W = \int dW = -\frac{1}{\mu_0} \int B_M \, dB_0 = \frac{1}{\mu_0} \int B_0 \, dB_0 = \frac{B_0^2}{2\mu_0} \tag{29.16}$$

---

* In our sketchy treatment, we are glossing over many important experimental details, such as the geometry of the superconducting sample.

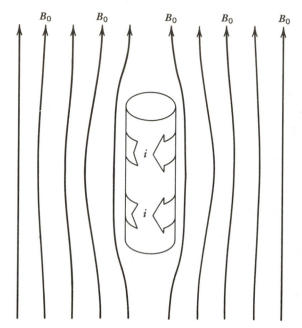

B_0 B_0 B_0 B_0 B_0

Figure 29.17 When placed in a magnetic field, the induced current around the perimeter of a superconductor will oppose the imposed field perfectly. Since there is no electrical resistance, the induced current will flow forever.

Just as the zero viscosity of a superfluid was only displayed up to a certain velocity before excitations from the ground state could occur, so is superconductivity good only up to a certain maximum current density before electrons can be knocked out of the ground state via interaction with the stationary lattice. Since the current density around the perimeter of the superconductor must increase as we increase the field, B_0, a certain critical value of the field, B_c, can be reached where the current density is sufficiently large for the superconductivity to break down—that is, for electrons to get knocked out of the lowest states.

The work per unit volume done on the system to bring it from its superconducting state, at zero magnetic field, to the point of transition to the nonsuperconducting state is given by Eq. 29.16.

$$W_s(0 \rightarrow H_c) = \frac{B_c^2}{2\mu_0}$$

At the critical field, B_c, where the transition occurs, both superconducting and normal states of the system are equally stable, so in a zero external field, the superconducting state is more stable by an energy of

$$\text{(energy of stabilization)} = V\left(\frac{B_c^2}{2\mu_0}\right) \qquad (29.17)$$

where V is the volume of the sample.

We have seen that it becomes easier for electrons to jump the gap in a superconductor as the temperature increases, due to both the narrowing of the gap and the widening of the tail of the occupation number. Therefore, as the temperature of a

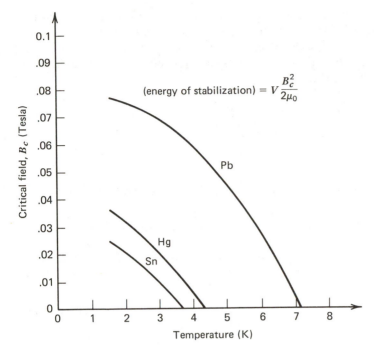

Figure 29.18 Plot of critical field versus temperature for superconducting lead (Pb), mercury (Hg), and tin (Sn). As the temperature rises, the superconducting state becomes less stable.

superconductor rises, we expect the energy of stabilization to decrease. This is depicted for some superconductors in Figure 29.18.

SUMMARY

Many metals become superconducting at a temperature on the order of a few degrees Kelvin. Superconductors display zero electrical resistance, caused by a finite energy gap between the electronic ground state and the first excited states, which must be due to mutually attractive forces among the electrons.

In the BCS theory, these mutually attractive forces are generated by very highly correlated behaviors among the electrons in the ground state, so that the phonon-mediated interactions are predominantly attractive. Thermally generated phonons tend to disrupt the highly correlated behaviors of electrons in the BCS ground state, so as temperatures rise, the width of the gap decreases.

As the temperature of a superconductor rises, the narrowing of the gap and the spreading of the "tail" in the fermion occupation number make it increasingly probable for electrons to jump the gap. Lower current densities are needed in order that collisions of electrons with atomic lattice points become sufficiently forceful as to knock them into excited states. These currents are produced by induction through the imposition of external fields. As the temperature rises, smaller critical magnetic fields, B_c, are required to produce current densities sufficient to cause the superconductor to break down. The energy of stabilization of a superconducting sample of volume V is given by

$$(\text{energy of stabilization}) = V\left(\frac{B_c^2}{2\mu_0}\right) \qquad (29.17)$$

PROBLEMS

29-26. Consider an electron moving in one dimension confined to some region of length L.

(a) If the size of one quantum state is given by $\Delta x \, \Delta p_x = h$, how much momentum separates two neighboring states if the electron is confined to a sample measuring 1 cm on a side?

(b) How much energy separates the ground state ($p = 0$) from the first excited state?

(c) At about what temperature would the transition from ground state to first excited state start occurring?

29-27. The critical field, B_c, for a certain metal is 0.1 T. If the typical separation of conduction electrons in this metal is 5 Å, what is the energy of stabilization per electron, in eV?

INDEX

A

Absolute zero, 157, 228-229, 498
Acceptors, 493
Accessible states:
  and diffusive interactions, 182-183
  and thermal interactions, 182-183
  and work, 162-165
Adiabatic cooling, 242, 500
Adiabatic demagnetization, 502
Adiabatic expansion, 266, 269
Adiabatic processes, 232, 233, 241-
  249
  in air masses, 236-237
  in earth, 247-249
  in gases, 236-237, 242-245
  in solids and liquids, 245-249
Air, 236-237, 281
Air molecules, distribution of, 41-42, 52,
  58-59
Ammonia, 60, 63-64, 105
Angular momentum:
  orbital, 19-21
  quantization of, 18-21
  spin, 19-21
  total, 18-19, 21, 29

B

Band edge, 487, 488
Band gaps, 425
Band structure, 482-486
Bardeen, 519
Barrier constraints, removal of, 261-262
BCS theory, 519
Binomial coefficient, 39
Binomial distribution, 35-41
Blackbodies, 449
Blackbody radiation, 443-454
  energy flux, 451
Bohr magneton, 337
Boltzmann's constant, 128, 142-143
Bose-Einstein statistics, 406
Boson gases, 432, 433
Bosons, 19, 405, 407

  degenerate, 438
Brown, Robert, 306
Brownian motion, 306-307

C

Canonical ensembles, 281-283
Cards, 3, 4, 43, 44
Carnot cycle, 263-267
  p-V diagrams for, 269
  work done, 270, 271
Carnot efficiency, 274
Carnot engine, components of, 264-265
Celcius temperature scale, 143
Charge, fundamental unit of, 8
  of electron, 9
  of proton, 9
  quantization of, 8-9
Chemical equilibrium, 373-383
Chemical potential, 84, 88, 92, 138, 147,
  163, 174-176, 374, 424, 438
  and accessible states, 182-183
  dependence on (T,p), 376-378
  dependence on (T,p,N), 177-178
  dependence on (T,V,N), 90, 91
  for identical particles, 360, 380, 383
  magnitude of, 90-92
  measuring changes in, 218-219
  and number of particles, 298-299
  from the partition function, 351
  and phase equilibrium, 396, 397
  for photons, 445
  in semiconductors and insulators, 487
  for water, 178
Classical statistics, 279-400
  comparison with quantum, 409-419
  limits of, 419-421
Clausius-Chapyron equation, 396-399
Coefficient of linear expansion, 171
Coefficient of thermal conductivity, 327,
  329, 330, 335
Coefficient of utility, 275
Coefficient of viscosity, 331, 332, 335
Coefficient of volume expansion, 169,
  222, 239

Coins, 2, 3, 5, 6, 33, 47, 48, 49-51, 59
  heads-tails states for, 112, 115
Collision fragments, 9
Collision frequency, 319-320, 322-323, 325
Collision between particles, 79-80
Collisions, elastic, 82
Combustion, 87
Compressibility, isothermal, 171-172
Conduction band, 426
  bandwidth and temperature, 489
  distribution of electrons in, 485, 486
Conduction electrons, 63, 200-201, 482
  degrees of freedom of, 455-457
  energy of, 472-474
  heat capacity of, 478-479
  particle distribution, 472
Conductivity:
  and electric field, 489
  and pressure, 490-491
  and temperature, 489, 490
Configurations for small systems, 37, 39
Constraints, 189-278
  imposed, 232-262
  natural, 209-231
  second law, 210-214
  third law, 228-229
Convection:
  in atmosphere, 248
  in mantle, 247-248
  stability against, 248
Cooling, adiabatic, 242, 500
Cooper, 519
Coulomb potential, 26
Critical field, for superconductors, 522, 523
Critical point, 386, 388
Curie law, 345, 346

D

Debye frequency, 481
Debye model, 461-470
Debye model notation, 481
Debye temperature, 481
Degeneracy, 109, 291
Degenerate electron gas, 470
Degenerate systems, 438, 498

Degrees of freedom, 71-73, 76
  average energy per, 300-301
  quantum effects, 456-457
Density of states, 17, 110-115, 423
Detailed balance, principle of, 448-450
Deuterium, 87-88
Diamagnetism, 338-340
Diatomic gas molecules, 363-370
Dice, 2, 33, 34, 36-37, 48, 52, 59
Differentials, exact and inexact, 94-97, 160-161
Diffraction, 10
Diffusion coefficient, 326, 327, 335
Diffusion in gases, 326-327
Diffusive cooling, 501, 502
Diffusive interactions, 83-92, 174-189, 213
Distributions:
  gaussian, 53-58
  skewed, 55
Domains, magnetic, 340
Donors, 492, 493
Duong-Petit law, 305

E

E, 8
Efficiency of engines, 273-277
Einstein model:
  for lattice vibrations, 459-461
  for superfluid helium, 513-516
Elastic waves in solids, 461
  polarizations of, 463
Electrical charge, 8, 166, 167
Electromagnetic radiation, 443-454
Electron affinities, 219
Electron gas, 484
Energy:
  conservation of, 78
  internal, 70-102
  quantization of, 25
Energy bands, 294-296, 425-426
Energy flux, 328
Energy gap:
  in semiconductors and insulators, 425
  in superconductors, 518-519
  in superfluid helium, 512
Engines, 263-367

coefficient of utility, 275
  efficiency, 273-277
  internal combustion, 266-267
  work done by, 270, 271
Engines and refrigerators, 263-278
Ensembles, 48, 280-283
Enthalpy, 214-215, 238
Entropy, 103-132
  and accessible states, 284, 286
  changes in, 163, 164, 226
  dependence on internal energy, 134,
    146
  and identical particles, 262
  and the partition function, 354-355
  and probabilities, 285, 286, 401
  and reversibility, 249-250
  and work, 164-165, 168
Equations of state, 190
Equilibrium, 104-106, 126, 179-180
  approaching, 180-181
  between phases, 384-399
Equilibrium constant, 381
Equipartition, 73-76, 300-308, 327
Equipartition theorem, 74, 84, 85, 301
  violations of, 223-224
Exact differentials, 96, 97, 215-216
Excitations between semiconductor
    levels, 494-495
Excitation temperature, 290-291, 363-
    364, 498
  in semiconductors and insulators, 483
Expansion, linear, 171
  volume, 169

F

Farenheit temperature scale, 143
Fermi-Dirac statistics, 405
Fermi level, 471
Fermion gases, 432, 433
Fermions, 19, 405, 407
  degenerate, 438, 470-471
Fermi surface, 471
  in doped semiconductors, 494, 495
  in intrinsic semiconductors, 487
Ferromagnetism, 340, 341
First law, 92-93, 162
Fluctuations, 4, 47-53
  relative, 49, 51, 63

Flux, of gas molecules, 316-319
Force, generalized, 166, 167
Fourier analysis, 11-12
Free expansion of gases, 256-261
  and temperature changes, 259-260
Friction, 251-252
Frictional drag in gases, 330
Fundamental postulate, 106-108

G

Gases:
  accessible states, 312
  cooling, 253-256
  degrees of freedom, 193, 195
  diffusion in, 326-327
  expansion of, 161-162, 165
  heat capacity of, 302-305
  ideal, 191-195
  internal energy of, 73
  models for, 191-197
  molecular diffusion, 326-327
  particle flux, 316-319
  partition function for, 362-370
  pressure exerted by, 318-319, 441-
    442
  quantum, 429-437
  real, 195-197
  rotation of molecules, 73
  spectra of accessible states, 424-425,
    429-431
  thermal conductivity, 327-330
  transport processes in, 324-335
  velocity distributions for, 309-324
  viscosity, 330-333
Gas laws, 193, 197
  for quantum gases, 436-437, 439-441
Gas molecules, relative motion of, 322-
    323
Gaussian distribution, 49, 53-59
Gibbs free energy, 213, 376-377
  and phases transitions, 391-395
Grand canonical ensembles, 281-283
Gyromagnetic ratios, 23, 337

H

H, 10
ħ, 11

Harmonic oscillators, 24-25, 455-457, 459-460
  energy levels in, 25
  equilibrium, 24
Heat and entropy, 148-149
Heat capacity, 155-157, 202-206, 222
  at absolute zero, 228, 304
  of conduction electrons, 478-479
  of gases, 204-206, 243-244, 302-305
  of ideal gases, 204
  of metals, 478-479
  molar, 155, 157
  near absolute zero, 158
  of solids, 305-306, 476-479
  of van der Waals gases, 204-206
  variations with volume, 225
Heat flux, 327, 329
Heat function, 238
Heat reservoirs, 148-154
Heat transfer, 78-80
  as inexact differential, 94
Helium, 87
  interatomic forces, 509
  isotopes of, 510-511
  phase diagrams, 570
  phase transitions in, 385, 510, 511
Helium II, 511
Helmholtz free energy, 214-215, 239, 355
Holes, 64, 200, 201
  conduction by, 483-484
  occupation number for, 485, 486
Hydrogen atom, 20, 109
  excitation of, 292-293

I

Ice crystals, 184-186
Ideal gases, 191-195
  accessible states, 192
  entropy, 192
  equation of state, 193
  heat capacity, 243-244
  internal energy, 192
  pressure, 193-194
Ideal gas law, 193
  derivation, 318-319
Identical particles:
  and the counting of states, 409-418

  partition function for, 359
Imposed constraints, 232-262
Impurities, in semiconductors, 492-497
Independent variables, 99-100
  and imposed constraints, 232
Insulators, 482-491
Integrals, standard, 322
Interactions, 133-189
  chemical, 373-383
  and independent variables, 190-191
  mechanical, 160-173
  thermal, 134-159
  types of, 76-77, 78, 133
Internal energy, 69-101
  changes in, 222-223
  dependence on temperature, 137
  as exact differential, 94
  fluctuations in, 152-153
  of gases, 72, 73
  of iron bar, 70, 94-95
  from partition function, 351
  of quantum gases, 435-436
  of solids, 71-73
Intrinsic conduction, transition temperature, 496-497
Intrinsic materials, 482
Isobaric processes, 232, 233, 237-239
  in gases, 237-239
  in solids and liquids, 239
Isochoric processes, 232, 233
Isothermal compressibility, 171-172, 206-208, 222, 240-241
  definition, 206
  for ideal gases, 207
  for van der Waals gases, 207-208
Isothermal processes, 239-241, 270-272
  in gases, 239-240
  in solids and liquids, 240-241

J

Joule-Thompson process, 253-256

K

Kelvin temperature scale, 143

L

Lambda point, 511, 512
Latent heat, 390, 392
Lattice vibrations, 459-470
Law of mass action, 378-382
Liquid helium, 501, 502
Liquids, models for, 198-199
Low temperature physics, 498-524
Low temperatures:
  attaining, 500-503
  measuring, 503-506

M

Magnetic domains, 340
Magnetic energy, 342
Magnetic field, magnetic moment in, 23
Magnetic moments, 22-23, 28-29, 107-
   108, 166
  and angular momentum, 22, 29, 336
  and diamagnetism, 338
  energy of, 23
  mean values of, 342
  and paramagnetism, 342-348
  and spin, 337
Magnetic work, 166
  in superconductors, 521
Magnetism, 336-348
  cause of, 336
Mantle, convection in, 247-248
Mars, water on, 387, 388
Maxwell distribution, 309-324
Maxwell, James Clerk, 10, 309
Maxwell's relations, 214-221, 229-231
  applications, 222-228
Mean free path, 321, 325, 326
Mean values, 32-35, 48
  from partition function, 350-353
  of velocities, 314-315, 325
Mechanical interactions, 80-83, 160-
   173
Metals, 200-201, 470-475
  conduction electrons in, 200-201,
   470-475
Microcanonical ensembles, 281-283
Models, 190-208
Mole, 196
Molecular diffusion, 326-327

Molecules, excited states of, 53
Momentum, and wavelength, 11
Momentum flux, 331, 332

N

Natural constraints, 209-231
  types of, 210
Neutron, charge distribution in, 337,
  338
Nitrogen molecule, 73
Nonequilibrium processes, 253-262
N-type semiconductors, 494
Nuclear magneton, 337
Nucleus, atomix, 8

O

Occupation number, 404-409
  for bosons, 406, 407
  for fermions, 405, 406
Occupation of quantum states, 403-421
Oven, photons in, 443-454

P

Paramagnetism, 339-340, 342-348
  strong field limit, 345
  weak field limit, 343-345
Parameters, interdependence among, 209,
  212, 214-231
Particle flux, 326, 328, 331
Particles, 10
Particle transfer, 83-92
Particle waves, 10
Partition function, 349-373
  definition, 349
  for a gas, 362-370
  for identical subsystems, 358-361
  for many subsystems, 357-361
  and mean values, 350-353
  rotational part, 364-365, 368
  translational part, 364
  vibrational part, 365-366, 368
Phase change, 75
Phase diagrams, 386-389
  for helium, 510
  for water, 387
Phase equilibrium, 396-399

Phase space, 15-17
  constraints on, 16
  coordinates, 16
Phase transitions, 146-148, 384-399
  and attractive forces, 385
  gas-liquid, 389-395, 398
  model for, 389-395
  pressure and temperature dependence,
    396-399
  second order, 386, 511
Phonon gas, 461-470
  density of states for, 464-465
  energy distribution, 463, 466-467
  particle distribution, 462-463, 465
Phonons, 200, 461
  maximum energy of, 463-465
  polarizations of, 463
  in superconductors, 500
Photon gas, 443-450
  energy density, 448
  energy distribution, 444-445
  particle distribution, 444
Photons, 288-289
Piezoelectric effect, 491
Plasma, 385
Polarization, of molecules, 197
  of phonons, 463
Potential energy, 83-92, 174-176
  of gases, 259-260
Pressure, 160, 163, 164
  and entropy, 164-165
  exerted by a gas, 318-319
  from the partition function, 351
  from quantum gases, 437, 441-442
  thermal, 390
Probabilities, 2, 3, 32-67, 106-108, 118,
    349
  of being in a state, 280, 284-288, 401-
    402
  in classical and quantum statistics, 287-
    289
  for energy of a small system, 150-154
  of exitation, 291
  for identical elements, 37
  for two criteria, 41
Probability amplitude, 13
Properties, microscopic and macroscopic,
    4, 5, 6, 8
P-V diagrams, 268-272

  for phase transitions, 394, 397
  for van der Waals gas, 393
P-type semiconductors, 494

Q

Quantum effects, 8, 279, 303, 498-
    500
  at low temperatures, 427
Quantum gases, 429-437
  energy spectrum, 434
  gas law for, 436-437
  internal energy of, 435-436
  particle distribution, 431
Quantum mechanics, 11
  and the harmonic oscillator, 24
Quantum states, 13-18, 103, 104
Quantum statistics, 401-524
  comparison with classical, 409-419
  survey of applications, 422-442
Quasistatic processes, 105, 253

R

Random walk, 59-64
Real gases, 195-197
  degrees of freedom, 195
  equation of state, 196-197
  pressure, 196-197
  van der Waals model, 197
Refrigeration, 500, 501
Refrigeration cycle, 266, 267-268
Relaxation time, 105
Reservoirs, 280-286
Reversibility, 249-253
Root mean square, 52
Root mean square velocity, 315
Rotation, diatomic molecules, 73
Roulette wheel, 4

S

Salt, dissolution in water, 183-184
Schrieffer, 519
Second law, 126-132
  violation of, 127
Second law constraints, 210-214
Semiconductors, 201, 482-497
  changes in conductivity, 489-491

doped, 492-497
intrinsic, 487-492
n-type, 494
p-type, 494
Shear stress, 330-331
Simple harmonic motion, 24
Sine waves, 11, 12
Small systems, 31-67, 280-288
  energy of, 150-154
Snowflakes, formation of, 184-186
Solids:
  degrees of freedom, 71-73, 199
  heat capacity of, 305-306, 476-479
  models for, 71, 199-201
  potential energy, 199
  spring constants, 199-200
  thermal properties of, 455-481
  vibrational waves in, 200
Specific heat, 155, 157
  for common substances, 156
Spin and statistics, 409
Spin-orbit coupling, and paramagnetism,
    344
Spin states, 107, 109
Spin-up and spin-down, 18
States:
  accessible, 403, 404, 422-429
  accessible to air molecules, 103, 110,
    114
  accessible to combined system, 130,
    131
  counting of, 412-415
  degenerate, 109
  density of, 423
  dependence on internal energy, 110-
    111, 116-117, 118, 133, 134
  discrete and continuous, 422-426
  for distinguishable particles, 412
  for identical particles, 412-413
  multiple occupancy, 412-415
  spacing of, 109-111
  spectra of, 422-431
  summation over, 423-424
  of a system, 104-117, 128
  for two interacting subsystems, 120-
    125
Statistical mechanics, 3, 5
Statistically independent behaviors, 41-
    44

Statistics:
  classical, 279-400, 402
  classical and quantum, 286-288, 409-
    419
  quantum, 401-524
Step, in random walk, 59-61
Stephan-Boltzmann constant, 451
Stirling's formula, 40, 45
Stoichiometric coefficients, 379-380
Sublimation, 387
Sulfuric acid, 85-87
Sun, 64
  radiation from, 451
Superconductivity, 427, 500, 518-524
  stability of, 520-523
Superfluid helium, 385
  excitations in, 511, 516
Superfluidity, 427, 500, 508-517
  Einstein model for, 513-516
System variables, 126
Systems, size of, 31

T

Taylor series expansion, 54, 65-67
Temperature, 74-75, 135-136, 163
  and heat flow, 140-143
  and internal energy, 73-76, 137-139,
    298
  and magnetic moments, 504
  and occupation of states, 506
Temperature scales, 141-143
Thermal conductivity of gases, 327-
    330
Thermal energy, 84, 85, 86, 87, 88, 90,
    138, 147, 175
Thermal equilibrium, 135
Thermal expansion, 169-173
Thermal interactions, 78-80, 134-159
Thermodynamical potential, 213, 374-
    375
  and phase transitions, 391-395
Thermodynamics, 5, 7
Thermometers, 143-145
  standard, 145
Thermometric parameters, 143-144
Thermonuclear reactions, 87-88
Third law, 157-158
Third law constraints, 228-229

Throttling process, 253-256
  and temperature changes, 256
Transition temperature:
  intrinsic-impurity, 496-497
  for superconductivity, 518
  for superfluidity, 516
Transport processes, 324-335

U

Uncertainty principle, 11-13, 14, 15

V

Valence band, 426
  distribution of holes in, 485, 486
Valence electrons, 201, 482
Van der Waals model, 197
  for liquids, 198
  and phase transitions, 389-395
Variables, independent, 99-100, 190
Viscosity of gases, 330-333
Volume, change in, 160-162

W

Water:
  boiling point, 386, 387, 399

condensation, 87
dissociation of molecule, 293-294
freezing point, 387
phases of, 386, 387
triple point, 143
Wave function, 11-13
  boundary conditions, 19
Wavelength, and momentum, 11
Wave nature of particles, 9-11
Wave number, 11
Waves, 9
Work, 80-83, 160-173
  and accessible states, 162-165
  electrostatic, 166
  gravitational, 166
  as inexact differential, 94
  kinds of, 165-169
  magnetic, 166, 521
  on microscopic scale, 81, 82-83
Work function, 239

Z

Zero-energy reference level, 83-84, 138, 147, 213
Zero-point oscillations, 509-510
Zeroth law, 137